ORGANIC LIGHT EMITTING DIODE – MATERIAL, PROCESS AND DEVICES

Edited by **Seung Hwan Ko**

Organic Light Emitting Diode – Material, Process and Devices
Edited by Seung Hwan Ko

Published by InTech
Janeza Trdine 9, 51000 Rijeka, Croatia

Publishing Process Manager Iva Lipovic
Technical Editor Teodora Smiljanic
Cover Designer Jan Hyrat

First published July, 2011
Printed in Croatia

A free online edition of this book is available at www.intechopen.com
Additional hard copies can be obtained from orders@intechweb.org

Organic Light Emitting Diode – Material, Process and Devices, Edited by Seung Hwan Ko
p. cm.
ISBN 978-953-307-273-9

For EU product safety concerns:
IN TECH d.o.o., Prolaz Marije Krucifikse Kozulić 3, 51000 Rijeka, Croatia,
info@intechopen.com or visit our website at intechopen.com.

Meet the editor

Professor Seung Hwan Ko got the Ph.D. degree from University of California, Berkeley under Prof. Costas Grigoropoulos in 2006. Later that year, he started a post-doctoral fellow position at UC Berkeley and a project researcher at Lawrence Berkeley National Lab till 2009 and is currently an assistant professor at Korea Advanced Institute of Science and Technology (KAIST). His main research interests are next generation, unconventional micro/nano fabrication approach development as an alternative to the conventional photolithography and vacuum deposition based processes. The novel micro/nano fabrication process development includes the new material development, the high resolution direct patterning process development, and specific application development. We concentrate on applying developed direct patterning technologies to realization of the flexible electronics such as flexible display, sensors and energy devices.

Contents

Preface

Organic light-emitting diodes (OLED) are playing a major role in information technology (IT) by providing the promise of further expanding the use of digital displays through making display fabrication technology lower in cost and higher in performance to replace liquid crystal displays (LCD). Due to various attractive features such as high contrast, high brightness, large color gamut and thin structure, various sized OLED displays from small-sized mobile phone display to large-sized TV display have already begun to be mass-produced.

This book is a collection of state-of-the-art works intended to cover theoretical and experimental aspects of OLED from material synthesis and characterization (Chapter 1-6) to actual process development and devices applications (Chapter 7-12). Each chapter features remarkable breakthrough on OLED and provides latest scientific knowledge and leading-edge technology. They offer research agenda and accelerate the research, development and distribution of OLED. I expect that this book will be useful to encourage further experimental and theoretical research in OLED.

In closing, I wish to express my sincere gratitude to the contributing authors of each chapter, publishing process manager Ms. Iva Lipovic, and the publishing staffs. In particular, I am deeply grateful to Prof. Costas P. Grigoropoulos (UC Berkeley), Dr. Hee K. Park (AppliFlex), Dr. Ming-Tsang Lee (Lawrence Berkeley National Laboratory), Dr. Heng Pan (Applied Materials. Inc) for valuable suggestions. I dedicate this book to my parents and my wife, Ms. Hyun Jung Kim.

Dr. Seung Hwan Ko
Applied Nano Technology and Science (ANTS) Lab
Korea Advanced Institute of Science and Technology (KAIST),
Daejeon, Korea

Part 1

OLED Materials

1

Synthesis, and Photo- and Electro-Luminescent Properties of Phosphorescent Iridium- and Platinum-Containing Polymers

Yuji Koga and Kouki Matsubara
Department of Chemistry, Faculty of Science, Fukuoka University
Japan

1. Introduction

Development of polymer light-emitting diode (PLED) has been attracted considerable attentions,[1] because polymeric materials could be applied to low-cost production of electro-luminescent (EL) devices exhibiting efficient luminescence for flat-panel displays. As polymeric property of the materials enables the solution processes, such as spin-coating, screen printing, and ink-jet printing (Figure 1), large-area and fine-pixel displays could be easily developed in comparison with the vapor deposition process. In addition to such easy preparation, it is of significant that it requires a fewer number of layers in PLED devices, which enables low driving voltage, even though PLED still has a drawback in lower luminescence efficiency than that of the organic light-emitting diode (OLED) in general.

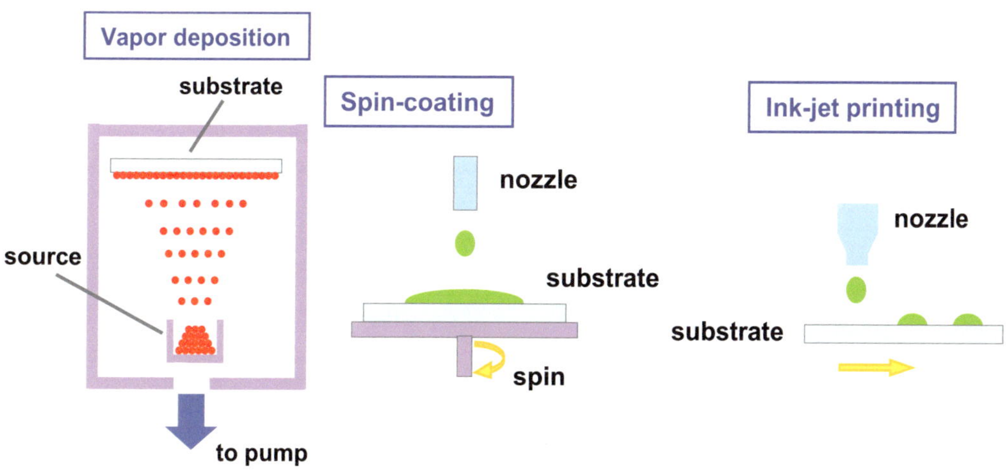

Fig. 1. Vapor deposition process, spin-coating, and ink-jet printing

In contrast to widely developed fluorescent π-conjugated polymers, such as polyfluorenes and polyphenylene vinylenes (PPVs), as polymeric EL materials (Figure 2),[2] researches for phosphorescent polymers are still now in progress, because it follows the development of the phosphorescent metal complexes, which are also in progress. Two types of phosphorescent PLED materials are known: (1) host polymers such as poly(vinylcarbazole) (PVK) and poly(9,9-di-*n*-octyl-2,7-fluorene) (PFO), into which phosphorescent small molecules are doped,[1a-f] and (2) polymers having phosphorescent pendant units in the side chain. In the former polymer, phase separation and crystallization of the small molecules in the polymer matrix may reduce the luminescence efficiency due to self-quenching mechanism and prevent uniform emission all over the films. Thus, several studies were focused on the latter phosphorescent polymer. Lee et al.[3] and Tokito et al.[4] independently developed non-conjugative copolymers in which monomers having luminescent cyclometalated iridium pendant units copolymerized (Figure 3), whereas Chen et al. reported preparation of a conjugative fluorene copolymer from a cyclometalated iridium-suspended co-monomer (Figure 4).[5]

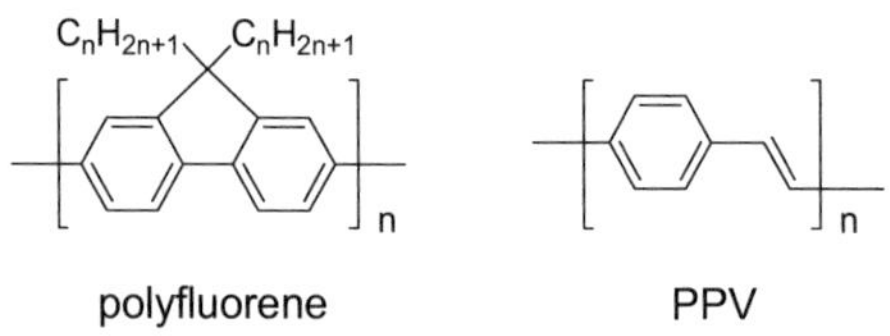

Fig. 2. Structure of polyfluorene and PPV

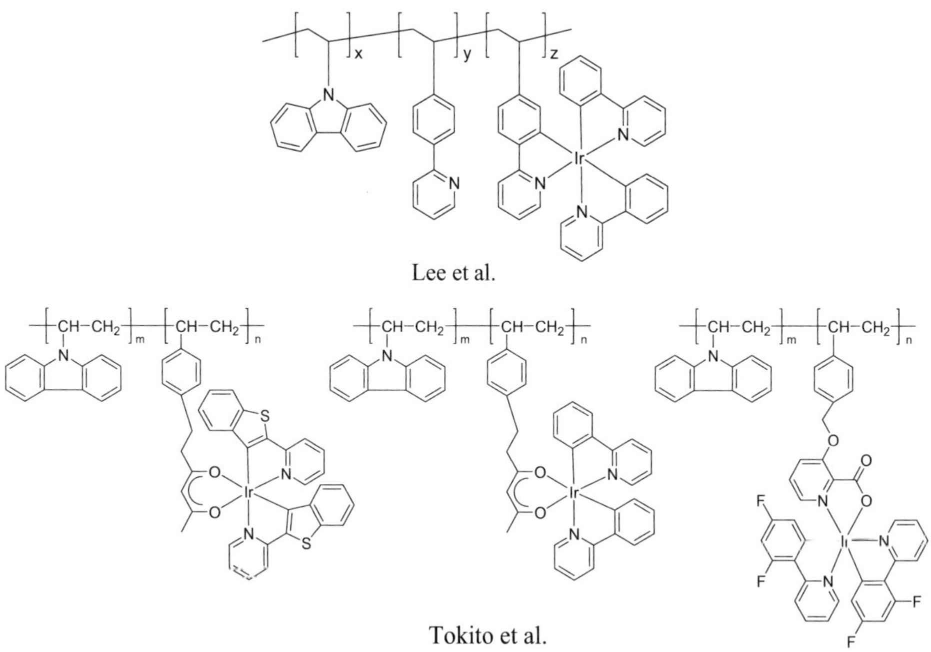

Fig. 3. Structure of non-conjugated copolymers with luminescent cyclometalated iridium pendant units

Scheme 1 depicts the general synthetic methodologies for the metallopolymers including the metal complex side chain. Almost all of the luminescent metallopolymers ((D) in Scheme 1) have been synthesized via monomers containing metal-complex pendant units (B) (Scheme 1, *Method A*).[6] However, polymerization of these metal-containing monomers led to the

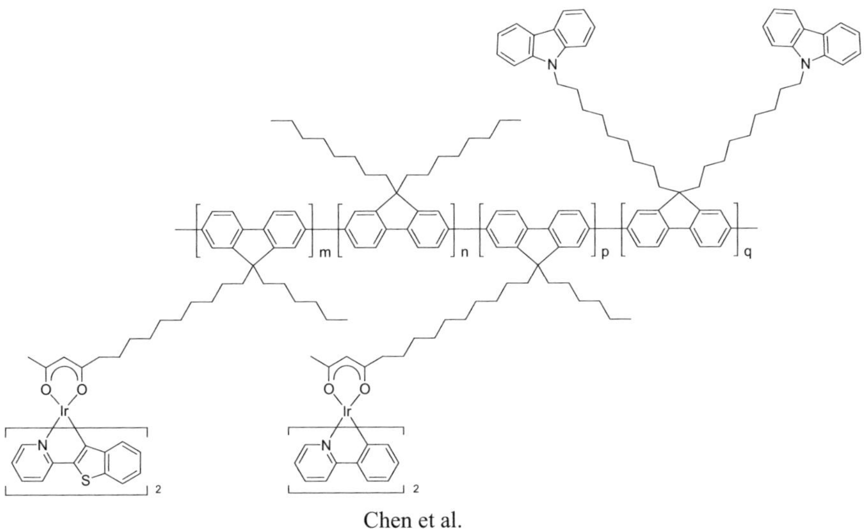

Fig. 4. Structure of conjugated fluorene copolymer with luminescent cyclometallated iridium pendant units

elimination of metal fragments to some extent or failure of polymer weight control. Alternatively, the polymer (D) can be synthesized by polymerization of a ligand-containing monomer followed by the reaction of the copolymer ligand (C) with a metal pendant unit (Scheme 1, *Method B*). *Method B* is seemed to be favored against *Method A* in the preparation of phosphorescent metallopolymers, because various kinds of ligand monomers can be easily copolymerized in desirable content with previously developed radical copolymerization processes in metal-containing polymer chemistry.[7] However, in the early reports, synthesis of luminescent metallopolymers via *Method B* was conducted only under severe conditions, such as Lee, Schulz or Fréchet reported (Scheme 2, 3, 4).[3, 7g]
Furthermore, there are few examples that the alpha and/or omega ends of the polmers are capped eith phosphorescent units for EL materials that can be provided by the final combination of the ligand unid units in the polymer ends with metal precursors.

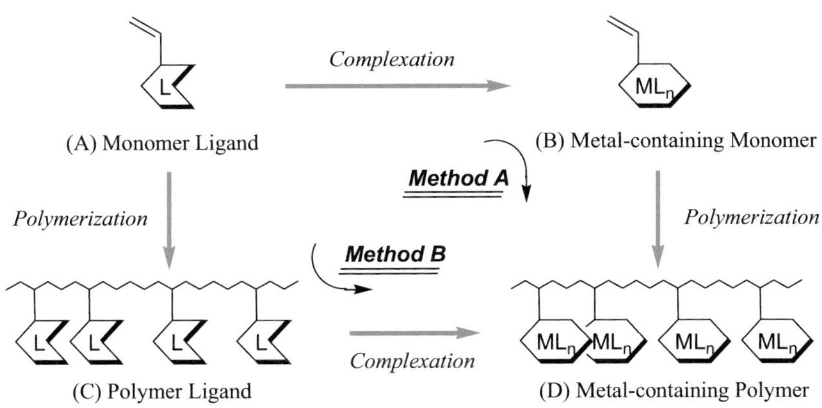

Scheme 1. Synthetic methodology of metal-containing polymer.

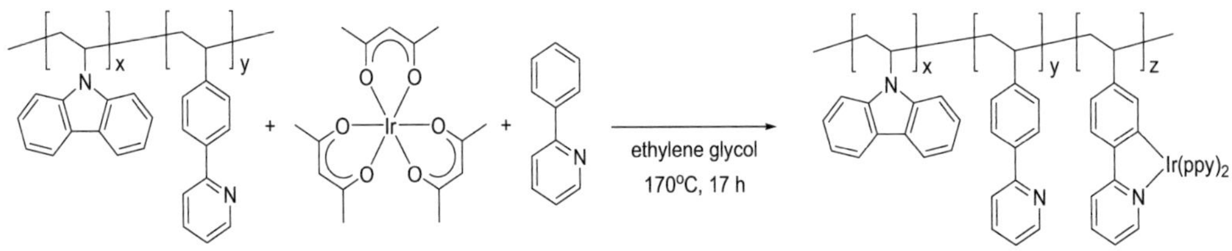

Scheme 2.

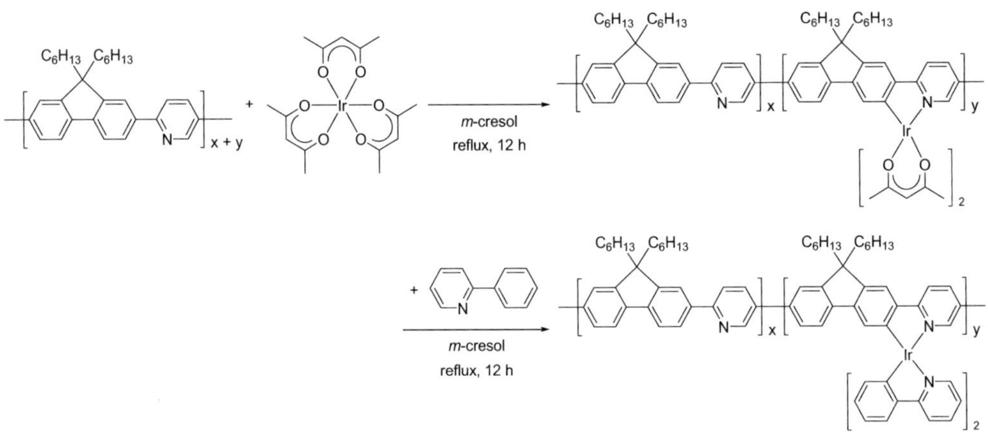

Scheme 3.

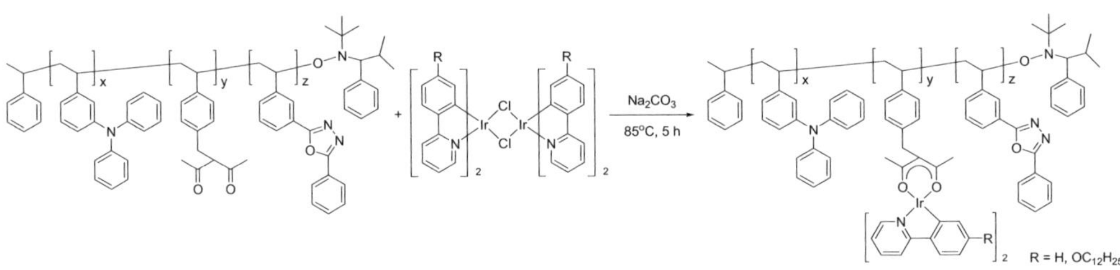

Scheme 4.

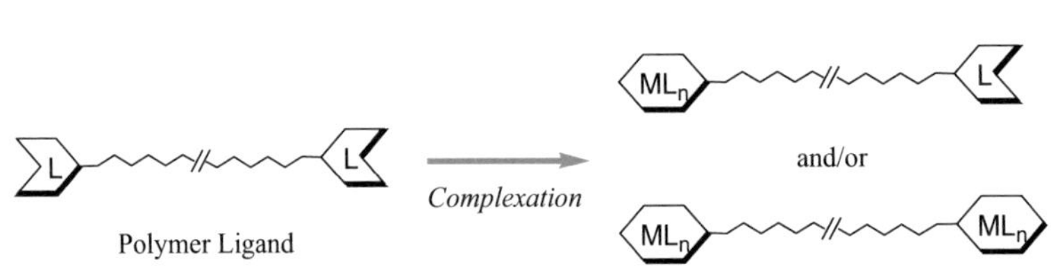

Scheme 5.

We previously developed metallopolymers produced from methyl methacrylate (MMA) and 4-styryldiphenylphosphine copolymers with an iridium precursor. The iridium polymers performed both photo- and electroluminescence, and showed characteristic features depending upon the content ratios of the iridium and phosphorus in the copolymers.[8] However, the devices containing these polymers exhibited rather low luminescent efficiency, because of its low charge-transporting ability caused by the nonconductive polyolefin backbone.

Here, we developed several series of iridium or platinum-containing metallopolymers under mild conditions in *Method B*, where pyridine was used as a linker module between the polymer main chain and the luminescent metal unit. Metal-incorporation effects of the metallopolymers on their luminescent properties were investigated in order to develop devices with high luminescent efficiency. Two independent types of the metallopolymers are shown in the following chapters: one is that composed of polyolefin main chain and metal units in its side chain, and the other is π-conjugated polymers, which has the luminescent metal groups in the chain ends.

2. Preparation of metallopolymers containing polyolefin main chain

In the previous paper, we have reported that luminescent polymers can be easily prepared by the reaction of the phosphorus ligand copolymers derived from MMA and 4-styryldiphenylphosphine with the iridium precursor under mild conditions.[8] Unfortunately, we found that the phosphorus side chain was easily oxidized to form oxide, probably leading to desorption of metal species from the metallopolymer. Alternatively, to improve the luminescent polymers, we used 4-vinylpyridine (Vp) as a linker module comonomer. It has been generally used as a ligand in metallopolymers.[9] In this chapter, we have reported preparation of a series of new luminescent metallopolymers. First, pyridine-containing polymers were produced as ligands for metal modules. Then, some amount of iridium complex precursor, $[IrCl(piq)_2]_2$ (**3**), where piq is 1-phenylisoquinoline, was added to the ligand polymer.

Scheme 6 depicts the synthetic procedure for the ligand copolymer **1** and the iridium-containing polymer **5**. MMA and Vp were copolymerized in the presence of benzoylperoxide (BPO) at 80 °C for 10 h to form the random copolymer **1**.[10] The number-averaged molecular weight of **1** was 963000 g/mol, determined by size-exclusion chromatography (SEC) which was calibrated using polystyrene standards. The ^1H NMR spectroscopy of **1** revealed that the content of Vp was 23 mol%. These copolymers reacted with $[IrCl(piq)_2]_2$ (**3**) in CH_2Cl_2 efficiently to form **5**, as a red compound. The expected quantity of the vinylpyridine iridium unit, [IrCl(piq)(Vp)], in metallopolymer **5** was 49 wt%, which was calculated by the feed ratios of **3** and Vp in **1**. The luminescent polymer of platinum analog **6** was also prepared efficiently by the reaction of **1** with a platinum precursor, $[PtCl(piq)]_2$ (**4**), in CH_2Cl_2.

In these luminescent polymers, MMA were used as a comonomer in order to reveal the properties of luminescent modules. However, as noted below, the EL efficiencies of the devices including such metallopolymers were extremely low, because of the poor conductivity of MMA copolymer. Therefore, we used *N*-vinylcarbazole (Vc) as an alternative comonomer instead of MMA to improve the conductivity and luminescence property. Poly(*N*-vinylcarbazole) (PVK) has been known as a host material for OLED component, performing high conductivity due to closely arranged π conjugated sites which hang from the polyolefin backbone as a pendant group.[11] Scheme 7 depicts the synthetic procedure for the iridium-containing polymers **7a** and **7b** from ligand copolymers **2a** and **2b**, which have different Vp contents, 4.7 and 15 mol%, respectively. The content of Vp was determined by absorption coefficient ratio for PVK at 345 nm in $CHCl_3$ (Table 1). The ligand copolymers **2a** and **2b** were prepared in the presence of AIBN.[12] The reaction of these copolymers with $[IrCl(piq)_2]_2$ (**3**) in CH_2Cl_2 gave red solutions, similar to the prior experimental result forming **5** (Table 2). The quantities of the iridium unit, $[IrCl(piq)_2(Vp)]_{y2}$, content in the metallopolymers **7a** and **7b** were 8.4 and 25 wt%, which were determined by the feed ratios of **3** and Vp in **2a-b** ·

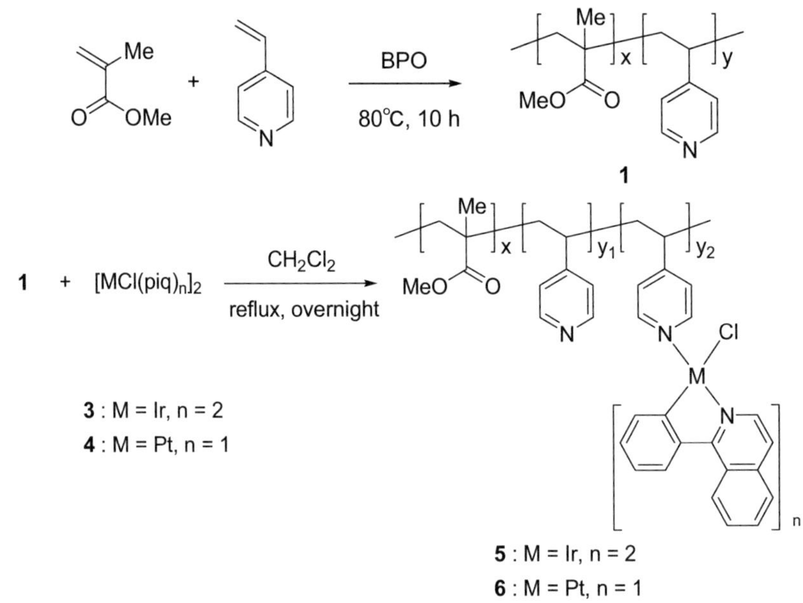

Scheme 6. Preparation of **1**, **5**, and **6**.

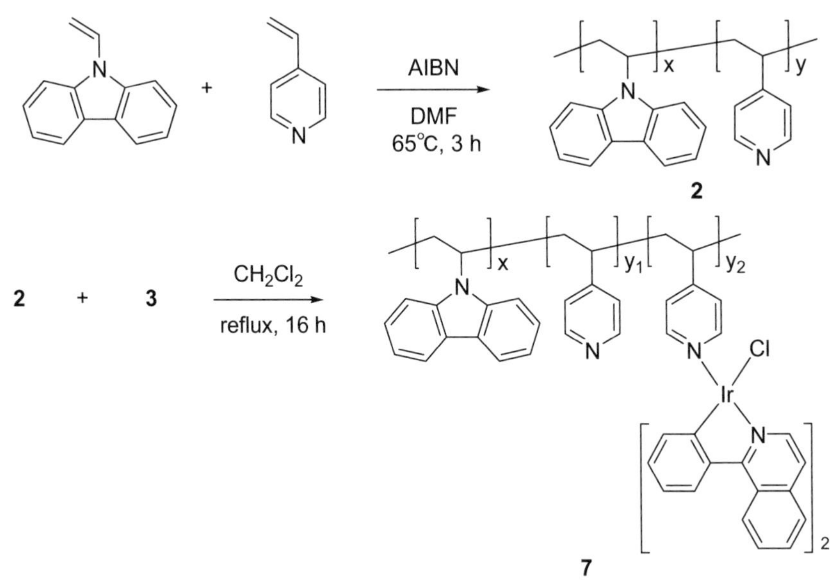

Scheme 7. Synthesis of copolymers **2a-b**, and metallopolymers **7a-b**.

The MMA-copolymerized metallopolymers **5** and **6** were readily soluble in several organic solvents, such as CH_2Cl_2, and $CHCl_3$, whereas the Vc-copolymerized metallopolymers **7** had poor solubility toward these solvents. Figure 5 shows the ^{1}H NMR spectra for the monomeric complex **14**, **1**, and **5**. Broadened signals due to the aromatic protons of **1** appeared from δ 8.5 to 8.2 and from δ 7.1 to 6.6 (Figure 5 (b)), whereas new broad resonances were observed from δ 10.1 to 6.1 (c), assigned as aromatic protons of the incorporated iridium unit in **5**, which provided the similar set of signals to those corresponding to **14** (a). The result suggested that the iridium unit in **5** has the same chemical structure as that of **14**. The spectra for **6** were similarly observed. Several broadened signals assigned as aromatic groups of the iridium unit in **7** were also observed as shown in Figure 6 (b) and (d). These signals were shifted to the higher field when these signals were compared with those due to the monomeric analog **14** (Figure 6 (e)), probably due to the shielding effect of the surrounding aromatic groups of the carbazole side chain.

Ligand Polymer	Comonomer			Initiator (mmol)		Yield (%)	Vp Content (mol%)	*Mn* (×104 g/mol)	PDI
	Vp (mmol)	Other (mmol)							
1	0.93	MMA	9.3	BPO	0.050	100	23	96	2.7
2a	1.0	Vc	20	AIBN	0.10	82	4.7	5.0	2.1
2b	2.0	Vc	20	AIBN	0.10	20	15	3.8	3.5

Table 1. Preparation of Ligand Polymers

Polymer Complex	Feed Polymer Ligand (Pyridine Content / mmol)		Feed Precursor Complex (Metal Content / mmol)		Yield (%)	$[IrCl(piq)_2(Vp)]_{y2}$ Content (wt%)
5	**1**	0.40	**3**	0.10	94	49
6	**1**	0.041	**4**	0.010	94	41
7a	**2a**	0.20	**3**	0.049	95	8.4
7b	**2b**	0.20	**3**	0.050	92	25

Table 2. Preparation of Metallopolymers

3. Preparation of metallopolymers containing conjugated polymer main chain

EL materials containing small molecules as doping luminescent compounds and conjugated polymers, such as PPV and PFO, have been developed as EL materials. Those performing more efficient luminescence have also been developed directly binding chromophores in the side chain of the conjugated polymers (Figure 4 and Scheme 3). [5,7f,7g] However, it is unknown that the luminescent iridium or platinum unit directory combines to the end of the conjugated polymers without breaking the π-conjugation, except one example.[13] The conjugated binding between the host polymer and the guest chromophore is expected that intramolecular electron transfer occurs easily. Here we developed new

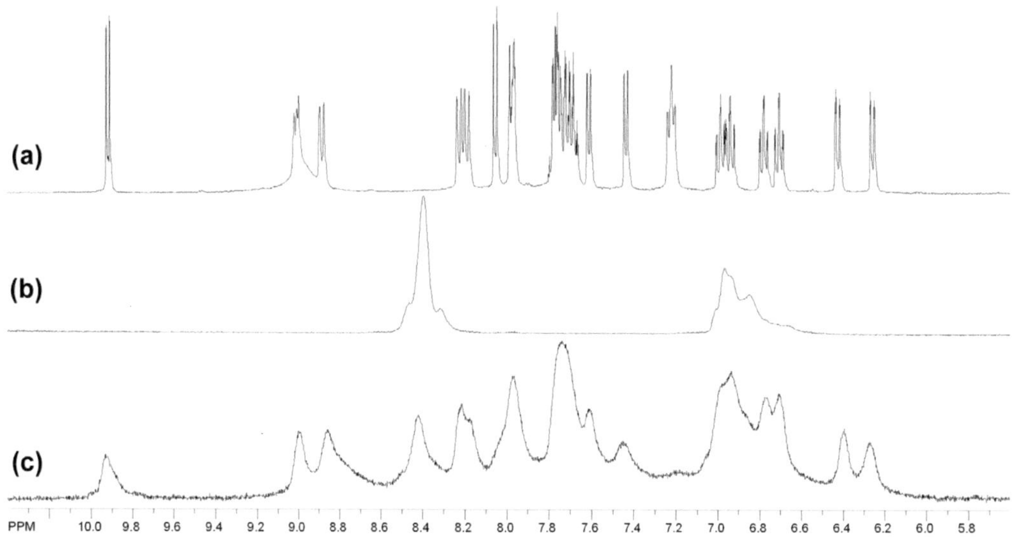

Fig. 5. ^{1}H NMR spectra in CD_2Cl_2 for (a) **14**, (b) ligand polymer **1**, and (c) metallopolymer **5** in the aromatic region.

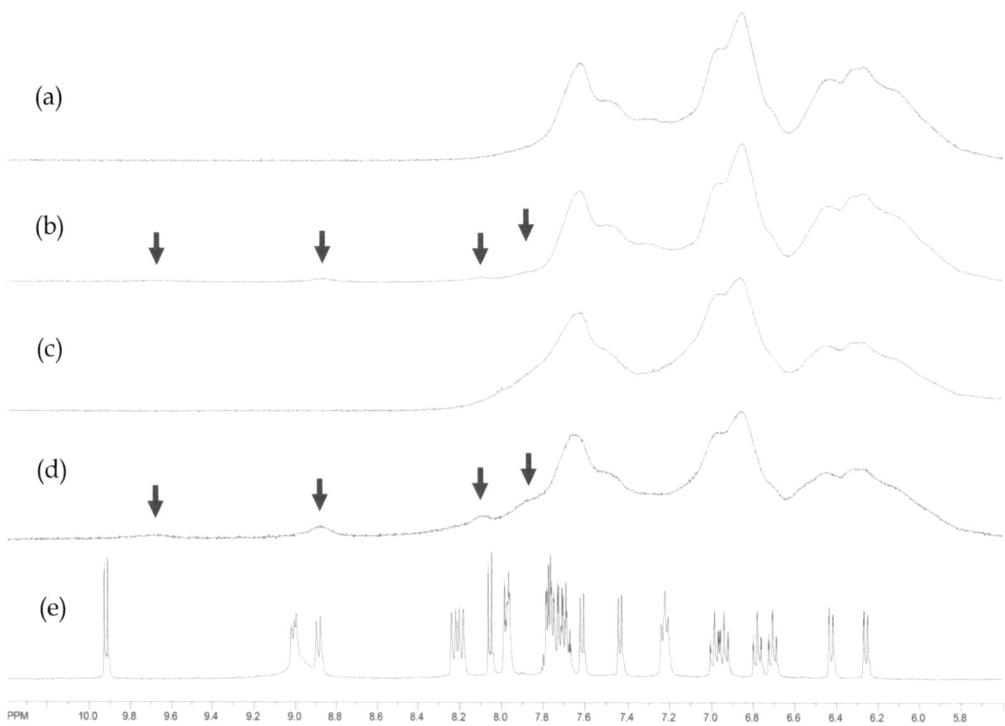

Fig. 6. ^{1}H NMR spectra in CD_2Cl_2 for (a) **2a**, (b) **7a**, (c) **2b**, (d) **7b** and (e) **14** in the aromatic region.

series of conjugated ligand polymers where a pyridyl group was introduced in the both ends of the polymers. As the previous preparation for the polyolefin metallopolymers, the luminescent 1-phenylisoquinoline iridium module was efficiently combined to the ligand polymers after polymerization. The procedures of the ligand polymers **8** and **9** and metallopolymers **10-13** are shown in Schemes 8 and 9. The conjugated polymer was prepared by the polycondensation reaction of two comonomers, 9,9-dioctylfluorene dibromide and bisboronic ester, mediated by a the catalyst system using palladium acetate, tri(*o*-methoxyphenyl)phosphine, and sodium carbonate. Subsequent reaction of the obtained mixture with pyridylboronic acid pinacol ester or pyridylboronic acid formed the end-capped ligand polymers **8a** and **8b** in 34 or 53% yield. The content of the pyridyl group was 2.9 or 2.0 mol%, which was determined by ^{1}H NMR spectrum. On the other hand, a comonomer, 4-*sec*-butylphenyl-*N,N*-bis(4-bromophenyl)amine, was added to the polycondensation, resulted in the formation of triarylamine-containing ligand copolymers **9a** and **9b**. The yields, averaged moleculer weights, and pyridyl-group contents are listed in Table 3. The reactions of the ligand polymers **8a** and **9a** with the iridium precursor **3** efficiently formed the metallopolymers **10** and **12**. The theoretical ratios of the iridium unit, [IrCl(piq)$_2$(py)] in **10** and **12** were 1.7 and 1.7 wt%. The platinum analogs **11, 13** were similarly synthesized from **8b, 9b** and the platinum precursor **5**. The yields and the content ratios of iridium and platinum are listed in Table 4.

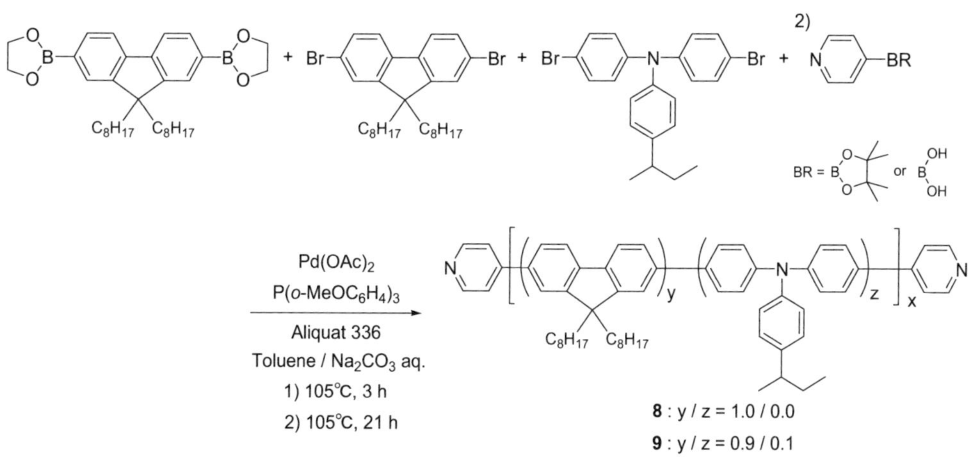

Scheme 8. Preparation of the conjugated ligand polymers **8** and **9**.

The metallopolymers **12** and **13**, in which the 4-*sec*-butylphenyl-diphenyleneamine group was included in the main chain, were readily soluble in several organic solvents, such as CH_2Cl_2, and $CHCl_3$, whereas the metallopolymers **10, 11** had poor solubility toward these solvents. The ^{1}H NMR spectra for **10** and **12** demonstrated no existence of the precursor **3** but small amount of the iridium unit binding to the terminal as shown in Figure 7. In addition to the signals around δ 8.7 and from 7.0 to 8.5, small new signals, most of which agreed with those due to the monomeric iridium analog **14**, appeared, suggesting that the iridium unit in **10** and **12** has the similar structure to **14**. The appearance of the signal at δ 8.7 in the spectra for the products indicated that the free pyridyl group at the end of the polymers still remained in part, even after combination of the iridium module. The spectra for the analogous polymers **11** and **13** containing platinum unit also showed the successful formation of the metallopolymers, in which the platinum unit has the similar structure to **15**.

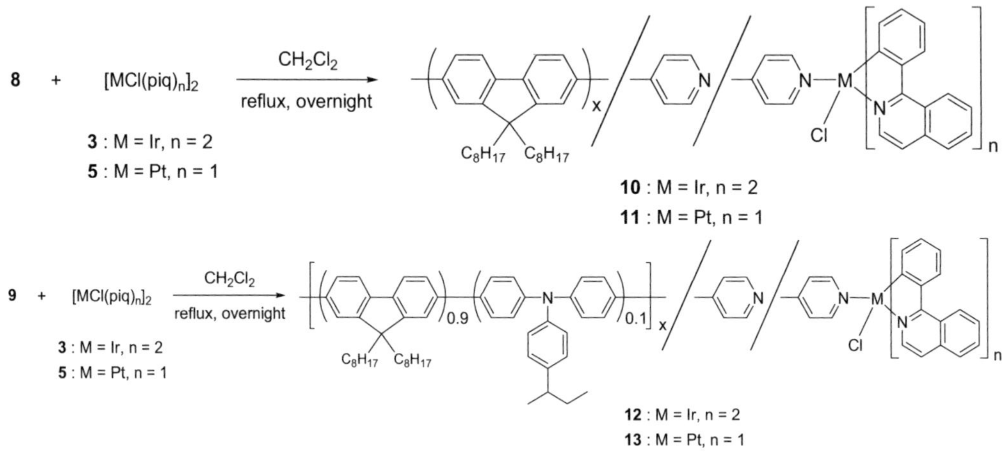

Scheme 9. Preparation of the conjugated metallopolymers **10-13**.

Ligand Polymer	Comonomer				Yield (%)	Pyridine Content (mol%)	*Mn* (×104 g/mol)	PDI
	FlBO (mmol)	FlBr (mmol)	PABr (mmol)	PyBO (mmol)				
8a[a]	5.1	6.2	0.0	1.8	34	2.9	2.9	1.4
8b	5.2	6.2	0.0	1.8	53	2.0	3.4	1.6
9a[a]	5.2	5.0	1.2	1.8	28	3.5	3.1	1.3
9b	5.4	5.0	1.2	1.8	52	2.6	3.9	1.4

[a] Pinacol ester instead of boronic acid was used as a coupling reagent in the synthesis of **8a** and **9a**.

Table 3. Preparation of the conjugated ligand copolymers.

Polymer Complex	Ligand Copolymer		Metal Precursor		Yield (%)	$[IrCl(piq)_2(py)]$ Content (wt%)
		Pyridine Content (mmol)		Metal Content (mmol)		
10	**8a**	0.029	**3**	0.0051	80	1.7
11	**8b**	0.032	**5**	0.0056	76	1.1
12	**9a**	0.016	**3**	0.0030	68	1.7
13	**9b**	0.029	**5**	0.0055	34	1.0

Table 4. Preparation of the conjugated metallopolymers.

4. PL behavior of metallopolymers

The corresponding pyridine-substituted monomeric complexes of iridium and platinum (Figure 8 (a)) were also prepared for the first time in order to estimate the luminescent behavior of the metallopolymers containing the similar metal units,[14] because polymer backbone sometimes influences luminescence of the metal units significantly both in the solid and solution states.[8, 15] The structures of these complexes were well defined and the PL behaviors were compared with those of the metallopolymers.[14] All the iridium and platinum complexes and polymers were irradiated at 462 and 433 nm, respectively, to observe the luminescence. As shown in Figure 8 (b), both the luminescent bands for the iridium and platinum complexes **14** and **15** were observed in almost the same area from 570 to 850 nm, although the low-energy absorption bands appeared in the different area (c).

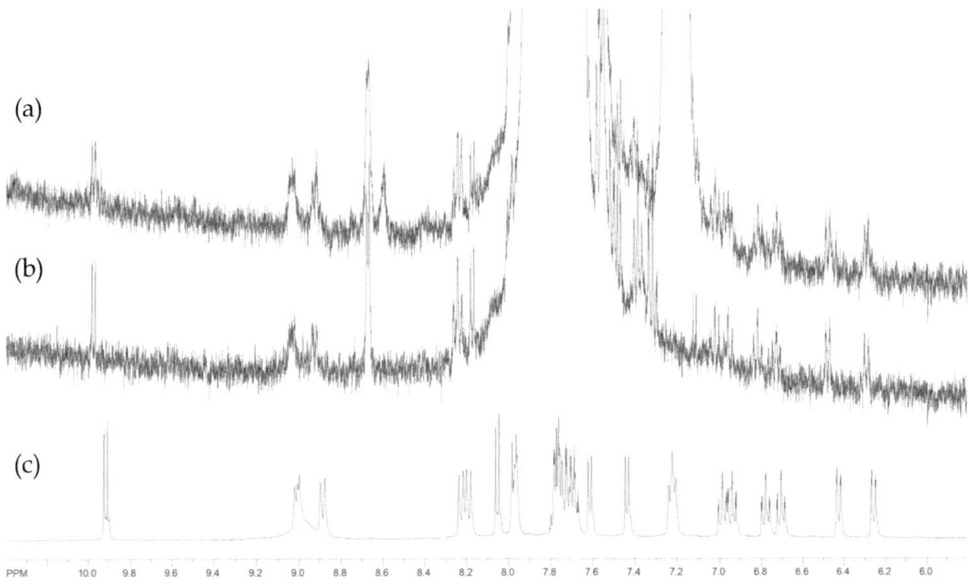

Fig. 7. ^1H NMR spectra in CD_2Cl_2 for (a) **12**, (b) **10** and (c) **14** in the aromatic region.

Figure 9 (a) and (b) show the PL spectra of the metallopolymers **5**, **6**, **7a** and **7b** in the CH_2Cl_2 solutions. Similar photoluminescent properties of the metallopolymers in the solutions to those of the monomeric metal analogs were observed when the concentrations of the metal units in the solutions were comparable. For example, red luminescence from **5** was observed and the wavelength at the maximum intensity was 620 nm. The PL spectra for **7** showed the emissions at ~620 nm from the incorporated [IrCl(piq)$_2$] unit. The intensity of the spectrum for the iridium polymer **5** was more than four times stronger than that for the platinum analog **6**. Lower intensity for **7a** than **7b** was coincident with the content ratios of the iridium unit in the metallopolymers (Table 2).

The PL spectra for the conjugated metallopolymers were shown in Figure 10. The spectra indicated that each structure of the metal unit was almost the same as that of the monomeric metal complex. The iridium-containing polymers **10** and **12** exhibited the similar luminescent spectra to that for the monomeric analog **14**, whereas the spectra for the platinum polymers **11** and **13** showed a characteristic strong band due to the conjugated main chain under 570 nm in addition to the luminescence emitted by the platinum unit. The result of the latter suggested that the low-energy absorption band of the conjugated main chain, which was irradiated simultaneously, lies on almost the same wavelength area as that of the platinum unit. It was confirmed by the electronic spectra for these compounds.

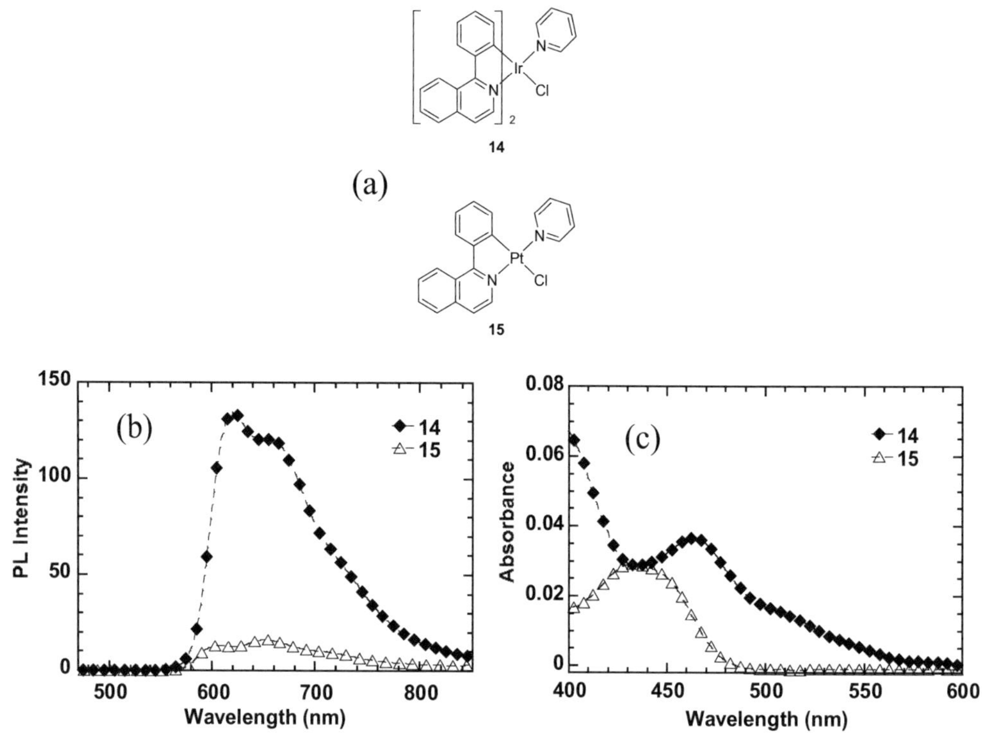

Fig. 8. (a) Structures, (b) PL and (c) UV-vis spectra of the monomeric complexes of iridium and platinum. (concentrations of the both complexes were 1×10^{-5} M.)

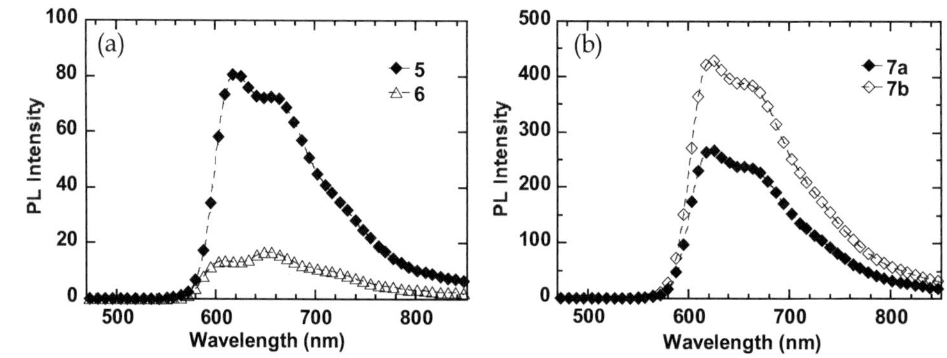

Fig. 9. PL spectra for (a) **5**, **6** and (b) **7**. Iridium polymers **5** and **7** and platinum polymer **6** were irradiated at 462 and 433 nm, respectively (metallopolymers **5**, **6** were 0.05 g/L and metallopolymers **7** were 0.5 g/L).

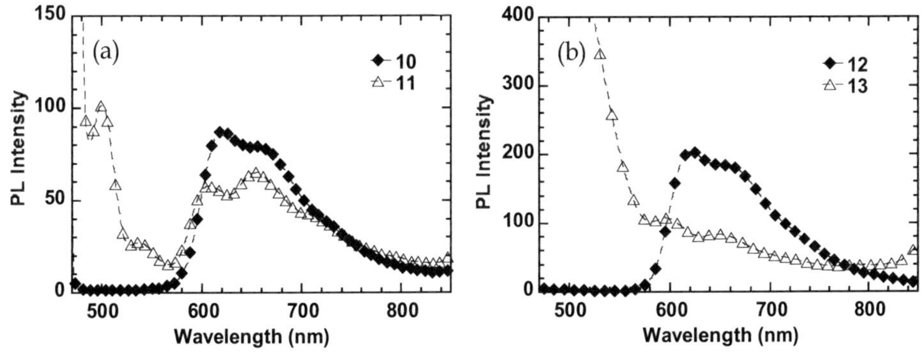

Fig. 10. PL spectra for (a) **10, 11**, (b) **12** and **13** in CH_2Cl_2. Iridium polymers **10** and **12** and platinum polymers **11** and **13** were irradiated at 462 and 433 nm, respectively (at 0.5 g/L).

5. EL behavior of the metallopolymers

EL behavior of devices containing the metallopolymers and the monomeric metal complexes was investigated. The metallopolymers composed of the conductive conjugated main chain exhibited the best performance among them, indicating that the structures of the polymer backbone closely affect the luminescence behavior, compared with those in the PL measurements.

We prepared several EL devices containing the luminescence layer of the metallopolymers as shown in Table 5. The first **device A** in entry 1 has only the iridium polymer **5** made from MMA / Vp copolymer, as the luminescence layer. The device performed 7.4 V of threshold voltage at 1 cd/m^2 and 0.026 cd/A (at 17.4 V) of maximum current efficiency, better than those of the previously reported iridium polymer made from 4-styryldiphenylphosphine copolymer, 13.2 V of threshold voltage at 1 cd/m^2 and 0.06 cd/A (at 20.0 V) of maximum current efficiency.[8] The better performance in **device A** than the phosphine-based copolymer is probably derived from improved conductivity of the pyridine-based copolymer.

When a host polymer **16** was added to the luminescent layer of **5** to make **devices B-D**, the threshold voltage and maximum current efficiency were improved to some extent. The threshold voltage decreased particularly to the comparable level to the **devices Q** and **R**, which were based on the monomeric iridium analog **14**. However, the dominant luminescence in these devices was from the host polymer **16** at 435 nm, indicating that energy transfer from the host to the guest polymer may be disrupted by the non-conductive polyolefin main chain and methacrylate side chain. The **device E** did not emit at least up to 20 V, whereas the emission was slightly observed from the platinum unit but mainly from the host polymer **16** in **device F**. The behavior was different from the **devices T** and **U**, in which the luminescence from the doped platinum complex (605 and 650 nm) into **16** was detected. These results strongly suggest the presence of non-conductive copolymers involved in the metallopolymers lead to poor EL performance.

The **devices G** and **J**, in which the Vc copolymer was involved instead of MMA copolymer, showed unexpected high threshold voltage, despite PVK was known as conductive host polymer in EL devices.[11] The **device S**, containing the monomeric **14**-doped PVK, also performed high threshold voltage, suggesting that PVK does not perform enough conductivity in the absence of conductive polymers. This high voltage was improved, when the conductive host polymer **16** was added to the luminescent layer. However, the maximum current efficiencies were not so different among **devices H, I** and **K, L**, in spite of the different iridium unit content ratios in these metallopolymers **7a** and **7b** (**7a** < **7b**, see Tables 1-3). Although the total performances of these devices based on the Vc copolymer were still not satisfactory, the energy transfer from the host polymer **16** to the metallopolymers occurred smoothly, leading to decrease of luminescence at 435 nm from the host **16**, in comparison with copolyMMA-based devices.

entry	Device[a]	Emitting Layer			Metal Unit Content[b]		V_{th}[c] (V)	$\eta_{c\,max}$[d] (cd/A, V)	λ_{max}[e] (nm)
		Host[f]	Guest	Feed Ratio (Host / Guest)		(wt%)			
1	**A**	—	**5**	0 / 100	Ir	49	7.4	0.026, 17.4	635
2	**B**	**16**	**5**	80 / 20	Ir	9.8	5.0	0.063, 7.4	430
3	**C**	**16**	**5**	90 / 10	Ir	4.9	5.6	0.15, 8.2	435
4	**D**	**16**	**5**	95 / 5	Ir	2.5	4.8	0.091, 12.6	435
5	**E**	—	**6**	0 / 100	Pt	41	—	—	—
6	**F**	**16**	**6**	95 / 5	Pt	2.1	5.6	0.096, 9.6	435
7	**G**	—	**7a**	0 / 100	Ir	8.4	19.2	0.026, 20.0	625
8	**H**	**16**	**7a**	60 / 40	Ir	3.4	4.2	0.13, 6.8	430
9	**I**	**16**	**7a**	80 / 20	Ir	1.7	4.6	0.14, 7.6	430
10	**J**	—	**7b**	0 / 100	Ir	25	11.0	0.082, 20.0	630
11	**K**	**16**	**7b**	60 / 40	Ir	10	4.4	0.12, 6.4	430
12	**L**	**16**	**7b**	80 / 20	Ir	5	4.0	0.097, 5.8	430
13	**M**	—	**10**	0 / 100	Ir	1.7	4.0	1.14, 4.0	625
14	**N**	—	**11**	0 / 100	Pt	1.1	5.4	0.14, 7.6	605, 650
15	**O**	—	**12**	0 / 100	Ir	1.7	3.4	0.47, 3.6	620
16	**P**	—	**13**	0 / 100	Pt	1.0	5.8	0.36, 10.0	605, 650
17	**Q**	**16**	**14**	90 / 10	Ir	10	6.4	0.31, 9.2	625
18	**R**	**16**	**14**	95 / 5	Ir	5.0	5.0	0.30, 8.2	615
19	**S**	PVK	**14**	95 / 5	Ir	5.0	8.0	0.014, 19.8	630
20	**T**	**16**	**15**	90 / 10	Pt	10	10.0	0.024, 16.2	605, 650
21	**U**	**16**	**15**	95 / 5	Pt	5.0	6.8	0.048, 9.8	435

[a] Device structure: ITO/PEDOT:PSS/Emitting layer/Ba/Al
[b] Metal unit is [MCl(piq)$_n$(Py-)] (n = 2, Ir; n = 1, Pt) or the monomeric complex in the emitting layer.
[c] Threshold voltage at 1 cd/m^2.
[d] Maximum current efficiency.
[e] The λ_{max} values correspond to the highest intensity peak in the EL spectrum at maximum current efficiency.

[f] 16:

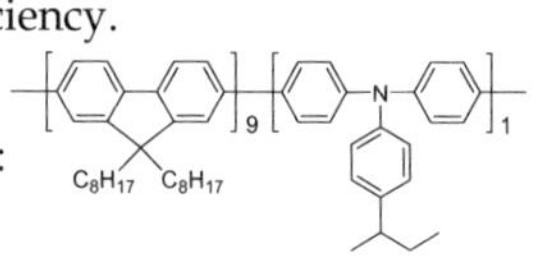

Table 5. EL properties of the devices containing the metallopolymers

The **devices M, N, O,** and **P** containing metal end-capped conjugated polymers provided satisfactory luminescence performances, compared with the other devices. As shown in Figure 11, negligible luminescence around 435 nm derived from the conjugated main chain was observed in the **devices M** and **O** containing iridium-capped polymers **10** and **12**, whereas considerable luminescence from the conjugated main chain appeared in the platinum-based **devices N** and **P**. We can conclude that iridium-based devices are superior to platinum-based ones in energy-transfer ability in this EL device system. The **device O** showed the highest performance as a red EL device among all the devices. It is of interest that the performances of the **devices M, O, N, P** excelled those of the **devices Q, R, T, U,** which contained the layer of the monomeric complex **14**- or **15**-doped copolymer **16**. We found that these **devices M, O, N, P** showed more than 1 V lower threshold voltages than those of the **devices Q, R, T, U**. These devices have the same structure except whether the metal chromophore is bound to the end of the host polymer (**M, O, N, P**) or exists independently (**Q, R, T, U**). We considered that direct combination of the conductive polymer and the metal unit led to facile electron transfer to the metal unit, resulting in low threshold voltages and high current efficiency of these devices. As for the iridium unit-containing devices, additional easy energy transfer from the host polymer to iridium caused the highest performance.

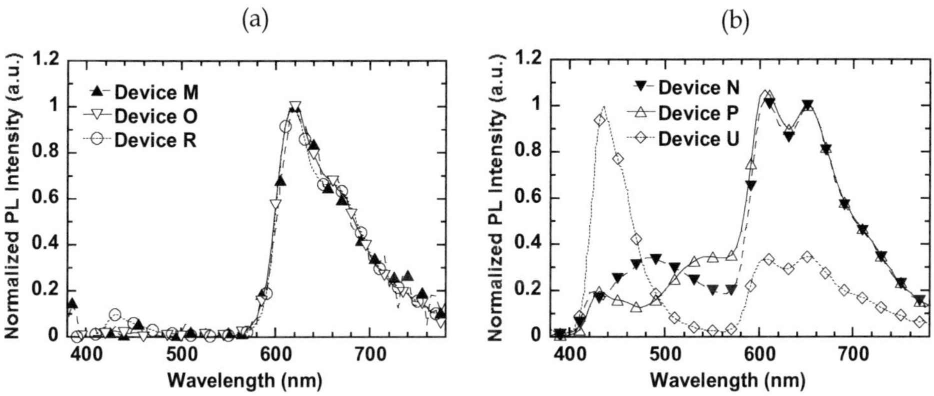

Fig. 11. EL spectra for (a) **devices M, O** and **R**, (b) **devices N, P** and **U**, of which the structures are shown in Table 5. (at 4.0, 4.0, 8.0, 8.0, 10.0, and 10.0 V, respectively) The origin of the small luminescnet bands from 480 to 570 nm in (b) is not identified.

6. Conclusion

One of the most important factors to design new devices that contain complicated organic/inorganic/polymeric compounds is how to prepare the compounds easily and efficiently. Here we described the successful preparation of several luminescent polymer materials in a few steps, that contained the simple coordination of the metal module precursor to the pyridine-bound ligand polymers under mild conditions. After several attempts to investigate the EL behavior of the devices containing the obtained metallopolymers, we found that structure of backbone host polymer is quite important for efficient luminescence and low driving voltage in these devices. We also demonstrated that the good EL performance was provided when the guest unit directly bound to the host polymer.

7. Experimental details

7.1 Synthesis of pyridine-capped conjugated copolymers

As a typical example, into a 200-mL three-necked flask equipped with a condenser, 2.77 g (5.2 mmol) of 9,9-dioctylfluorene-2,7-bis(boronic acid ethylene glycol ester), 2.72 g (5.0 mmol) of 9,9-dioctyl-2,7-dibromofluorene, 0.551 g (1.2 mmol) of 4-(1-methylpropyl)-*N*,*N*-bis(4-bromophenyl)aniline, 0.79 g of methyltrioctylammonium chloride (Aliquat 336, made by Sigma-Aldrich Corporation), and 60 mL of toluene were placed. Under a nitrogen atmosphere, 2.2 mg of palladium diacetate and 12.9 mg of tris(2-methoxyphenyl)phosphine were added to the solution, and the solution was heated to 95°C. While a 17.5 wt% sodium carbonate aqueous solution (16.5 mL) was dropped to the obtained solution over 30 minutes, the solution was heated to 105°C, and subsequently stirred at 105°C for 3 hours. Then, 369 mg of 4-(4,4,5,5-tetramethyl-1,3,2-dioxaborolan-2-yl)pyridine dissolved in toluene (30 mL) was added, and the mixture was stirred at 105°C for 21 hours. After the aqueous layer was removed, 3.65 g of sodium *N*,*N*-diethyldithiocarbamate trihydrate and 36 mL of water were added, and the solution was stirred at 85°C for 2 hours. An organic layer was separated and washed with water (78 mL, twice), a 3 wt% aqueous acetic acid (78 mL, twice), and then water (78 mL, twice). The organic layer was dropped to methanol to form precipitates, which were filtrated and dried to obtain a solid. The residual solid was dissolved in toluene (186 mL), and the solution was passed through a silica gel / alumina column, where toluene was passed in advance. The filtrate was concentrated under reduced pressure and dropped into methanol, and a precipitate was filtered to obtain ligand polymer **9a** (1.26 g). The number-averaged molecular weight M_n was 3.1×10^4 g/mol, which was determined by SEC calibrated with polystyrene standards.

7.2 Synthesis of conjugated iridium polymers

As a typical example, under an inert-gas atmosphere, a mixture of $[IrCl(piq)_2]_2$ (**3**) (0.0038 g, 0.0030 mmol) and pyridine-capped copolymer **9a** (0.243 g, containing 0.016 mmol of pyridine) in CH_2Cl_2 (6 mL) was refluxed for 16 h. After cooling to room temperature, the resulting solution was poured into hexane to afford a precipitate, which was filtered and washed with hexane and dried under reduced pressure to obtain light orange powder **10** in 80 % yield ($M_n = 3.3\times10^{4}$l g/mol).

8. Acknowledgement

The EL experiments in this study were conducted with kind supports of Sumitomo Chemical Co., Ltd.

9. Abbreviations

PL: photo-luminescent
EL: electro-luminescent
PLED: polymer light-emitting diode
OLED: organic light-emitting diode
PPV: polyphenylene vinylene
PVK: poly(vinylcarbazole)
PFO: poly(9,9-di-n-octyl-2,7-fluorene)

MMA: methyl methacrylate
Vp: 4-vinylpyridine
piq: 1-phenylisoquinoline
SEC: size-exclusion chromatography
Vc: *N*-vinylcarbazole
AIBN: azobisisobutylonitrile
BPO: benzoylperoxide
FlBO: 9,9-dioctylfluorene-2,7-bis(boronic acid ethylene glycol ester)
FlBr: 9,9-dioctyl-2,7-dibromofluorene
PABr: 4-*sec*-butylphenyl-*N,N*-bis(4-bromophenyl)amine
PyBO: 4-(4,4,5,5-tetramethyl-1,3,2-dioxaborolan-2-yl)pyridine boronic acid

10. References

[1] (a) Lee, C.-L.; Lee, K. B.; Kim, J.-J. Appl Phys Lett 2000, 77, 2280–2282; (b) Negres, R. A.; Gong, X.; Ostrowski, J. C.; Bazan, G. C.; Moses, D.; Heeger, A. J. Phys Rev B 2003, 68, 115209; (c) Zhu, W. G.; Ke, Y.; Wang, F.; Liu, C. Z.; Yuan, M.; Cao, Y. Synth Met 2003, 137, 1079–1080; (d) van Dijken, A.; Bastiaanssen, J. J. A. M.; Kiggen, N. M. M.; Langeveld, B. M. W.; Rothe, C.; Monkman, A.; Bach, I.; Stössel, P.; Brunner, K. J Am Chem Soc 2004, 126, 7718–7727; (e) Tanaka, I.; Tabata, Y.; Tokito, S. Jpn J Appl Phys 2004, 43, L1601–L1603; (f) Xia, H.; Zhang, C.; Qiu, S.; Lu, P.; Zhang, J.; Ma, Y. Appl Phys Lett 2004, 84, 290–292; (g) Gong, X.; Robinson, M. R.; Ostrowski, J. C.; Moses, D.; Bazan, G. C.; Heeger, A. J. Adv Mater 2002, 14, 581–585; (h) Jiang, C.; Yang, W.; Peng, J.; Xiao, S.; Cao, Y. Adv Mater 2004, 16, 537–541; (i) Adachi, C.; Kwong, R.; Forrest, S. R. Org Electron 2001, 2, 37–43; (j) Lee, C.-C.; Yeh, K.-M.; Chen, Y. J Polym Sci Part A: Polym Chem 2008, 46, 7960–7971; (k) Wang, P.; Chai, C.; Yang, Q.; Wang, F.; Shen, Z.; Guo, H.; Chen, X.; Fan, X.; Zou, D.; Zhou, Q. J Polym Sci Part A: Polym Chem 2008, 46, 5452–5460; (l) Yuan, M.-C.; Shih, P.-I.; Chien, C.-H.; Shu, C.-F. J Polym Sci Part A: Polym Chem 2007, 45, 2925–2937.
[2] Kraft, A.; Grimsdale, A. C.; Holmes, A. B. Angew Chem Int Ed 1998, 37, 402.
[3] Lee, C.-L.; Kang, N.-G.; Cho, Y.-S.; Lee, J.-S.; Kim, J.-J. Opt Mater 2002, 21, 119–123
[4] Tokito, S.; Suzuki, M.; Sato, F.; Kamachi, M.; Shirane, K. Org Electron 2003, 4, 105–111.
[5] Chen, X.; Liao, J.-L.; Liang, Y.; Ahmed, M. O.; Tseng, H.-E.; Chen, S.-A. J Am Chem Soc 2003, 125, 636–637.
[6] (a) Xu, Y.; Guan, R.; Jiang, J.; Yang, W.; Zhen, H.; Peng, J.; Cao, Y. J Polym Sci Part A: Polym Chem 2008, 46, 453–463; (b) Mei, C.; Ding, J.; Yao, B.; Cheng, Y.; Xie, Z.; Geng, Y.; Wang, L. J Polym Sci Part A: Polym Chem 2007, 45, 1746–1757.
[7] (a) Yamada, Y. M. A.; Takeda, K.; Takahashi, H.; Ikegami, S. Org Lett 2002, 4, 3371–3374; (b) Yamada, Y. M. A.; Takeda, K.; Takahashi, H.; Ikegami, S. J Org Chem 2003, 68, 7733–7741; (c) Bianchini, C.; Frediani, M.; Vizza, F. Chem Commun Commun 2001, 479–480; (d) Borbone, F.; Caruso, U.; Maria, A. D.; Fusco, M.; Panunzi, B.; Roviello, A. Macromol Symp 2004, 218, 313–321; (e) Marin, V.; Holder, E.; Hoogenboom, R.; Schubert, U. J Polym Sci Part A: Polym Chem 2004, 42, 4153–4160; (f) Deng, L.; Furuta, P. T.; Garon, S.; Li, J.; Kavulak, D.; Thompson, M. E.; Fre´chet, J. M. J Chem Mater 2006, 18, 386–395; (g) Schulz, G. L.; Chen, X.; Chen, S.-A.; Holdcroft, S. Macromolecules 2006, 39, 9157–9165; (h) Aamer, K. A.; Tew, G. N. J Polym Sci Part A: Polym Chem 2007, 45, 1109–1121.

[8] Koga, Y.; Yoshida, N.; Matsubara, K. J Polym Sci Part A: Polym Chem 2009, 47, 4366–4378.
[9] (a) Daniel, S.; Gladis, J. M.; Rao, T. P. Anal Chim Acta 2003, 349, 173–182; (b) Sumi, V. S.; Kala, R.; Praveen, R. S.; Rao, T. P. Int J Pharm 2008, 349, 30–37; (c) Fan, P.; Wang, B. J Appl Polym Sci 2010, 116, 258–266.
[10] Salahuddin, N. J. Appl. Polym. Sci. 2007, 104, 3317-3323
[11] (a) Xiao, J.; Yao, Y.; Deng, Z.; Wang, X.; Liang, C. J Lumin 2007, 122–123, 639–641; (b) Ye, T.; Chen, J.; Ma, D. Phys. Chem. Chem. Phys. 2010, 12, 15410–15413; (c) Song, M.; Park, J. S.; Yoon, M.; Kim, A. J.; Kim Y. I.; Gal, Y.-S.; Lee, J. W.; Jin, S.-H. J Organomet Chem 2011, 696, 2122-2128.
[12] Zhang, H.; Zhou, Z.; Liu, K.; Wang, R.; Yang, B. J. Mater. Chem. 2003, 13, 1356-1361.
[13] Zhang, K.; Chen, Z.; Yang, C.; Zou, Y. ; Gong, S.; Tao, Y.; Qin, J.; Cao, Y. J Mater Chem 2008, 18, 3366–3375.
[14] Koga, Y.; Matsumoto, T.; Matsubara, K. unpublished results.
[15] Koga, Y.; Ueno, K.; Matsubara, K. J Polym Sci Part A: Polym Chem 2006, 44, 4204–4213.

2

Synthesis and Photophysical Properties of Pyrene-Based Multiply Conjugated Shaped Light-Emitting Architectures: Toward Efficient Organic-Light-Emitting Diodes

Jian-Yong Hu[1,2] and Takehiko Yamato[1]
[1]*Department of Applied Chemistry, Faculty of Science and Engineering, Saga University*
[2]*Department of Organic Device Engineering, Yamagata University*
Japan

1. Introduction

Since the pioneering works on the first double-layer thin-film Organic electroluminescence (EL) devices (OLEDs) by C. W. Tang and co-workers in the Kodak Company in 1987 (Tang & Vanslyke, 1987), OLEDs have attracted enormous attentions in the scientific community due to their high technological potential toward the next generation of full-color-flat-panel displays (Hung & Chen, 2002; Wu et al., 2005; Geffroy et al., 2006) and lighting applications (Duggal et al., 2007; So et al., 2008). In today's developments of OLED technologies, the trends of organic EL devices are mainly focusing both on optimizations of EL structures and on developing new optoelectronic emitting materials. Obviously the key point of OLEDs development for full-color-flat display is to find out materials emitting pure colors of red, green and blue (RGB) with excellent emission efficiency and high stability. Numerous materials with brightness RGB emission have been designed and developed to meet the requirements toward the full-color displays. Among them, organic small molecules containing polycyclic aromatic hydrocarbons (PAHs) (e. g. naphthalene, anthracene, perylene, fluorene, carbazole, pyrene, etc.) are well known and are suitable for applications in OLEDs. Recently, naphthalene, anthracene, perylene, fluorene, carbazole, pyrene and their derivatives have been widely used as efficient electron-/hole-transporting materials or host emitting materials in OLED applications. In this chapter an overview is presented of the synthesis and photophysical properties of pyrene-based, multiply conjugated shaped, fluorescent light-emitting materials that have been disclosed in recent literatures, in which several pyrenes have been successfully used as efficient hole-/electron-transporting materials or host emitters or emitters in OLEDs, by which a series of pyrene-based, cruciform-shaped π-conjugated blue-light-emitting architectures can be prepared with an emphasis on how synthetic design can contribute to the meeting of promising potential in OLEDs applications.

2. Pyrene and pyrene derivatives

Pyrene is an alternant polycyclic aromatic hydrocarbon (PAH) and consists of four fused benzene rings, resulting in a large, flat aromatic system. Pyrene is a colorless or pale yellow solid, and pyrene forms during incomplete combustion of organic materials and therefore can be isolated from coal tar together with a broad range of related compounds. Pyrene has been the subject of tremendous investigation. In the last four decades, a number of research works have been reported on both the theoretical and experimental investigation of pyrene concerning on its electronic structure, UV-vis absorption and fluorescence emission spectrum. Indeed, this polycyclic aromatic hydrocarbon exhibits a set of many interesting electrochemical and photophysical attributes, which have results in its utilization in a variety of scientific areas. Some recent advanced applications of pyrene include fluorescent labelling of oligonucleotides for DNA assay (Yamana et al., 2002), electrochemically generated luminescence (Daub et al., 1996), carbon nanotube functionallization (Martin et al., 2004), fluorescence chemosensory (Strauss & Daub, 2002; Benniston et al., 2003), design of luminescence liquid crystals (de Halleux et al., 2004), supermolecular self-assembly (Barboiu et al., 2004), etc.. On the other hand, as mentioned above, PAHS (e. g. naphthalene, anthracene, perylene, fluorene, carbazole, etc.) and their derivatives have been developed as RGB emitters in OLEDs because of their promising fluorescent properties (Jiang et al., 2001; Balaganesan et al., 2003; Shibano et al., 2007; Liao et al., 2007; Thomas et al., 2001). In particular, these compounds have a strong π-electron delocalization character and they can be substituted with a range of functional groups, which could be used for OLEDs materials with tuneable wavelength. Similarly, pyrene has strong UV-vis absorption spectra between 310 and 340 nm and emission spectra between 360 and 380 nm (Clar & Schmidt, 1976), especially its expanded π-electron delocalization, high thermal stability, electron accepted nature as well as good performance in solution. From its excellent properties, it seems that pyrene is suitable for developing emitters to OLEDs applications; however, the use of pyrene molecules is limited, because pyrene molecules easily formed π-aggregates/excimers and the formation of π-aggregates/excimers leads to an additional emission band in long wavelength and the quenching of fluorescence, resulting in low solid-state fluorescence quantum yields. Recently, this problem is mainly solved by both the introduction of long or big branched side chains into pyrene molecules and co-polymerization with a suitable bulky co-monomer. Very recently, it was reported that pyrene derivatives are useful in OLEDs applications (Otsubo et al., 2002; Thomas et al., 2005; Ohshita et al., 2003; Jia et al., 2004; Tang et al., 2006; Moorthy et al., 2007) as hole-transporting materials (Thomas et al., 2005; Tang et al., 2006) or host blue-emitting materials (Otsubo et al., 2002; Ohshita et al., 2003; Jia et al., 2004; Moorthy et al., 2007). To date, various pyrene-based light-emitting materials have been disclosed in recent literatures, which can be roughly categorized into three types of materials: (1) *Functionalized pyrene-based light-emitting monomers;* (2) *Functionalized pyrene-based light-emitting dendrimers*; and (3) *Functionalized pyrene-based light-emitting oligomers and polymers.*

3. Functionalized pyrene-based light-emitting monomers

Because of its extensive π-electron delocalization and electron-accepted nature, pyrene is a fascinating core for developing fluorescent π-conjugation light-emitting monomers. In those compounds, pyrene was used as a conjugation centre core substituted by some functionalized groups or as function substituents introduced into others PAHs rings. In this section, the synthesis and photophysical properties of two types of functionalized pyrene-based light-emitting monomers, namely, *pyrene-cored organic light-emitting monomers* and *pyrene-functionalized PAHs-cored organic light-emitting monomers* were fully presented. In particular, the use of these light-emitting monomers as efficient emitters in OLEDs will be discussed in detail.

3.1 Pyrene-cored organic light-emitting monomers

Although pyrene and its derivatives have been widely used as fluorescence probes in many applications, there are two major drawbacks using pyrene as a fluorescence probe: One is the absorption and emission wavelengths of the pyrene monomer are confined to the UV region of 310-380 nm, and the other is pyrene can easily forms an excimer above concentrations of 0.1 mM. In order to probe biological membranes using fluorescence techniques it is desirable to have a fluorophore probe that absorbs and emits in the long wavelength region, preferably in the visible region of the electromagnetic spectrum in order to minimize the spectral overlap of the intrinsic fluorescence of the bio-molecules that occur in the UV region. Furthermore, molecular systems that are light emitters in the visible region are potentially useful in the fabrication of organic light emitting diodes (OLEDs). Therefore, it is desirable to design molecules that have emission in the visible region. Consequently, the most common method to bathochromically shift the absorption and emission characteristics of a fluorophore is to extend the π-conjugation by introducing unsaturated functional groups (e. g. acetylenic group) or rigid, bulky PAHs moieties (e. g. phenylene, thiophene, bithiophene, thienothiophene, benzothiadiazole-thiophene, pridine, etc.) to the core of the fluorophore. In recent papers, using pyrene as a conjugation centre core, the synthesis, absorption and fluorescence-emission properties of the 1,3,6,8-tetraethynylpyrenes and its derivatives have been reported (Venkataramana & Sankararaman 2005, 2006; Fujimoto et al., 2009), and monomers of 1-mono, 1,6-bis-, 1,8-bis-, 1,3,6-tris-, and 1,3,6,8-tetrakis-(alkynyl)pyrenes have also been prepared (Maeda et al., 2006; Kim et al., 2008; Oh et al., 2009). On the other hand, 1,3,6,8-tetraarylpyrenes as fluorescent liquid-crystalline columns (de Halleux et al., 2004; Sienkowska et al., 2004) or organic semiconductors for organic field effect transistors (OFETs) (Zhang et al., 2006) or efficient host blue emitters (Moorthy et al., 2007; Sonar et al., 2010) or electron transport material (Oh et al., 2009) have recently been reported. The starting point for the above-mentioned synthesis was 1-mono (**2a**), 1,6-di-(**2b**), 1,8-di-(**2c**), 1,3,6-tris-(**2d**), and 1,3,6,8-tetrabromopyrenes (**2e**), which is readily prepared by electrophinic bromination of pyrene (**1**) with one to excess equivalents of bromine under the corresponding reaction conditions, respectively (Grimshaw et al., 1972; Vollmann et al., 1937) (Scheme 1). These materials were consequently converted to the corresponding alkynylpyrenes (pyrene-C≡C-R) or arylpyrenes (pyrene-R) by Sonogashira cross-coupling reaction or Suzuki cross-coupling reaction, respectively.

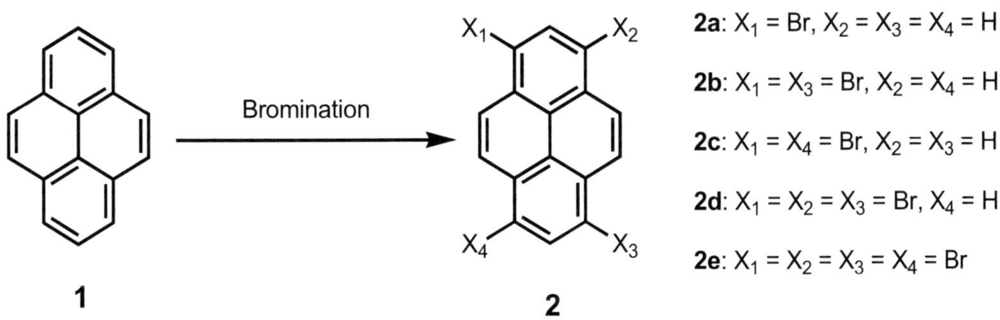

Scheme 1. Electrophilic bromination of pyrene (**1**)

3.1.1 Alkynyl-functionalized pyrene-cored light-emitting monomers

Acetylene has been widely applied for linking π-conjugated units and for effectively extending the π-conjugation length. The progress of such π-conjugated materials by means of acetylene chemistry has strongly dependent on the development of Sonogashira coupling reaction. Thereby, many attractive acetylene-linked molecules have emerged such as for semiconducting polymers (Swager et al., 2005; Swager & Zheng, 2005), macrocyclic molecules (Kawase, 2007; Hoger et al., 2005), helical polymers (Yamashita & Maeda, 2008) and energy transfer cassettes (Loudet et al., 2008; Han et al., 2007; Jiao et al., 2006; Bandichhor et al., 2006). Accordingly, the use of acetylene group for extending the conjugation of the pyrene chromophore is one of the most common methods. Sankararaman *et al.* (Venkataramana & Sankararaman, 2005) reported the synthesis, absorption and fluorescence-emission of 1,3,6,8-tetraethynylpyrene derivatives **3a-f**, which were prepared by the Sonogashira coupling of tetrabromomide (**2e**) with various terminal acetylenes yielded the corresponding tetraethynylpyrenes. Significant bathochromic shifts of absorptions band were observed in the region of 350-450 nm for **3a-d**, 375-474 nm for **3e-f**, respectively, compare with that of pyrene (**1**) in dilute THF solutions due to the extended conjugation of the pyrene chromophore with the acetylenic units. Similarly, the fluorescence emission bands of **3a-f** are also bathochromically shifted in region of 420-550 nm in comparison of pyrene in THF. The quantum efficiency of fluorescence emission for **3a-d** was in the rang of 0.4-0.7; these values are comparable to that of pyrene, while **3e** and **3f** are low due to the deactivation of the excited state resulting from the free rotation of the phenyl groups. The results suggest these derivatives are potentially useful as emitters in the fabrication of organic light emitting diodes (OLEDs). A pyrene octaaldehyde derivative **4** and its aggregations through π-π and C-H···O interactions in solution and in the solid state probed by its fluorescence emission and other spectroscopic methods are also prepared by Sankararaman *et al.* (Venkataramana & Sankararaman, 2006) In view of its solid-state fluorescence, this octaaldehyde **4** and its derivatives might find applications in the field of molecular optoelectronics. Similarly, Fujimoto and co-workers (Maeda et al., 2006) have synthesized a variety of alkynylpyrene derivatives **5a-d** from mono- to tetrabromo-pyrenes (**2a-2e**) and arylacetylenes using the Sonogashira coupling, and comprehensively examined their photophysical properties. The alkynylpyrenes **5a-h** thus prepared showed not only long absorption (365-434 nm, 1.0×10^{-5} M, in EtOH) and fluorescence emission (386-438 nm, $1.6\text{-}2.5 \times 10^{-7}$ M, in EtOH) wavelengths but also high fluorescence quantum yields (0.55-0.99, standards used were 9,10-diphenylanthracene) as compared with pyrene itself. Additionally, the alkynylpyrene skeletons could be applied to practically useful fluorescence probes for proteins and DNAs. Fujimoto *et al.* (Fujimoto et al., 2009) recently also prepared a series of 1,3,6,8-tetrakis(arylethynyl)pyrenes **6a-e** bearing electron-donating or electron-withdrawing groups. Their photophysical properties analysis demonstrated that the donor-modified tetrakis(arylethynyl)pyrene **6a-c** showed fluorescence solvatochromism on the basis of intramolecular charge transfer (ICT) mechanism, while the acceptor-modified ones **6d-e** never did. Furthermore, the donor-modified tetrakis(arylethynyl)pyrene **6a-c** were found to be stable under laboratory weathering as compared with that of coumarin. Thus, the tetrakis(arylethynyl)pyrenes **6** are expected to be applicable to bioprobes for hydrophobic pockets in various biomolecules and photomaterials.

More recently, Kim *et al.* prepared a series of alkynylpyrenes **7a-e** that bear peripheral [4-(*N,N*-dimethylamino)phenylethynyl] (DMA-ethynyl) units using pyrene as the π-center and their two-photon absorption properties (Kim et al., 2008) and electrogenerated chemiluminescence (ECL) properties (Oh et al., 2009) were investigated in detail, respectively. These alkynylpyrenes **7a-e** showed unique patterns in photophysical and electrochemical properties. For example, compound **7e**, which has four peripheral DMA-ethynyl moieties, exhibits a marked enhancement in ECL intensity compared to the other compounds **7a-7d**; this is attributable to its highly conjugated network that gives an extraordinary stability of cation and anion radicals in oxidation and reduction process, respectively. The result is a promising step in the development of highly efficient light-emitting materials for applications such as organic light-emitting diodes (OLEDs).

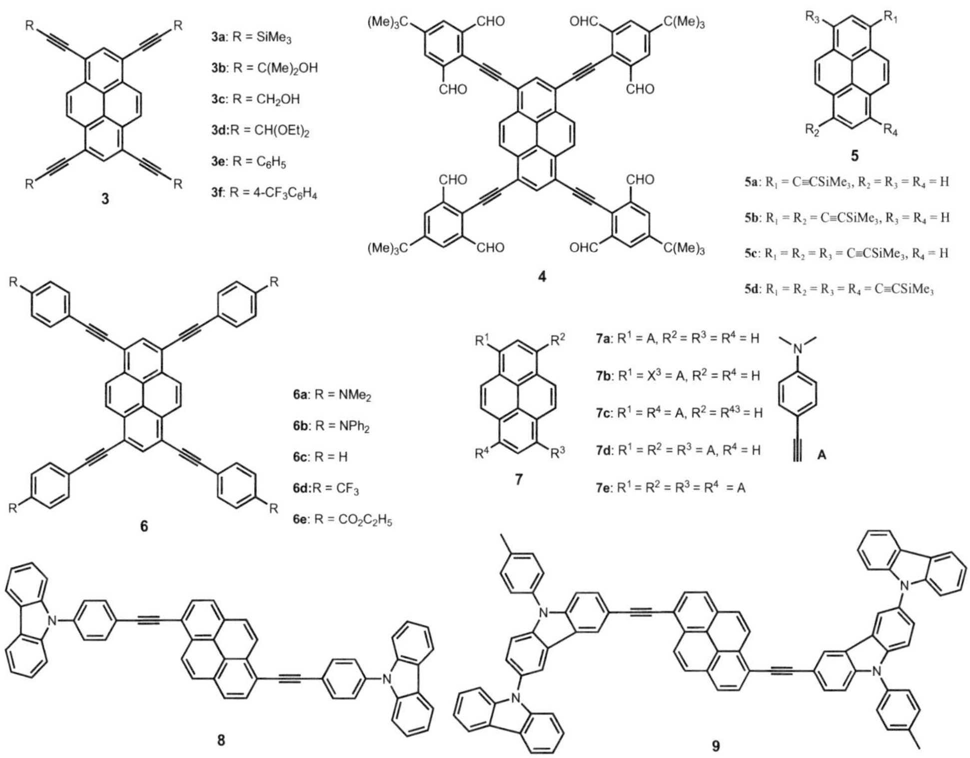

Fig. 1. Alkynyl-functionalized pyrene-cored light-emitting monomers (**3-9**).

Despite various alkynyl-functionalized pyrene-based light-emitting monomers with excellent efficiency and stability have been designed and studied by many research groups, there are very few examples of alkynylpyrenes-based OLED materials. Xing *et al.* (Xing et al., 2005) synthesized two ethynyl-linked carbazole-pyrene-based organic emitters (**8** and **9**, Figure 1) for electroluminescent devices. Both **8** and **9** show extremely high fluorescence quantum yield of nearly 100% because of the inserting of pyrene as electron-acceptor. Due to its higher solubility and easier fabrication than those of **8**, they fabricated a single-layer electroluminescence device by doping **9** into PVK. The single-layer device (ITO/PVK: **9** (10: 1, w/w)/Al) showed turn-on voltage at 8 V, the maximum luminance of 60 cd/m^2 at 17 V, and the luminous efficiency of 0.023 lm/W at 20 V. the poor performance of the device is probably due to the unbalance of electrons and holes in PVK. To improve the device performance, an additional electron-transporting layer (1,3,5-tri(phenyl-2-

benzimidazole)benzene (TPBI) was deposited by vacuum thermal evaporation in the structure of device: ITO/PVK : **9** (10 : 1, w/w) (60 nm)/TPBI (30 nm)/Al (100 nm). Physical performance of the device appeared to be improved: turn-on voltage 11 V, maximum luminance reached 1000 cd/m^2, external quantum efficiency was found to 0.85% at 15.5 V, and luminous efficiency was 1.1 lm/W at 15.5 V. The molecular structures of these alkynyl-functionalized pyrene-cored light-emitting monomers (**3-9**) are shown in Figure. 1.

3.1.2 Aryl-functionalized pyrene-cored light-emitting monomers

Recently, due to their extended delocalized π-electron, discotic shaped, high photoluminescence efficiency, and good hole-injection/transport properties, 1,3,6,8-tetrafunctional pyrene-based materials (i. e. 1,3,6,8-tetra-alkynylpyrenes and 1,3,6,8-tetraarylpyrenes) have the potential to be very interesting class of materials for opto-electronic applications. All the tetraarylpyrenes were mainly synthesized starting from the 1,3,6,8-tetrabromopyrene (**2e**). Suzuki coupling reaction between the tetrabromopyrene **2e** and the corresponding arylboronic acids or esters under Pd-catalyzed conditions afforded the corresponding tetraarylpyrenes. The first example of tetraarylpyrenes is 1,3,6,8-tetraphenylpyrene (TPPy, **10)**. TPPy is a highly efficient fluorophore showing strong blue luminescence in solution (quantum yield Φ = 0.9 in cyclohexane) (Berlamn, 1970), and the organic light emitting field-effect transistor devices (OLEFET) based on TPPy have been shown to exhibit electroluminescence (EL) with an external quantum efficiency of only 0.5% due to aggregation (Oyamada et al., 2005). In view of its high fluorescence quantum yield in solution and ease of substitution by flexible later side chains, TPPy has recently been selected as a discotic core to promote liquid-crystalline fluorescent columns. Greets and co-workers synthesized and studied several new derivatives of pyrenes (**11**) (de Halleux et al., 2004); the pyrene core has been substituted at the 1,3,6,8-positions by phenylene rings bearing alkoxy, ester, thioether, or tris(alkoxy)benzoate groups on the *para* positions. In order to generate liquid-crystalline phases, they varied the nature, number, and size of the side chains as well as the degree of polarity around the TTPy core, however, the desired liquid-crystalline behavior has not been observed. Kaszynski *et al.* (Sienkowska et al., 2007) also prepared and investigated series 1,3,6,8-tetraarylpyrenes **12** on their liquid crystalline behavior by using thermal, optical, spectroscopic, and powder XRD analysis. No mesogenic properties for these tetraarylpyrenes exhibited. Zhang and co-workers (Zhang et al., 2006) recently reported the synthesis and characterization of the first examples of novel butterfly pyrene derivatives **13** and **14**, in which thienyl and trifluoromethylphenyl aromatic groups were introduced in the 1-, 3-, 6- and 8-positions of pyrene cores through Suzuki coupling reactions of 2-thiopheneboronic acid and 4-trifluoromethylphenylboronic acid with 1,3,6,8-tetrabromopyrene (**2e**) in refluxing dioxane under a nitrogen atmosphere in good yields, respectively. The physical properties of **13** and **14** were investigated. The absorption maximum of **13**, containing electron-donating thienyl units has double absorption maximum at 314 nm and 406 nm, while **14**, with electron-withdrawing groups of trifluoromethylphenyl is located at 381 nm. The optical band gaps obtained from the absorption edges are 2.58 eV for **13** and 2.84 eV for **14**. The lower band gap for **13** is probably attributable to intramolecular charge transfer from thienyl ring to the pyrene core. Furthermore, compounds **13** and **14** exhibit strong green (λ_{max} = 545 nm) and blue (λ_{max} = 452 nm) fluorescence emission at longer wavelengths in the solid state than in solution (λ_{max} = 467 nm for **13**; λ_{max} = 425 nm for **14**; 27-78 nm red shift), indicating strong intermolecular interaction in the solid state. The field effect transistors (FETs) device based on **14** did not show any FET performance, while the FET device using **13** as active material exhibited p-type performance. The mobility was 3.7 × 10^{-3} $cm^2V^{-1}s^{-1}$ with an on/off ratio of 10^4, and the threshold voltage was -21 V. This is the first example of a p-type FET using a butterfly pyrene-type moleculae (**13**) as the active material. More recently, a typical example of piezochromic luminescence material **15** based on TPPy was designed and prepared by Araki *et al.* (Sagara et al., 2007), in which to the *para* position of the phenyl groups of this parent molecule TPPy, four hexyl amide units were introduced as the multiple hydrogen-bonding sites. The addition of methanol to a chloroform solution of **15** resulted in precipitation of a white powder (B-form), interestingly; this blue-emitting white solid (B-form) was converted to a yellowish solid showing a strong greenish luminescence (G-form) simply by pressing it with a spatula. The absorption and fluorescence bands of **15** in chloroform solution showed structureless features at 392 and 439 nm (Φ = 0.7, life time τ = 1.3 ns), respectively, which are not much different form those of TPPy (Raytchev et al., 2003). In the solid state, the emission band of the B-form (Φ = 0.3, τ = 3.1 ns) appeared at a position similar to that in solution, but the G-Form solid showed considerable red-shifted emission at 472 nm (Φ = 0.3, τ = 3.2 ns). To clarify the different spectroscopic properties of these two solids, their solid-state structures were studied by IR spectra analysis and powder X-ray diffraction (XRD), respectively.

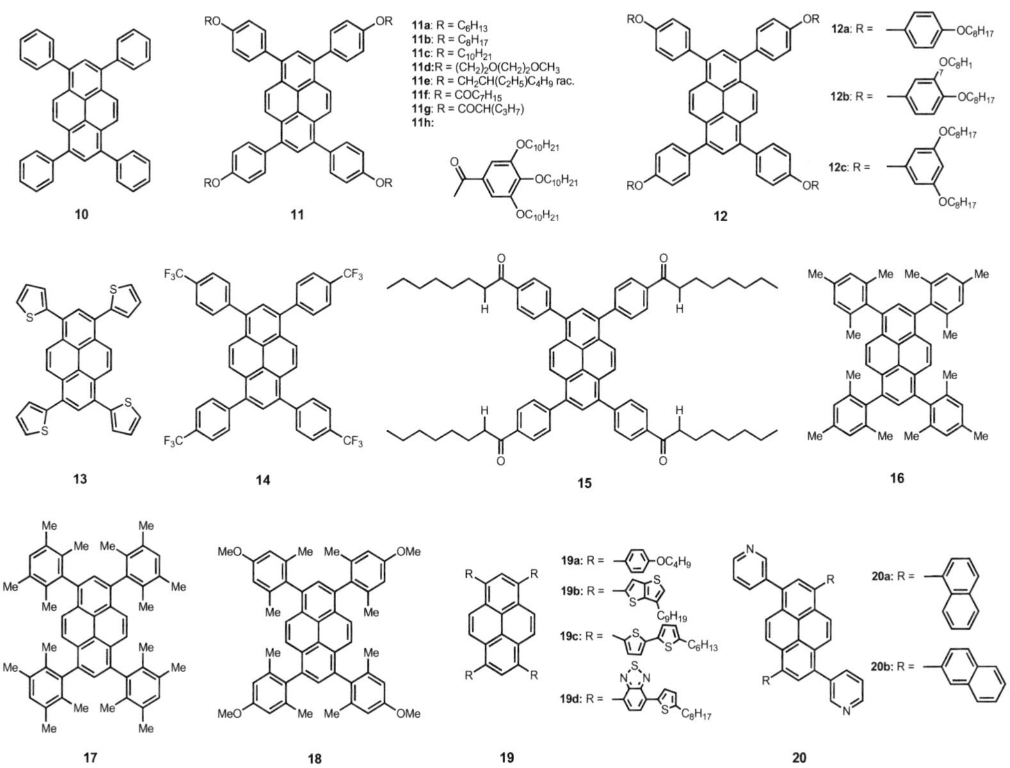

Fig. 2. Aryl-functionalized pyrene-cored light-emitting monomers (**10**-**20**).

Although the IR spectra of **15** in the B- and G-form were essentially the same, and the lower-shifted peak of the amide NH stretching at 3282 cm^{-1} indicated the formation of strong hydrogen bonds, however, closer examination of the spectra revealed that the NH stretching peak of the G-form solid was apparently broader and extended to the higher wave-number side, indicating the presence of weakly hydrogen-bonded amide units. The powder X-ray diffraction pattern of the B-form solid showed clear reflection peaks, indicating that **15** molecules in the B-form were packed in a relatively well-defined microcrystalline-like structure. On the other hand, the G-form solid did not show any noticeable diffraction in the XRD profile. The results indicated that the crystalline-like ordered structure of the B-form was disrupted in the G-form solid, agreeing well with the IR results. Therefore, the design principle in the studies could be widely applicable to other molecular systems.

As described above, numerous aryl-functionalized pyrene-cored light emitting materials have been developed, but very few of these offer OLED devices. Quite recently, several 1,3,6,8-tetraaryl-functionlized pyrenes as efficient emitters in organic light emitting diodes (OLEDs) have been reported. Moorthy *et al.* (Moorthy et al., 2007) prepared three sterically congested tetraarylpyrenes **16-18**, which can be readily accessed by Suzuki coupling between the 1,3,6,8-tetrabromopyrene (**2e**) and the corresponding arylboronic acids in isolated yields. The UV-vis absorption spectra of these arylpyrenes revealed a vibronic feature that is characteristics of unsubstituted parent pyrene with short wavelength absorption maximum at ca. 289 nm. The long wavelength absorption maximum for **16**, **17**, and **18** occurred at 363, 364, and 367 nm, respectively, which points to only a marginal difference. The photoluminescence (PL) spectra for **16-18** show that their emission maximum lie in the region between 400 and 450 nm with maximum bands at 411 nm for **16**, 412 nm for **17**, 411 nm for **18** in solutions, and at 435 nm for **16**, 434 nm for **17**, 442 nm for **18** in the solid-state, respectively. The absence of palpable red-shifted emission in all **16-18** attests to the fact that the molecules do not aggregate in the solid state to form excimers, which is also evidenced from X-ray crystal structure determination of **16**. The PL quantum yields of **16-18** in cyclohexane solution and vacuum-deposited films were found to be in the range of 0.28-0.38 and 0.24-0.44, respectively. The HOMOs and LUMOs for **16-18** vary from 5.75 to 5.80 eV and 2.56 to 2.62 eV, respectively. Thermal properties of **16-18** were gauged by thermogravimetric analysis (TGA) and differential scanning calorimetry (DSC), which were found to exhibit decomposition temperatures (T_d) above 300 °C and a broad endothermic melting transition (T_m) above 240 °C, respectively. The functional behavior of **16-18** as pure blue host emitting materials in OLEDs was investigated by fabricating devices for capturing electroluminescence (EL) as following: ITO/NPB (40 nm)/**16-18** (10 nm)/TPBI (40 nm)/LiF (1 nm)/Al (150 nm), where ITO (indium tin oxide) was the anode, NPB (*N,N'*-bis-(naphthlen-1-yl)-*N,N'*-bis(phenyl)benzidine) served as a hole-transporting layer, **16/17/18** as an emitting layer, TPBI (1,3,5-tri(phenyl-2-benzimidazole)-benzene) as an electron-transporting layer, and LiF:Al as the composite cathode. The luminance and external quantum efficiencies of the devices constructed for **16** were 1.85 cd/A and 2.2%, respectively, at a current density of 20 mA/cm^2 (9.48 V). The maximum luminance achieved was 3106 cd/m^2 with CIE of (0.15, 0.10). In contrast, the device constructed for **18** exhibited much better performance yielding a maximum external quantum efficiency of ca. 3.3% at 6.5 V. The maximum luminance efficiency achieved was 2.7 cd/A at a current density of 5.25 mA/cm^2 (6.5) with a maximum luminance of 4730 cd/m^2 with CIE of (0.14, 0.09). Thus, the maximum external efficiency achieved for the non-doped blue emitting device fabricated for **18** is 3.3%; this value is comparable to commonly used blue emitting materials based on spirofluorene (3.2 cd/A) (Kim et al., 2001), diarylanthracences (2.6-3.0 cd/A) (Kim et al., 2005; Tao et al., 2004), diphenylvinylbiphenyls (1.78 cd/A) (Xie et al., 2003), biaryls

(4.0cd/A) (Shih et al., 2002), etc. The maximum luminance efficiency of 2.7 cd/A achieved in **18** underscores the fact that the attachment of sterically hindered substituents to the pyrene does indeed lead to suppression of face-to-face aggregation.

Small molecules advantageous because they can be i) purified by column techniques such as recrystalliztion, chromatography, and sublimation and ii) vacuum-deposited in multilayer stack both important for device lifetime and efficiency (Anthony et al., 2008). However, vacuum-deposition techniques require costly processes that are limited to practical substrate size and relatively low yields in the manufacture of high volume products using masking technologies (Shtein et al., 2004). On the other hand, polymers are generally of lower purity than small molecules but can be used to achieve larger display sizes at much lower costs using technologies such as inkjet and screen printing (Krebs, 2009; Loo & McCulloch, 2008; Sirringhaus & Ando, 2008). By combining the advantages of both small molecules and polymers, specially, 1,3,6,8-tetraarylpyrenes (**19**) with high purity and solution processability for application in organic electronics was recently reported by Sellinger and co-workers (Sonar et al., 2010), which were synthesized by both Stille and Suzuki-Miyaura cross coupling, respectively. All compounds **19a-d** are readily soluble in common organic solvents such as $CHCl_3$, CH_2Cl_2, THF and toluene, which allows for purification by column chromatography and solution processing. The photophysical properties of **19** were measured by UV-vis absorption and photoluminescence (PL) spectroscopy in chloroform and in thin films. Compounds **19a** and **19d** show red-shifted wavelength absorption maximum (λ_{max}) at 451 and 452 nm in comparison to **19b** and **19c** (λ_{max} = 394 and 429 nm, respectively) due to their slightly more extended conjugation lengths. In thin film absorption, compounds **19a**, **19c**, and **19d** show red shifts of ca. 13-19 nm, while **19b** shows similar absorbance compared to their respective dilute solutions. Solution PL spectra of **19b** and **19c** show deep blue and sky blue emission, respectively, at 433 and 490 nm, whereas **19a** and **19d** exhibit green and orange emission at 530 and 541 nm, respectively. In thin film PL, all compounds **19** are red-shifted 29-95 nm compared their corresponding solutions due to aggregation in the solid state. The calculated HOMO values for **19** are in range of 5.15 to 5.33 eV, these energy levels match quite with commonly used hole injection/transport layer and anodes such as PEDOT: PSS (5.1 eV) and ITO (4.9 eV), indicating the materials are suitable for application in OLEDs. The strong PL emission, tunable energy levels, excellent solubility, enhanced thermal properties, and good film-forming properties make these materials promising candidates for application in solution-processed devices. Using **19b** as a potential deep blue emitting material, a structure of the OLED is fabricated as following: indium tin oxide (ITO)/PEDOT: PSS (50 nm)/**19b** (50 nm)/1,3,5-tris(phenyl-20benzimidazolyl)-benzene (TPBI) (20 nm)/Ca (20 nm)/Ag (100 nm) where PEDOT/PSS and TPBI act as hole-injecting/transport and electron-injecting/transport layers, respectively. The maximum brightness and luminance efficiency are 5015 cd/m^2 (at 11 V) and 2.56 cd/A (at 10 cd/m^2) with CIE coordinates (0,15, 0.18), respectively. The efficiency numbers are quite promising for unoptimized small molecule solution processed blue OLEDs (Zhang et al., 2010; Wang et al., 2009). The turn-on voltage for the device of around 3 V is quite low, suggesting that the barrier for hole injection from PEDOT: PSS is low, which is expected from the measured HOMO level of **19b**.

For OLEDs, to achieve maximum device efficiency is highly depended on the balance of carrier recombination, because the hole mobility is usually much higher than the electron mobility under the same electric field (Chen et al., 1999; Chu et al., 2007). And thus, emitting and charge-transporting materials with a high ionization potential values such as oxadiazole

(Tokito et al., 1997), benzimidazole (Shi et al., 1997), diarylsilole group materials (Uchida et al., 2001) and electron transport materials (ETMs) (Kulkarni et al., 2004; Strohriegl & Grazulevicius, 2002) were synthesized and applied to OLEDs. On the other hand, the polycyclic aromatic hydrocarbons such as naphthalenes, anthracenes, and pyrenes, compared with hetero-aromatic compounds, are known to be not suitable for ETMs due to their high reduction potentials, but they have good thermal stability and no absorption at longer wavelengths than 430 nm (Tonzola et al., 2003). Among them, pyrene has relatively high electron affinity values and better thermal stability. Lee an co-workers (Oh et al., 2009) synthesized new kinds of pyrene-based electron transport materials (ETMs): 1,6-di(pyridin-3-yl)-3,8-di-(naphthlen-1-yl)pyrene (**20a**) and 1,6-di(pyridin-3-yl)-3,8-di(naphthlen-2-yl)pyrene (**20b**) *via* Suzuki coupling reaction starting from 1,6-dibromopyrene (**2b**). Three blue OLEDs (**1-3**) were fabricated by high-vacuum thermal evaporation of OLED materials on to ITO-coated glass as following: ITO/DNTPD (60 nm)/NPB (30 nm)/AND: TBP 3wt% (25 nm)/Alq_3 (device **1**) or **20a** (device **2**) or **20b** (device **3**) (25 nm)/LiF (0.5 nm)/Al (100 nm), where 4,4′-bis[*N*-[4-{*N,N*-bis(3-methylphenyl)amino}-phenyl]-*N*-phenylamino]biphenyl (DNTPD) and 4,4′-bis[*N*-(1-naphthyl)-*N*-phenylamino] biphenyl (NPB) act as hole-injecting/transport layers (HTL), AND : TBP 3wt% act as emitting layers (EML), and Alq_3 or **20a** or **20b** act as electron-injecting/transport layers (ETL), respectively. The external quantum efficiencies of the devices **2-3** with the newly-developed pyrene-based molecules **20a/20b** as electron transport materials increase by more than 50% at 1 mA cm^{-2} compared with those of the device **1** with representative Alq_3 as an electron transport material. The enhanced quantum efficiencies are due to the balanced charge recombination in an emissive layer. Electron mobilities in **20a** and **20b** films are 3.7×10^{-5} cm^2 $(Vs)^{-1}$ and 4.3×10^{-5} cm^2 $(Vs)^{-1}$, respectively. These values are three times higher than that of Alq_3. Highly enhanced power efficiency is achieved at 1.4 lm/W for device **1** with Alq_3, 2.0 lm/W for device **2** with **20a**, and 2.1 lm/W for device **3** with **20b** at 2000 cd/m^2 due to a low electron injection barrier and high electron mobility. All structures for these 1,3,6,8-tetraaryl-functionalized pyrene-cored light-emitting monomers (**10-20**) are shown in Figure. 2.

3.2 Pyrenyl-functionalized PAHs-cored light-emitting monomers

Due to the most attractive features of its excimer formation, delayed fluorescence, rather fluorescence lifetimes, etc., pyrene is also a fascinating subchromophores for constructing highly efficient fluorescent light-emitting monomers for OLEDs applications. Recent literatures survey revealed that there are many number of investigations concerning the attachment of pyrene to other aromatic fluorophores such as benzene, fluorene, and carbazole, etc. as highly efficient emitters in OLEDs. In this section, the synthesis and photophysical properties of three types of pyrenyl-functionalized PAHs-cored light-emitting monomers were summarized. Especially, several these light-emitting monomers as emitters in efficient OLEDs will be fully discussed.

3.2.1 Pyrenyl-functionalized benzene-cored light-emitting monomers

Hexaarylbenzenes have received much attention in material science in recent years (Rathore et al., 2001; Rathore et al., 2004; Sun et al., 2005) for application as light emitting and charge-transport layer in OLEDs (Jia et al., 2005). Lambet *et al.* (Rausch & Lambert, 2006) designed a synthetic route to the first hexapyrenylbenzene **21** starting from 4,5,9,10-tetrahydropyrene, in which six pyrenyl substituents are arranged in a regular manner and held together by a central benzene core. The absorption spectrum of **21** in dichloromethane shows the typical allowed bands at 465 and 398 nm, and the much weaker forbidden bands at 347 and 280 nm, respectively, which display vibronic structure and are distinctly shifted to lower energy region. The emission spectrum of **21**, peaked at 415 nm, also shows a vibronic fine structure but is much less resolved than that of the parent pyrene. At the low-energy side, there is a broad and unresolved shoulder at 483 nm, which is more intense in polar solvents than in moderately or apolar solvents. The emission spectrum is independent of the concentration such that excimer formation between two molecules of **21** could be excluded, which is both from locally excited pyrene states and from the excitonic states of the aggregate. The others pyrenyl-functionalized benzenes such as 1,3,5-tripyrenyl-functionalized benzenes (**22**) and dipyrenylbenzens (**23**), as organic luminescence (EL) lighting materials are recently disclosed by many patents (Charles et al., 2005; Cheng & Lin, 2009). For example, 1,3,5-tripyrenylbenzene **22**, this arrangement results in a good blue emissive material with a peak emission at 450 nm. However, the compound still has a minor aggregation problem in its solid state, resulting in a shoulder emission at 482 nm and reduced blue color purity. More recently, Sun and co-workers (Yang et al., 2007) reported the synthesis of dipyrenylbenzenes (**24** and **25**) as the light emissive layer for highly efficient organic electroluminescence (EL) diodes. The UV-vis absorption of **24** and **25**, in chloroform, shows the characteristic vibration pattern of the pyrene group at 280, 330 and 349 nm for **24**, 281 and 352 nm for **25**, respectively. Upon excitation, the PL spectra with λ_{max} = 430, 426 nm are observed and the full-widths at half-maximum (FWHM) of **24** and **25** are 63 and 64 nm, respectively. To study the EL properties of **24** and **25**, multilayer devices with the configuration of ITO/NPB (50 nm)/**24** or **25** (30 nm)/BCP (10 nm)/Alq_3 (30 nm)/LiF (1 nm)/Al were fabricated. For device with **24**, the maximum intensity is located at 488 nm with the CIE coordinates of (0.21, 0.35). The color of the emission is bluish green, covering the visible region of 420-600 nm, which probably due to the injected charge carriers are recombined at NPB layer, results in broadening the EL spectrum for this device. For device with **25**, EL emission is centered at 468 nm and the CIE coordinates are (0.19, 0.25). The best power efficiency obtained for the **24** device and the **25** device was 4.09 lm/W at a voltage, current density, and luminance of 5.6 V, 20 mA/cm^2, and 1459 cd/m^2, and 5.18 lm/W at a voltage, current density, and luminance of 5.2 V, 20 mA/cm^2, and 1714 cd/m^2, respectively. Another examples of dipyrenylbenzene derivatives, 1-(4-(1-pyrenyl)phenyl)pyrene (**PPP**, **26a**), 1-(2,5-dimethoxy-4-(1-pyrenyl)-phenyl)pyrene (**DOPPP**, **26b**), and 1-(2,5-dimethyl-4-(1-pyrenyl)phenyl)pyrene (**DMPPP**, **26c**) have been recently reported by Cheng *et al.* (Wu et al., 2008), which was synthesized by the Suzuki coupling reaction of aryl dibromides with pyreneboronic acid. These compounds exhibit high glass-transition temperatures (T_g) at 97 °C for **PPP**, 135 °C for **DOPPP**, and 137 °C for **DMPPP**, respectively. The lower T_g of **PPP** is probably because of the low rotations barrier of the central phenylene group in **PPP** compared with that of substituted phenylene group in **DOPPP** and **DMPPP**. Single-crystal X-ray analysis revealed that these dipyrenylbenzenes adopt a twisted conformation with inter-ring torsion angles of 44.5°-63.2° in the solid state. Thus, The twisted structure is responsible for the low degree of aggregation in the films that leads to fluorescence emission of the neat films at 446-463 nm, which is shorten than that of the typical pyrene excimer emission, and also conductive for the observed high fluorescence quantum yields of 63-75 %. The absorption spectra for **PPP**, **DOPPP** and **DMPPP** in dilute dichloromethane solutions (< 10^{-4} M) showed vibronic structures typical for the pyrene moiety at 300-350 nm.

The PL spectra of **PPP** and **DOPPP** have lost vibronic structure and red-shifted to 428-433 nm with respect to the pyrene monomer emission, while the **DMPPP** emission centered at 394 nm still maintains a weak vibronic feature. On the other hand, the absorption spectra of these thin-film samples show a broad band at 354-361 nm, which is red-shifted by approx. 10 nm relative to those in solution. The PL spectra of **PPP**, **DOPPP**, and **DMPPP** are red-shifted by 18-52 nm to 463, 451, and 446 nm, respectively and become broader with a FWHM of 68-78 nm from dichloromethane solution to the thin-film state. Such red-shifts could be attributed to i) the aggregation of pyrene groups and ii) the extension of π-delocalization caused by the more coplanar configuration (An et al., 2002) of these dipyrenylbenzenes (**26**) in the neat film. A bilayer device using PPP as the hole transporter and Alq_3 as the emitter emits green light at 513 nm from the Alq_3 emitter, which can comparable to the common Alq_3-based devices using NPB as the hole transporter, indicating **PPP** is an excellent hole transporter. Furthermore, three devices consists of [ITO/CuPc (10 nm)/NPB (50 nm)/**PPP** or **DOPPP** or **DMPP** (30 nm)/TPBI (40 nm)/ Mg:Ag/Ag] were fabricated, where CuPc act as a hole injecter, TPBI act as a hole blocker and electron transporter, respectively. **PPP**-based device emits blue light at 474 nm efficiently with a maximum η_{ext} of 4.5 % and CIE coordinates of (0.14, 0.20); at a current density of 20 mA/cm^2, the luminance and the η_{ext} are 1300 cd m^{-2} and 4.2 %, respectively. In particular, **DOPPP**-based device emits blue light at 455 nm efficiently with a maximum η_{ext} of 4.3 % and CIE coordinates of (0.15, 0.16); at a current density of 20 mA/cm^2, the luminance and the η_{ext} are 980 cd m^{-2} and 3.7 %, respectively. **DMPPP**-based device emits deep-blue light at 446 nm with CIE coordinates of (0.15, 0.11), the maximum η_{ext} and luminance reach as high as 5.2 % and 40400 cd m^{-2} (14 V), respectively, at 20 mA cm^{-2}, the luminance and η_{ext} are 902 cd m^{-2} and 4.4 %, respectively. All chemical structures of these pyrenyl-functionalized benzenes are show in Figure. 3.

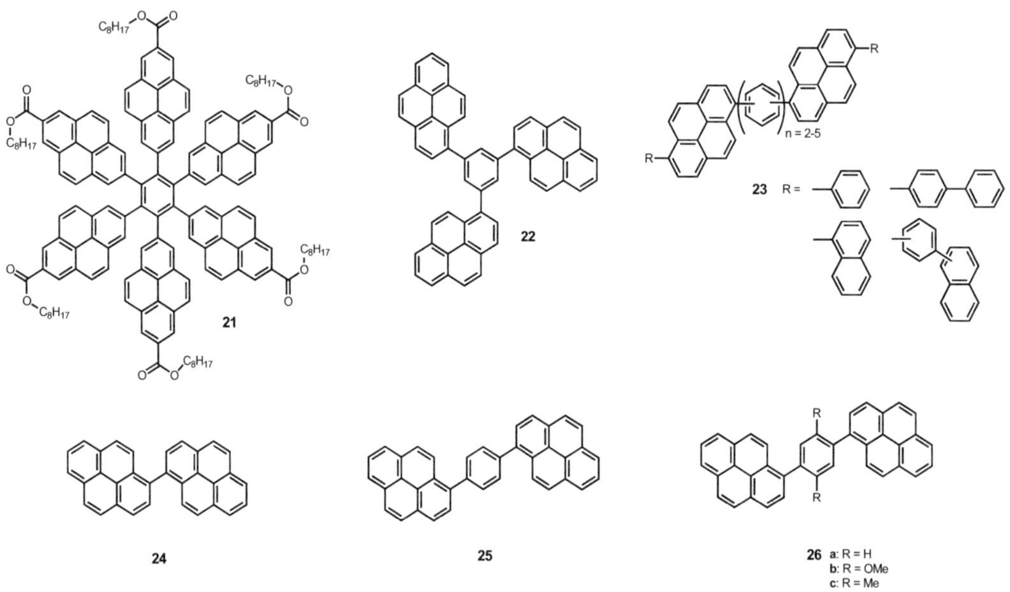

Fig. 3. Pyrenyl-functionalized benzene-cored light-emitting monomers (**21-26**).

3.2.2 Pyrenyl-functionalized fluorene-cored light-emitting monomers

Because of their high photoluminescence (PL) efficiency, much research into blue-emitting materials has focused on conjugated fluorene derivatives (Yu et al., 2000; Wong et al., 2002; Kim et al., 2001), and further the introduction of aryl groups at C9 position of fluorene could improve the stability of the materials. On the other hand, as a large conjugated aromatic ring, pyrene has the advantages of high PL efficiency, high carrier mobility, and the much improved hole-injection ability than oligofluorenes or polyfluorene (Tao et al., 2005). Thus, the combination of the high thermal stability of diarylfluorene with the high efficiency and hole-injection ability of pyrene is expected to develop new blue-light-emitting materials for OLEDs applications. Recently, some fluorene derivatives that functionalized by pyrenyl groups at 2-, and 7-positions have been used in OLEDs. Tao *et al.* (Tao et al., 2005) reported the synthesis and characterization of a series of fluorene detivatives, 2,7-dipyrene-9,9′-dimethylfluorene (**27**, **DPF**), 2,7-dipyrene-9,9′-diphenylfluorene (**28**, **DPhDPF**), and 2,7-dipyrene-9,9′-spirobifluorene (**29**, **SDPF**), in which pyrenyl groups are introduced because they are highly emissive, bulky, and rigid, and thus expected to improve the fluorescence quantum yields and thermal stability of the fluorene derivatives. The fluorene derivatives (**27-29**) have high fluorenscence yields (0.68-0.78, in CH_2Cl_2 solution), good thermal stability (stable up to 450 °C in air), and high glass-transition temperatures in the range of 145-193 °C. All UV-vis absorption spectra of **DPF**, **DPhDPF**, and **SDPF** in dilute CH_2Cl_2 solution exhibits the characteristic vibration pattern of the isolated pyrene groups with maximum peaks at 360 nm for **DPF**, 362 nm for **DPhDPF**, and 364 nm for **SDPF**, respectively. The PL spectrum of the **DPF**, **DPhDPF**, and **SDPF** and films shows blue emission peaks at 421, 422, and 422 nm in solution, and 460, 465, and 465 nm in films, respectively. The shifts are probably due to the difference in dielectric constant of the environment (Salbeck et al., 1997). Using the three derivatives as host emitters, Blue-light-emitting OLEDs were fabricated in the configuration of ITO/CuPc (15 nm)/NPB (50 nm)/**DPF** or **DPhDPF** or **SDPF** (30 nm)/Alq_3 (50 nm)/Mg:Ag (200 nm). Among them, the turn-on voltage of the **DPF**-based device is 5.8 V and the device achieves a maximum brightness of 14300 cd/m^2 at a voltage of 16 V and a current density of 390 mA/cm^2 with CIE coordinates of (0.17, 0.24). The maximum current efficiency of the blue OLEDs made with **DPF**, **DPhDPF**, and **SDPF** host layers is 4.8, 5.0, and 4.9 cd/A, respectively. To confine and enhance electron-hole recombination in the EML and thus to increase device efficiency, a hole-blocking layer, such as a layer of 2,2′,2″-(benzen-1,3,5-triyl)tris(1-phenyl-1*H*-benzimidazole) (TPBI) is commonly used between the EML and ETL. Thus, a modified device with a configuration: ITO/CuPc (15 nm)/NPB (50 nm)/**DPF** (nm)/TPBI (50 nm)/Mg:Ag was fabricated Compared to the Alq_3-based device, the TPBI-based device indeed shows a higher efficiency of 5.3 cd/A and 3.0 lm/W, and better CIE coordinates of (0.16, 0.22) with a lower turn-on voltage of 5.2 V.

Along this line, Huang and co-workers (Tang et al., 2006) also designed and synthesized two highly efficient blue-emitting fluorene derivatives, 2-pyrenyl-9-phenyl-9-pyrenylfluorene (**30a**, $\mathbf{P_1}$) and 2,7-dipyrenyl-9-phenyl-9-pyrenyl-fluorene (**30b**, $\mathbf{P_2}$). They fabricated devices of ITO/TCTA (8 nm)/$\mathbf{P_1}$ or $\mathbf{P_2}$ (30 nm)/BCP (40 nm)/Mg: Ag, where the TCTA (4,4′,4″,-tri(*N*-carbazolyl)triphenylamine) was used as both the buffer layer and hole-transporting layer, and BCP (2,9-dimethyl-4,7-diphenyl-1,10-phenanthroline) as both the buffer layer and electron-transporting layer. The devices have low turn-on voltages of 4 and 3.5 V, with high current efficiencies of 2.56 (9.5 V) and 3.08 cd/A (9 V), high power efficiencies of 0.85 and 1.17 lm/W (7.5 V), and high brightness of 16664 (15 V) and 19885 cd/m^2 (13 V) for $\mathbf{P_1}$-based device and $\mathbf{P_2}$-based device, respectively. The peaks of the blue EL spectra were all at 454 nm with CIE coordinates of (0.17, 0.17) and (0.17, 0.19) for **P$_1$**-based device and **P$_2$**-based device, respectively. In order to obtain a better solubility and a low tendency to crystallize in devices, a long chain alkyloxy group was introduced into C9 phenyl at *para*-position, two new efficient blue-light-emitting materials, 2-pyrenyl-9-alkyloxyphenyl-9-pyrenylfluorene (**31a**) and 2,7-dipyrenyl-9- alkyloxyphenyl-9-pyrenylfluorene (**31b**) have also been reported by Huang's group (Tang et al., 2006). A preliminary simple three layer blue-light-emitting diodes with a configuration of ITO/TCTA (8 nm)/**31a** (30 nm)/BCP (45 nm)/Mg:Ag, was fabricated and obtained without the need for a hole-injection layer, with high luminance of 11620 cd/m^2 (14.6 V), turn-on voltage of 4.0 V, current efficiency of 3.04 cd/A (8.8 V) and CIE coordinates of (0.18, 0.23). More recently, two solution-processable, pyrenyl-functionalized, fluorene-based light emitting materials, 2-(1-ethynylpyrenyl)-9-alkyloxyphenyl-9-pyrenylfluorene (**32a**) and 2,7-di(1-ethynyl-pyrenyl)-9- alkyloxyphenyl-9-pyrenylfluorene (**32b**) were reported by Huang's group (Liu et al., 2009). They emit blue light in solution and green light in film at 412, 439 nm and 514, 495 nm, respectively. Because of the good thermal stability and excellent film-forming ability, **32a** was chosen as the active material in solution processed devices. Two single layered devices with the configurations of [ITO/PEDOT: PSS (40 nm)/**32a** (80 nm)/Ba (4 nm)/Al (120 nm)] (Device 1) and [ITO/PEDOT: PSS (40 nm)/**32a** (80 nm)/CsF (4 nm)/Al (120 nm)] (Device 2) were fabricated. For device 1, it showed bright green emission with peak at 522 nm with CIE coordinate of (0.36, 0.54). The turn-on voltage was 4.2 V, the maximum brightness was 3544 cd/m^2 and the maximum current efficiency reached 0.9 cd/A. For device 2, it showed bright green emission with peak at 528 nm with CIE coordinate of (0.39, 0.54). The turn-on voltage was 3.2 V, the maximum brightness was 8325 cd/m^2 and the maximum current efficiency reached 2.55 cd/A. The results suggest that for pyrene-based materials, CsF was a more efficient cathode than barium in electron injection. This was because the Ba was still an injection limited cathode for **32a**, While the presence of Al capping cathode, free low work function alkali metal would be generated at the **32a**/CsF interface and the CsF layer produced an interfacial dipole. Another solution processable, pyrenyl-functionalized

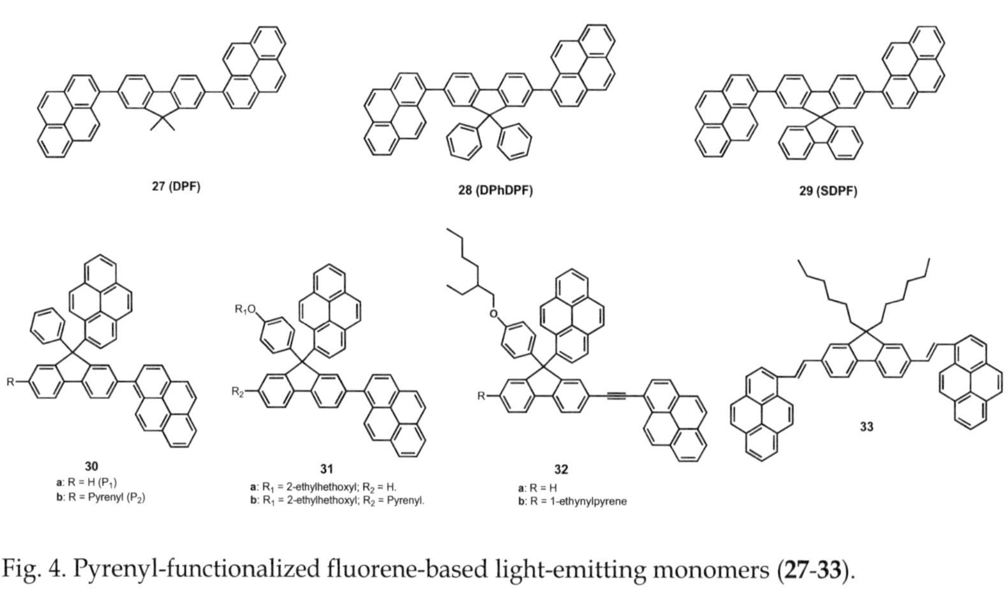

Fig. 4. Pyrenyl-functionalized fluorene-based light-emitting monomers (**27-33**).

Quite recently, Tao *et al.* (Tao et al., 2010) reported the synthesis and characterization of a new carbazole/fluorene/pyrene-composed organic light emitting hybrids, 9,9-bis-(3-9-phenyl-carbazoyl)-2,7-dipyrenylfluorene (**39**). Using 39 as a emitter, a nondoped device with typical three-layer structure of ITO/NPB (50 nm)/**39** (20 nm)/TPBI (30 nm)/LiF (0.5 nm)/MgAg (100 nm) was fabricated. The device exhibits deep-blue emission with a peak centered at 458 nm and CIE coordinates of (0.15, 0.15). The devices shows a turn-on voltage (at 1 cd/m^2) of < 3.5 V and achieves a maximum brightness of 7332 cd/m^2 at a voltage of 9 V and a current density of 175 mA/cm^2, the maximum current efficiency of 4.4 cd/A (at 3.1 lm/W). The results indicate that introduction of carbazole units at the 9-position of fluorene is an efficient means for reducing red shift of the emission from molecular aggregation. Furthermore, the chemical structures for the carbazole/arylamine/ fluorene/pyrene-composed hybrids light-emitting monomers are show in Figure. 5.

4. Functionalized pyrene-based light-emitting dendrimers

Dendrimers are hyperbranched macromolecules that consist of a core, dendrons and surface groups (Newkome et al., 1996). Generally light-emitting dendrimers consist of a fluorescent light-emitting core to which one or more branched dendrons are attached. Furthermore, light-emitting dendrimers possess many potential advantages over conjugated polymers and small molecule materials. First, their key electronic properties, such as light emission, can be finely tuned by the selection of the core drawing from a wide range of luminescent chromophores, including fluorescent groups and phosphorescent groups. Second, by selecting the appropriate surface groups, solubility of the molecule can be adjusted. Finally, the level of intermolecular interactions of the dendrimers can be controlled by the type and generations of the dendrons employed, that are vital element to OLEDs performance. Recently, several types of fluorescent light-emitting dendrimers (Wang et al., 1996; Halim et al., 1999; Freeman et al., 2000; Adronov et al., 2000; Lupton et al., 2001; Kwok & Wong, 2001) and phosphorescent light-emitting dendrimers (Lo et al., 2002, 2003; Markham et al., 2004; Ding et al., 2006) have been disclosed in recent literatures that successfully used in the fabrication of OLEDs by means of solution process. For example, Ding and co-workers have developed a class of phosphorescent iridium dendrimers based on carbazole dendrons (Ding et al., 2006), with a device structure of ITO/PEDOT: PSS/neat **dendrimers**/TPBI/LiF/Al, a maximum external quantum efficiency (EQE) of 10.3 % and a maximum luminous efficiency of 34.7 cd A^{-1} are realized. By doping these dendrimers into carbazole-based host, the maximum EQE can be further improved to 16.6%.

On the other hand, substitution in the pyrene core at the most active centers (i.e. 1-, 3-, 6-, and 8-positions) exclusively by the dendrons can lead to interesting dendritic architectures with a well-defined number of chromophores in a confined volume. Thus, pyrene is a fascinating core for constructing fluorescent-conjugated
light emitting dendrimers. In addition, several types of chromophoric dendrimers backbone such as polyphenylene (Gong et al., 2001; Xu et al., 2002; Kimura et al., 2001), poly(phenylacetylene) (Xu & Moore, 1993; Meliger et al., 2002), and poly(benzyl ether) (Jiang & Aida, 1997; Harth et al., 2002) have been widely used as light absorbers, and the energy was efficiently funnelled to the core accepter. In this section, we presented the synthesis and photophysical properties of pyrene-cored light-emitting dendrimers that have been investigated for the preparation of optoelectronically active solution-processable light emitting dendritic materials and concentrate on the potential applications in OLEDs, in

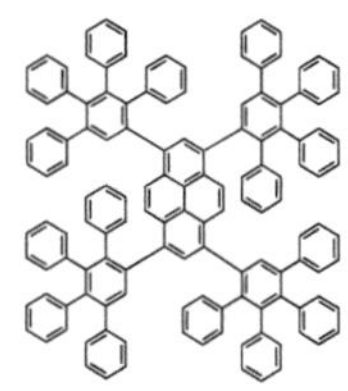

40

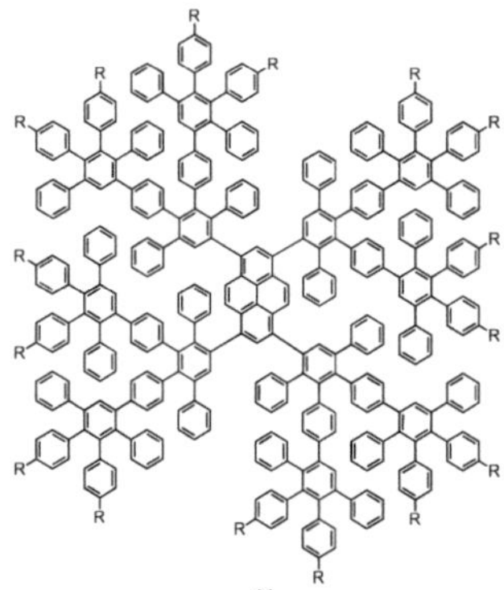

41
a: R = H; **b**: R = $C_{14.10}$

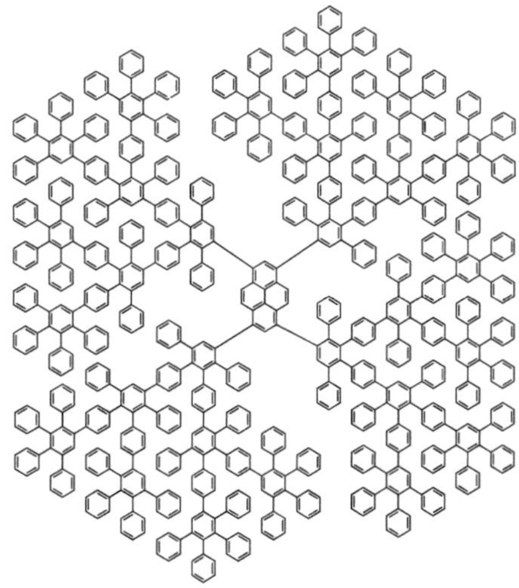

42

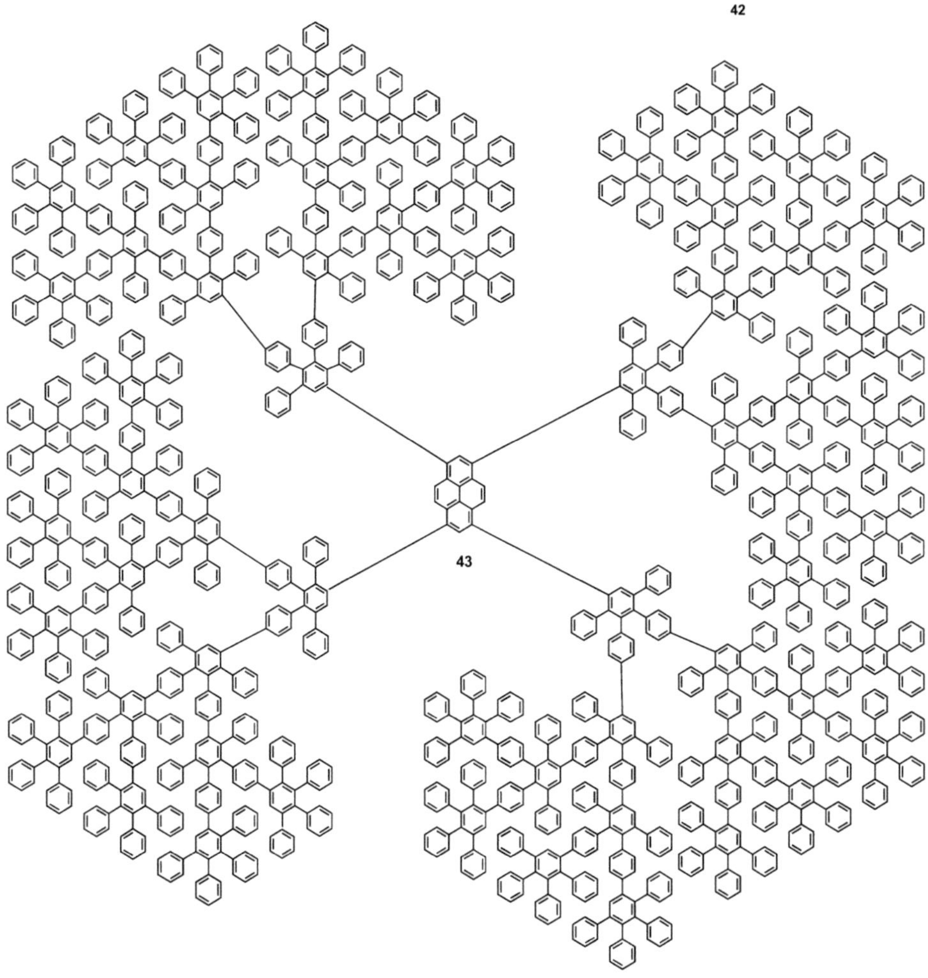

43

Fig. 6. Polyphenylene-functionalized pyrene-cored light emitting dendrimers (**40-43**).

which some light emitting dendrimers have been used. Due to its highly stiff, rigid, and shaped-persistent dendritic backbone (Wind et al., 2001; Rosenfeldt et al., 2003, 2004), recently, Mullen and co-workers (Bernhardt et al., 2006) successfully prepared a series of encapsulation (**40-43**, Figure 6) in a rigid polyphenylene shell using pyrene as chromophore and electrophore core. Around the fluorescent pyrene core, a first-generation (**40**), a second-generation (**41**), a third-generation (**42**), and a fourth-generation (**43**) polyphenylene dendritic environment consisting pyrene building blocks are constructed starting from the fourfold ethynyl-substituted chromophore, 1,3,6,8-tetraethynylpyrene by combining divergent and convergent growth methods. All UV-vis absorption spectra of the first- to fourth-generation dendrimers **40, 41a**, **42**,and **43** in $CHCl_3$ showed two distinct bands, one in the visible region at ca. 395 nm and the other in the UV region at ca. 280-350 nm. The absorption in the visible region is due to the π-π^* transition of the pyrene core and showed a red shift of up to 55 nm compared with unsubstituted parent pyrene (337 nm). Additionally, the fine structure of the pyrene absorption spectrum was lost due to substitution with the phenyl rings. The absorption band in the UV region can be predominantly attributed to the polyphenylene dendrons (Liu et al., 2003), as indicated by the linear increase in theextinction coefficients $\varepsilon(\lambda)$ with increasing number of attached phenylene moieties. The emission spectra of **40, 41a**, **42**, and **43** displayed a broad emission band at 425 nm upon excited the pyrene core at 390 nm. No change was observed in emission maximum or fluorescence intensity of the pyrene core with the changes of the dendrimers generation. Excitation of the polyphenylene dendrons at 310 nm resulted in strong emission of the pyrene core at 425 nm, which indicates efficient energy transfer from the polyphenylene dendrons to the pyrene core. By using 9,10-diphenylanthracene as reference chromophore, the fluorenscence quantum yieldsΦ_f of **40**, **41a**, **42**, and **43** in $CHCl_3$ were determined at 0.92-0.97. Furthermore, the Stern-Volmer quenching experiments and temperature-dependent fluorescence spectroscopy indicated that a second-generation dendrimers shell (**41**) is sufficient for efficiently shielding the pyrene core and thereby suppressing aggregation. In order to investigate the solid-state photophysics of polyphenylene-dendronized pyrenes (**40-43**), the absorption and emission spectra of alkyl-chain-decorated second-generation dendromers **41b** were recorded. Films of good optical quality were obtained by Simple drop casting and spin coating from toluene solution onto quartz substrates. Thin-film of the second-generation 41b displayed an absorption maximum of at 393 nm, almost unshifted compared with solution spectra. The absorption band in the UV region (ca. 260 nm) should be assigned to the polyphenylene dendrons. The emission spectra of **41b** occurred at 449 nm, that is, a bathochromic shift of only 20 nm compared with solution spectra. Thus, these pyrene-cored dendrimers (**40-43**) are exciting new light-emitting materials that combine excellent optical features and good film-forming ability, which make them promising candidates for several applications in electronic devices such as OLEDs.

The synthesis and characterization of a new class of dendrimers (**44** (PyG0), **45** (PyG1), and **45** (PyG2), Figure 7) consisting of a polysulfurated pyrene core, namely, 1,3,6,8-tetra-(arylthio)pyrene moiety, with appended thiophenylene units, recently, reported by Gingras *et al.* (Gingras et al., 2008). The UV-vis absorption spectra of the compounds **44** (PyG0), **45** (PyG1), and **45** (PyG2) in CH_2Cl_2 solutions showed broad absorption band in the visible region with a maximum at 435 nm, which is essentially the identical for the three dendrimers, can be straightforwardly assigned to the pyrene core strongly perturbed by the four sulfur substituents. The band with a maximum around 260 nm can be assigned to the dendrons of thiophenylene units, which increase in intensity with the dendrimers generation. The emission bands for these three dendrimers in $CHCl_3$ solutions

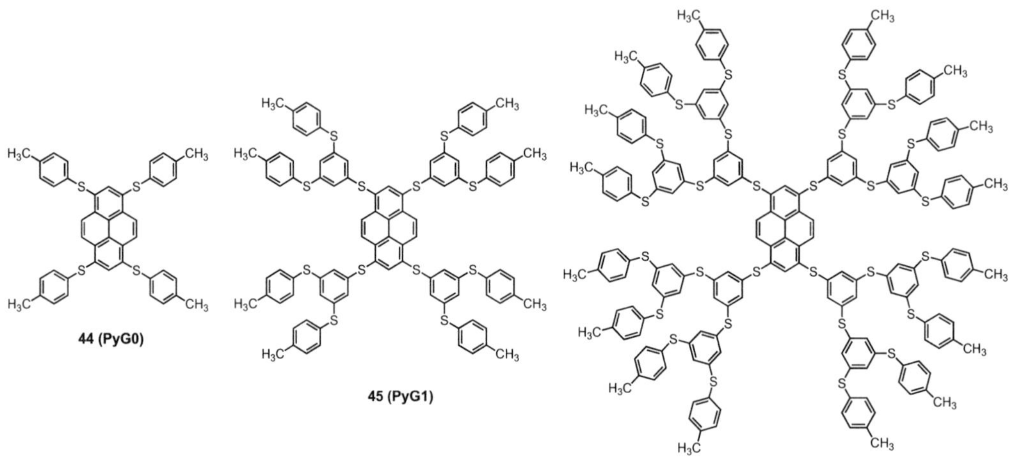

Fig. 7. Polythiophenylene-functionalized pyrene-cored light emitting dendrimers (**44-46**).

strongly red-shifted with maximum at 448 nm for **44**, 452 nm for **45**, and 457 nm for **46**, respectively, compared to that of pyrene (375 nm). Compared to pyrene (277 ns), a remarkable difference is that the excited state lifetime is so short at 1.4-2.4 ns that it is unaffected by the presence of oxygen. The strong fluorescence is also present in the solid state (powder) as a broader band at lower energy with the same lifetime. The fluorescence anisotropy and redox properties for the three dendrimers **44-46** were also investigated in cyclohexane solution at 293 K and in CH_2Cl_2 solutions, respectively. Thus, these newly developed dendrimers with unique photophysical properties might be exploited for optoelectronic and electrochromic applications.

Examples of light emitting dendrimers including pyrene both at the core and periphery are very rare (Mondrakowski et al., 2001), and multichromophoric dendrimers consisting exclusively of pyrene units have recently been reported by Mullen group (Figueira-Duarte et al., 2008). They designed and characterized a new type of light emitting dendrimers, polypyrene dendrimers, represented by the first-generation dendrimer (**49**, **Py(5)**) and the second-generation dendrimers (**50**, **Py(17)**), consisting of five and seventeen pyrene units, respectively, as well two model compounds, **47** (**Py(2)**), and **48** (**Py(3)**), comprising two and three pyrene chromophores, respectively (Figure 8). The UV-vis absorption and fluorescence spectra of the polypyrene dendrimers **Py(*n*)** (n = 2, 3, 5, and 17) were recorded in toluene at 25 °C. In the absorption spectra, from **47 (Py(2)** to **50 (Py(17)**, the broad red sifted from the structural part of the spectrum increases in relative intensity. This broad band reflects the intramolecular interaction between the pyrene units in the **Py(*n*)** compounds, which becomes stronger from **47** to **50**. The fluorescence spectra of **Py(2)** consists of a single band, with some vibrational structure. The fluorescence band shifts to red from 23300 cm^{-1} for **47** (**Py(2)**) to 20660 cm^{-1} for **50** (**Py(17)**), with a simultaneous loss of structure. The fluorescence quantum yields (Φ_f) are 0.72 for **47**, 0.72 for **48**, 0.70 for **49**, and 0.69 for **50**, respectively. The fluorescence decay times of **47** (**Py(2)**), **48** (**Py(3)**), and **49** (**Py(5)**) are 1.76, 1.86, and 1.51 ns, respectively, while, for **50** (**Py(17)**), a triple-exponential decay is found, with τ_1 = 1.75 ns as the major component. The results indicates that the intermolecular excimer formation in the polypyrene dendrimers **47-50** can be excluded due to their small concentrations (< 10^{-5} mol L^{-1}) and the short monomer-fluorescence lifetime τ_1 of around 2 ns. In addition, the computed structures (AM1) analysis of **49** (**Py(5)**) and **50** (**Py(17)**) revealed that the calculated dihedral angle between the core and the first branch is 65-66° for **Py(5)** and 71-73° for **Py(17)**, with the angle between the first and the second branch in **Py(17)** around 84-89°. Thus, the rigid and strongly twisted 3D structure allows a precise spatial arrangement in which each unit is a chromophore. Furthermore, the results on photophysical properties and molecular structure design make these dendrimers model compounds or attractive candidates for use as fluorescence labels or optoelectronics applications.

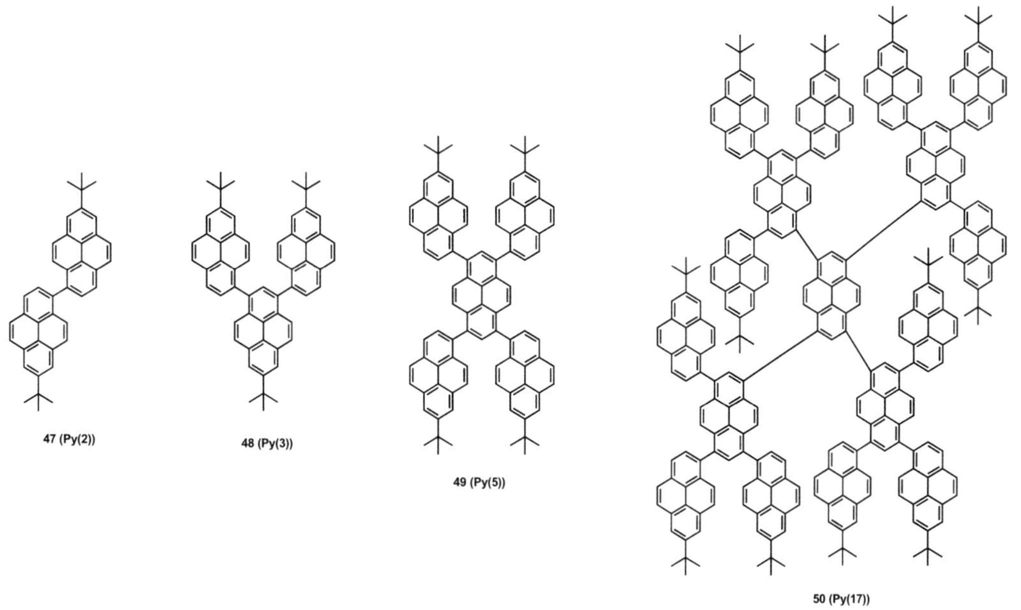

Fig. 8. Polypyrene light emitting dendrimers (**47-50**).

In recent years, one-dimensional self-assembly of functional materials has received considerable interest in the fabrication of nanoscale optoelectronic devices (Lehn, 1995). Research reports (Hill et al., 2004; Kastler et al., 2004; Balakrishnan et al., 2006) suggest that the aromatic organic molecules and large macromolecules are prone to one-dimensional self-assembly through strong π-π interactions. For example, the self-assembly of stiff polyphenylene dendrimers with pentafluorophenyl units has reported by Mullen group (Bauer et al., 2007), in which the driving force for nanofiber formation is attribute to the increase in intermolecular π-π stacking and van der Waals interactions among dendrons by pentafluorophenyl units. On the other hand, for the acetylene-linked dendrimers, their stretched and planar structures may enable facial π-π stacking, resulting in efficient intermolecular electronic coupling. More recently, Lu and co-workers (Zhao et al., 2008) reported two new solution-processable, fluorinated acetylene-linked light emitting dendrimers (**51**a (**TP1**) and **51b** (**TP2**), Figure 9) composed of a pyrene core and carbazole/fluorene dendrons. The strong electron-withdrawing groups of tetrafluorophenyl are introduced at the peripheries of the dendrimers may enhanced electron transportation (Sakamoto et al., 2000), thus balancing the number of holes and electrons in LEDs devices. Both dendrimers are highly soluble in common organic solvents. Their thermal stability is investigated by differential scanning calorimetry (DSC) and thermogravimetric analysis

(TGA) in N_2 at a heating rate of 20 °C/min. dendrimers **TP1** and **TP2** exhibit high glass-transition temperatures (T_g's) at 142 and 130 °C, respectively, and decomposition temperatures (T_d's, corresponding to a 5 % weight loss) at 456 and 444 °C, respectively. The UV-vis absorption spectra of the dendrimers in CH_2Cl_2 solutions exhibit two prominent absorption bands: the first band is attributed to the π-π* transition of the core (pyrene with a certain extension) with a maximum peak at ca. 501 nm, which reveals that the dendrimers are highly conjugated; the second bands is should assigned to the dendrions with a maximum peaks at ~390 nm for **TP1** and ~399 nm for **TP2**. In the case of thin neat films, similar absorption spectra for both dendrimers are observed except for a slight red shift and a loss of fine structures. Upon excitations, both dendrimers **TP1** and **TP2** exhibit emission peaks located at 522 nm with a shoulder at ~558 nm, which is attributed to the emission of the core. There is only a trace emission from the dendrons in the range of 400~450 nm, which indicates efficient photon harvesting and energy transfer from dendrons to the core.

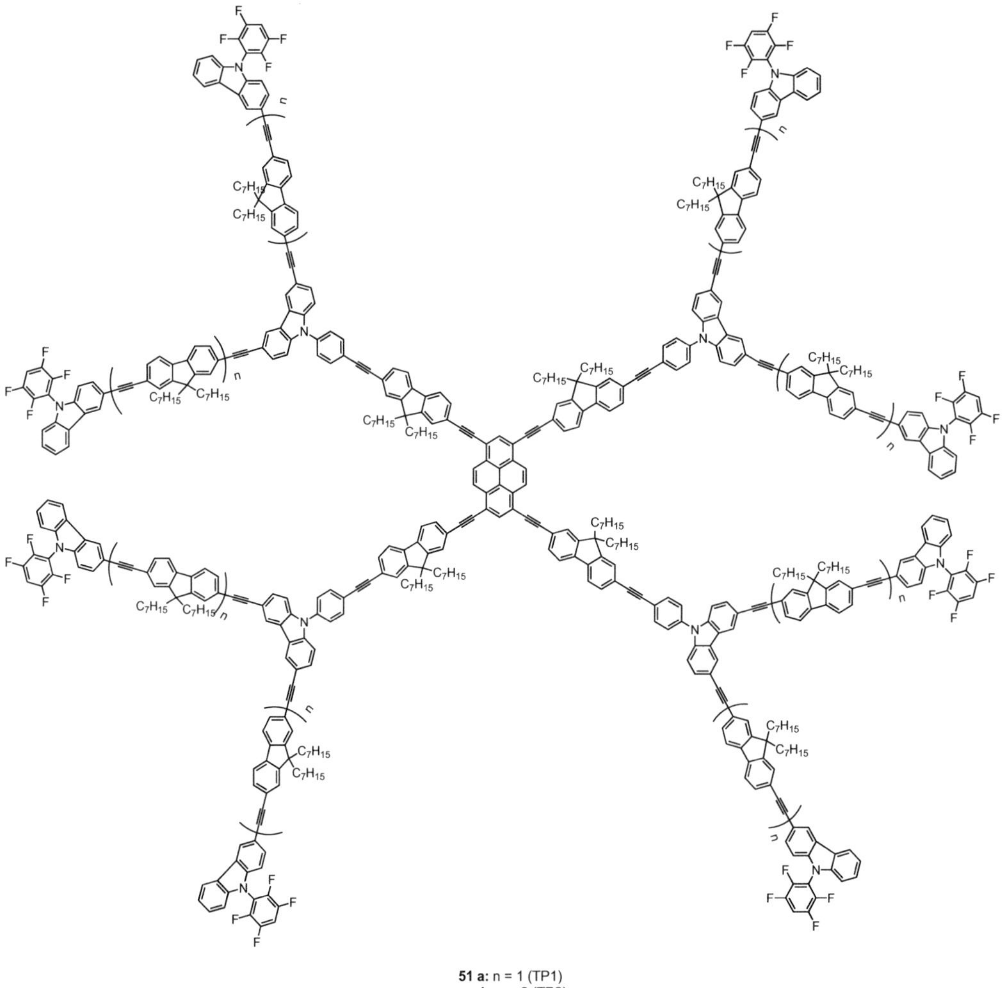

51 a: n = 1 (TP1)
b: n = 2 (TP2)

Fig. 9. Fluorinated acetylene-linked pyrene-cored light emitting dendrimers (**51**).

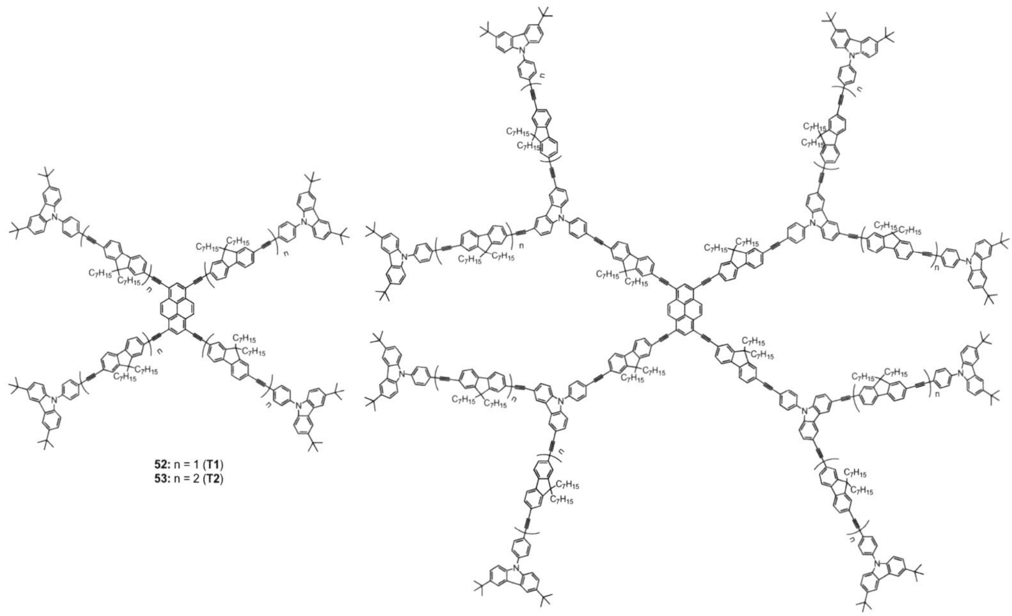

52: n = 1 (**T1**)
53: n = 2 (**T2**)

54: n = 1 (**T3**)
55: n = 2 (**T4**)

Fig. 10. Aacetylene-linked pyrene-cored light emitting dendrimers (**52-56**, **T1-T5**).

In the thin films, **TP1** and **TP2** exhibit strong yellow emission with peaks at 532 nm and 530 nm and relatively peak at 568 nm, respectively, which are ascribed to aggregate formation in the solid states. A nanofibrous suspension was obtained in CH_2Cl_2 solution of **TP1** due to its facile one-dimensional self-assembly property. Interestingly, the nanofiber suspension exhibits a green emission with a peak at 498 nm, which is blue-shifted by 34 nm with respective to that of thin neat film. This blue-shifted emission is somewhat abnormal because, generally, aggregations of molecules through intermolecular π-π interaction should result in a red-shifted emission (Balakrishnan et al., 2005; Hoeben et al., 2005). Thus, these finding suggests that the self-assembly process occurs in a nonhomocentric way. Atomic force microscopy (AFM) detected that compounds **TP1** and **TP2** exhibited good film-forming ability despite its rigid and hyperbranched structures. The EL properties of **TP1** and **TP2** were fabricated with the configuration of ITO/PEDOT (25 nm)/**TP1** or **TP2**/Cs_2CO_3 (1 nm)/Al (100 nm) by spin-coating with 1500 rpm from their 2% (wt%) *p*-xylene solutions. Two dendrimers exhibit yellowish green with main peaks at 532 nm and shoulder peaks at 568 nm and CIE coordinates of (0.38, 0.61) for **TP1** and (0.36, 0.62) for **TP2**, respectively. The devices exhibits a maximum efficiency of 2.7 cd/A at 5.8 V for **TP2**, 1.2 cd/A at 6.4 V for **TP1**, and a maximum brightness of 5300 cd/m^2 at 11 V for **TP2**, 2530 cd/m^2 at 9 V for **TP1**, respectively. These obtained results indicated that the dendrimers with fluorinated terminal groups are promising candidates for optoelectronic materials. Quite recently, Lu and co-workers (Zhao et al., 2009) reported another series of acetylene-linked,solution-processable stiff dendrimers (**52-56**, **T1-T5**, Figure 10) consisting of a pyrene core, fluorene/carbazole-composed dendrons. The dendrimers **52-56** show good thermal stability, strong fluorescence, efficient photo-harvesting, and excellent film-forming properties. The single-layer devices with a configuration of ITO (120 nm)/PEDOT (25 nm/**dedrimer**/Cs_2CO_3 (1 nm)/Al (100 nm) are fabricated and fully investigated. The dendrimer films are fabricated by a spin-coating speed ranging from 800 to 3500 rpm from their *p*-xylene solutions. For example, at a speed of 1500 rpm, the **T3**-based LED exhibits yellow EL (CIE: 0.49, 0.50) with a maximum brightness of 5590 cd/m^2 at 16 V, a high current efficiency of 2.67 cd/A at 8.6 V, and a best external quantum efficiency of 0.86%. These results indicate the constructive one offsets the distinctive effect of intermolecular interaction.

5. Functionalized pyrene-based light-emitting oligomers and polymers

In recent years, organic materials with π-conjugated systems, such as conjugated polymers (Kraft et al., 1998) and monodisperse conjugated oligomers (Mullen & Wenger, 1998) have been intensively studied due to their potential applications in photonics and optoelectonics, such as field-effect transistors (FETs) (Tsumura et al., 1986), OLEDs (Burroughes et al., 1990), solar cells (Brabec et al., 2001), and solid-state laser (McGehee & Heeger, 2000), and the academic interest on the structure-property relationship of molecules. To date, many π-conjugated oligomers and polymers possessing benzene, naphthalene, thiophene, and porphyrin as a conventional core. Although pyrene is a fascinating core in fluorescent π-conjugated light-emitting monomers and dendrimers, the use of pyrene as central core for the construction of oligomers or polymers is quite race.

Purified by precipitated, conjugated polymers are typically characterized by chemical composition and distribution in chain length. However, the polydispersity in chain length leads to complex structural characteristics of the thin films, and make it very difficult for researchers to establish a proper structure-property relationship. In contrast, monodisperse conjugated oligomers are strucrally uniform with superior chemical purity accomplished by recrystallization and column chromatography. Thus, oligomers generally possess more predictable and reproducible properties, facilitating systematic investigation of structure-property relationship and optimization. Recently, some pyrene-based conjugated light-emitting oligomers and polymers have been reported. For instance, pyrene-cored crystalline oligopyrene nanowires (**57**, Figure 11) exhibiting multi-colored emission have been reported by Shi *et al.* (Qu & Shi, 2004). Inoue and co-workers reported the synthesis and photophysical properties of two types of acetylene-linked π-conjugated oligomers based on alkynylpyrene skeletons (Shimizu et al., 2007). The chemical structures of these alkynylpyrene oligomers **58** and **59** are also show in Figure 11, and the structural difference between **58** and **59** is only the linkage position of terminal acetylene groups on the benzene rings, i.e., *para* for **58** and *meta* for **59**. The optical properties of the oligomers **58** and **59** were investigated by using $CHCl_3$ as a solvent at dilute concentrations (1.0 x 10^{-6} M) under degassed conditions, respectively. Both absorption maximum and its corresponding coefficient (log ε) of the *para*-linked oligomers **58** are varied from 436 nm to 454 nm, and 4.84 M^{-1} cm^{-1} to 5.58 M^{-1} cm^{-1}, with increasing of oligomer length. In the case of *meta*-linked oligomers **59** only a slight bathochromic shift was observed that varied from 440 nm to 444 nm with increasing of oligomer length, which probably because of partial insulation of the π-conjugation on these oligomers. The fluorescence spectra of the oligomers were also measured in degassed $CHCl_3$ solutions. Two strong emission bands were observed in the visible region in all spectra. The emission maxima for the *para*-linked oligomers **58** shifted to longer wavelength from 448 nm to 473 nm, in a manner similar to their absorption maximum. On the other hand, for the *meta*-linked oligomers **59**, the fluorescence spectra varied from 455 nm to 461 nm in agreement with the electronic absorption spectra. The fluorescence quantum yields (Φ) were found in the range of 0.35-0.74 in $CHCl_3$ and 0.44-0.79

in THF, respectively. Thus, the newly developed π-conjugated oligomers will facilitate the synthesis of alkynylpyrene polymers and the useful to optical devices.

More recently, Lu and co-workers reported (Zhao et al., 2007) a series of highly fluorescent, pyrene-modified light-emitting oligomers, namely, pyrene-end-capped oligo(2,7-fluorene ethynylenes)s (**60-62**) and pyrene-centered oligo(2,7-fluorene ethynylenes)s (**63-65**) (Figure 11). The absorption spectra of the oligomers were investigated in both dilute CH_2Cl_2 solutions and in thin neat films. For the pyrene-end-capped oligomers **60**, **61**, and **62**,

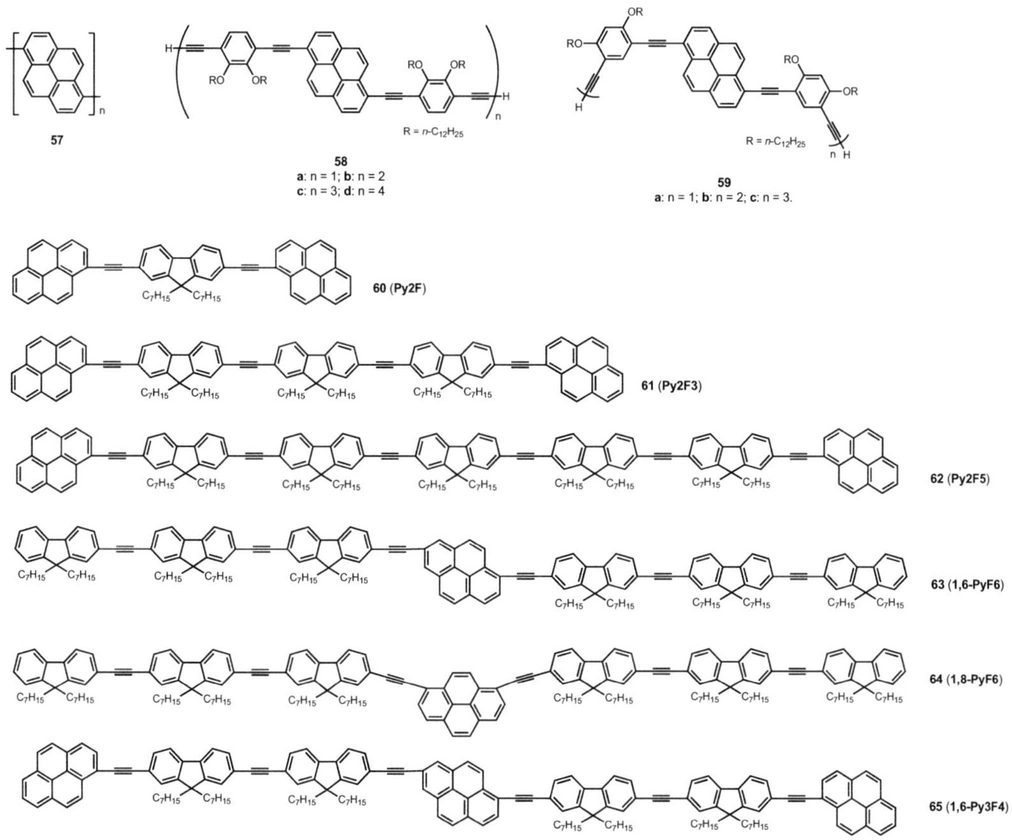

Fig. 11. Functionalized pyrene-based light-emitting oligomers (**57-65**).

the maximum absorption peaks were located at 426, 421, and 418 nm, respectively, which could be attributed to the π-π* transition of the molecular backbone. A interesting blue-shift was observed as the molecular chain length increased, which might be due to the complicated intramolecular conformation such as the two pyrene units might not conjugate to the whole molecular backbone efficiently at one time, thus lead to a weak influence. Compared to that of pyrene-end-capped oligomer **Py2F5** (**62**), the pyrene-centered 1,6-PyF6 (**63**) and 1,6-Py3F4 (**65**) show red-shifted by ~33 nm located at ~451 nm. 1,8-PyF6 (**64**) exhibited a similar maximum absorption peak in comparison to that of 1,6-PyF6, but the relative absorption intensity changed, which might be due to the interruption of delocalization of the π-electrons along the oligomer backbone by the 1,8-pyrene linkage. All absorption spectra in solid-state for these oligomers were almost identical, but had a slightly bathochromic shift (2-10 nm) compared to the corresponding solutions, which indicated that these oligomers exhibited very similar conformations in both states (Chen et al., 2005). In CH_2Cl_2 solutions, the PL spectra of the pyrene-end-capped oligomers **60**, **61**, and **62** showed a main emission peak at 436, 430, and 429 nm, respectively, with a shoulder peak at 464, 456, and 455 nm, respectively. The blue-shift emissions were attributed to the same reason for blue-shifted absorption spectra of them. On the other hand, the PL spectra of the pyrene-centered oligomers 63-65 exhibited quite similar main emission peaks at ~465 nm with shoulder peaks at ~495 nm, actually emanating from disubstituted pyrene. In thin neat films, the disappearance of the fine structures of spectra were observed with main peaks at 492 nm for **60**, 489 nm for **61**, and 476 nm for **62**, respectively. Emission of Py2F (**60**) was strongly red-shifted by 56 nm compared to the emission in solution which should be due to the facile formation of excimers between pyrene units. All the oligomers were highly fluorescent. The PL quantum yields of these oligomers were in the range of 0.78-0.98 in degassed cyclohexane solutions using 9,10-diphenylanthracence (DPA, Φ = 0.95) as a standard (Melhuish, 1961). Moreover, Py2F5 (**62**) exhibitedhigher quantum yields than the pyrene-centered oligomers **63-65** with similar chain length, which might be that excitons were well confined to the whole backbone of Py2F5 (**62**). By using these oligomers as emitters, the devices with the same configurations of ITO/PEDOT: PSS (30 nm)/ **oligomers** (50 nm)/TPBI (20 nm)/Al (100 nm) were fabricated. For the pyrene-end-capped oligomers **60-62**, the EL emissions were observed from green (532 nm) to blue (468 nm) with the increment of the fluorene moieties. The EL emission of **60** (Py2F) was significantly red-shifted (40 nm) comparison with that of PL emission in film, while the EL emission of **62** (Py2F5) was slightly blue-shifted (8 nm). Since **60** (Py2F) had the shortest chain length among the pyrene-end-capped oligomers, the highest chain mobility was suggested. Results have pointed out that materials with repeating fluorene units should be underwent a process of alignment in an electric field, and molecules with the high chain mobility more easily formed excimers than molecules with low chain mobility (Weinfurtner et al., 2000). Due to the higher chain mobility of **60** compared to that of **61** and **62**, it is was more possible for **60** molecules to align under the electric field. Thus, the pyrene groups on one Py2F (**60**) molecule could be close to the pyrene groups on the neighbouring molecules, and when the distance between the two fluorophores was appropriate, excimers were formed under the electric excitations. On the other hand, for the pyrene centered oligomers **63-65**, the EL spectra showed green emissions from 472 to 504 nm, which similar to their corresponding PL emission in films except for slight red shifts. The results indicate that both PL and EL emission originated from the same radiative decay process of singlet excitons. The turn-on voltages of the oligomers-based devices were in the range of 4.3-5.4 V. the **Py2F**-based device exhibited maximum brightness at 2869 cd/cm^2 at 10.5 V and a highest external quantum efficiency of 0.64%. While with the increase of the fluorene moiety, the device based on **Py2F3** and **Py2F5** exhibited a substantial decrease of maximum brightness from 918 cd/cm^2 at 9.0 V to 207 cd/cm^2 at 8.0 V as well as the external quantum efficiency of 0.41% for **Py2F3** and 0.15% for **Py2F5**. The pyrene-end-capped-based devices exhibited comparable brightness, 493 cd/cm^2 at 8.5 V for **63**, 520 cd/cm^2 at 8.5 V for **64**, and 340 cd/cm^2 at 6.5 V for **65**, respectively, as well as an external quantum efficiency, 0.22% (**63**), 0.22% (**64**) and 0.14% (**65**), respectively. Obviously, as chain length elongated, the performance of the devices was decreased. One possible explanation for this phenomenon was that the oligomrs with more fluorene moieties were more easily crystallized than the oligomers with fewer fluorene units. It was well known that crystallization was disadvantageous to the electroluminescence properties of organic materials. As a result, the good performance of the pyrene-modified oligomers-based devices indicated that they were promising light-emitting materials for efficient OLEDs.

In comparison of small molecules, conjugated polymers have the advantageous of being applicable in larger display sizes and lighting devices at much lower manufacturing costs via solution-based deposition techniques. Conjugated polymers such as polyphenylvinylene (PPV) and its derivatives are known as visible light emitters and have been widely used in the fabrication of organic light-emitting diodes (OLEDs) (Son et al., 1995). Only a few numbers of investigations concerning on the attachment of pyrene to the polymeric chain (Rivera et al., 2002) or the use of pyrene along the polymeric backbone (Ohshita et al., 2003; Mikroyannidis et al., 2005; Kawano et al., 2008; Figueira-Duarte et al., 2010) were reported as model systems or new materials for molecular electronics.

Giasson and co-workers reported (Rivera et al., 2002) the synthesis and photoproperties of four different polymers (**66** (**PEP**), **67** (**PTMSEP**), **68** (**PBDP**), and **69** (**PTMSBDP**), Figure 12) by the W and Ta-catalyzed polymerization of 1-ethynylpyrene, 1-(trimethylsilylethynyl)-pyrene, 1-(buta-1,3-diynyl)pyrene, and 1-(4-trimethylsilylethynyl)pyrene, respectively, in which pyrene as functional group attached in the polymeric chain. For comparison, the dimmer of 1-ethynylpyrene (**DEP**) was prepared. The absorption spectra of the polymers and **DEP** are recorded in THF. For **DEP**, three peaks were observed, the peak at 336 nm can be attributed to the pyrene moieties, and the peak at 346 nm and shoulder peak at 390 nm should have their origin in intramolecular interactions (complexation) between the pyrene units present in the dimer. The absorption spectrum of **PEP** is significantly different from that of **DEP**. The shoulder peak around 390 nm in **DEP** disappeared in the absorption spectra of **PEP**. This suggests that the intramolecular interactions between adjacent pyrene units in the polymer are weaker than those in **DEP**. Moreover, a broad band is observed around 580 nm in the absorption of **PEP**, which should be caused by the polyacetylene chain. The result indicates that the effective electronic conjugation is relatively long for this polymer. The absorption spectra of **PTMSEP**, **PBDP**, and **PTMSBDP** are relatively similar to each other. However, the bands of **PTMSBDP** and **PBDP** are broader than that of **PTMSEP** suggesting that stronger interactions between pyrene units are present in the former polymers. Thus, two facts can be demonstrated that the distortion of the polymer backbone caused by the presence of a trimethylsilyl group significantly weakens the electronic interactions between pyrene moieties and the incorporation of triple bond into the polymeric chain permits better interactions between the pyrene units. On the other hand, the band around 580 nm observed in **PEP** is not observed for these polymers, which indicates that the effective conjugation is much shorter. In the fluorescence spectra of **DEP** and **PEP** in THF, both compounds show a band in the range of 360-465 nm arising from non-associated pyrene moieties. **DEP** also shows a broad band around 480 nm, which should due to the molecular interactions between pyrene units present in this molecule. Surprisingly, such a distinct band is not observed in the case of PEP that might be caused by an inner-filter effect involving the main chain. However, the fluorescence intensity of **PEP** near 480 nm is significant. This strongly suggests that a complex between pyrene units is also formed in the polymer. The fluorescence spectra of **PTMSBDP** and **PBDP** show two distinct bands similar to the ones observed in the fluorescence spectra of **DEP**. These results are consistent with the absorption spectra of these two polymers showing that strong interactions exist between pyrene moieties in the conjugated chain. However, the fluorescence intensity around 480 nm is much reduced in the case of **PTMSEP**, further confirming the above results that the incorporation of trimethylsilyl groups into the polymeric backbone decreases the interactions between the pyrene units. On the other hand, the band around 580 nm observed in **PEP** is not observed for these polymers, which indicates that the effective conjugation is much shorter. In the fluorescence spectra of **DEP** and **PEP** in THF, both compounds show a band in the range of 360-465 nm arising from non-associated pyrene moieties. **DEP** also shows a broad band around 480 nm, which should due to the molecular interactions between pyrene units present in this molecule. Surprisingly, such a distinct band is not observed in the case of PEP that might be caused by an inner-filter effect involving the main chain. However, the fluorescence intensity of **PEP** near 480 nm is significant. This strongly suggests that a complex between pyrene units is also formed in the polymer. The fluorescence spectra of **PTMSBDP** and **PBDP** show two distinct bands similar to the ones observed in the fluorescence spectra of **DEP**. These results are consistent with the absorption spectra of these two polymers showing that strong interactions exist between pyrene moieties in the conjugated chain. However, the fluorescence intensity around 480 nm is much reduced in the case of **PTMSEP**, further confirming the above results that the incorporation of trimethylsilyl groups into the polymeric backbone decreases the interactions between pyrene units. On the other hand, by using pyrene as the polymeric backbone, pyrene-based polymers have been studied by several research groups. For example, Ohshita *et al.* prepared (Ohshita et al., 2003) two organosilanylene-diethynylpyrene polymers **70** and **71** (Figure 12) by the reactions of 1,6-di(lithioethynyl)pyrene and the corresponding dichloroorganosilanes. The hole-transporting properties of the polymers were evaluated by the performance of electroluminescent (EL) devices with the configuration of ITO/polymer **70** or **71** (70-80 nm)/Alq_3 (60 nm)/Mg-Ag, in comparison with those of an organosilanylene-9,10-diethynyl-anthracene alternating polymer, reported previously (Adachi et al., 1997; Manhart et al., 1999). Among them, the device with polymer **70** (device I) exhibited the best performance with a maximum luminescence of 6000 cd/cm^2. This is presumably due to the favored inter- and intra-molecular π-π interactions in the solid states by reducing the volume of the

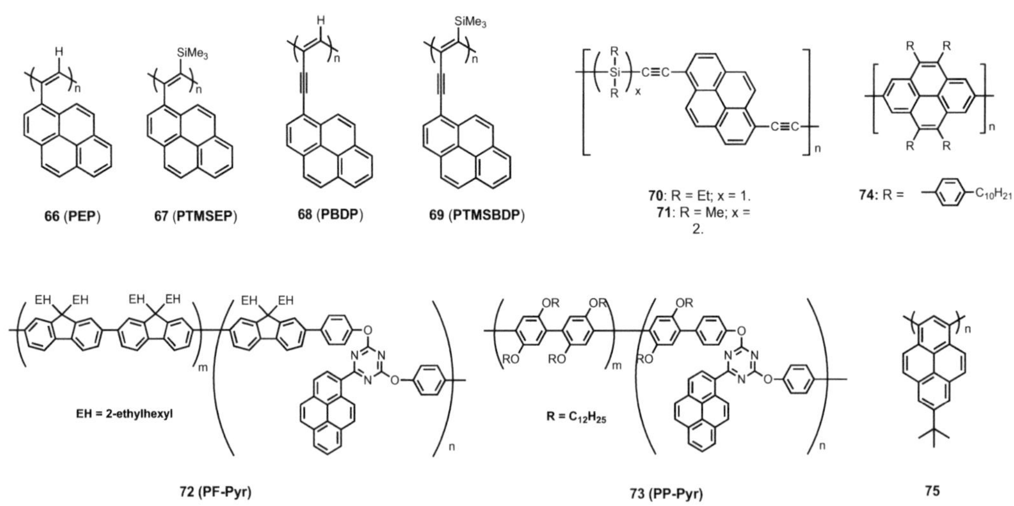

Fig. 12. Functionalized pyrene-based light-emitting polymers (**66-69** and **70-75**).

silicon units. Further improvement of the performance of the device with polymer **70** was realized by introducing a TPD (*N,N'*-diphenyl-*N,N'*-di(*m*-tolyl)-1,1-biphenyl-4,4'-diamine) layer as electron-block with the structure of ITO/**70** (40 nm)/TPD (10 nm)/Alq_3 (60 nm)/Mg-Ag (device II). The optimized device emitted a maximum brightness of 16000 cd/cm^2 at the bias voltage of 14-16 V. when compared with that of the device of ITO/**TPD** (50 nm)/Alq_3 (60 nm)/Mg-Ag (device III), the device II showed lower turn-on voltage (4-5 V for device II and 6 V for device III) and higher current density. These results clearly indicate the excellent hole-transporting properties of polymer **70** films. Mikroyannidis and co-workers recently reported (Mikroyannidis et al., 2005) the synthesis, characterization and optical properties of two new series of soluble random copolymer **72** (**PF-Pyr**) and **73** (**PP-Pyr**) (Figure 12) that contain pyrenyltriazine moieties along the main chain by Suzuki coupling. The photophysical properties of these polymers were fully investigated in both solutions and thin films. For the copolymer **PF-Pyr** (**72**), blue emissions in solutions with PL maximum at 414-444 nm (PL quantum yields 0.42-0.56) and green emissions in the thin films with PL maximum around 520 nm were observed, respectively. The green emission in solid state of these random copolymers **72** was a result of the energy transfer from the fluorene to the pyrenyltriazine moieties. For the copolymer **PP-Pyr** (**73**), blue light both in solution and in thin film with PL maximum at 385-450 nm were observed, respectively. More specially, the copolymers **PF-Pyr** (**73**) showed outstanding color stability since their PL trace in thin film remained unchanged with respect to the PL maximum and the spectrum pattern even following annealing at 130 °C for 60 h. The color stability of the polymer **PF-Pyr** is an attractive feature regarding the high temperature developed during the device operation.

More recently, Mullen group described (Kawano et al., 2008) the synthesis and photophysical properties of the first 2,7-linked conjugated polypyrenlene, **74** (Figure 12), tethering four aryl groups by Yamamoto polycondensation (Yamamoto, 2003). Although composed of large π-units, the polymer **74** is readily soluble in common organic solvent due to the unique substitution with bulky alkylaryl groups at the 4-, 5, 9-, and 10-positions in pyrene ring. The polymer **74** shows a blue fluorescence emission with a maximum band at 429 nm in solution, fulfilling the requirements for a blue-emitting organic semiconductor. However, the fluorescence spectra of **74** exhibit a remarkable long-wavelength tailing as well as additional emission bands with maximum at 493 and 530 nm. To recognize and verify the most probable explanation for the substantially red-shifted band in the case of **74**, concentration dependence of the fluorescence, solvatochromic shifts of the emission maximum (Jurczok et al., 2000; Fogel et al., 2007), and time-resolved measurements of the fluorescence are investigated. These facts together indicated that the red-shifted broad emission bands are not caused by aggregation, but by intramolecular energy redistribution between the vibrational manifold of the single polymer chain (VandenBout et al., 1997; Becker et al., 2006). Furthermore, the additional red-shifted emission (green color) of the polymer **74** in the solid state could be strongly reduced by blending with a non-conjugated polymer such as the polystyrene. Thus, these properties of the polymer **74** could have application in materials processing, for example, as a surrounding media sensor or optoelectronics.

Quite recently, Mullen research group reported (Figueira-Duarte et al., 2010) the suppression of aggregation in polypyrene **75** (Figure 12) with a highly twisted structure of the polymeric chain. The use of *tert*-butyl groups was crucial for selectively affording substitution at the 1,3-positions in the monomer synthesis, and also for both attaining sufficient solubility and avoiding the use of long alkyl chains. The UV-vis absorption and

PL spectra of the polypyrene **75** exhibit very similar spectra in the diluted THF solution and the thin film. The absorption spectra show a π-π* transition at ca. 357 nm and a higher energy absorption band at ca. 280 nm. In contrast, the emission in both solution and thin film showed a broad unstructured band with a maximum at 441 nm in solution and a slight bathochromic shift to 454 nm in the solid state, respectively. Both a classical concentration dependence analysis (in toluene at different concentration ranging from 0.1 to 1000 mg/L) and the calculated molecular structure of a linear 1,3-pentamer model compound (AM1) for the polypyrene **75** provided good evidence for the absence of excimer and aggregation emission. It is well known that the morphological stability at high temperature is a critical point for device performance. Thermal characterization of the polypyrene **75** was made using differential scanning calorimetry (DSC) and thermogravimetric analysis (TGA), and the influence of thermal treatment on its optical properties was investigated. The high morphological stability and glass transition temperature, T_g, could be attributed to the presence of the rigid pyrene unit in the main chain of the polymer. Thus, the device with structure of ITO/PEDOT: PSS/polypyrene **75**/ CsF/Al was fabricated. The device showed bright blue-turquoise electroluminescence with a maximum at 465 nm and a profile very similar to the PL in the solid state. Brightness values at 300 cd/m^2 were obtained at 8 V with CIE coordinates of (0.15, 0.32). The devices show remarkable spectral stability over time with only minor changes in the spectra as a consequence of a thermal annealing under device operation. The OLEDs display a detectable onset of electroluminescence at approximately 3.5 V and maximum efficiencies of ca. 0.3 cd/A. The performance of the presented devices is comparable to devices fabricated without evaporated transport layers from similar poly(*para*-phenylene)-type based materials with respect to the overall devices efficiency and brightness (Pogantsch et al., 2002; Jacob et al., 2004; Tu et al., 2004). Thus, the simple chemical route and the exciting optical features render this polypyrene a promising material toward high-performance polymer blue light-emitting diodes.

6. Pyrene-based cruciform-shaped π-conjugated blue light-emitting architectures: promising potential electroluminescent materials

In recent years, carbon-rich organic compounds with a high degree of π-conjugation have attracted much attention due to their unique properties as ideal materials for modern electronic and photonic applications, such as organic light-emitting diodes (OLEDs), liquid-crystal displays, thin-film transistors, solar cells and optical storage devices (Meijere, 1998, 1999; Haley & Tykwinski, 2006; Mullen & Weger, 1998; Mullen & Scherf, 2006; Kang et al., 2006; Seminario, 2005; Van der Auweraer & De Schryer 2004). Among them, functionalized, cruciform-shaped, conjugated fluorophores are well-known because they exhibit interesting optoelectronic properties due to their special, multi-conjugated-pathway structures. Examples of cruciform-shaped phores are the 1,2,4,5-tetrasubstituted(phenylethynyl) benzenes of Haley *et al.* (Marsden et al., 2005), the X-shaped 1,2,4,5-tetravinyl-benzenes of Marks *et al.* (Hu et al., 2004), the 1,4-bis(arylethynyl)-2,5-distyrylbenzenes of Bunz *et al.* (Wilson & Bunz, 2005), and other cross-shaped fluorophores developed by Nuckolls *et al.* (Miao et al., 2006) and Scherf *et al.* (Zen et al., 2006). Therefore, their seminal studies on the structure-property relationships for those materials provided valuable information for the molecular design of material as model systems or promising candidates toward high-performance optoelectronic devices.

Accordingly, our previous report (Yamato et al., 1993; Yamato et al., 1997) on the synthesis of 4,5,9,10-tetrabromo-2,7-di-*tert*-butylpyrene prompted us to explore 4,5,9,10-tetrakis(phenylethynyl)pyrenes as emissive materials. We surmised that i) The presence of the sterically bulky *t*Bu groups in pyrene rings at the 2- and 7-positions would play important roles for both inhibiting undesirable face-to-face π stacking in solution and the solid state (Bennistom et al., 2007) and attaining sufficient solubility; ii) The ready synthetic accessibility by the Sonogashira coupling, the phenylacetylenicgroups were a priori anticipated to facilitate the construction of the cruciform-shaped structure and further extend the conjugation length of the pyrene chromophore, resulting in a shift of the wavelength of absorption and fluorescence emission into the visible region of the electromagnetic spectrum. Along this lines, as our efforts on the construction of extended π-conjugation compounds based on pyrene (Hu et al., 2009; Hu et al., 2010), we recently succeed to prepare a new series of pyrene-based, cruciform-shaped, π-conjugated, blue-light-emitting monomers (**79**) with a low degree of aggregations in the solid state and pure-blue emission by various spectroscopic techniques (Hu et al., 2010).
The simple chemical route to the cruciform-shaped conjugated pyrenes **79** is shown in scheme 2. The Lewis-acid-catalyzed bromination of 2,7-di-*tert*-butylpyrene (**76**) (Yamato et al., 1993; Yamato et al., 1997) readily afforded the 4,5,9,10-tetrabromo-2,7-di-*tert*-butylpyrene **77** in high yiled of 90%. The modified Sonogashira coupling of the tetrabromide **77** with various phenylacetylenes **78** produced the corresponding 2,7-di-*tert*-butyl-4,5,9,10-tetrakis(*p*-R-phenylethynyl0pyrenes **79** in excellent yields. As a comparison, 1,3,6,8-tetrakis(4-methoxyphenylethynyl)pyrene **80** is prepared according to literature procedure (Venkataramana & Sankararaman, 2005). The chemical structures of these new pyrenes **79** and **80** were fully confirmed by their $^{1}H/^{13}C$ NMR spectra, FT-IR spectroscopy, mass spectroscopy as well as elemental analysis. All results were consistent with the proposed cruciform-shaped structures.

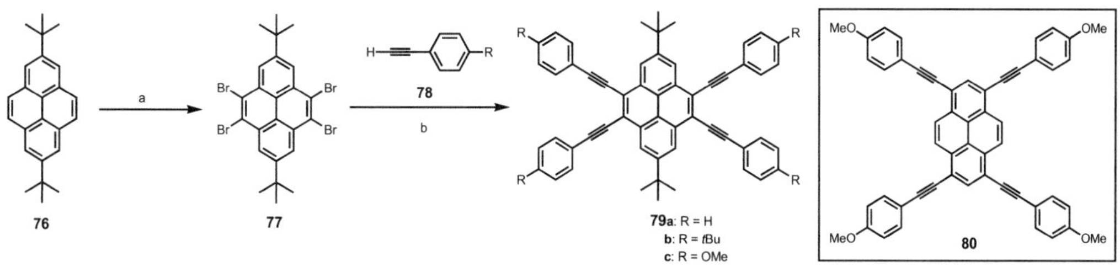

Scheme 2. Synthesis of 4,5,9,10-tetraksi(phenylethynyl)pyrene derivatives **79a-c**. Reagents and conditions: (a) Br_2, Fe powder, CH_2Cl_2, r. t., for 4 h, 90%; (b) $[PdCl_2(PPh_3)_2]$, CuI, PPh_3, Et_3/DMF (1:1), 24-48 h, 100 °C.

The performance of the organic compounds in optoelectronic devices strongly relies on the intermolecular order in the active layer. Small single crystals of **79c** are suitable for X-ray structural determination under the synchrotron. Both the X-ray crystal-structures diagram and packing diagram of **79c** are shown in Figure 13, respectively. As revealed from this analysis, there is a herringbone pattern between stacked columns, but the π-π stacking average distance of adjacent pyrene units was not especially short at ca. 5.82 Å in this crystal lattice. The results strongly indicate that the two bulky *t*Bu groups attached to the pyrene rings at the 2- and 7-positions play an important role in suppressing the aggregations

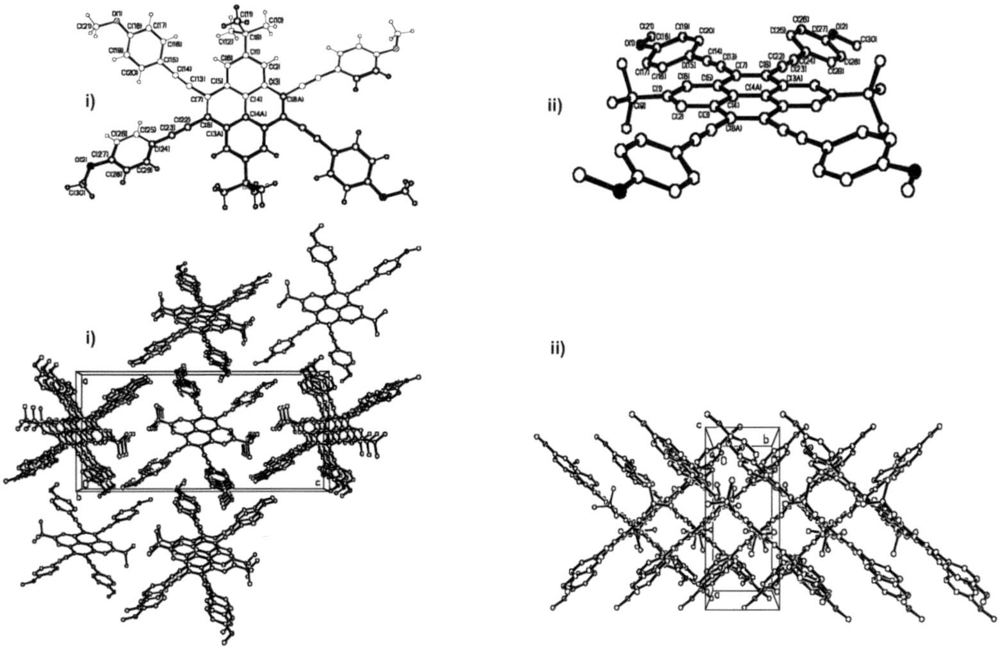

Fig. 13. X-ray crystal-structure diagram of **79c** (i) top view; (ii) side view. down); Packing diagram of **79c** (i) view parallel to *b*, highlighting the π-π stacking; (ii) view parallel to *c*, showing the herringbone packing motif.

in the solid state. Hence, the newly developed cruciform-shaped pyrenes with both the unique intermolecular order of stacked column and low degree of π stacking suggest that they might be advantageous to high charge-carrier transport and robust blue-light emitting materials in optoelectronic devices (Naraso et al., 2005; Wu et al., 2007; Gao et al., 2008).

The UV-vis absorption spectra of **79** are shown in Figure 14, that together with those of pyrenes **76** and **80**. Compared with that of pyrene **76**, the absorption spectra of both **79** and **80** were broad and less well-resolved, and the longest-wavelength, π-π* transition absorption maximum of **79** and **80** occurred at ca. 410-415 nm and 477 nm, respectively, due to the extended conjugation length of the pyrene chromophore with the four phenylethynylenic units. Interestingly, although the vibronic features of **79a-c** were more similar to those of **80** than to those of **76** (Figure 14), the spectra of **79** were less red-shifted than that of **80**, despite the presence of the two electron-donating *t*Bu groups in **79**. A reasonable explanation for these different shifts between **79** and **80** is their quite different conjugation pathway. For **79**, the four phenylacetylenic units are connected with the central pyrene moieties at the nearby 4-, 5-, 9-, and 10-positions to afford a short, cruciform, π-conjugated molecular structure, hence, short, cruciform π-conjugation occurs; for **80**, however, these four phenylacetylenic units are connected with pyrene rings at the more distant 1-, 3-, 6-, and 8-positions, resulting in a longer cruciform, π-conjugated structure. Hence, the conjugation length of **80** is larger than that of **79**, which leads to a larger red shift to ~ 500 nm. Upon excitation, a dilute solution of **79** and **80** in CH_2Cl_2 showed pure-blue and green emission (Figure 14) with a maximum band at 441 nm for **79a**, 448 nm for **79b**, 453 nm for **79c**, and 496 nm for **80**, respectively, which are systematically varied in agreement with the electronic absorption spectra. High quantum yields of **79** in solution were found to in the range of 0.66-0.98.

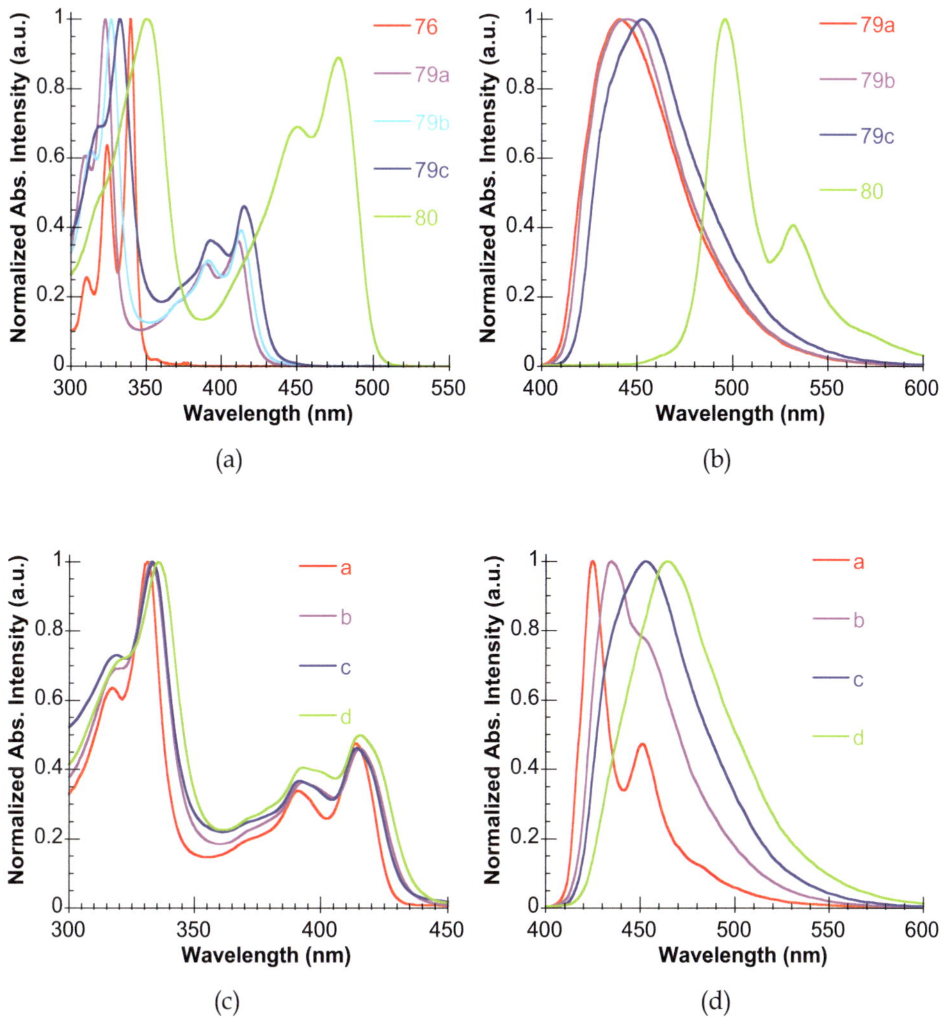

Fig. 14. Normalized absorption (A) and fluorescence emission spectra (B) of **79** and **80** recorded in CH_2Cl_2; down) Normalized absorption (C) and fluorescence emission spectra (D) of **79c** recorded in (a) cyclohexane, (b) THF, (c) CH_2Cl_2, and (d) DMF.

In order to obtain more insight into the photophysical properties of these new cruciform-shaped, conjugated pyrenes, both concentration dependence of the fluorescence and solvatochromic shifts of the absorption and emission spectra of **79c** are investigated, respectively. By increasing the concentration from 1.0×10^{-8} M to 1.0×10^{-4} M, the intensity of this emission band to gradually increase, only the monomer emission at 453 nm was observed. The result further indicate that the two sterically bulky *t*Bu groups at the 2- and 7-positions can prevent two molecules of **79** from getting close enough to result in excimer emission at high concentrations. For **79c**, a change of solvent from nonpolar cyclohexane to polar DMF caused only a very slight, positive, batnochromic shift in the π-π* absorption band from 413 to 417 nm. However, in the case of the emission of **79c**, a substantial positive bathochromism with a maximum peak from 425 nm to 464 nm was observed fromcyclohexane to DMF (Figure 14). The results suggest that these new pyrenes **79** are more solvated in the excited state than in the ground state (Chew et al., 2007). Thus, these molecules emit very bright, pure-blue fluorescence and have good solubility in common organic solvents and high stability, which make them potential candidates as blue organic light-emitting materials for the fabrication of OLED devices, and further exploration into this area is underway.

7. Conclusions

In this Chapter, we have given an overview of the recent work on the synthesis and photophysical properties of pyrene-based light-emitting architectures and their application to molecular optoelectronic devices. We have demonstrated here some concrete examples that pyrene can be appropriately modified with electro- or photo-active chromosphores such as phenyl, phenylethynyl, fluorene, carbazole, and pyrene, and the resulting functional pyrenes can be successfully applied to the fabrication of EL devices. Thus, we expect that further research work on the functional pyrenes will promote a basic understanding of molecular design and optoelectronic properties, and their potential applications to molecular devices such as organic light-emitting diodes (OLEDs).

8. Acknowledgments

The authors wish to acknowledge financial support, respectively, from the CANON Company, the Royal Society of Chemistry and the Cooperative Research Program of "Network Joint Research Center for Materials and Devices (Institute for Materials Chemistry and Engineering, Kyushu University)". The authors wish to thank Dr. Yong-Jin Pu (Department of Organic Device Engineering, Yamagata University) for fruitful discussions.

9. References

Adachi, A.; Manhart, S. A.; Okita, K.; Kido, J.; Ohshita, J. & Kunai, A. (1997). *Synth. Met.*, 91, p. 333.

Adronov, A. & Frechet, J. M. J. (2000). *Chem. Commun.*, p. 1701.

An, B.-K.; Kwon, S.-K.; Jung, S.-J. & Park, S. Y. (2002). *J. Am. Chem. Soc.,* 124, p. 14410.

Anthony, J. E.; Heeney, M. & Ong, B. S. (2008). *MRS Bulletin,* 33, p. 698.

Anthopoulos, T. D.; Frampton, M. J.; Namdas, E. B.; Burn, P. L. & Samuel, I. D. W. (2004). *Adv. Mater.*, 16, p. 557.

Balaganesan, B.; Shen, W.-J. & Che, C.-H. (2003). *Tetrahedron Letters,* 44, p. 5747.

Balakrishnan, K.; Datar, A.; Oitker, R.; Chen, H.; Zuo, J. & Zang, L. (2005). *J. Am. Chem. Soc.,* 127, p. 10496.

Balakrishnan, K.; Datar, A.; Zhang, W.; Yang, X.; Naddo, T.; Huang, J.; Zuo, J.; Yen, M.; Moore, J. S. & Zang, L. (2006). *J. Am. Chem. Soc.*, 128, p. 6576.

Bandichhor, R.; Petrescu, A. D.; Vespa, A.; Kier, A. B.; Schroeder, F. & Burgess, K. (2006). *J. Am. Chem. Soc.*, 128, p. 10688.

Barboiu, M.; Prodi, L.; Montalti, M.; Zaccheroni, N.; Kyritsakas, N. & Lehn, J.-M. (2004). *Chem.-Eur. J.*, 10, p. 2953.

Bauer, R.; Liu, D. V.; Heyen, A.; Schryver, F. D.; Fryter, S. D.; & Mullen, K. (2007). *Macromolecules,* 40, p. 4753.

Becker, K.; Lupton, J. M.; Feldmann, J.; Setayesh, S.; Grimsdale, A. C.; & Mullen, K. (2006). *J. Am. Chem. Soc.*, 128, p. 680.

Bennistom, A. C.; Harriman, A.; Howell, S. L.; Sams, C. A. & Zhi, Y.-G. (2007). *Chem.-Eur. J.*, 13, p. 4665.

Benniston, A. C.; Harriman, A.; Lawrie, D. J.; Mayeux, A.; Rafferty, K.; & Russel, O. D. (2003). *Dalt. Trans.*, p. 4762.

Berlamn,I. B. (1970). *J. Phys. Chem.*, 74, p. 3085.

Bernhardt, S.; Kastler, M.; Enkelmann, V.; Baumgarten, M. & Mullen, K. (2006). *Chem.-Eur. J.*, 12, p. 6117.

Brabec, C. J.; Sariciftci, N. S. & Hummelen, J. C. (2001). *Adv. Funct. Mater.*, 11, p. 15.

Burroughes, J. H. Burroughes, J. H.; Bradley, D. D. C.; Brown, A. R.; Marks, R. N.; Mackay, K.; Friend, R. H.; Burnss, P. L. & Holmes, A. B. (1990). *Nature,* 347, p. 539.

Charles Li, X.-C.; Okamura, Y.; Ueno, K.; Tashiro, M.; & Surya Prakash, G. K. *US Patent* 6852429 2005.

Chen, B. J.; Lai, W. Y.; Gao, Z. Q.; Lee, C. S. & Gambling, W. A. (1999). *Appl. Phys. Lett.*, 75, p. 4010.

Chen, C. H. & Shi, J. (1998). *Coord. Chem. Rev.*, 171, p. 161.

Cheng, C.-H.; & Lin, C. –S. *US Patent* 0169921 2009.

Chen, S.; Xu, X.; Liu, Y.; Yu, G.; Sun, X.; Qiu, W.; Ma, Y. & Zhu, D. (2005). *Adv. Funct. Mater.*, 15, p. 1541.

Chew, S.; Wang, P.; Hong, Z.; Kwong, H. L.; Tang, J.; Sun, S.; Lee, C. S. & Lee, S.-T. (2007). *J. Lumin.*, 124, p. 221.

Chu, T.-Y. & Song, O.-K. (2007). *Appl. Phys. Lett.*, 90, p. 203512.

Clar, E. & Schmidt, W. (1976). *Tetrahedron,* 32, p. 2563.

Daub, J.; Beck, M.; Knorr, A. & Spreitzer, H. (1996). *Pure Appl. Chem.*, 68, p. 1399.

de Halleux, V.; Calbert, J.-P.; Brocorens, P.; Cornil, J.; Declercq, J.-P.; Bredas, J.-L. & Greets, Y. (2004). *Adv. Funct. Mater.*, 14, p. 649.

Ding, J. Q.; Gao, J.; Cheng, Y. X.; Xie, Z. Y.; Wang, L. X.; Ma, D. G.; Jing, X. B. & Wang, F. S. (2006). *Adv. Funct. Mater.*, 16, p. 575.

Duggal, A. R.; Heller, C. M.; Shiang, J. J.; Liu, J. & Lewis, L. N. (2007). *J. Displ. Tech.*, 3, p. 184.

Figueira-Duarte, T. M.; Del Rosso, P. G.; Trattnig, R.; Sax, S.; List, E. J. W. & Mullen, K. (2010). *Adv. Mater.*, 22, p. 990.

Figueira-Duarte, T. M.; Simon, S. C.; Wager, M.; Druzhinin, S. I.; Zachariasse, K. A. & Mullen, K. (2008). *Angew. Chem. Int. Ed.*, 47, p. 10175.

Fogel, Y.; Kastler, M.; Wang, Z.; Rienko, D.; Bodwell, G. J. & Mullen, K. (2007). *J. Am. Chem. Soc.*, 129, p. 11743.

Freeman, A. W.; Koene, S. C.; Malenfant, P. R. L.; Thompson, M. E. & Frechet, J. M. (2000). *J. Am. Chem. Soc.*, 122, p. 12385.

Fujimoto, K.; Shimizu, H.; Furusyo, M.; Akiyama, S.; Ishida, M.; Furukawa, U.; Yokoo, T. & Inouye, M. (2009). *Tetrahedron,* 65, p. 9357.

Gao, B.; Wang, M.; Cheng, Y.; Wang, L.; Jing, X. & Wang, F. (2008). *J. Am. Chem. Soc.*, 130, p. 8297.
Geffroy, B.; le Roy, P. & Prat C. (2006). *Poly. Inter.*, 55, p. 572.
Gingras, M.; Placide, V.; Raimundo, J.-M.; Bergamini, G.; Ceroni, P. & Balzani, V. (2008). *Chem.-Eur. J.*, 14, p. 10357.
Gong, L.-Z.; Hu, Q.-S. & Pu, L. (2001). *J. Org. Chem.*, 66, p. 2358.
Grimshaw, J. & Grimshaw, J. T. (1972). *J. Chem. Soc. Perkin Trans 1*, p. 1622.
Haley, M. M.; Tykwinski, R. R. (Eds), *Carbon-Rich Compounds: From Molecules to Materials*, Wiley-VCH: Weinheim, Germany, 2006.
Halim, M.; Pillow, J. N. G.; Samuel, I. D. W. & Burn, P. L. (1999). *Adv. Mater.*, 11, p. 371.
Han, J.; Jose, J.; Mei, E. & Burgess, K. (2007). *Angew. Chem. Int. Ed.*, 46, p. 1684.
Harth, E. M.; Hecht, S.; Helms, B.; Malmstrom, E. E.; Frechet, J. M. J. & Hawker, C. J. (2002). *J. Am. Chem. Soc.*, 124, p. 3926.
Hill, J. P.; Jin, W.; Kosawa, A.; Fukushima, T.; Ichihara, H.; Shimomura, T.; Ito, K.; Hashizime, T.; Ishii, N. & Aida, T. (2004). *Science*, 304, p. 1481.
Hoeben, F. J. M.; Jonkheijim, P.; Meijer, E. W. & Schenning, A. P. H. J. (2005). *Chem. Rev*, 105, p. 1491.
Hoger, S. *In Acetylene Chemistry*; Diederich, F.; Stang, P. J.; & Tykwinski, R. R. Eds.; Willey-VCH: Weinheim, Germany, 2005.
Hu, J.-Y.; Era, M.; Elsegood, M. R. J. & Yamato, T. (2010). *Eur. J. Org. Chem.*, 1, p. 72.
Hu, J.-Y.; Paudel, A. & Yamato, T. (2009). *J. Chem. Res.*, p. 109.
Hu, K.; Zhu, P. W.; Yu, Y.; Facchetti, A. & Marks, T. J. (2004). *J. Am. Chem. Soc.*, 126, p. 15974.
Hung, L. S. & Chen, C. H. (2002). *Mater. Sci. Engi. R. Rep.*, 39, p. 143.
Jacob, J.; Sax, S.; Plok, T.; List, E. J. W.; Grimsdale, A. C. & Mullen, K. (2004). *J. Am. Chem. Soc.*, 126, p. 6987.
Jiang, D. L.; & Aida, T. (1997). *Nature* (London) 388, p. 454.
Jiang, X.-Y.; Zhang, Z.-L.; Zheng, X.-Y.; Wu, Y.-Z. & Xu, S.-H. (2001). *Thin Solid Films*, 401, p. 251.
Jiao, G.-S.; Thoresen, L. H.; Kim, T. G.; Haaland, W. C.; Gao, F.; Topp, M. R.; Hochstrasser, R. M.; Metzker, M. L. & Burgess, K. (2006). *Chem.-Eur. J.*, 12, p. 7816.
Jia, W.-L.; Mccormick, T.; Liu, Q.-D.; Fukutani, H.; Moltala, M.; Wang, R.-Y.; Tao, Y. & Wang, S. (2004). *J. Mater. Chem.*, 14, p. 3344.
Jia, W. L.; Wang, R. Y.; Song, D. T.; Ball, S. J.; McLean, A. B. & Wang, S. N. (2005). *Chem.-Eur. J.*, 11, p. 832.
Joule, J. A. (1984). *Adv. Heterocycl. Chem.*, 35, p. 83.
Jurczok, M.; Plaza, P.; Rettig, W. & Martin, M. M. (2000). *Chem. Phys.*, 256, p. 137.
Kang, H.; Evrenenko, G.; Dutta, P.; Clays, K.; Song, K. & Marks, T. J. (2006). *J. Am. Chem. Soc.*, 128, p. 6194.
Kastler, M.; Pisula, W.; Wasser-fallen, D.; Pakula, T. & Mullen, K. (2004). *J. Am. Chem. Soc.*, 126, p. 5234.
Kawano, S.-I.; Yang, C.; Ribas, M.; Baluschev, S.; Baumgarten, M. & Mullen, K. (2008). *Macromolecules*, 41, p. 7933.
Kawase, T. (2007). *Synlett*, p. 2609.
Kido, J. & Lizumi, Y. (1997). *Chem. Lett.*, 26, p. 963.
Kim, H. M.; Lee, Y. O.; Lim, C. S.; Kim, J. S. & Cho, B. R. (2008). *J. Org. Chem.*, 73, p. 5127.

Kim, Y.-H.; Jeong, H.-C.; Kim, S.-H.; Yang, K. & Kwon, S.-K. (2005). *Adv. Funct. Mater.*, 15, p. 1799.

Kim, Y. H.; Shin, D. C.; Kim, S. H.; Ko, C. H;. Yu, H. S.; Chae, Y. S. & Kwon, S. K. (2001). *Adv. Mater.*, 13, p. 1690.

Kimura, M.; Shiba, T.; Yamazaki, M.; Hanabusa, K.; Shirai, H. & Kobayashi, N. (2001). *J. Am. Chem. Soc.*, 123, p. 5636.

Koene, B. E.; Loy, D. E. & Thompson, M. E. (1998). *Chem. Mater.*, 10, p. 2235.

Kraft, A.; Grimsdale, A. & Holmes, A. B. (1998). *Angew. Chem. Int. Ed.*, 37, p. 402.

Krebs, F. C. (2009). *Solar Energy Materials and Solar Cells,* 93, p. 394.

Kulkarni, A. P.; Tonzola, C. J.; Babel, A. & Jenekhe, S. A. (2004). *Chem. Mater.*, 16, p. 4556.

Kuwabara, Y.; Ogawa, H.; Inada, H.; Noma, N. & Shirota, Y. (1994). *Adv. Mater.*, 6, p. 677.

Kwok, C. C. & Wong, M. S. (2001). *Macromolecules,* 34, p. 6821.

Law, M. Goldberger, J.; & Yang, P. D. (2004). *Ann. Rev. Mater. Res.*, 34, p. 83.

Lee, S. H.; Nakamura, T. & Tsutui, T. (2001). *Org. Lett.*, 3, p. 2005.

Lehn, J.-M. *Supramolecular Chemistry-Concepts and Perspectives*; VCH: Weiheim, Germany, 1995.

Liao, Y.-L.; Lin, C.-Y.; Wong, K.-T.; Hou, T.-H. & Hung, W.-Y. (2007). *Org. Lett.*, 10, p. 4511.

Liu, D. J.; De Feyter, S.; Cotlet, M.; Stefan, A.; Wiesler, U. M.; Herrmann, A.; Grebel-Koehler, D.; Qu, J. Q.; Mullen, K. & De Schryver, F. C. (2003). *Macromolecules,* 36, p. 5918.

Liu, F.; Tang, C.; Chen, Q.-Q.; Li, S.-Z.; Wu, H.-B.; Xie, L.-H.; Peng, B.; Wei, W.; Cao, Y. & Huang, W. (2009). *Organic Electronics,* 10, p. 256.

Loo, Y. L. & McCulloch, I. (2008). *MRS Bull.*, 33, p. 653.

Lo, S.-C.; Male, N. A.; Markham, J. P. J.; Magennis, S. W.; Burn, P. L.; Salata, O. V. & Samuel. I. D. W. (2002). *Adv. Mater.*, 14, p. 975.

Lo, S.-C.; Namdas, E. B.; Burn, P. L. & Samuel, I. D. W. (2003). *Macromolecules,* 36, p. 9721.

Loudet, A.; Bandichhor, R.; Wu, L. & Burgess, K. (2008). *Tetrahedron,* 64, p. 3642.

Lupton, J. M.; Samuel, I. D. W.; Beavington, R.; Frampton, M. J.; Burn, P. L. & Bassler, H. (2001). *Phys. Rev. B,* 63, p. 155206.

Maeda, H.; Maeda, T.; Mizuno, K.; Fujimoto, K.; Shimizu, H. & Inouye, M. (2006). *Chem.-Eur. J.*, 12, p. 824.

Manhart, S. A.; Adachi, A.; Sakamaki, K.; Okita, K.; Ohshita, J.; Ohno, T.; Hamaguchi, T.; Kunai, A. & Kido, J. (1999). *J.Organomet. Chem.*, 592, p. 52.

Markham, J. P. J.; Samuel, I. D. W.; Lo, S.-C.; Burn, P. L.; Weiter, M. & Bassler, H. (2004). *J. Appl. Phys.*, 95, p. 438.

Marsden, J. A.; Miller, J. J.; Shirtcliff, L. D. & Haley, M. M. (2005). *J. Am. Chem. Soc.*, 127, p. 2464.

Martin, R. B.; Qu, L.; Harruff, B. A.; Bunker, C. E.; Gord, J. R.; Allard, L. F. & Sun, Y.-P. (2004). *J. Phys. Chem. B,* 108, p. 11447.

McGehee, M. D. & Heeger, A. J. (2000). *Adv. Mater.,* 12, p. 1655.

Meijere, A. de. (Ed.), *Topics in Current Chemistry, Carbon Rich Compounds,* I, Spinger: Berlin, Germany, 196 1998.

Melhuish, W. H. (1961). *J. Phys. Chem.,* 65, p. 229.

Melinger, J. S.; Pan, Y.; Kleiman, V. D.; Peng, Z.; Davis, B. L.; McMorrow, D. & Lu, M. (2002). *J. Am. Chem. Soc.*, 124, p. 12002.

Miao, Q.; Chi, X. L.; Xiao, S. X.; Zeis, R.; Lefenfeld, M.; Siegrist, T.; Steigerwald, M. L. & Nuckolls, C. (2006). *J. Am. Chem. Soc.*, 128, p. 1340.

Mikroyannidis, J. A.; Fenenko, L. & Adachi, C. (2006). *J. Phys. Chem. B*, 110, p. 20317.

Mikroyannidis, J. A.; Persephonis, P. G. & Giannetas, V. G. (2005). *Synth. Met.*, 148, p. 293.

Modrakowski, C.; Flores, S. C.; Beinhoff, M. & Schluter, A. D. (2001). *Synthesis*, p. 2143.

Moorthy, J. N.; Natarajin, P.; Venkatakrishnan, P.; Huang, D.-F. & Chow, T. J. (2007). *Org. Lett.*, 9, p. 5215.

Mullen, K.; & Wegner, G. Eds: *Electronic Materials: The Oligomer Approach*, Wiley-VCH: Weinheim, Germany, 1998.

Mullen, K.; Scherf, U. Eds: *Organic Light Emitting Devices: Synthesis Properties and Applications* Wiley-VCH: Weiheim, Germany, 2006.

Naraso; Nishida, J.-I.; Ando, S.; Yamaguchi, J.; Itaka, K.; Koinuma, H.; Tada, H.; Tokito, S. & Yamashita, Y. (2005). *J. Am. Chem. Soc.*, 127, p. 10142.

Newkome, G. R.; Moorefield, C. N.; & Vogtle, F. *Dendritic Molecules: Comcepts, Synthesis, and Perspectives*, Wiley-VCH, Weinheim, Germany 1996.

O'Brien, D. F.; Burrows, P. E.; Forrest, S. R.; Koene, B. E.; Loy, D. E. & Thompson, M. E. (1998). *Adv. Mater.*, 10, p. 1108.

Oh, H.-Y.; Lee, C.-H. & Lee, S.-H. (2009). *Organic Electronics*, 10, p. 163.

Oh, J.-W.; Lee, Y. O.; Kim, T. H.; Ko, K. C.; Lee, J. Y.; Kim, H. & Kim, J. S. (2009). *Angew. Chem. Int. Ed.*, 48, p. 2522.

Ohshita, J.; Yoshimoto, K.; Tada, Y.; Harima, Y.; Kunai, A.; Kunugi, Y. & Yamashita, K. (2003). *J. Organomet. Chem.*, 678, p. 33.

Otsubo, T.; Aso, Y. & Takamiya, K. (2002). *J. Mater. Chem.*, 12, p. 2565.

Oyamada, T.; Uchiuzou, H.; Akiyama, S.; Oku, Y.; Shimoji, N.; Matsushige, K.; Sasaba, H. & Adachi, C. (2005). *J. Appl. Phys.*, 98, p. 074506.

Pogantsch; Wenzl, F. P.; List, E. J. W.; Leising, G.; Grimsdale, A. C.; & Mullen, K. (2002). *Adv. Mater.*, 14, p. 1061.

Pu, Y.-J.; Higashidate, M.; Nakayama, K. & Kido, J. (2008). *J. Mater. Chem.*, 18, p. 4183.

Qu, L. & Shi, G. (2004). *Chem. Commun.*, p. 2800.

Rathore, R.; Burns, C. L. & Abdelwahed, S. A. (2004). *Org. Lett.*, 6, p. 1689.

Rathore, R.; Burns, C. L. & Deselnicu, M. I. (2001). *Org. Lett.*, 3, p. 2887.

Rausch, D. & Lambert, C. (2006). *Org. Lett.*, 8, p. 5037.

Raytchev, M.; Pandurski, E.; Buchvarov, I.; Modrakowski, C. & Fiebig, T. (2003). *J. Phys. Chem. A*, 107, p. 4592.

Rivera, E.; Belletete, M.; Zhu, X. X.; Durocher, G. & Giasson, R. (2002). *Polymer*, 43, p. 5059.

Rosenfeldt, S.; Dingenouts, N.; Potschke, D.; Ballauff, M.; Berresheim, A. J.; Mullen, K.; & Linder, P. (2003). *Angew. Chem.*, 115, p. 111; (2004). *Angew. Chem. Int. Ed.*, 43, p. 109.

Sagara, Y.; Mutai, T.; Yoshikawa, I. & Araki, K. (2007). *J. Am. Chem. Soc.*, 129, p. 1520.

Sakamoto, Y.; Suzuki, T.; Miura, A.; Fujikawa, H.; Tokito, S. & Taga, Y. (2000). *J. Am. Chem. Soc.*, 122, p. 1832.

Salbeck, J.; Weissortel, F.; Yu, N.; Baner, J. & Bestgen, H. (1997). *Synth. Met.*, 91, p. 209.

Seminario, J. M. (2005). *Nature Materials*, 4, p. 111.

Shibano,Y.; Umeyama, T.; Matano, Y. & Imahori, H. (2007). *Org. Lett.*, 10, p. 1971.

Shih, H. T.; Lin, C.-H.; Shih, H.-H.; & Cheng, C.-H. (2002). *Adv. Mater.*, 14, p. 1409.

Shi, J.; Wang, C. W.; & Chen, C. H. *US Patent* 5646948, 1997.

Shtein, M.; Peumans, P.; Benziger, J. B. & Forrest, S. R. (2004). *Adv. Mater.*, 16, p. 1615.
Sienkowska, M. J.; Farrar, J. M.; Zhang, F.; Kusuma, S.; Heiney, P. A. & Kaszynski, P. (2007). *J. Mater. Chem.*, 17, p. 1399.
Shimizu, H.; Fujimoto, K.; Furusyo, M.; Maeda, H.; Nanai, Y.; Mizuno, K. & Inouye, M. (2007). *J. Org. Chem.*, 72, p. 1530.
Sirringhaus, H. & Ando, M. (2008). *MRS Bull.*, 33, p. 676.
So, F.; Kido, J. & Burrows, P. L. (2008). *MRS Bull.*, 33, p. 663.
Sonar, P.; Soh, M. S.; Cheng, Y. H.; Henssler, J. T. & Sellinger, A. (2010). *Org. Lett.*, 12, p. 3292.
Son, S.; Dodabalapur, A.; Lovinger, A. J. & Galvin, M. E. (1995). *Science*, 269, p. 376.
Strohriegl, P.; & Grazulevicius, J. V. (2002). *Adv. Mater.*, 14, p. 1439.
Strauss, J. & Daub, J. (2002). *Org. Lett.*, 4, p. 683.
Sun, D. L.; Rosokha, S. V. & Kochi, J. K. (2005). *Angew. Chem. Int. Ed.*, 44, p. 5133.
Swager, T. M. *In acetylene chemsity*; Diederich, F.; Stang, P. J.; & Tykwinski, R. R.; Eds; Willey-VCH: Weinheim, Germany, 2005.
Swager, T. M. & Zheng, J. (2005). *Advances in Polymer Science*, 177, p. 151.
Tang, C.; Liu, F.; Xia, Y.-J.; Lin, J.; Xie, L.-H.; Wei, A.; Li, S.-B.; Fan, Q.-L. & Huang, W. (2006). *J. Mater. Chem.*, 16, p. 4074.
Tang, C.; Liu, F.; Xia, Y.-J.; Lin, J.; Xie, L.-H.; Zhong, G.-Y.; Fan, Q.-L. & Huang, W. (2006). *Organic Electronics*, 7, p. 155.
Tang, C. W. & VanSlyke, S. A. (1987). *Appl. Phys. Lett.*, 51, p. 913.
Tao, S. L.; Hong, Z. R.; Peng, Z. K.; Ju, W. G.; Zhang, X. H.; Wang, P. H.; Wu, S. K. & Lee, S. T. (2004). *Chem. Phys. Lett.*, 397, p. 1.
Tao, S. L.; Peng, Z. K.; Zhang, X. H.; Wang, P. F.; Lee, C.-S. & Lee, S.-T. (2005). *Adv. Funct. Mater.*, 15, p. 1716.
Tao, S. L.; Zhou, Y.; Lee, C.-S.; Zhang, X.-H. & Lee, S.-T. (2010). *Chem. Mater.*, 22, p. 2138.
Thomas, K. R. J.; Velusamy, M.; Lin, J. T.; Chuen, C. H. & Tao, Y. T. (2005). *J. Mater. Chem.*, 15, p. 4453.
Thomas, K. R.; Lin, J T.; Tao, Y.-T.; & Ko, C.-W. (2000). *Adv. Mater.*, 12, p. 1949.
Thomas, K. R.; Lin, J T.; Tao, Y.-T. & Ko, C.-W. (2001). *J. Am. Chem. Soc.*, 23, p. 9404.
Tokito, S.; Tanaka, H.; Koda, N.; Okada, A. & Tago, Y. (1997). *Macromol. Symp.*, 125, p. 181.
Tonzola, C. J.; Alam, M. M.; Kaminsky, W. & Jenekhe, S. A. (2003). *J. Am. Chem. Soc.*, 125, p. 13548.
Tsumura, A.; Koezuka, K. & Ando, T. (1986). *Appl. Phys. Lett.*, 49, p. 1210.
Tu, G.; Zhou, Q.; Cheng, Y.; Wang, L.; Ma, D.; Jing, X. & Wang, F. (2004). *Appl. Phys. Lett.*, 85, p. 2172.
Uchida, M.; Izumizawa, T.; Nakano, T.; Yamaguchi, S.; Tamao, K. & Furukawa, K. (2001). *Chem. Mater.*, 13, p. 2680.
VandenBout, D. A.; Yip, W. T.; Hu, D. H.; Fu, D. K.; Swager, T. M. & Barbara, P. F. (1997). *Science*, 277, p. 1074.
Van der Auweraer, M. & De Schryer, F. C. (2004). *Nature Materials*, 3, p. 507.
Venkataramana, G. & Sankararaman, S. (2005). *Eur. J. Org. Chem.*, 19, p. 4162.
Venkataramana, G. & Sankararaman, S. (2006). *Org. Lett.*, 8, p. 2739.
Vollmann, H.; Becker, M.; & Correll, H. S. (1937). *Justus Liebigs Ann. Chem.*, 1, p. 531.

Wang, L.; Jiang, Y.; Luo, J.; Zhou, Y.; Zhou, J.; Wang, J.; Pei, J. & Cao, Y. (2009). *Adv. Mater.*, 21, p. 4854.
Wang, P. W.; Liu, Y. J.; Devadoss, C.; Bharathi, P. & Moore, J. S. (1996). *Adv. Mater.*, 8, p. 237.
Weinfurtner, K.-H.; Fujikawa, H.; Tokito, S. & Taga, Y. (2000). *Appl. Phys. Lett.*, 76, p. 2502.
Wilson, J. N. & Bunz, U. H. F. (2005). *J. Am. Chem. Soc.*, 127, p. 4124.
Wind, M.; Wiesler, U. M.; Saalwachter, K.; Mullen, K.; & Spiess, H. W. (2001). *Adv. Mater.*, 13, p. 752.
Wong, K. T.; Chien, Y. Y.; Chen, R. T.; Wang, C. F.; Lin, Y. T.; Chiang, H. H.; Hsieh, P. Y.; Wu, C. C.; Chou, C. H.; Su, Y. O.; Lee, G. P. & Peng, S. M. (2002). *J. Am. Chem. Soc.*, 124, p. 11576.
Wu, C.-C., Chen, C.-W.; Lin, C.-L. & Yang, C.-J. (2005). *J .Displ. Tech.*, 1, p. 248.
Wu, J.; Pisula, W. & Mullen, K. (2007). *Chem. Rev.*, 107, p. 718.
Wu, K.-C.; Ku, P.-J.; Lin, C.-S.; Shih, H.-T.; Wu, F.-I.; Huang, M.-J.; Lin, J.-J.; Chen, I.-C.; & Cheng, C.-H. (2008). *Adv. Funct. Mater.*, 18, p. 67.
Xie, W.; Hou, J. & Liu, S. (2003). *Semicond. Sci. Tech.*, 18, p. 142.
Xing, Y.-J.; Xu, X.-J.; Zhang, P.; Tian, W.–J.; Yu, G.; Lu, P.; Liu, Y.-Q. & Zhu, D.-B. (2005). *Chem. Phys. Lett.*, 408, p. 169.
Xu, M.-H.; Lin, J.; Hu, Q.-S. & Pu, L. (2002). *J. Am. Chem. Soc.*, 124, p. 14239.
Xu, Z. & Moore, J. S. (1993). *Angew. Chem. Int. Ed. Engl.*, 32, p. 1354.
Yamana, K.; Iwai, T.; Ohtani, Y.; Sato, S.; Namakura, M. & Nakano, H. (2002). *Bioconjugate Chem.*, 13, p. 1266.
Yamamoto, T. (2003). *Synlett*, 4, p. 425.
Yamashita, E. & Maeda, K. (2008). *Macromolecules,* 41, p. 3.
Yamato, T.; Fujimoto, M.; Miyazawa, A. & Matsuo, K. (1997). *J. Chem. Soc. Perkin Trans 1*, p. 1201.
Yamato, T.; Miyazawa, A. & Tashiro, M. (1993). *J. Chem. Soc. Perkin Trans 1*, p. 3127.
Yang, C.-H.; Guo, T.-F. & Sun, I.-W. (2007). *J. Lumine.*, 124, p. 93.
Yu, W.-L.; Pei, J.; Huang, W. & Heeger, A. J. (2000). *Adv. Mater.*, 12, p. 828.
Zen, A.; Bilge, A.; Galbrecht, F.; Alle, R.; Meerholz, K.; Grenzer, J.; Neher, D.; Scherf, U. & Farrell, T. (2006). *J. Am. Chem. Soc.*, 128, p. 3914.
Zhang, H.-J.; Wang, Y.; Shao, K.-Z.; Liu, Y.-Q.; Chen, S.-Y.; Qiu, W.-F.; Sun, X.-B.; Qi, T.; Ma, Y.-Q.; Yu, G.; Su, Z-.M. & Zhu, D.-B. (2006). *Chem. Commun.*, 2, p. 755.
Zhang, M.; Xue, S.; Dong, S.; Wang, Q.; Fei, T.; Gu, C. & Ma, Y. (2010). *Chem. Commun.*, 46, p. 3923.
Zhao, Z. J.; Li, J.-H.; Chen, X. P.; Lu, P. & Yang, Y. (2008). *Org. Lett.*, 10, p. 3041.
Zhao, Z. J.; Li, J.-H.; Chen, X. P.; Wang, X. M.; Lu, P. & Yang, Y. (2009). *J. Org. Chem.*, 74, p. 383.
Zhao, Z. J.; Xu, X. J.; Jiang, Z. T.; Lu, P.; Yu, G. & Liu, Y. Q. (2007). *J. Org. Chem.*, 72, p. 8345.
Zhao, Z. J.; Xu, X. J.; Wang, F.; Yu, G.; Lu, P.; Liu, Y. & Zhu, D. (2006). *Synth. Met.*, 156, p. 209.

3

Organometallic Materials for Electroluminescent and Photovoltaic Devices

Boris Minaev[1], Xin Li[2,3], Zhijun Ning[2], He Tian[3] and Hans Ågren[2]
[1]*Bogdan Khmelnitskij National University*
[2]*Royal Institute of Technology*
[3]*East China University of Science and Technology*
[1]*Ukraine*
[2]*Sweden*
[3]*People's Republic of China*

1. Introduction

Electroluminescent devices, solar energy conversion technologies and light-emitting electrochemical cells represent a promising branch of modern optoelectronic industry based on organic dyes and polymers as the main working materials. Elementary processes like energy flow through an organic-inorganic interface and voltage control at a molecular level with peculiar electronic properties are now well understood and used in fabrication of new efficient and sophisticated optoelectronic devices. Today, organic light emitting diodes (OLEDs) are used commercially in displays and various lighting applications providing high external quantum efficiency (up to 19%) and low power consumption (Nazeeruddin et al. 2009).

Electroluminescence of organic materials was observed for the first time by Martin Pope et al. (Pope et al. 1963) a half of a century ago. Twenty-four years later the pioneering work of Eastman Kodak Company (Tang & VanSylke 1987) provided the use of 8-hydroxyquinoline aluminum (Alq_3) as electron-transporting and emissive material (Scheme 1) in an OLED device. Since then a growing progress has been witnessed in the field of organic optoelectronics through incorporation of various combinations of organic polymers, dyes and organometallic complexes (Yersin & Finkenzeller 2008, Köhler & Bässler 2009).

Organic conjugated polymers, like poly(para-phenylene vinylene) (PPV) doped by various chromophores, are now used in OLEDs as they lend the possibility to create charge carrier recombination and formation of excitons with high efficiency of light emission. Typical OLEDs are fabricated by spin-coating, inkjet printing or by vacuum deposition of organic materials on an indium-tin-oxide (ITO)-coated glass and with a multilayer structure of the device including NPB (N,N′-Bis(naphthalene-1-yl)-N,N′-bis(phenyl)-benzidine) and Alq_3 as the hole transport layer (HTL) and electron transport layer (ETL), respectively. These materials are presented here as typical examples. In between there is a doped emission layer (EML). Usually some additional layers which protect the ETL from reactions with the cathode material, or reduce the injection barrier and electron-hole quenching, are incorporated into the device architecture. These OLEDs are thin, flexible, stable, and energy conserving devices; they have prompt response times (μs), high color purity and are suitable for large screen displays and even for illuminating wallpapers in the near future.

In this review we will mostly discuss a new OLED generation with triplet emitters based on cyclometalated transition metal complexes which have attracted extensive attention in the past decade since their external quantum efficiency exceeds that of usual organic materials. In purely organic EML polymers, like PPV or PPP, the energy stored in triplet states cannot be utilized in order to increase the emissive efficiency of OLEDs, since the transition from the lowest triplet (T_1) to the singlet ground (S_0) state is strictly forbidden; this energy thus is mostly spread non-radiatively into backbone phonons and heating of the sample. It has been proposed that that the use of the triplet states may improve an efficiency of solar cells to a large extent (Köhler & Bässler, 2009; Wong, 2008). Singlet state emitters, like Alq_3 (Scheme 1), as a key electron transporter in ETL and new nonmetallic complexes will also be analyzed in this review.

Introducing $Ir(ppy)_3$ in doped EMLs provided a new revolution in modern OLED optoelectronics (Baldo et al. 1999). With the Ir(III) complexes as dopants the electroluminescence is enhanced by harnessing both singlet and triplet excitons after the initial charge recombination, since spin-orbit coupling (SOC), being much stronger in heavy elements like iridium, removes the spin-forbidden character of the singlet-triplet (S-T) transitions. The search for OLED materials was initiated by the classic example $Ru(bpy)_3^{2+}$, which was used as a photocatalyst in solar-driven photoconversion processes. The task was to alter the excited state redox potential of similar metal complexes by several modifications; changing the central transition metal, replacing some of the ligands and modifying the ligands by adding suitable functional groups. For example, changing the metal center in $Ru(bpy)_3^{2+}$ to Ir(III) produces a complex, $Ir(bpy)_3^{3+}$, with excellent photo-oxidizing power. The neutral bpy ligand was changed by the negatively charged ppy. These changes provide a crucial improvement of the chromophore stability and finally became a lucky choice for implementation in modern OLEDs.

Similar problems of the dye optimization are present for the dye-sensitized solar cells (DSSC) based on TiO_2 nanocrystals. They have become of considerable interest as renewable power sources because of the ability to provide a high coefficient of light conversion to electricity (up to 10.4 percent). DSSCs commonly use sensitizers based on complexes of ruthenium with bipyridine (bpy) and other ligands. The most successful example of DSSCs is a Grätzel cell in which the ruthenium(II) bis(4,4'-dicarboxy-2,2'-bipyridyl)-bis(isothiocyanate), denoted by the N_3 dye, is used as sensitizer (Grätzel, 1990); other similar dyes have also served for this purpose. All these organometallic dyes absorb visible light and, being in the excited state, provide electron transfer to TiO_2 crystals on the surface of which they are adsorbed. After that, the oxidized chromophore is reduced by the electrolyte and the cycle is repeated. The requirements to a sensitizing chromophore are universal: high light absorption coefficient in the entire intense solar spectrum, ability to inject an electron into the conduction band of TiO_2, and fast reduction by the electrolyte. The choice and optimization of the chromophore are highly important for the DSSC technology.

To make systematic choices, it is necessary to know the relationship between structures and optical properties of dye molecules. For this purpose, a number of quantum-chemical calculations of electron absorption spectra of the most important sensitizers based on complexes of ruthenium(II) and iridium(III) with polypyridines have been carried out in recent years. Theoretical studies of phosphorescent dopants for OLEDs are of similar importance. Nowadays, almost all new dyes for DSSCs and OLEDs become subjects for comprehensive photophysical, electrochemical and quantum-chemical density functional theory investigations. The aim of the present review is to describe new synthesis of organometallic complexes based on Ir, Ru, Pt and other ions which are prospects for effective OLEDs and solar cell fabrication and to present the theoretical background that is important for these dyes.

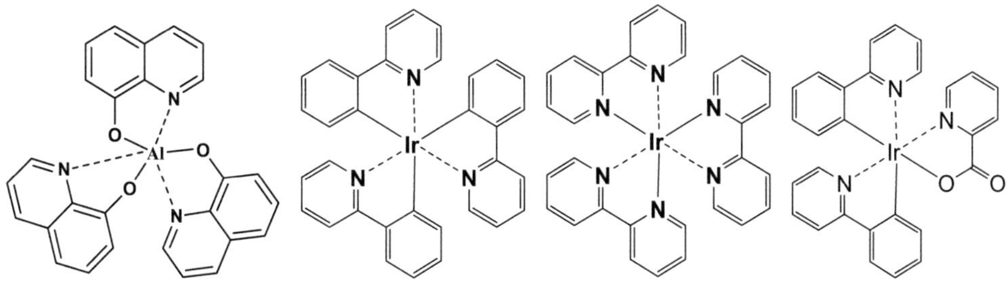

Alq_3 $Ir(ppy)_3$ $Ir(bpy)_3^{3+}$ $Ir(ppy)_2(pic)$

Scheme 1.

We will discuss Ir-complexes of the type shown in Scheme 1, our own synthesized dyes and a number of those which are known from the literature. The main subject of this Review is a theoretical analysis of the structural factors which determine high efficiency of new OLED materials, which we have provided by calculations of various CICs and other dyes. Our unique ab initio calculations of spin-orbit coupling effects and phosphorescent lifetimes in these heavy metal compounds permit us to derive some general ideas about the synthesis of the best phosphorescent sensitizing chromophores for OLEDs and DSSCs.

2. Principles of optoelectronics based on organic materials

2.1 Electroluminescence with low power consumption

OLEDs are based on polymers multi-layer structures. Inside each layer the electronic excitations in the repeating molecular units, being linked by covalent bonds, represent rather complicated excitons. Such excitations in a polymer chain are extended over several molecular links and are associated with the distortion of the polymer. The exciton energy decreases when the size of the excited area increases, but the Coulomb attraction in the electron-hole pair (EHP) can confine the exciton. Principles of electro-luminescence in organic polymers are described in a number of references (e.g. Pope & Swenberg, 1999; Yersin & Finkenzeller, 2008; Köhler & Bässler, 2009). In this review we concentrate our attention on the theoretical description of molecular states at the microscopic level; thus the macroscopic picture of the structural elements of OLEDs and photovoltaic devices will be shortly presented in the form of main concepts of their architecture, connected with organic and inorganic materials in different aggregate states.

Organic materials used in modern optoelectronics can occur in the form of polymer layers, molecular crystals or supramolecular assemblies. Optical characteristics of these materials are intrinsically determined by their molecular properties, which we here will generalize in a simplified manner (Fig. 1). Their conductivity and charge carrier properties are also determined by molecular orbital (MO) properties, but mainly by external perturbations;

injection from the electrodes in the OLED devices or through the dissociation of EHPs created by the incident light in the solar cells, which also require some simple model descriptions. In organic molecular crystals weak van der Waals interaction provides very narrow (≤0.5 eV) valence and conduction bands; in polymers these bands are wider.
In Fig. 1 we present a scheme that illustrates the creation and propagation of excitons in a conjugated polymer chain, which is simulated by four molecular blocks. In real polymers there are thousands of blocks; furthermore, in OLEDs there are few different polymer layers, where the molecular blocks have different MO energy levels. We omit these details for simplicity in Fig. 1.

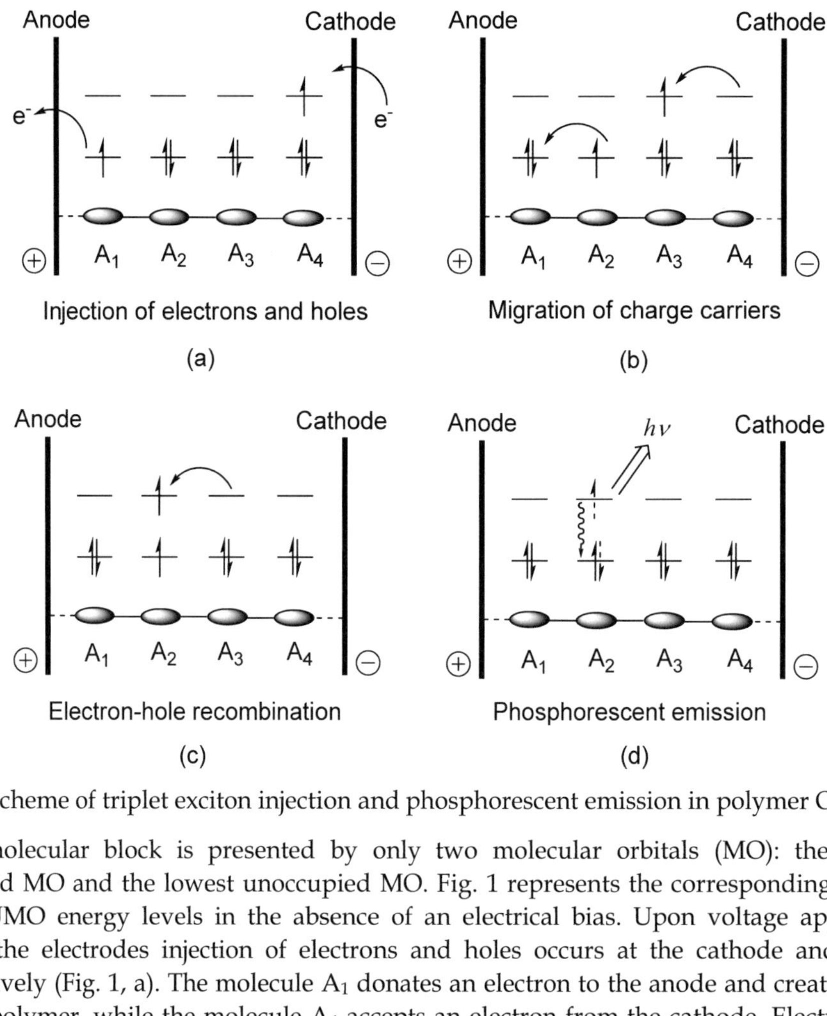

Fig. 1. Scheme of triplet exciton injection and phosphorescent emission in polymer OLEDs

Each molecular block is presented by only two molecular orbitals (MO): the highest occupied MO and the lowest unoccupied MO. Fig. 1 represents the corresponding HOMO and LUMO energy levels in the absence of an electrical bias. Upon voltage application across the electrodes injection of electrons and holes occurs at the cathode and anode, respectively (Fig. 1, a). The molecule A_1 donates an electron to the anode and creates a hole in the polymer, while the molecule A_4 accepts an electron from the cathode. Electrons and holes migrate through the polymer chain in the opposite directions because of the applied voltage; getting closer they start to attract each other until they occur at the neighboring blocks A_2 and A_3 (Fig. 1, b).
The critical radius R_C for the mutual interaction inside the EHP is defined as the distance at which the Coulomb attraction is equal to the thermal energy:

$$kT = e^2 / 4\pi\varepsilon\varepsilon_0 R_C \ , \tag{1}$$

where $\varepsilon, \varepsilon_0$ represent the dielectric constants of the polymer material and the vacuum (Köhler & Bässler, 2009). The low dielectric constant for a typical organic polymer ($\varepsilon = 3$) provides at room temperature quite a big Coulomb capture radius of about 19 nm. In organic polymer films the repeating molecular units (A_n) are linked through covalent chemical bonds and electrons and holes are not strictly localized; in some polymers the charge carrier wave function can extend over several molecular blocks. If we return back to the simple picture in Fig. 1, it is clear that electrons and holes localized at the neighboring blocks A_2 and A_3 have overlapping wave functions. This means a high probability of an electron "jump" from A_3 to A_2, or electrons-hole recombination. After their recombination an electronically excited molecule A_2 is obtained (Fig. 1, c), which can emit light (Fig. 1, d).
In Fig. 1 the triplet state of the electron-hole pair (EHP) is shown, but it can be created in the singlet state as well. Annihilation of the singlet EHP leads to the singlet excited state S_1 of molecule A_2 which can emit light in the spin allowed singlet-singlet transition to the ground state S_0 (Fig. 2). Spontaneous emission $S_1 \rightarrow S_0$ is usually characterized by a short lifetime (ns) because of the large electric dipole transition moment for the spin allowed singlet-singlet transition. This is a typical mechanism of electroluminescence in pure organic light emitting diodes. Radiative rate constants of photofluorescence (k_3) in such polymers is usually a million times larger than those of phosphorescence (k_4), which cannot compete with nonradiative quenching (k_5) in pure organic polymers at room temperature.

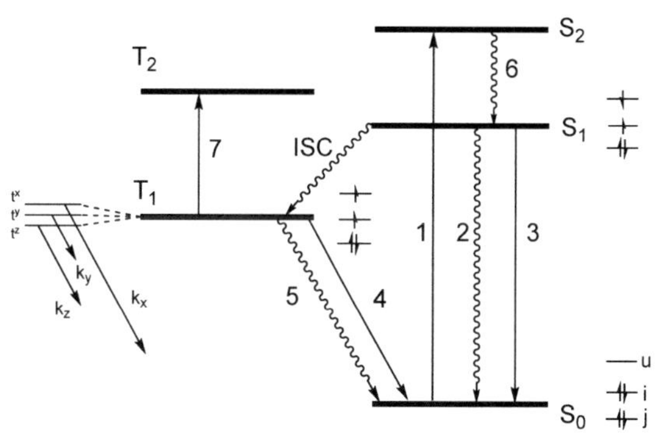

Fig. 2. Main photoprocesses following light absorption by an organic molecule (1). Wavelines - nonradiative quenching; ISC – intersystem crossing; 3 – fluorescence (rate constant $k_3 \sim 10^9$ s^{-1}, radiative lifetime $\tau_3^r = 1/k_3 \approx$ ns); 4 – phosphorescence (k_4 is in the range 0.01-10^6 s^{-1}); 7 – triplet-triplet absorption.

In organic solids solvents, excited state levels are shifted with respect to free molecules in gas phase due to inhomogeneous broadening. In an amorphous film or glass, each molecule has its own particular environment with different orientations and distances to its neighbors. This random variation of electronic polarization of the neighborhood leads to a particular shift of the excited state energy for each molecule (inhomogeneous broadening of spectral lines). For the triplet excited state this broadening is smaller than for the singlets, because electron correlation is stronger for the former (Pope& Swenberg, 1999).
Modern OLEDs often contain an organic polymer in the emission layer doped by a transition-metal complex with heavy iridium or platinum ions, which provide a strong SOC

that makes the $T_1 \rightarrow S_0$ transition to be effectively allowed ($k_4 \sim 10^6$ s^{-1}). In this case the triplet excitons also produce useful work in the OLED (Fig. 1, d). Dopants in EMLs not only collect the S and T excitons by the EHP recombination, but can also be used for regulation of the OLED color. In particular, iridium complexes, containing large π-conjugated ligands (Scheme 1), such as 2-phenylpyridine anions (ppy$^-$) and neutral 2,2′-bipyridines (bpy), have the advantage that their emission wavelength can be tuned from blue to red by peripheral substitution in the rings by electron-withdrawing and electron-donating substituents or by replacement of chelating ligands. The S_1-T_1 splitting (Fig. 2) is determined by the double exchange integral, $2K_{i-u}$, for φ_i and φ_u orbitals (typically HOMO and LUMO)

$$K_{i-u} = \int\int \varphi_i(1)\varphi_u(1)(e^2/r_{1,2})\varphi_i(2)\varphi_u(2)dv_1 dv_2 \,, \qquad (2)$$

which is large for the π-π^* states of conjugated molecules (about 1 eV); for the σ-π^* or n-π^* and charge-transfer states the exchange integral is rather small (about 0.1 eV); numbers in brackets (1) and (2) denote coordinates of two electrons in Eq. (2), $r_{1,2}$ is the interelectron distance.

The rate of intersystem crossing in most conjugated molecules and polymers is apparently very low with some exceptions like fullerene and anthracene. Deuteration of organic molecules often suppresses the k_5 rate; the C-H vibrational frequency is much higher than that of the C-D bond vibration and higher overtones should be excited in the deuterated species in order to accept the excess energy $E(T_1)$-$E(S_0)$ and transfer it into vibrational relaxation. From this example one can realize that the nonradiative energy transfer is determined by electron-vibrational (vibronic) interaction to a large extent. This notion can also be applied to the electron-hole injection, migration and recombination processes, and electron transfer in DSSCs (Minaev et al, 2009b).

Since ISC is a spin-forbidden T-S quantum transition, its rate constant also depends on SOC, and on the relative positions of nearby electronic and vibronic states of different symmetry and spin-vibronic interactions (Minaev & Ågren, 2006). Calculations of SOC and radiative rate constants are very important for understanding the function of modern OLEDs. This will be considered in next chapter with explanation of the left part in Fig. 2, where the splitting of spin sublevels of the T_1 state is exaggerated.

Here we need to elucidate some principles of electron-hole migration in more detail. Organic semiconductors have low conductance due to disorder in the amorphous or polycrystalline body; electron and hole mobilities are typically 10^{-8} - 10^{-3} cm^2/V s. In contrast, the perfect molecular single crystal of pentacene has a hole mobility as large as 1.5 cm^2/V s at room temperature (Köhler & Bässler, 2009). All these organic materials have very narrow conduction and valence bands (CVBs), since the molecules are weakly bound by van der Waals interactions. Narrow CVBs imply a mean scattering length of charge carriers to be comparable with intermolecular distances (0.4 nm). Photoexcitation creates predominantly the excited state on an individual molecule (Fig. 2) in such a crystal. Because of translational invariance this excited state may likely reside on any neighboring molecular block in the crystal. It can move through the crystal and is treated as a quasiparticle (exciton). In polymers the exciton wave function can extend over two molecules depending on geometry distortion of the excited state in the chain (charge-transfer exciton). In π-conjugated polymers, like PPV, the electron-hole distance is about 1 nm in the singlet state and about 0.7 nm in the triplet state (Köhler & Bässler, 2009). The difference is determined by exchange interaction of the type presented in Eq. (2). The notations of molecular orbitals (i, u) can refer to HOMO and LUMO inside one molecule, or to different molecules (even to different polymer chains in the case of an inter-chain exciton). The total wave function may be presented in a general form which includes charge-transfer and local molecular excitations:

$$\Psi = c_1\Psi(A^-B^+) + c_2\Psi(A^+B^-) + c_3\Psi(A^*B) + c_4\Psi(AB^*), \quad (3)$$

where A and B refer to different polymer blocks, A* indicates the excited A molecule. If the ionic terms dominate $(c_1, c_2 >> c_3c_4)$ the wave function in Eq. (3) describes a charge-transfer exciton. The opposite case $(c_1, c_2 << c_3c_4)$ corresponds to an exciplex or an excimer. For an excimer the two molecules are the same (A=B) and this is a model for a Frenkel exciton in molecular crystals. In OLEDs the excited states are formed by recombination of two injected charges in EHP ($c_2 \approx 0.99$ in Eq. (3) and $A^+ = A_2$, $B^- = A_3$ in Fig. 1, b). If the positive and negative charges on two molecules are bound by Coulombic interaction one can speak about a geminate polaron pair. For real polymers all coefficients in Eq. (3) are nonzero and their ratio depends on the the A-B distance.

2.2 Solar energy conversion

The mechanism of electric power generation in solar cells is opposite to the mechanism of OLED operation, presented in Fig. 1. The incident light produces an electronic excitation of a dye unit or of a polymer/inorganic crystal followed by charge separation with the subsequent need for the EHP to reach some heterojunction. In solar cells based on crystalline silicon an exciton is created by photoexcitation in one material and the singlet (or triplet) excited state diffuses to the interface with the other material, where dissociation to an electron and a hole takes place. If the energy gained exceeds the exciton binding energy, and if the percolation path for the separated charges affords them to reach the respective electrodes, a voltage occurs. Similar principles of bulk heterojunctions are used in organic semiconductors, when two solutions of polymers with different electronegativity are mixed and spinned on a film (Köhler et al, 1994). The morphology of the film can be optimized by the annealing conditions and the choice of solvent. Solar cells operating both with the singlet and triplet excited states (like in Fig. 1, d) are known. The triplet excitons have longer diffusion length compared to the singlets and this could be used as advantage for such organic solar cells. Despite of the slow Dexter mechanism for the triplet exciton transfer (Forrest, 2004), the large lifetime provides a triplet diffusion length ranging to 140 nm in amorphose organic films, while for singlet excitons it is typically in the range 10-20 nm (Köhler et al, 1994). Polymer-based solar cells operating by triplet excitons also have some advantages, like the triplet emitters in OLEDs, but with completely different physical origin. Inorganic semiconductors, like crystalline silicon, have wider valence and conductive bands than organic solids and also larger dielectric constants ε_r (in silicon ε_r =12, in anthracene ε_r =3). A wide band implies that the mean scattering length of the charge carriers is much larger than the lattice site and exceeds the capture radius R_C of Coulombic attraction for an electron-hole pair (EHP). When an incident light creates an EHP, both charge carriers are delocalized in their wide bands and are not bound by Coulomb attraction. Any scattering event occurs at some distance outside the Coulomb capture radius, R_C, thus the created charge carriers are free in the valence and conduction bands (Köhler & Bässler, 2009). As they are independent of each other, the mutual orientation of their spins is arbitrary; the singlet and triplet states in such EHPs are degenerate, since there is no overlap between the electron and hole wave functions and their exchange energy is zero. This situation for the spin dynamics is similar to that earlier considered in organic chemical reactions of radicals in solvents. At room temperature the excitons in crystallin silicon are similar to separating radical pairs in the solvent cage. At low temperature, the capture radius R_c in Eq. (1) increases and the EHP bound by Coulomb attraction can exist as a Wannier-type exciton. The binding energy of Wannier excitons in silicon is only 1.42 kJ/mol and the electron-hole separation is about 50 Å (Köhler & Bässler, 2009). The exchange integral, Eq. (2), at this distance is of the order 0.1 kJ/mol, so the S-T splitting of Wannier excitons is marginal. In silicon crystals the exciton wave function, Eq. (3), is presented by the first term $c_2 \approx 0.99$ and the Si atoms are bound by σ-bonds inside the A and B moieties. The direct SOC matrix element between such S and T excitons is equal to zero, since their spatial wave functions are identical; but one-center SOC integrals can contribute in the second order through SOC mixing with the intermediate σσ* states. Since the S-T mixing of excitons is an important problem for both OLEDs and solar cells we will consider here spin-dependent exciton recombination, light emission and other photophysical phenomena starting with spin statistics of a geminate radical pair.

2.3 Spin dynamics in organic solvents and its relation to OLED excitons

Interest in spin-statistics problems in organic chemistry was initiated during studies of radical recombination reactions and chemically induced dynamic nuclear polarization (CIDNP) (Salikhov et al, 1984). CIDNP was detected as a non-equilibrium absorption intensity and emission in NMR spectra of radical recombination products in organic chemical reactions in solvents. It was recognized that the radical pair in the triplet spin state cannot recombine and that it dissociates after the first collision in the cage of a solvent. Only the singlet state pair can recombine and produce a product. After the first collision the triplet radical pair (RP) has a large probability for a new reencounter in the solvent cage. Between the two collisions the separated radical pair can provide a triplet-singlet (T-S) transition and then produce a product of recombination, which is enriched by the nuclei with a particular nuclear spin orientation. The T-S transition is induced by hyperfine interactions (HFI) between the magnetic moment of an unpaired electron in the radical and the particular magnetic moment of the nearby nucleus. The HFI provides a "torque" that promotes the electron spin flip in one radical, which means that a T-S transition takes place in the RP. This, the most popular RP mechanism of CIDNP, has also been applied to chemically induced dynamic electron polarization (CIDEP) in EPR spectra of radical products in photochemistry as well as to magnetic field effects (MFE) in chemistry (Salikhov et al, 1984; Hayashi & Sakaguchi, 2005). In non-geminate radical pairs, produced from different precursors, all possible spin states are equally probable. There are three triplet sub-states and one singlet for each RP; by statistics the number of non-reactive triplet collisions is three times larger than the number of reactive singlets. Thus the T-S transitions in the separated RP between reencounter sequences can increase the rate and the yield of the radical recombination reaction in the solvent. The splitting of triplet sublevels and the rate of the T-S transitions depends on the external magnetic field and this is the reason for MFE in radical reactions. The radical-triplet pair mechanism was later developed for explaining the MFE in radical-triplet interactions. It takes into account MFE for the quartet and doublet states mixing in such interactions. This mechanism has to be used for the treatment of the polaron-triplet annihilation, which is now considered as a reason for triplet state quenching by charge carriers in OLEDs (Köhler & Bässler, 2009).

Similar ideas have to be applied for electron-hole recombination in OLEDs in order to compel the triplet excitons to do useful work in electroluminescent devices. The Wannier excitons are quite similar to the separated radical pairs in the solvent cage if comparison with the CIDNP theory is relevant. Unfortunately CIDEP and MFE theories were not utilized in OLED technology during long time until the first application of the triplet emitters in doped polymers (Baldo et al, 1999), and magnetic field effects are still not used in electroluminescent applications for electron-hole recombination, though it could have some technological applications in organic polymers. The T_1 sublevels are usually depopulated with different rates (k_i, $i = x,y,z$, Fig. 2). In 1979, Steiner reported MFE due to the depopulation type triplet mechanism (d-type TM) on the radical yield of electron transfer reactions between a triplet-excited cationic dye ($^3A^{+*}$) and Br-substituted anilines (D) in methanol at 300 K (Hayashi & Sakaguchi, 2005). Steiner proposed that a triplet exciplex $^3(A^*D^+)$ is generated by charge transfer in this reaction and that the sublevel-selective depopulation is induced by strong SOC at a heavy Br atom during decomposition $^3(A^*D^+) = A^\bullet + D^{\bullet+}$. Similar reactions with triplet exciplexes were found to produce CIDEP and MFE due to d-type TM. The corresponding theory of magnetic field effects due to spin-orbit coupling in transient intermediates and d-type TM has been proposed (Hayashi & Sakaguchi, 2005; Serebrennikov & Minaev, 1987). Its application for charge-transfer excitons in phosphorescent OLEDs is ongoing. First we need to consider the main elementary processes, which occur within the close vicinity of the emitting center in the polymer layer, but general principles of the charge carrier migration and their spin statistics are also discussed.

2.4 Spin statistics of excitons in OLEDs and spin-dependent optoelectronics

As shown in Fig. 1, organic-conjugated polymers are used in OLEDs as they lend the possibility to create charge carrier recombination and formation of excitons with high efficiency of light emission. The typical OLED device consists of a layer of a luminescent organic polymer sandwiched between two metal electrodes. Electrons and holes are first injected from the electrodes into the polymer layer. These charge carriers migrate through the organic layer and form excitons when non-geminate pairs of oppositely charged polarons capture each other. The colliding charge pairs origin from different sources, so they have random spin orientation. Thus the singlet and triplet colliding pairs are equally probable. According to statistical arguments the excitons are created in an approximate 1:3 ratio of singlet to triplet. Fluorescence occurs from the singlet states, whereas the triplets are non-emissive in typical organic polymers, which do not contain heavy metal ions. The triplet-singlet (T_1 - S_0) transitions in organic polymers are six to eight orders of magnitude weaker than the spin-allowed singlet-singlet (S_1 - S_0) fluorescence. The phosphorescence gains the dipole activity through spin-orbit coupling (SOC) perturbation. SOC is very weak in organic polymers because the orbital angular momentum between the π-π^* states of conjugated chromophores is almost quenched. The other reason is that the SOC integrals inside the valence shell of the light atoms are relatively small (for carbon, nitrogen and oxygen atoms they are 30, 73 and 158 cm^{-1}, respectively). These integrals determine the fine-structure splitting of the ^3P_J term into sublevels with different total angular momentum (J). In light atoms such splitting obeys the Lande interval law and can be described in the framework of the Russell-Saunders scheme for the angular momentum summation.

Thus the emission from triplet states of organic chromophores has very low rate constant and cannot compete with non-radiative quenching at room temperature. Consequently, it has been assumed that the quantum yield has an upper statistical limit of 25 per cent in OLEDs based on pure organic polymers. In order to compel the triplet excitons to emit light and to do useful work in OLEDs one needs to incorporate special organometallic dyes containing heavy transition-metals into the organic polymers, which will participate in the charge carrier recombination and provide strong SOC in order to overcome spin prohibition of the T_1 - S_0 transition. Incorporation of $Ir(ppy)_3$ into a polymer leads to an attractive OLED material by two reasons: the high rate of electron-hole recombination on the $Ir(ppy)_3$ dye and relatively strong SOC at the transition-metal center induces a highly competitive T_1 - S_0 transition probability and quantum efficiency of the OLED. The cyclometalated photocatalytic complexes of the Ir(III) ion fit these conditions quite well. Involvement of such a heavy atom into metal-to-ligand charge transfer (MLCT) states of different symmetries increases configuration interaction between them and the π-π^* states of the ligands, which finally leads to a strong singlet-triplet SOC mixing in the cyclometalated Ir complexes.

While the ppy ligands are structurally similar to bipyridines, it has been earlier recognized that the metal-carbon bonds which they form with transition-metal ions provide a specific influence on their complex properties that are quite distinct from those of the N-coordinated bpy analogues. Replacing bpy in $Ir(bpy)_3^{3+}$ by 2-phenylpyridine produces a very strong photoreductant, $Ir(ppy)_3$. The enhanced photo-reducing potential of such complexes is attributed to the increase in electron density around the metal due to the stronger donor character of the coordinating carbon atoms. Species containing both bpy and ppy ligands, such as $[Ir(ppy)_2bpy]^+$, have intermediate photoredox properties and can operate as either photo-oxidants or photoreductants. Use of cationic complexes in OLEDs provides some advantages since they do not require complicated fabrication of multilayer structure for charge injection and recombination, which is promising for large-area lighting applications (De Angelis et al. 2007). The presence of mobile cations and negative counter-anions (PF_6^-) makes the ionic complexes more efficient than the neutral cyclometalated iridium complexes (CIC). The ions create high electric fields at the electrode interfaces, which enhances the electron and hole injection into the polymer and also the exciton formation at the dopant metal complexes. Electrons and holes are injected at a voltage just exceeding the potential to overcome the HOMO-LUMO energy gap in the active material of the OLED, irrespective of the energy levels of the electrodes.

The SOC effects on the T_1 - S_0 transition in the $[Ir(ppy)_2(bpy)]^+$ (PF_6^-) and other ionic and neutral iridium complexes have been theoretically studied in order to interpret the high efficiency of the corresponding OLED materials (Minaev et al. 2006; Jansson et al, 2007; Minaev et al. 2009; Baranoff et al. 2010). This affords to foresee new structure-property relations that can guide an improved design of organic light-emitting diodes based on phosphorescence. Modern density functional theory (DFT) permits to calculate the optical phosphorescence properties of such complexes because of their fundamental significance for OLED applications. First principle theoretical analysis of phosphorescence of organometallic compounds has recently become a realistic task with the use of the quadratic response (QR) technique in the framework of the time-dependent density functional theory (TD DFT) approach. These DFT calculations with quadratic response explain a large increase in radiative phosphorescence lifetime when going from the neutral Ir(ppy)3 to cationic $[Ir(bpy)_3]^{3+}$ compounds and other trends in the spectra of tris-iridium(III) complexes. Calculations show the reason that some mixed cationic dyes consecutively improve their T_1 - S_0 transition probabilities and unravel the balance of factors governing the quantum emission efficiency in the corresponding organic light-emitting devices.

In order to present connections between main features of electronic structures and photo-physical properties including phosphorescence efficiency and energy transfer mechanisms we have to consider spin properties and the SOC effect in atoms and molecules in detail. Since the SOC description in atoms and the multiplet splitting in the framework of the Russell-Saunders scheme is a crucial subject for the new OLED generation of triplet-type emitters, we will pay proper attention to atomic and molecular SOC with special attention to the Ir atom and CIC spectra.

2.5 Spin-orbit coupling

The electron spin wave function Ψ satisfies the equation for the spin square operator of: $\vec{s}^2\Psi = s(s+1)\hbar^2\Psi$, where $s = 1/2$ is a spin quantum number, $\hbar = (h/2)\pi$ is the Planck constant. Two types of spin wave functions Ψ which satisfy this requirement (α, β) and all components of the spin operator are:

$$\alpha = \begin{pmatrix} 1 \\ 0 \end{pmatrix}, \beta = \begin{pmatrix} 0 \\ 1 \end{pmatrix}; s_x = \frac{\hbar}{2}\begin{pmatrix} 0 & 1 \\ 1 & 0 \end{pmatrix}, s_y = \frac{i\hbar}{2}\begin{pmatrix} 0 & 1 \\ -1 & 0 \end{pmatrix}, s_z = \frac{\hbar}{2}\begin{pmatrix} 1 & 0 \\ 0 & -1 \end{pmatrix} \quad (4)$$

Spin was first postulated in order to explain the fine structure of atomic spectra and formulated by Pauli in matrix form, Eq. (4); then it was derived by Dirac in the relativistic quantum theory. In many-electron systems - atoms, molecules, polymers - the electron spins are added by quantum rules into the total spin $\vec{S} = \sum_i \vec{s}_i$, which plays an important role as a fundamental conservation law

$$\vec{S}^2\Psi = S(S+1)\hbar^2\Psi \quad (5)$$

For the even number of electrons the total spin quantum number can be equal $S = 0$ (singlet state), $S = 1$ (triplet state), $S = 2$ (quintet state), which are the most important states in organic chemistry and quantum theory of OLEDs. For odd number of electrons (holes, radicals) the total spin quantum number is usually equal $S = 1/2$ as for one electron, but excited states could have high spin quartet ($S = 3/2$) and sextet ($S = 5/2$) spin. Multiplicity in general is equal to $2S+1$, which determines a number of spin sublevels in an external magnetic field.

Before calculation of efficiency of triplet emitters in OLEDs one has to analyze quantization of the orbital angular momentum $\vec{L}$ in atoms, which is determined by quantum number L; it needs to be added to spin in order to determine the total angular momentum of atom $\vec{J}$:

$$\vec{L}^2\Psi = L(L+1)\hbar^2\Psi \qquad \vec{J}^2\Psi = J(J+1)\hbar^2\Psi, \text{ where } \vec{J} = L + \vec{S} \quad (6)$$

In relativistic theory all atomic states with L ≠ 0 acquire additional correction to the total energy which is equal to the expectation value of the SOC operator; thus a splitting of atomic terms with different J occurs. Calculation of fine structure is easy to illustrate for a one-electron atom. The SOC operator for the hydrogen-like atom with nuclear charge Z is obtained by Dirac:

$$H_{so} = \frac{e^2\hbar^2}{2m^2c^2}\frac{Z}{r^3}\vec{l}\vec{s} \quad (7)$$

The operators $\vec{l}\vec{s}$ here are given in $\hbar$ units. The scalar product of two operators $\vec{l}\vec{s}$ can easily be calculated by the definition $\vec{J}^2 = (\vec{L}+\vec{S})^2 = \vec{L}^2 + 2\vec{L}\vec{S} + \vec{S}^2$ with account of Eqs. (5) - (6), which applies also to the single electron case:

$$\vec{L}\vec{S} = \frac{1}{2}[J(J+1) - L(L+1) - S(S+1)] \tag{8}$$

A simple generalization of the SOC operator for a many-electron atom can be summarized in the forms:

$$H_{so} = \zeta\sum_i \vec{l}_i\vec{s}_i = \lambda\vec{L}\vec{S}, \text{ where } \lambda = \pm\frac{\zeta}{2S}, \ \zeta_{np} = \frac{e^2\hbar^2}{2m^2c^2}\left\langle\frac{Z}{r^3}\right\rangle_{np} \tag{9}$$

In Eq. (9) Z is a semi-empirical parameter; the "plus" sign corresponds to the open shell, which is "less-than-half" occupied, "minus" – to the "more-than-half" occupied open shell. Using this semi-empirical constant one can calculate SOC in organic molecules. The Ir(III) ion has a $(5d)^6$ configuration: thus its ground state is a quintet ^5D which is split in five sublevels. According to the third Hund's rule the lowest one is $^5D_4$ since the open shell $(5d)^6$ is "more-than-half" occupied and the "minus" sign is used in Eq. (9); thus λ is negative in this case. The maximum J=4 provides SOC energy 4λ, next levels with J=3 has zero correction, and J=2,1 and 0 have positive SOC corrections -3λ, -5λ and -6λ, respectively. The Ir(III) ion is a rather difficult example of SOC treatment in atoms (Koseki at el. 2001). In the neutral Ir atom the ground state 1^4F $(5d)^7(6s)^2$ splitting is more complicated because of non-diagonal SOC mixing with the excited configuration 2^4F $(5d)^8(6s)^1$. In our SOC calculations of iridium complexes we use effective core potential (ECP) and basis set for the Ir atom, augmented with a set of f polarization functions, proposed in Refs. (Cundari & Stevens, 1993; Koseki at el. 2001). The valence orbitals of this ECP are already adjusted for relativistic contractions and expansions, but 5d AOs are nodeless (even though they should have two inner nodes). Instead of the full Breit-Pauli operator (Ågren et al. 1996) we use for the CIC and Pt compounds an effective one-electron SOC operator with effective nuclear charge for each atom A (Koseki at el. 1998)

$$H_{so} = \frac{e^2\hbar^2}{2m^2c^2}\sum_i\sum_A \frac{Z_{eff}(A)}{r_{iA}^3}\vec{l}_{iA}\vec{s}_i \tag{10}$$

This operator was widely used for SOC calculations in molecules and charge-transfer complexes with semi-empirical self-consistent field (SCF) configuration interaction (CI) methods (Minaev & Terpugova, 1969; Minaev, 1972; Minaev, 1978) and also in ab initio approaches (Koseki at el. 1998). For the ECP basis set in heavy elements the effective nuclear charge in Eq. (10) loses its physical meaning and becomes a rather large fitted parameter, since the 5d AO is nodeless. Koseki at el. have obtained $Z_{eff}(Ir)$ =1150.38, $Z_{eff}(Pt)$ =1176.24. For the first row transition metals and for the lighter elements these parameters have the usual meaning and are close to the values found earlier (Minaev & Terpugova, 1969), since the 3d and 2p functions lack nodes. Multiconfiguration (MC) SCF method with account of second order CI and SOC (Koseki at el. 1998) provides moderate agreement with the observed spectra of Ir and Pt atoms. For the ground state of the Pt atom ^3D $(5d)^9(6s)^1$ the MC SCF + SOCI calculations predict negative excitation energy to the excited ^1S $(5d)^{10}$ configuration which leads to disagreement with experiment when SOC is included in the CI

matrix (Minaev & Ågren, 1999). A multi-reference (MR) CI + SOC calculation improves the results (Table 1). The SOC-induced splitting of the ^3D_J sub-levels deviates rather much from the Lande interval rule but is semiquantitatively reproduced by MRCI+SOC calculations (Table 1) with the parameter Z_{eff}(Pt) =1312 (Minaev & Ågren, 1999). One needs to stress that the experimental S-T energy gap between the $^3D_3$ and $^1S_0$ states (6140 cm^{-1} =0.76 eV) is very far from non-relativistic CI results (0.03 eV) and is determined mostly by SOC. That is why many attempts to reproduce this S-T gap in non-relativistic CI methods have failed (Minaev & Ågren, 1999). This is in a large contrast to the Pd atom with the ^1S $(5d)^{10}$ ground state, where the S-T energy gap is well reproduced in simple CI calculations.
Account of $^3F_4$ $(5d)^8(6s)^2$ state does not influence the old results (Minaev & Ågren, 1999) because the ^3F state energy is rather large in MRCI calculations. But the $^1D_2$ singlet state strongly interacts with the $^3D_2$ and $^3F_2$ triplets, which leads to a low-lying level with J=2. A study of the Pt complexes used in OLEDs indicates that ligand fields strongly influence the S-T energy gap and SOC splitting of the multiplets. The orbital angular momentum of the Pt atom is almost quenched by ligands such as porphine and acetylides (Minaev at el. 2006/a,b) and the zero-field splitting (ZFS) is strongly reduced. ZFS can be reliably estimated by second order perturbation theory, and depends on the square of the SOC matrix elements. The S-T mixing is determined by first order perturbation theory and it is still large in Pt complexes used in OLED; thus the SOC-induced by the Pt atom strongly influences the $T_1 \rightarrow S_0$ emission (phosphorescence) rate in platinum acetylides (Minaev at el. 2006.a) and platinum porphyrines (Minaev at el. 2006.b).

State	MRCI	MRCI+SOC	Expim.	Degener.
(configurat.)	a.u.	cm^{-1}	cm^{-1}	
$^3D_3$ $(5d)^9(6s)^1$	-0.823370	0.00	0.00	7
$^3D_2$ $(5d)^9(6s)^1$	-0.823370	2066.54	775.9	5
^1S $(5d)^{10}$	-0.822295	4646.12	6140.0	1
$^3D_1$ $(5d)^9(6s)^1$	-0.823370	11025.067	10132.0	3
$^1D_2$ $(5d)^9(6s)^1$	-0.807364	12471.36	13496.3	5
$^3F_4$ $(5d)^8(6s)^2$	-0.791214	945.32	823.7	9

Table 1. Splitting of the low-lying states in the Pt atom; from Ref. (Minaev & Ågren, 1999) with some additions; -118.0 a.u. should be added to MRCI column.

The treatment of SOC in the iridium atom is also complicated (Koseki at el. 2001). Account of all electrons with the Breit-Pauli SOC operator definitely improves the SOC splitting of the two low-lying ^4F states (Koseki at el. 2001), but the ECP basis set with an effective single-electron operator, Eq. (10), and the Z_{eff}(Ir) value also give reliable results (Koseki at el., 1998). Our calculations with this approximation of SOC and phosphorescence lifetime in cyclometalated iridium complexes, used in OLED emissive layer, provide good agreement with experimental measurements for radiative characteristics. This is important for a comprehensive understanding of the electronic mechanisms in order to formulate chemical requirements for OLED materials.

2.6 Triplet-singlet transitions and zero-field splitting of the triplet state

Spin-orbit coupling can mix the triplet (T) and singlet (S) states in atoms, molecules and solids. Before studying SOC mixing between excitons one has to analyze the electric dipole operator ($\vec{m}$ =e$\sum_i \vec{r}_i$) and its transition moment $T_1 \rightarrow S_0$ for a typical molecule or cyclometalated complex with a ground S_0 state (Fig. 2). Let us consider first order perturbation theory for the T_1 and S_0 states:

$$\tilde{T}_1^{\alpha} = T_1^{\alpha} + \sum_n \frac{\langle S_n | \hat{H}_{SO} | T_1^{\alpha} \rangle}{E(T_1) - E(S_n)} S_n \; ; \; \tilde{S}_0 = S_0 + \sum_k \frac{\langle T_k^{\alpha} | \hat{H}_{SO} | S_0 \rangle}{E(S_0) - E(T_k)} T_k^{\alpha} \tag{11}$$

The perturbed wave function of the first excited triplet state is denoted here as $\tilde{T}_1$; it is mixed with all singlet states S_n wave functions, including the ground state, n=0. In a similar way the ground state perturbed wave function $\tilde{S}_0$ has admixtures of all triplet states, including k=1. The triplet state wave function T_k^{α} can be represented as a product of the spatial part ${}^3\Psi_k = \sum {}^3A_{i-u,k} {}^3\Phi_{i-u}$ and the spin part t^{α}. In the TD DFT method the ${}^3\Phi_{i-u}$ configurations are presented as two-component matrices, which include single excitations above the closed shell of the type: ${}^3\Theta_{i-u} = \frac{1}{\sqrt{2}}[\varphi_i(1)\varphi_u(2) + \varphi_i(2)\varphi_u(1)]$. Spin functions of the ZFS sub-levels have a general form (Vahtras et al. 2002):

$$t^x = \frac{1}{\sqrt{2}}[\beta(1)\beta(2) - \alpha(1)\alpha(2)] \,; t^y = \frac{i}{\sqrt{2}}[\beta(1)\beta(2) + \alpha(1)\alpha(2)] \,; t^z = \frac{1}{\sqrt{2}}[\alpha(1)\beta(2) + \alpha(2)\beta(1)] \tag{12}$$

In organic π-conjugated molecules the i - u orbitals, HOMO - LUMO, are of π-type. Zero-field splitting in the T_1 state of such molecules and in organic π-conjugated polymers is determined by weak spin-spin coupling, which usually does not exceed 0.1 cm^{-1}. The SOC contribution to ZFS in these cases is negligible; it occurs in the second order of perturbation theory:

$$|H_{\alpha\beta} - E\delta_{\alpha\beta}| = 0, \quad \alpha, \beta = x, y, z \tag{13}$$

where

$$H_{\alpha\beta} = H_{\alpha\beta}^{(1)} + H_{\alpha\beta}^{(2)} = \langle T_1^{\alpha} | H_{ss} | T_1^{\beta} \rangle + \sum_k \sum_{\alpha\beta} \langle T_1^{\alpha} | H_{so} | {}^{\lambda}\Psi_k^{\beta} \rangle \langle {}^{\lambda}\Psi_k^{\alpha} | H_{so} | T_1^{\beta} \rangle / ({}^3E_1 - {}^{\lambda}E_k) \tag{14}$$

Here $\lambda = 2S + 1$ means multiplicity of the perturbing state. Summation in Eq. (14) includes S=0, 1, 2, that is SOC mixing of the lowest triplet T_1 with all singlet, triplet and quintet states in the spectrum. If the SOC mixing with the triplet state ${}^3\Psi_k$ produces down-shift of the T_1^x and T_1^y spin-sublevels, then the corresponding singlet state ${}^3\Psi_k$ produces a similar shift of the ${}^zT_1$ sub-level. If the T_1 state is of π-π^* nature, the perturbing states are of σ-π^* (or π-σ*) nature. In this case the S-T splitting ${}^3E_{\pi\pi*} - {}^1E_{\sigma\pi*}$ and T-T splitting ${}^3E_{\pi\pi*} - {}^3E_{\sigma\pi*}$ are almost the same. The corresponding SOC integrals between T-T and S-T states are also very similar. Thus the SOC contribution to ZFS from the analogous singlet and triplet counterparts is negligible. It is less than 10^{-5} cm^{-1} in the benzene and naphthalene molecules, thus the ZFS is completely determined by weak spin-spin coupling. One can see that the SOC contribution to ZFS strongly depends on the S-T splitting of the perturbing states. If the lowest triplet is of n-π nature, like in pyrazine or benzoquinone, the perturbing S and T states are of π-π^* type. The exchange integral, Eq. (2), for π-π^* orbitals is usually rather large, thus one can expect an appreciable SOC contribution, Eq. (14), to ZFS of the T_1(n-π^*) state. Similar analysis has been presented for the $Ir(ppy)_3$ complex (Jansson et al. 2006; Yersin &

Finkenzeller, 2008), which shows that the SOC splitting of the ^{3}MLCT state can be relatively large.
Let us use the perturbed states, Eq. (11), in order to calculate the triplet-singlet transition:

$$<\tilde{T}_1^{\alpha} | m_{\gamma} | \tilde{S}_0 >= \sum_n \frac{\langle S_n | \hat{H}_{SO} | T_1^{\alpha} \rangle^*}{E(T_1) - E(S_n)} < S_n | m_{\gamma} | S_0 > + \sum_k \frac{\langle T_k^{\alpha} | \hat{H}_{SO} | S_0 \rangle}{E(S_0) - E(T_k)} < T_1^{\alpha} | m_{\gamma} | T_k^{\alpha} >$$

Since SOC integrals are imaginary and hermitian, $\langle S_n | \hat{H}_{SO} | T_1^{\alpha} \rangle^* = \langle T_1^{\alpha} | \hat{H}_{SO} | S_n \rangle$, the last equation can be presented in the form

$$<\tilde{T}_1^{\alpha} | m_{\gamma} | \tilde{S}_0 >= \sum_{n(\neq 0)} G_{1,n}^{\alpha} < S_n | m_{\gamma} | S_0 > - \sum_{k(\neq 1)} G_{k,0}^{\alpha} < T_1^{\alpha} | m_{\gamma} | T_k^{\alpha} > + G_{1,0}^{\alpha} ({}^1m_{0,0} - {}^3m_{1,1}), \quad (15)$$

$$G_{k,0}^{\alpha} = \frac{\langle T_k^{\alpha} | \hat{H}_{SO} | S_0 \rangle}{E(T_k) - E(S_0)}$$

and $({}^1m_{0,0} - {}^3m_{1,1})$ is the difference of the permanent dipole moments of the ground singlet state and the lowest triplet state; its contribution to the phosphorescence k_4 rate constant requires special attention and will be analyzed later.

3. Iridium(III) complexes in OLED materials

Iridium as heavy metal center can provide large SOC and therefore allows the spin-forbidden S_0-T_1 transition which facilitates the utilization of triplet emission energy in OLED materials. The first prototype of iridium-containing dyes used in OLED was tris(2-phenylpyridine)iridium, i.e. the $Ir(ppy)_3$ complex, which was found to improve OLED devices. Nowadays iridium complexes constitute an important class of dopants for organic polymers used in OLEDs in order to increase the efficiency of electroluminescence. Iridium complexes have advantages such as strong phosphorescence in the visible region and tunable emission wavelengths through peripheral functionalization of the ligands.
Heteroleptic iridium complexes have advantage that functions of different groups can be integrated into one molecule. Such complexes usually consist of two cyclometalating ligands (C^N) and one ancillary ligand. By changing the functional groups in the ancillary ligand or introducing a novel ancillary ligand, the photophysical properties of the complex can be tuned. For example, fluorine substitutions are often introduced into the ligand in order to lower the HOMO energy level and to obtain a blue-shifted emission wavelength. Interestingly, some iridium complexes containing switching units can respond to external electric or photo stimuli, leading to controllable and modulatable phosphorescence emission.

3.1 Spin-orbit coupling in cyclometalated iridium complexes

Modification of a CIC by modulating ligands for enhancement of their phosphorescence and tuning of its wavelength from blue to green and red colors is an important task for both theoretical and applied research. A theoretical background for the chemical and photophysical properties of transition metal complexes with polypyridyl ligands was developed a long time ago in the framework of crystal field theory and ligand field theory using quasi-octahedral symmetry (Nazeeruddin at el. 2009). High symmetry of the coordination sphere and relatively weak perturbation of d-AOs of the metal center by a ligand field implies that the orbital angular momentum of the metal ion is not completely quenched in the complex. Though an expectation value of L is zero in polyatomic systems, and Eq. (8) provides zero SOC correction to the nonrelativistic energy, non-diagonal terms of the SOC operators in Eq. (9) and (10) can generate large coefficients $G_{k,n}$ in Eq. (15) and even corrections to the expectation value of L (Minaev, 1978). The Ir atom is in the group VIII B, and lies below Co and Rh. The splitting of d-orbitals is rather specific in this series. The Ir(III) ion is characterized by relatively strong ligand field splitting between the occupied t_{2g} MO group and the unoccupied e_g pair of the 3d orbitals compared to other ions, thus it is easier to tune CIC emission by ligand modulation. Because of the larger nuclear charge of Ir, the SOC splitting and multiplet mixing is much stronger in CIC than in cobalt and rhodium complexes, thus enhanced S-T transitions and ISC is expected in CIC compounds. That is why the efficient quantum yield of the T_1 states and intense phosphorescence distinguish the photophysics of heavy metal complexes from those of organic and light metal compounds.

The photophysics of polypyridyl iridium complexes can be understood accounting for three types of excited state configurations: metal-centered (MC) excited dd* states of the t_{2g} - e_g type, ligand-centered (LC) excited π-π^* states, and metal-to-ligand charge transfer (MLCT) states. The TD DFT calculations indicate that the lowest triplet T_1 state is a mixture of the MLCT and LC excited state configurations (Minaev et al. 2006, Minaev et al. 2009, Nozaki 2007). In $Ir(ppy)_3$ the HOMO is a mixture of 5d-AO (t_{2g}) and the phenyl ring π-orbitals, in contrast the LUMO is a pure π^*-orbital of the pyridine moiety. In this case the $G_{1,0}$ value (Eq. 17) is negligible because the SOC integral includes a HOMO-LUMO angular momentum matrix element which does not contain one-center integrals at the metal. With this as background one can explain the low rate constant (k_5, Fig. 2) for the $T_1 \sim\rightarrow S_0$ non-radiative quenching of the phosphorescent emission. This is in a general agreement with a high phosphorescence quantum yield (φ_p) of CIC compounds. Some variations in φ_p are explained by SOC calculations of the $G_{1,0}$ coefficient, Eq. (15) (Li et al. 2011).

Analysis of Eq. (15) in the framework of TD DFT quadratic response calculations reveals general reasons for the high radiative rate constant (k_4, Fig. 2) for the $T_1 \rightarrow S_0$ phosphorescent emission. Intensity borrowing from the $T_1 \rightarrow T_k$ electric dipole transitions (last sum in Eq. (15)) provides the largest contribution to the phosphorescence intensity. The metal-centered (MC) excited triplet ^{3}dd* states of the $t_{2g} \rightarrow e_g$ type represent the higher triplets, T_k, which have strong SO coupling with the ground singlet state, S_0, and simultaneously – a large $T_1 \rightarrow T_k$ electric dipole transition moments (last sum in Eq. (15)). The reason is obvious; the $\langle T_k | H_{so} | S_o \rangle$ matrix elements include one-center SOC integrals at the metal, which are determined by a relatively large $\zeta_{5d}(\mathrm{Ir})$ value. The $T_1 \rightarrow T_k$ electric dipole transition moments do not depend on SOC and include transitions between LUMO (pyridine π^* MO) and e_g ($5d_{x2-y2}$ and $5d_{z2}$) orbitals, which are allowed, though they are not intense. Besides, there are LUMO+1 contributions which provide more efficient overlap with 5d-AOs and higher dipole moments. Substitution of ligands can influence HOMO and LUMO energies and their mixing with metal 5d-AOs, thus modulating the phosphorescence lifetime and tuning of its wavelength. A series of TD DFT calculations with SOC treatment by quadratic response provide a very good explanation of emission tuning in various CIC compounds and illustrate the physical reasons for OLED architecture and design (Li et al., 2011, Minaev et al., 2009, Janson et al., 2007, Nozaki 2007).

3.2 Cationic Ir(III) complexes

It is known that ionic cyclometalated complexes of the type $[Ru(bpy)_3]^{2+}(PF_6^-)_2$ do not need complicated fabrication of multilayer devices for charge injection and recombination (Nazeeruddin et al, 2009). These systems are used now in electrochemical devices, which are promising for large-area lighting. Only a single-layer of such ionic complexes operates at a low voltage and these devices are shown to be rather insensitive to the choice of electrode material, allowing the use of air-stable anodes and cathodes. The presence of mobile ions, which carry two net positive and negative charges makes such ionic materials quite different from the neutral organic semiconductors typically used in OLEDs. Upon application of a voltage the anions and cations move toward the anode and cathode, respectively, creating high electric fields at the electrode interfaces, which enhances charge injection into the polymer layer and exciton formation at the metal complexes (Nazeeruddin et al, 2009).

Unfortunately, the ionic systems provide a low quantum yield compared to the neutral complexes; the reason was established by the recent SOC calculation of ionic CIC (Minaev et al. 2009). Until recently, the majority of ionic chromophores used in the single-layer devices have been Ru-based complexes (Nazeeruddin et al, 2009). They emit light in the orange-red region, while for OLED displays white light is usually needed, which can be obtained by mixing blue with red and green colors. Such systems were synthesized in a form of mixed ligand cationic iridium complexes: the green-blue emitting [Ir(2-phenylpyridine)2(4,4′-dimethyl amino-2,2′-bipyridine]($PF6^-$) complex, labeled as N926, and the [Ir(2,4-difluorophenylpyridine)2 (4,4′-dimethyl amino-2,2′-bipyridine]($PF6^-$) complex, labeled as N969. Both show bright emission with high phosphorescence quantum yield (80-85%) at room temperature in an argon-degassed solution of CH_2Cl_2(Nazeeruddin et al, 2009). TD DFT calculations of these systems together with the pure ionic $[Ir(bpy)_3]^{3+}$ complex (Scheme 1) reveal the nature of the $T_1 \rightarrow S_0$ transition efficiency of the corresponding CICs (Minaev et al. 2009).

The spin density distribution and hyperfine constants in the optimized T_1 excited state of the $[Ir(bpy)_3]^{3+}$ complex indicates the biradical "quinoid" structure in one ligand. In this particular bpy ligand the ring bonds, being parallel to the C-C bridge, are getting shorter and the other bonds are elongated upon $S_0 \rightarrow T_1$ excitation. Thus the lowest T_1 state in the pure ionic $[Ir(bpy)_3]^{3+}$ complex is a local $\pi \rightarrow \pi^*$ excitation in one bipyridine moiety. Because of this the $T_1 \rightarrow S_0$ transition is not intense and the calculated phosphorescence lifetime, τ_p, is relatively large (0.1 ms), in agreement with experiment (0.054 ms). In mixed cationic Ir(III) systems the lifetime is much lower and close to the neutral *fac*-$Ir(ppy)_3$ complex: for the latter dye our theory and measurements provide the same value τ_p = 2 μs (Jansson et al. 2007). Our TD DFT calculations of τ_p include SOC between thousands of S,T states and are rather complicated. Thus a good agreement for both τ_p values seems to be a miracle. But it is not, since for the mixed $[Ir(ppy)_2(bpy)]^+$ complex the calculation (Minaev et al. 2009) provides τ_p = 4.83 μs in a perfect agreement with τ_p measurements in solid glass (4.4-5.2 μs). For N926 complex the calculated and experimental τ_p values are equal to 2.94 and 3.04 μs, respectively. The DFT method also provides an explanation for the high phosphorescence quantum yield; a direct SOC between S_0 and T_1 states is negligible in these systems, which explains the low rate constant (k_5, Fig. 2) for the T_1 state quenching. This SOC integral enters the last term in Eq. (15). It is not important for the radiative $T_1 \rightarrow S_0$ transition dipole moment, Eq. (15), since there are other big contributions at n=4-6. Tuning of the colors in cationic CICs is explained by the energy shifts of the π*(bpy) LUMO in N926 and by strong HOMO stabilization (Ir-ppy) in the N969 complex upon fluorine introduction (Nazeeruddin et al, 2009).

De Angelis et al. (2007) have reported a combined experimental and theoretical study on cationic Ir(III) complexes for OLED applications. The authors also described a strategy to tune the phosphorescent emission wavelength and to improve the quantum yields by modulating the electronic structures the iridium complexes through selective ligand functionalization. The newly synthesized cationic Ir(III) complex, [Ir(2,4-difluorophenylpyridine)$_2$(4,4'-dimethylamino-2,2'-bipyridine)](PF6) or N969 is observed to exhibit blue-green emission at 463 nm with a high quantum yield of 85% in acetonitrile solution at ambient temperature. DFT and TD DFT calculations with solvent effects taken into account were carried out to characterize the electronic structures of the ground state and the excited states. This work shows the possibility of tuning the electronic structures and the excited-state properties as useful for the design of new iridium(III) emitters with specific characteristics.

Ladouceur et al. (2010) have synthesized a family of other cationic iridium(III) complexes containing 4'-functionalized 5,5'-diaryl-2,2'-bipyridines ligands as triplet emitters for OLEDs. Most of the complexes show intense and long-lived phosphorescent emission in both 2-MeTHF and acetonitrile at 77 K and at ambient temperature. Quantum chemical calculations suggest that the emission arises from an admixture of the ^{3}LLCT (π(ppy) --> π*(bpy*)) and the ^{3}MLCT (dπ(Ir) --> π*(bpy*)) states. TD DFT calculations also provide insight into the origin of the electronic transitions. Moreover, the introduction of the peripheral aryl groups in the bpy* ligand is expected to enhance the shielding of the iridium center and therefore to increase the stability of the device.

Cationic bis-cyclometalated iridium(III) phenanthroline complexes with pendant fluorenyl substituents have been described by Zeng et al. (2009). These complexes consist of two 2-phenylpyridine ligands and one substituted phenanthroline ligand, which provides extended π-conjugation. Single-crystal X-ray diffraction measurements reveal that the iridium center adopts an octahedral coordination structure. Two of the complexes display reversible cyclic voltammetric waves which are assigned to the Ir(III)/Ir(IV) couple. Broad bands are observed in the photoluminescence spectra of all the complexes, corresponding to the mixed ^{3}MLCT and $^3\pi$-π* states. The lifetimes in the microsecond time-scale indicate the phosphorescent character of the luminescence, and it is found that larger conjugation length in the ligand leads to longer lifetime. DFT calculations show that the HOMOs are localized on the iridium center and the benzene rings of the phenylpyridine ligand, while the LUMOs are mainly located on the phenanthroline ligand. The light-emitting cells fabricated through the spin-coating approach exhibit maximal brightness efficiency of 9 cd A^{-1} and show very good stability in air.

3.3 Fluorine substitution in the ligands

Fluorine substitution is usually introduced into CIC to obtain intense blue emission. In the TD DFT study carried out by Li et al. (2011), linear and quadratic response approaches are employed to investigate the absorption and luminescence spectra of several facial and meridional iridium complexes with fluorine-substituted phenylpyridine ($F_n$ppy) ligands, as shown in Fig. 3. Similar to other Ir(III) complexes, the HOMOs are mainly localized on the metal center and the phenyl ring of the ppy ligands while the LUMOs are delocalized mostly on the pyridine part of the ppy ligands. The computations also suggest that the presence of the fluorine atoms in the ppy ligand will enlarge the HOMO-LUMO energy gap and result in blue-shifted emission. Moreover, the SOC strength and the radiative rate constant are diminished by the introduction of fluorine substitutions. Linear response calculations reveal that the S_0-T_1 SOC matrix element is smaller in the *fac*-isomer than in *mer*-complexes, which means that the nonradiative quenching of the T_1 state is faster in the latter complexes. Therefore in the meridional isomer the SOC matrix element together with the difference between the permanent dipole moments of the T_1 and S_0 states, Eq.(15), provide destructive contribution to the total $S_0 \rightarrow T_1$ transition moment. This study has shown the effects of the fluorine substitutions and the facial to meridional isomerization to the photophysical properties of the iridium complexes.

Avilov et al. (2007) have studied a series of Ir(III)-based heteroleptic complexes with phenylpyridine (*ppy*) and 2-(5-phenyl-4H-[1,2,4]triazol-3-yl)-pyridine (*ptpy*) derivatives as coordinating ligands through a number of experimental and theoretical approaches. The presence of the fluorine and trifluoromethyl substituents is found to affect both the emission energy and the localization of the lowest excited triplet states, which are characterized as local excitations of the chromophoric ligands (*ppy* or *ptpy*) by DFT calculations. The admixture between metal-to-ligand charge-transfer (MLCT) and ligand-to-ligand charge-transfer (LLCT) is small and their contributions are strongly dependent on the energy gaps between the relevant molecular orbitals.

The sky-blue emitting phosphorescent compound Ir(4,6-dFppy)$_2$(acac) has been doped into matrices and studied under ambient conditions as well as at low temperatures by Rausch et al. (2009). The emitting triplet state is found to be of MLCT character, and the polycrystalline and amorphous hosts are found to show distinct influence on the emissive properties. A clear difference is found through comparison with the similar Ir(4,6-dFppy)$_2$(pic) complex, which could be explained by the different effects of acac and pic ligands on the iridium d-orbitals, leading to different zero-field splittings, radiative emission rates and phosphorescence quantum yields. Highly resolved spectra reveal the importance of the spin-orbit coupling effect related to the emission from individual triplet sub-states. The authors emphasized that the complex symmetry and matrix effects are important factors that affect the performance of OLED devices.

Byun et al. (2007) have synthesized a number of bis-cyclometalated iridium(III) complexes with a common ancillary ligand ZN (3,5-dimethylpyrazole-N-carboxamide), which emit in the sky blue region. DFT calculations show that the cyclometalating ligands contribute negligibly to the HOMO while the ZN ligand is the main contributor together with the iridium d-orbitals. Moreover, it is found that the Ir(MeOF$_2$ppy)$_2$ZN complex possesses the largest phosphorescence quantum efficiency and the lowest nonradiative emission rate. The solution-synthesized organic light emitting device (OLED) of Ir(F$_2$ppy)$_2$ZN doped in a blend of polystyrene and m-bis(N-carbazolylbenzene) has shown an efficiency of 7.8 cd A^{-1} (Byun et al. 2007).

Takizawa et al. (2007) have prepared and systematically studied a series of new blue-phosphorescent iridium(III) complexes containing 2-phenylimidazo[1,2-a]pyridine (pip) derivatives as ligands. Electron-withdrawing substituents on the pip ligands are found to lower the HOMO energy level and lead to blue-shifted emission wavelengths. Based on experimental data it is found that the HOMO of the iridium complex with pip ligands is mixed Ir-d, phenyl-π and pip-π in character. The pip ligand is able to shift the emission wavelengths into the blue region and the polymer light-emitting devices (PLEDs) suggest that the pip-based iridium complexes are good phosphorescent materials for OLED applications.

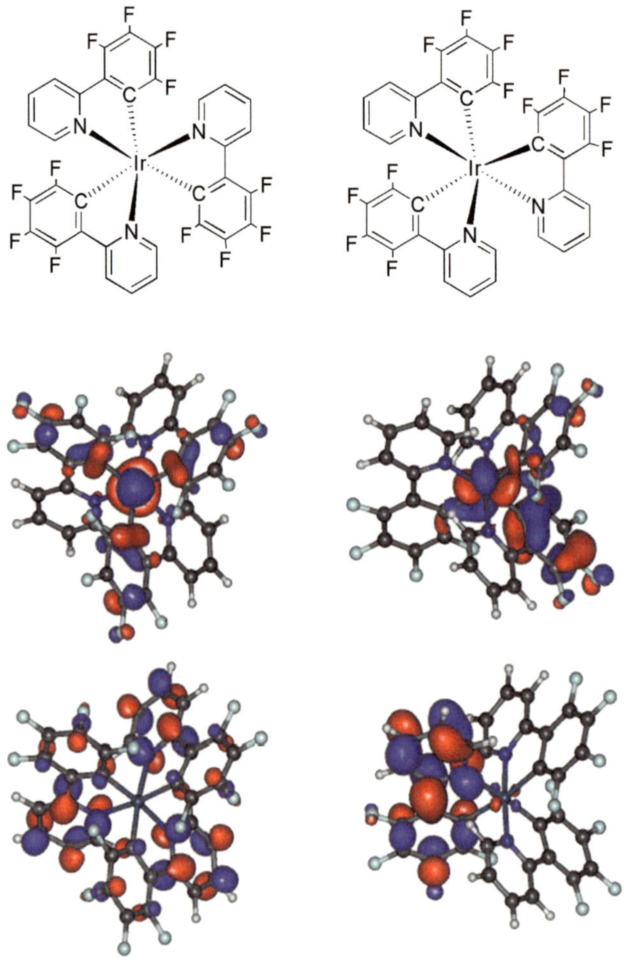

Fig. 3. Structures (top), HOMOs (middle) and LUMOs (bottom) of *fac-Ir(F_4ppy)$_3$* (left) and *mer*-Ir(F_4ppy)$_3$ (right).

3.4 Introduction of novel ligands

The introduction of novel ligands other than conventional ppy ligands provides the possibility of fine-tuning the spectra of the iridium complexes. For instance, imidazole-based ligands could lead to more delocalized frontier molecular orbitals. Baranoff et al. (2011) have studied two series of heteroleptic bis-cyclometalated Ir(III) complexes with phenyl-imidazole-based ligands, with phosphorescence emission ranging from greenish-blue to orange. The systematic study on these complexes shows that the photophysical and electrochemical properties could be tuned by changing the substitution pattern on the ligands. DFT calculations suggest that the highest occupied molecular orbital (HOMO) and the lowest unoccupied molecular orbital (LUMO) are more delocalized in complexes with phenyl-imidazole as ligands. Interestingly, the presence of chlorine substitution leads to an unexpected red shift in the emission energy, which could be explained by significant geometrical and electronic relaxation as confirmed by theoretical calculations.

Chang et al. (2007) reported the preparation of a series of new heteroleptic Ir(III) complexes chelated by two cyclometalated 1-(2,4-difluorophenyl)pyrazole ligands (*dfpz*)H and a third ancillary bidentate ligand (L^X). The cyclometalated *dfpz* ligands give rise to a larger π-π^* gap in the iridium complexes, and the lowest one-electron excitation are expected to accommodate the π^* orbital of the ancillary L^X ligands, which could be modified to fine-tune the phosphorescent emission. The reduction and oxidation reactions are found to occur mainly at the ancillary L^X ligands and the iridium metal site, respectively. The authors have shown a simple and straightforward approach to tune the color by varying the ancillary ligands only (Chang et al. 2007).

Volpi et al. (2009) have studied the cationic heteroleptic cyclometalated iridium complexes with 1-pyridylimidazo[1,5-alpha]pyridine (pip) ligands in order to provide exploitation of an efficient intersystem crossing in OLEDs. Blue luminescence is observed upon excitation of $[Ir(ppy)2(pip)]^+$ with lifetimes between 0.6 and 1.3 μs. TD DFT calculations with solvent effect taken into account reveal that the iridium center contributes significantly to most transitions. Furthermore, a photochemical reaction has been observed to give rise to a new class of cyclometalated iridium complexes with dipyridylketone and deprotonated amide as ligands.

Chou et al. have presented general concepts that have guided important developments in the recent research progress of room-temperature phosphorescent dyes. The authors elaborate on both the theoretical background for emissive metal complexes and the strategic design of the 2-pyridylazolate ligands, aiming to fine-tune the chemical and photophysical properties. The 2-pyridylazolate ligands are incorporated to give rise to the highly emissive transition-metal complexes, which show potential usefulness in the application as OLED dyes. Based on this family of metal complexes, the possibility of tuning the emission toward the near-IR region with for future applications in solar cell and near-IR imaging has been proposed.

Yang et al. (2008) studied neutral mixed-ligand Ir(N=C=N)(N=C)X complexes which are not emissive at room temperature but exhibit strong phosphorescence at 77 K. The 0-0 transition energies are located at around 450 nm with lifetimes of 3-14 μs. Through temperature-dependent lifetime measurements and unrestricted density functional theory calculations, the mechanism and pathway of thermal deactivation are analyzed in detail. The calculated activation energies of approximately 1800 cm^{-1} are in excellent agreement with the observed values.

Liu et al. (2007a) have presented calculations on geometries, electronic structures, and spectroscopic properties of a series of cationic iridium(III) complexes with C^N and PH_3 ligands. The geometries at the ground state and the excited state are optimized at the B3LYP/LANL2DZ and CIS levels of theory, respectively. The HOMOs are found to localize on the iridium atom and the C^N ligands, while the LUMOs are dominantly localized on the C^N ligand. TD DFT calculations with solvent effects taken into account by the polarized continuum model provide absorption and phosphorescence wavelengths in acetonitrile solution, and the low-lying absorptions are assigned as the $d_{yz}(Ir) + \pi(C\text{^}N)] \rightarrow$

π^*(C^N) transition. The computations also suggest that the phosphorescent emission wavelengths could be blue-shifted by introducing π electron-withdrawing groups and by suppressing the π-conjugation in the C^N ligand.

Liu et al. (2008) also investigated the photophysical properties of heteroleptic iridium complexes containing carbazole-functionalized β-diketonates. The authors have studied the influence of the triplet energy level of the ancillary carbazole-containing ligand on the photophysical and electrochemical behavior, and found that the superposition of the state density map of the triplet energy levels between the β-diketonate and the Ir(C^N)$_2$ fragment is the key factor to obtain strong ^{3}LC or ^{3}MLCT-based phosphorescence and high photophysical performance. DFT calculations reveal that the lowest excited state is mainly determined by the C^N ligand but not by β-diketonate when there is large difference between the triplet energy levels of the two parts, providing satisfactory explanation for experimental results.

The photophysical properties of facial and meridional tris-cyclometalated iridium(III) complexes containing 2-phenylpyridine and 1-phenylisoquinoline ligands have been reported by Deaton et al. (2010). The facial isomers show similar photophysical properties in 2-MeTHF solutions, indicating that the emission occurs based on the *piq* ligand(s). By comparing the photophysical properties between *fac*-Ir(*piq*)$_3$ and fac-Ir(*piq*)(*ppy*)$_2$ the effect of the *piq* ligand is revealed; it is found that the quantum yield is higher in *fac*-Ir(*piq*)$_3$ than that in *fac*-Ir(*piq*)(*ppy*)$_2$, suggesting a larger nonradiative rate in the latter compound. The meridional complexes have much lower quantum yields in solution comparing with their facial counterparts, and the difference between *mer*-Ir(*piq*)(*ppy*)$_2$ and *fac*-Ir(*piq*)(*ppy*)$_2$ is interpreted by more π-π^* character and less MLCT character in the former compound. The authors have shown that the phosphorescent decay is very efficient and may be used in OLEDs.

3.5 External modulation of Ir(III) phosphorescence

Considering the development of molecular switches that respond to external electric or photonic stimuli, it is interesting to introduce such a switching unit into the iridium complexes to realize a controllable phosphorescent emission. Zapata et al. (2009) have studied a heterobimetallic Ir(III) complex with a ferrocenyl azaheterocycle as ancillary ligand, which acts as a redox-fluorescent molecular switch. The ancillary ligand consists of a redox-acitve ferrocene unit and a 1,10-phenanthroline chelator coordinating with the iridium center. By tuning the oxidation state of the ferrocene through electrochemical stimuli, the emission intensity of the iridium complex can be modulated. This is an interesting example of the effective control of emission in iridium compounds. Besides, Tan et al. (2009) reported the photochromic iridium(III) complex (Py-BTE)$_2$Ir(acac) in containing two bis-thienylethene (BTE) switching units and one iridium(III) center. This molecule has shown distinct photo-reactivity and photo-controllable phosphorescence due to the combination of the photochromic BTE switch and the highly luminescent iridium(III) complex into one molecule. Through photo-induced isomerization, the phosphorescence is almost completely quenched by the closed-ring form of the BTE unit. Through DFT calculations, Li et al. (2010) have shown a monotonic relationship between the metal character of phosphorescence and the radiative deactivation rate constant function and rationalized the non-radiative deactivation rate using the energy gap law, leading to a theoretical interpretation of photochromic modulation of the iridium(III) phosphorescence.

4. Other new OLED materials

In the following we will consider our own experimental design of new materials for molecular electronics, electroluminescent and solar energy conversion devices. First we consider improvements of light-emitting and ETL materials, which do not include transition-metal complexes.

4.1 Modification of hole transport and electron transport layers

In Ref. (Xie et al, 2005) a new soluble 5-carbazolium-8-hydroxyquinoline Al(III) complex was synthesized and used in OLEDs as a dipolar luminescent material instead of Alq_3 (Scheme 1), which was the milestone emitting ETL material during two decades. An excellent capacity of electron transportation of Alq3 is determined by effective LUMO overlap between neighboring molecules (A_4 and A_3 in Fig. 1b). But the overlap of HOMOs, which governs the hole-transport (overlap between HOMO of molecules A_2 and A_1 in Fig. 1b) is very low. That is why we need to use an extra HTL material like NPB in order to obtain an effective OLED function. Carbazole derivatives are also widely used as hole-transport materials between the emitting layer and the anode. The idea to combine ETL and HTL properties and unite carbazole and Alq_3 moieties in a new luminescent material has been realized (Xie et al, 2005). The new soluble synthesized complex includes carbazolium substituted in a *para*-position to the oxygen in Alq_3. The highest spin density is at the HOMO of the ionized hole at this C-5 carbon atom position in the Alq_3 quinolate moiety. The electron-donating carbazolium substituent in the C-5 position causes a red-shift in the emission and absorption spectra of a new aluminum complex. The photoluminescence spectrum indicates an effective intramolecular singlet energy transfer from the carbazole groups to Alq_3 (no carbazole emission). The half oxidation potential of the new complex provides the HOMO energy (-5.51 eV) which is higher than that of Alq_3 (-5.9 eV). This significant improvement of the hole transport property is determined by the fact that the HOMO is mostly localized on the carbazole groups (Xie et al, 2005). The new complex being soluble is much better than Alq3, which must be vacuum deposited in fabrication of OLEDs. The main interest in soluble luminescent materials with high ETL and HTL properties lies in the scope for low-cost manufacturing, like spin coating, which is in line with the current trend to fabricate OLEDs from solutions.

4.2 Functionalization of nonmetallic photoluminescent complexes for red-emititng OLEDs

For full color displays red-emitting materials are required (besides green and blue emitters discussed above). Materials with red emission are usually achieved by doping red dyes (e.g., porphyrins) into a host matrix with a large band gap (Li et al. 2007). Because typical organic red dyes are large π-conjugated planar systems, they are prone to aggregate and quench their luminescence. Such organic dyes being highly emissive in dilute solutions become non-luminescent in the solid state. Many functional groups like oligo-fluorenes, truxene, indoles, which act as light-harvesting antenna and prevent aggregation, have been attached to porphyrins to obtain novel red-emitting materials for OLEDs. Bisindolylmaleimides provide wide luminescence bands in the range 550-650 nm and have also been found useful in fabrication of white color OLEDs (Ning et al. 2007a). In Ref. (Li et al. 2007) the bisindolylmaleimide (BIM) group, conjugated with tetraphenylporphin (TPP) in the form of dyad (PM-1) and pentamer (PM-2), have been sensitized and found to serve as good candidates of red-light emitting materials for OLEDs. These dendrimers have been prepared through imidization of bisindolylmaleic anhydride with aminoporphyrins. The long hexyl chains on the BIM groups improve solubility and suppress the aggregation in the solid state (Li et al. 2007). The new sensitized porphyrin dendrimers, PM-1 and PM-2, exhibit an intense Soret band (420 nm) and week Q-bands (500-650 nm) in the absorption spectra in dilute THF, which are typical for the TPP itself. The Soret band is slightly red shifted and broadened (compared with TPP) and new UV absorption occurs at 290 nm. The latter coincides with the BIM band and increases when the numbers of BIM groups increase in the dendrimers. Week additional BIM absorption occurs at 480 nm, which corresponds to charge transfer from the indolyl to the maleimide moiety. The luminescence spectra of all dendrimers exhibit characteristic emission of porphyrin, which consists of two vibronic Q_x bands: the strong 0-0 and week 0-1 peaks at 660 and 750 nm, respectively (Minaev et al. 2005). The dendrimer emission is much more intense than that of TPP, especially for the 0-0 band. This indicates an efficient singlet energy transfer from the BIM-antenna groups to the porphyrin ring. The through-bond energy transfer by the Förster intramolecular mechanism provides efficient fluorescence of the porphyrin moiety with the quantum yield in PM-2 being twice as large in comparison with TPP. The porphyrin dendrimer PM-2 with four BIM groups exhibits much stronger emission comparing to the PM-1 dye with only one BIM group due to enhanced antennae harvesting effects.

OLEDs made with solid film PM-2 spin-coated on quartz plate (ITO/PEDOT/PVK/PFO +PBD: PM-2 (5%)/Ba/Al) show pure red emission. The external quantum efficiency (0.13%) demonstrates effective EHP recombination and energy transfer in this EML. Another OLED device with 2.5% PM-2 doped within the PFO+PBD emissive layer exhibits higher external quantum efficiency (0.2%) and luminance maximum (101 cd m^2), but the chromaticity is not so pure (Li et al. 2007).

BIM derivatives themselves have also been used as non-doped red light-emitting materials (Ning et al. 2007a). Their solid state fluorescence quantum yield can be dramatically changed by introduction of different substituents on the non-conjugated linkage to the BIM skeleton. The OLED configuration (ITO/NPB/maleimide/TPBI/LiF:Al) with BIM derivative (1S) of 2,3-bis(N-benzyl-2′methyl-3′-indolyl)-N-methylmaleimide reaches the brightness 393 cd m-2 at 100 mA cm-2. Though the performance of such non-doped red organic light-emitting materials is not as good as the conventional doped ones, they are more promising for mass production (Ning et al. 2007a). OLEDs based on non-doped host EML can simplify the manufacturing significantly.

As indicated earlier, iridium(III) complexes are promising materials for applications in OLEDs, due to their strong spin-orbit coupling, intense phosphorescence, high quantum efficiency and tunable emission wavelengths. In most iridium complexes the HOMOs are found to locate on the metal center and on π-orbital of the ligands, and the LUMOs are delocalized on the π^*-orbital of the ligands. Electron-hole recombination in the doped emitting layer results in a mixed MLCT and ILCT triplet state, leading to efficient phosphorescence. By utilizing the triplet emission energy in OLED devices, iridium complexes are able to significantly enhance the efficiency of electroluminescence. However, cheaper organic molecules and supramolecular aggregates which utilize only singlet excited states for charge carrier recombination and energy transfer are still useful for low-cost OLED applications and for solar energy conversion. The position of HOMO and LUMO of the dye and their redox electrochemical parameters with respect to the electrode materials are crucial not only in OLEDs but also for photovoltaic devices. We shall in the coming section consider two promising types of such devices; dye-sensitized solar cells (DSSC) and organic semiconductor devices with triplet excitons.

5. Metal complexes for dye-sensitized solar cells

In order to see some common features in light emitters (OLED) and absorbers (DSSC) let us consider first a typical solar cell. DSSCs of the Grätzel type mainly consist of an optically transparent photoanode (lower Fluorine-doped tin oxide (FTO) glass sintered with TiO_2 nanocrystals), dyes adsorbed on mesoporous nanocrystalline TiO_2, electrolyte, and a cathode which consists of a platinum thin film layer sputtered on the upper FTO layer (Fig.4).

Fig. 5 shows the main carrier transport channels. At first, the incident light (hv) is absorbed by the sensitizer dye, the electrons of which are excited from the HOMO to the LUMO. Consecutively, the photogenerated electrons are injected from the LUMO of the dye to the conduction band (CB) of the TiO_2 (channel (a) in Fig. 5). The oxidized dye will later be reduced by the redox couple, channel (b), that recieves electrons from the counter electrode (cathode). Apart from these normal electron transfer channels, there are some other undesirable carrier transport channels, such as charge recombination of electrons from TiO_2-CB to the dye cations (c) and to the redox couples (d), and excited dye quenching by direct decay from LUMO to HOMO (e).

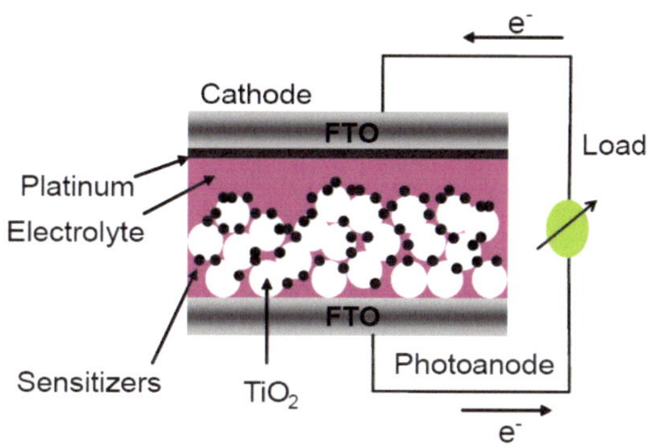

Fig. 4. Schematic structure diagram of DSSCs.

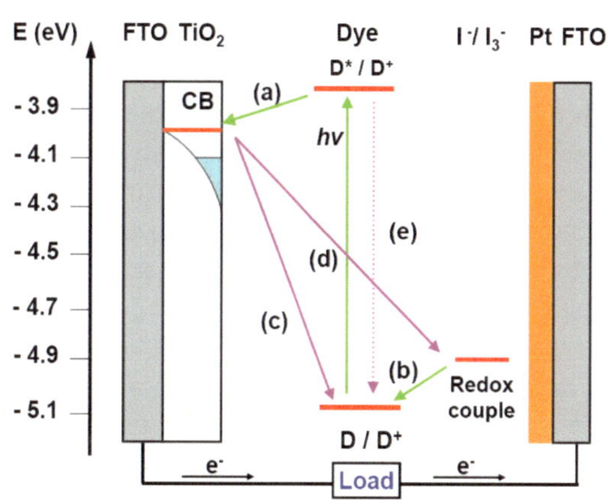

Fig. 5. Energy band structure and major electron transfer processes in DSCs.

Among all the constituent components in DSSCs (Fig. 4), the sensitizer, being charged with the task of the light absorption and electron injection, is generally regarded as the most crucial one for the overall efficiency. Since the first report by Grätzel and coworkers, the metal complex dyes are generally considered as the best sensitizers for DSSCs (O'Regan et al. 1991). During the development of DSSCs, a benchmark is given by the introduction of dye N3 (*cis*-dithiocyanato bis(2,2'-bipyridine-4,4'-dicarboxylate)ruthenium-(II)), which achieves efficiency over 10 % (Nazeeruddin et al. 1993. The other famous dye N719 is similar to N3, differing by the number of protons). A great deal of efforts were made to optimize the performance of metal complex dyes by molecular modification, meanwhile, the relationship between the sensitizer structure and performance has been extensively studied. In this section, we mainly focus on the development of metal complex sensitizers, which are similar to chromophores developed for OLEDs.

5.1 Extension of the absorption spectra

Although the efficiency of N3 is as high as 10 %, its absorption spectrum is mainly located in the visible region, and is quite weak in the near-infrared region of the Sun radiation. To improve the efficiency of DSSCs based on Ru complexes further, the absorption spectrum must be extended (Ning et al. 2009). For this purpose a number of new dyes have been sensitized.

5.1.1 Introduction of additional substitute on one of the bipyridine ligand in N3

The group of Wu and the Wang's group have reported a new kind of ruthenium complexes such as C101 (Scheme 2) (Gao et al. 2008), with high molar extinction coefficients by the addition of alkyl-substituted thiophene on the spectator ligands, with a motivation to enhance the optical absorptivity of the sensitizer. Along with the acetonitrile-based electrolyte, the C101 sensitizer achieved a strikingly high efficiency of 11.0-11.3%. The cells based on a low-volatility 3-methoxypropionitrile electrolyte and a solvent-free ionic liquid electrolyte, show conversion efficiency over 9.0%. In addition, this DSSC is highly stable under full sunlight soaking at 60°C during 42 days. It was speculated that alkyl chains (C_6H_{13}) can create a hydrophobic environment to improve the stability of the cells. Another possible reason might be that alky chains facilitate formation of a more compact sensitizer layer to prevent the approach of the electrolyte to the TiO_2 surface. Substitution of sulfur by selen in the dye C101 produces an effective sensitizer (C105) (Gao et al. 2009). TD DFT calculations of these dyes (Baryshnikov et al. 2010) indicate much more intense absorption of C105 with respect to the N3 dye and explain the negative solvatochromic effect, which is a sequence of the selenophene conjugation with the bpy ligand. Strong changes of the Ru-N and C-C bond lengths in the substituted bpy ligand also indicate the π-conjugation effect with selenophene. Besides, it supports the planar structure of the ligands (Baryshnikov et al. 2010). An intense absorption band of C105 at 746 nm is determined by two transitions: from HOMO to LUMO+2 with small admixtures of other excitations and from HOMO to LUMO+3. The highest occupied MOs in the C101 and C105 dyes are localized on the 3d orbitals of the Ru ion and on the N=C=S groups. Four quasi-degenerate vacant MOs in C105 are localized on the bpy ligands with some admixture from the metal. Thus the sun light induces the MLCT and NCS→bpy charge-transfer transitions. All types of bpy ligands are involved, something that provides an efficient electron injection through the carboxyl-groups.

Wu et.al have introduced ethylene-dioxythiophene groups on the bipyridine ligand, which further extended the absorption spectra compared with the thiophene ligand (Chen et al. 2007). The sensitizer SJW-E1 (Scheme 2) shows higher efficiency than N3 under the same conditions due to the extension of the electron donor ligand, which can uplift the HOMO energy edge and reduce the energy waste between the HOMO energy of the sensitizer and the redox couple. Based on the thiophen-substituted complex, Wu et. al further developed a new ruthenium-based dye (CYC-B6S, Scheme 2) in which alkyl-substituted carbazole moieties were incorporated in the thiophene-substituted bipyridine ligand (Chen et al. 2008). Compared with the N3 dye, the unique ancillary ligand in CYC-B6S is well-designed to enhance the light-harvesting capacity with the thiophene unit to further enrich the spectral response. Wang et.al developed a new Ru complex sensitizer K19 (Scheme 2) with a styryl unit attached to the bipyridly ligand (Kuang et al. 2006). The addition of this styryl ligand significantly enhanced the light absorbing capability. In addition, the DSS cells based on the K19 sensitizer also show an excellent photochemical stability. After 1000 h of light soaking at 60 °C, no drop in efficiency was observed for the cells covered with an ultraviolet absorbing polymer film.

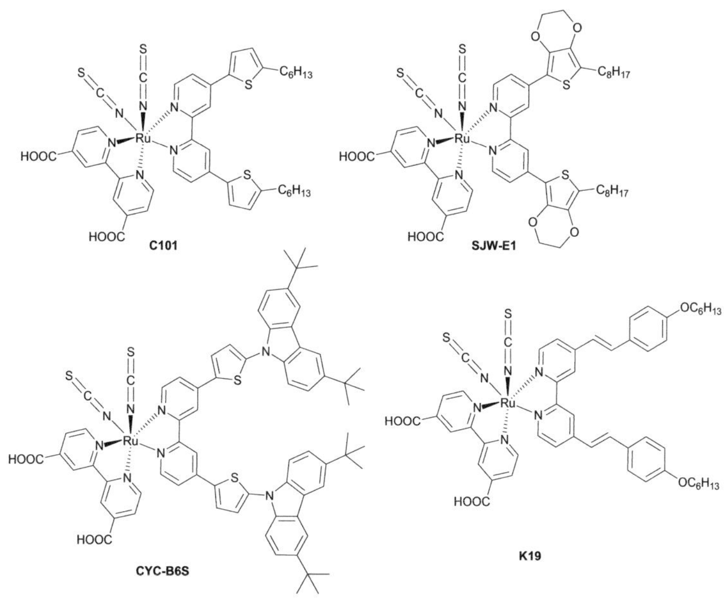

Scheme 2.

5.1.2 Modification of the ligand

A series of panchromatic ruthenium(II) sensitizers (black dye, Scheme 3) derived from carboxylated terpyridyl complexes of tris-thiocyanato Ru(II) has been developed by Grätzel's team (Nazeeruddin et al. 2001). Due to the presence of three thiocyanate ligands, the absorption spectrum is obviously red-shifted compared with complexes which have two thiocyanate ligands. The black dye, when anchored to nanocrystalline TiO_2 films, achieves very efficient sensitization over the whole visible range extending into the near-IR region up to 920 nm, yielding over 80% incident photon-to-current efficiencies (IPCE). Employing this dye the highest record conversion efficiency up to now, 11%, was achieved.

Jin et.al synthesized a novel kind of Ru complex sensitizer with a triarylamine-ligand (Jin et al. 2009). Under standard global AM 1.5 solar conditions, the J13 (Scheme 3)-sensitized solar cells demonstrate short circuit photocurrent densities of 15.6 mA/cm^2, open circuit voltages of 700 mV, fill factors of 0.71, and overall conversion efficiencies of 7.8%, which is comparable to the N719 dye under which identical measurement conditions gives 7.91%. DFT/TDDFT calculations indicate that the triarylamine ligand acts as an electron donor in a manner similar to the thiocyanato ligands and Ru metal. However, no obvious bathochromic shift of the absorption spectrum was observed.

Grätzel and coworkers developed a novel thiocyanate-free cyclometalleted ruthenium sensitizer (Ru-F, Scheme 3) with electron acceptor fluorine atoms substituted on one ligand (Bessho et al. 2009). Density functional theory (DFT) and time-dependent DFT (TDDFT) calculations show that the HOMO is located mostly on ruthenium and the cyclometalated ligands. Molecular orbital analysis confirms the experimental assignment of the redox potentials, and TDDFT calculations allow an assignment of the visible absorption bands. The DSSC based on Ru-F exhibits a remarkable IPSE value of 83%.

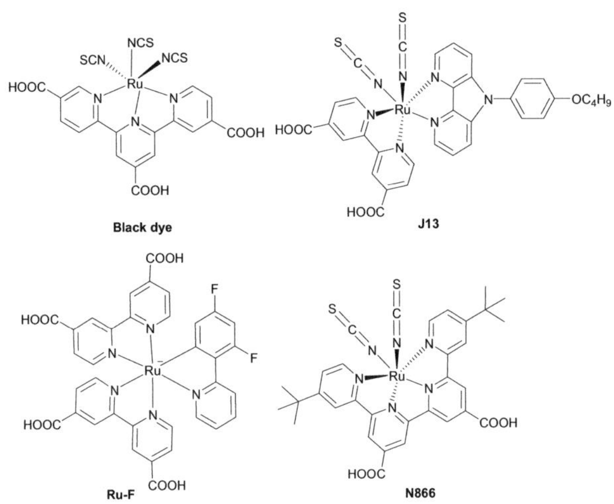

Scheme 3.

A ruthenium complex N886 (Scheme 3) with quarterpyridine as a ligand has been developed by Grätzel's group (Barolo et al. 2006). The absorption spectrum of the N886 complex shows metal-to-ligand charge-transfer transitions in the entire visible region. The TD DFT method qualitatively reproduces the experimental absorption spectra. The absorption bands were assigned to the mixed Ru/SCN-to-quaterpyridine charge-transfer transitions, which extend from the near-IR to the UV regions. Dramatic red shift of the absorption spectrum compared with N3 is observed. A DSSC the based on panchromatic sensitizer N886 complex shows an overall conversion efficiency of 5.85%.

5.2 Molecular modification of the Ru complex in order to reduce charge recombination

The charge recombination between the injected electrons in TiO2 and oxidized sensitizer or redox couple can significantly decrease the conversion efficiency of the DSSCs (Ning et al. 2010). Much efforts have been devoted to reduce these processes.

5.2.1 The task to reduce charge recombination between the oxidized sensitizer and TiO_2

In recent years, it was reported that the introduction of a triarylamine unit can increase the distance between the TiO_2 surface and the sensitizer electron-donor unit where the charge recombination normally occurs, and thus to reduce the charge recombination between sensitizer and TiO_2. Durrant et al. developed the sensitizer Ru-1 (Scheme 4) with a triphenylamine unit connected to the Ru complex, which showed higher efficiency than the sensitizer without the triphenylamine (Hirata et al. 2004). The observed suppression of the carriers recombination is attributed to an increase in the physical separation between the dye cation and the metal oxide surface. Bonhôte and coworkers have studied the charge recombination process in a series of Ru dyes connected with triphenylamine (Bonhôte et al. 1999). The lifetime of injected electrons in TiO_2 is enhanced by a significant factor of 100 times after the incorporation of those units in their model system (without I^-/I_3^- redox couples so that the carrier recombination mostly occurs between the oxidized dye and the injected electrons). In the model system, the sensitized nanocrystalline TiO_2 film employing the Ru-4 dye (Scheme 4) achieves a remarkably long lifetime of 4 s for injected electrons in TiO_2.

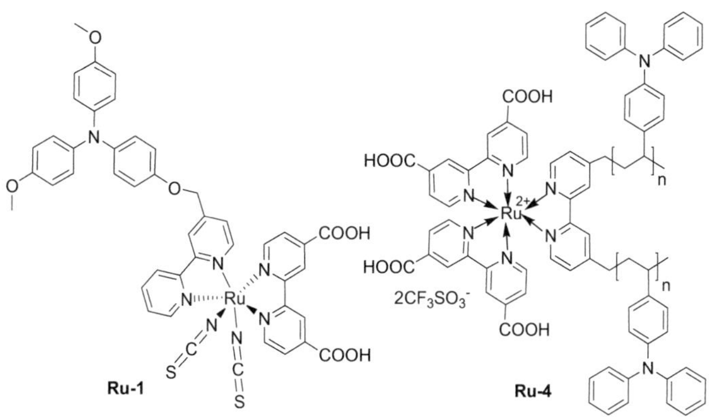

Scheme 4.

5.2.2 The task to reduce the charge recombination between oxidized sensitizer and redox couple

Except the charge recombination between TiO_2 and the oxidized sensitizer, the electron transfer from TiO_2 to the electrolyte will decrease the efficiency as well. It was reported that the starburst structure is also suitable for Ru dyes to reduce the charge recombination between TiO_2 and electrolyte. Haque and coworkers reported significantly reduced charge recombination between TiO_2 and redox couple by the connection of triphenylamine on the ligand (Haque 2005). Thelakkat and coworkers developed Ru dyes (Ru-TPD-NCS, Ru-TPA-NCS) with triarylamine substituents and applied them in solid state dye-sensitized solar cells (SDSCs) (Scheme 5) (Karthikeyan et al. 2007). The exterior starburst triarylamine can effectively reduce the carrier recombination between the injected electrons and redox couples, leading to higher V_{oc} and efficiency than N719. Kroeze and coworkers found that for Ru-complex-sensitizer-based DSCs, charge recombination can be reduced by connecting long alkyl chains (Fig. 4) (Schmidt-Mende et al. 2005). Snaith and coworkers obtained a much prolonged carrier lifetime by linking oxyethylene and/or diblock ethylene-oxide:alkane pendent groups to the Ru sensitizers (Snaith et al. 2007). By suppressing the carrier recombination, V_{oc} of SDSCs based on K68 reach as high as 0.93 V, and the total energy efficiency is 5.1% under AM 1.5 irradiation.

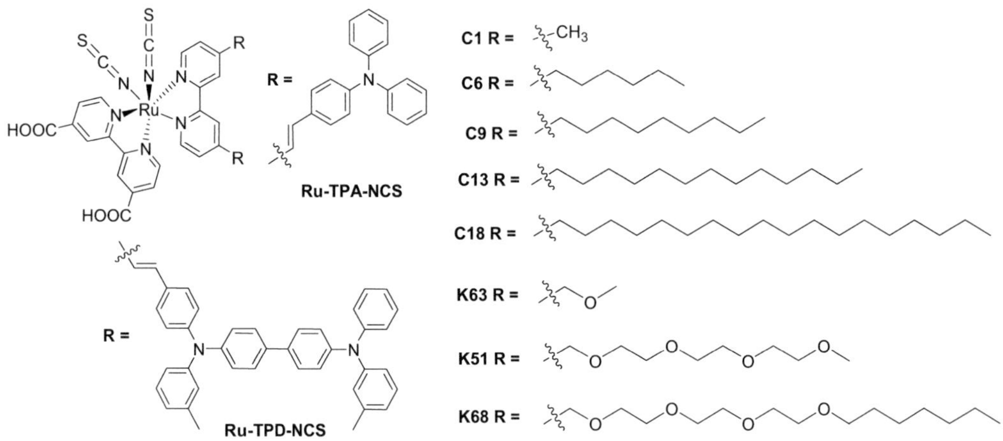

Scheme 5.

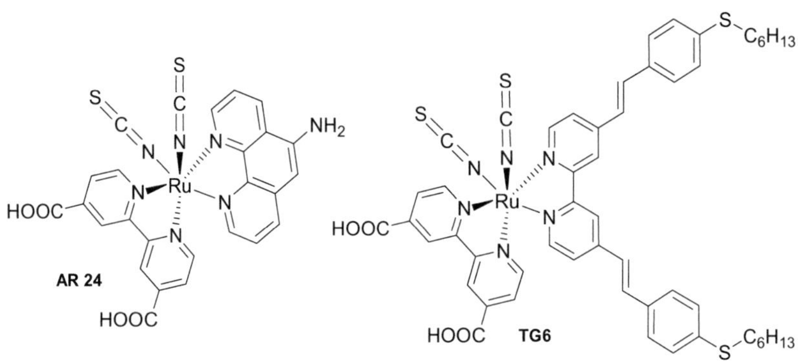

Scheme 6.

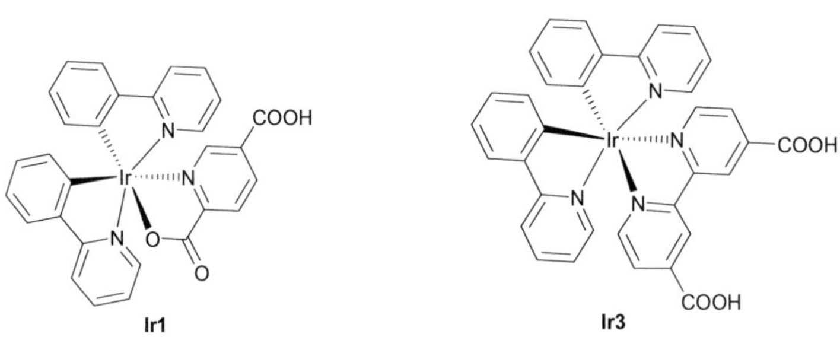

Scheme 7.

In addition to the blocking effect of the sensitizer, O'Reagan and coworkers recently found another potential factor that is crucial to determine the charge recombination, namely that the dye molecules can form complexes with the redox couple, and thus enhance the recombination reaction between electrons in TiO_2 and the electrolyte (O'Regan et al. 2009). They observed that the presence of an amine AR24 (Scheme 6) group in the sensitizer can significantly aggravate the charge recombination because of its strong iodide binding capability (Reynal et al. 2008). In addition, they found that the charge recombination of the sensitizer with an lkoxy group (K19) was clearly more serious than for the alky sulfide substitute (TG6) (O'Regan et al. 2009a). The difference was attributed to the different complexation capability with iodide of the sensitizer. However, up to now, the detailed mechanism of the complex is not clear.

5.3 The task to increase the electron injection efficiency

To increase the electron injection efficiency of DSSCs, it is critical to decrease the distance between the sensitizer acceptor and the TiO_2. An effective strategy might be the adoption of multi-anchor units. Tian et al. investigated a series of iridium sensitizers with one or two carboxyl anchor groups. It was found that the efficiency of a sensitizer with two carboxyl units (Ir3, Scheme 7) is pronouncedly higher than for a sensitizer with a single carboxyl unit (Ir1, Scheme 7) (Ning et al. 2009a).

Another factor that affects the electron injection efficiency is the non-radiative decay of the sensitizer, which results in energy loss. Tian et. al investigated the relationship between the emission quantum yield and the electron injection efficiency of sensitizers (Ning et al. 2009a). It was found that the electron injection efficiency is consistent with the luminescence quantum yield of the sensitizer. Since less non-radiative decay guarantees high luminescence quantum yield to enhance the electron injection efficiency, it is important to reduce the non-radiative decay which arises mainly from the molecular vibrations. The ethylene linkage is susceptible to isomerization upon irradiation, which leads to vibrational energy loss. For sensitizers with several ethylene units, the efficiencies are generally low (Ning et al. 2009).

The Ir1 complex (Scheme 7) synthesized recently for DSSC devices (Ning et al. 2009a) is very similar to $Ir(ppy)_2(pic)$ species (Scheme 1), used for OLEDs (Nazeeruddin et al. 2009, Minaev et al. 2009). The only difference is the presence of the COOH group in the 2-pyridinecarboxylate (picolinate) moiety, which is necessary for adsorption on the TiO_2 surface in DSSCs. The LUMO in both complexes is localized entirely on the picolinate ligand; in the Ir1 species the LUMO has a large contribution from the carboxyl group (Ning et al. 2009a). This is important for the LUMO overlap with the surface of the semiconductor and for the electron injection efficiency of the DSSC. The photocurrent action spectrum of the TiO_2 electrode sensitized by Ir1 dye indicates that the weak absorption at 490 nm (first HOMO→LUMO transition) produces electron injection, which is increased up to 80% IPCE at 440 nm ($S_0 \rightarrow S_2$ absorption). The S_2 state has no admixture of the carboxyl group, which means that injection occurs after the fast $S_2 \rightarrow S_1$ relaxation.

Introduction of the N,N-dimethylamino group into the *para*-position of the picolinate ligand provides a quite efficient CIC dopant (N984) for the emissive layer in OLEDs (Nazeeruddin et al. 2009). This is explained by SOC calculations and the large change in the T_1 state wave function (Minaev et al. 2009) of the $Ir(ppy)_2(pic)$ complex. In the absence of the dimethylamino group the antibonding π MO of picolinate ligand shifts down and becomes the LUMO which gets lower by 0.38 eV in comparison with the N984 complex. This is in agreement with the cyclic voltammogram of the N984 complex, which shows a reversible couple at 0.61 V versus ferrocene Cp_2Fe/Cp_2Fe^+ redox couple due to the Ir(III/IV) reduction-oxidation cycle. Such a reduction potential of N984 demonstrates that the LUMO is located on the 2-phenylpyridine ligand rather than on the aminopicolinate ancillary ligand, the lowest unoccupied MO of which is destabilized by the presence of the N,N-dimethylamino group. The changes of MO energy levels determine the differences in UV-vis absorption and phosphorescence spectra induced by the insertion of the N,N-dymethylamino group in the 4-position of the picolinate ancillary ligand (Minaev et al. 2009). One can thus see that common quantum-chemical studies of the similar chromophores used in OLED and DSSC devices (Minaev et al. 2009, Ning et al. 2009a) can help to understand the most essential electronic structure features responsible for emissive and electron injection properties of cyclometalated iridium complexes.

6. Organic solar cells based on a bulk heterojunction architecture

Organic solar cells (OSC) based on a bulk heterojunction architectures can be realized by mixing of two solutions of organic semiconductors with different electronegativities and subsequently spinning a film (Köhler & Bässler 2009). The photoexcited state in one material diffuses to the interface of the other where dissociation occurs. The size of the phase separation between the two materials should be on the same length scale as the exciton diffusion length. This also requires a percolation path for separated charges to be sufficient to reach the corresponding electrodes. Fabrication of the film can be optimized by proper annealing, solvent mixture, and by spin-coating a blend. In this way a solar cell based on a bulk heterojunction (fullerene/low-bandgap polymer) has been obtained recently with a PCE of 5.5% (Köhler & Bässler 2009). The triplet excitons have longer diffusion length compared to singlets and this could be used as advantage for such OSCs. Despite the slow Dexter mechanism for the triplet exciton transfer, the large lifetime provides a triplet diffusion length ranging from 20 to 140 nm in amorphous organic films, while for singlet excitons it is typically in the range 10-20 nm (Köhler & Bässler 2009, Köhler at el. 1994).
From the energetic point of view OSCs based on triplet excitons are less favorable than usual polymer solar cells based on singlets (Köhler et al. 1994). Triplet excitons are more tightly bound than singlet excitons (by two exchange integrals, $2K_{ij}$) and this increases the barrier for exciton dissociation. It can be overcome by suitable LUMO energy level matching. Anyway, this leads to waste of a fraction of the absorbed solar energy. The maximum possible PCE is predicted to be about 11% for OSCs based on singlets and is likely to be somewhat lower for triplet solar cells (Köhler & Bässler 2009). In the first produced triplet OSC the material used was a conjugated platinum(II)-containing polymer (Köhler et al. 1994) of the form trans-$[\text{-Pt}(PBu_3)_2C{\equiv}CRC{\equiv}C\text{-}]_n$, where R= phenylene. The efficiency of single-material OSCs based on such Pt-polymers with triplet excitons are comparable to that of analogously built solar cells with singlet excited states (Köhler et al. 1994). When the Pt-polymers with triplet excitons were incorporated in OSCs based on a bulk heterojunction architecture with fullerene the PCE increased up to 0.3% (Köhler et al. 1996). These Pt-polymers have blue absorption (Minaev et al. 2006; Lindgren et al. 2007), while solar light peaks in the red. Thus for practical applications other Pt- and Pd-containing polymers have been synthesized with conjugated spacers R which have strong

electron-acceptor character and various such heterojunction devices have been fabricated using this concept (Köhler & Bässler 2009).

7. Conclusions

In this review we have discussed the understanding and design of optimal organometallic chromophores for light-emitting layers in OLEDs and for light-absorbing dyes and charge separation in DSSC interfaces. As an illustrating example, electro-luminescence OLED devices based on cyclometalated Ir(III) complexes (CICs) are discussed in some detail with special attention to spin-orbit coupling effects and triplet state emission. In pure organic polymers, like PPV or PPP, the energy stored in triplet states cannot be utilized in order to increase the emissive efficiency of OLEDs. With CICs as dopants the electroluminescence is enhanced by harnessing both singlet and triplet excitons after the initial charge recombination. Because the internal phosphorescence quantum efficiency is high - as high as 100% can theoretically be achieved - these heavy metal containing emitters will be superior to their fluorescent counterparts in future OLED applications. That has spurred quantum theory research on internal magnetic perturbations in such heavy transition metal complexes. The spin conservation rule as well as its violation in modern phosphorescent OLEDs is of principal importance in optoelectronics and spintronics applications. Synthesis of new materials for OLEDs can be rationalized if proper understanding of spin quantization and spin-orbit coupling is taken into account. Moreover, since the manufacturing of a full color display requires the use of emitters with all three primary colors, *i.e.* blue, green and red, the rational tuning of emission color over the entire visible range has emerged as an important task. Similar tasks are met in dye optimization for DSSCs. We discussed in this review issues on DSSCs on the basis of electronic structure and excited states calculations. The main reason for strong phosphorescence in the studied Pt and Ir complexes is connected with the fact that the S_0 - S_1 transition moments are relatively low, but the "spin-forbidden" T_1 - S_0 transition "borrows" large intensity from the higher lying excited states. This is introduced by SOC at the metal ion, whose electrons are involved in relevant excitations through the metal to ligand charge transfer (MLCT) admixtures. Site-selective phosphorescence in solid matrices at low temperature has revealed that zero-field splitting and spin-sublevel activity can be changed in different sites of the matrix, which shows that the MLCT character of the T_1 state is rather sensitive to the intermolecular environment of the dye. This is an important message; electron-hole recombination also depends on similar factors and all of them should be taken into account in proper simulations of OLEDs.

8. References

Abe, T.; Miyazava, A.; Konno, H. & Kawanishi, Y. (2010). Deuteration isotope effect on nonradiative transition of fac-tris (2-phenylpyridinato) iridium (III) complexes. *Chemical Physics Letters*, Vol. 491, pp. 199-202.

Adachi, C.; Baldo, M.A.; O'Brien, D.F.; Thompson, M.E.; & Forrest, S.R. (2001). Nearly 100% internal phosphorescence efficiency in an organic light-emitting device. *Journal of Applied Physics*, Vol. 90, pp. 5048-5052.

Avilov, I.; Minoofar, P.; Cornil, J. & De Cola, L. (2007). Influence if substituents on the energy and nature of the lowest excited states of heteroleptic phosphorescent Ir(III) complexes: A joint theoretical and experimental study. *J. Am. Chem. Soc.* Vol. 129, pp. 8247-8258.

Baldo, M.A.; O'Brien, D.F.; Thompson, M.E.; & Forrest, S.R. (1999). Excitonic singlet-triplet ratio in a semiconducting organic thin films. *Physical Review B: Condensed Matter and Material Physics*, Vol. 60, pp. 14422-14428.

Baranoff, E.; Bolink, H. J.; De Angelis, F.; Fantacci, S.; Di Censo, D.;p Djellab, K.; Grätzel, M. & Nazeeruddin, Md. K. (2010) An Inconvenient Influence of Iridium(III) Isomer on OLED Efficiency, *Dalton Transactions*, Vol. 39(2010), pp. 8914-8918, DOI: 10.1039/C0DT00414F.

Baranoff, E.; Fantacci, S.; De Angelis, F.; Zhang, X.; Scopelliti, R.; Gratzrl, M. & Nazeeruddin, M.K. (2011). Cyclometalated Iridium(III) Complexes Based on Phenyl-Imidazole Ligand, *Inorganic Chemistry*, Vol. 50(2011), pp. 451-462, DOI: 10.1021/ic901834v.

Baryshnikov, G.V.; Minaev, B. F. & Minaeva, V. A (2011). Quantum-chemical study of effect of conjugation on structure and spectral properties of C105 sensitizing dye. *Optics and Spectroscopy*, Vol. 110 (3), pp. 393-400.

Barolo, C.; Nazeeruddin, Md. K.; Fantacci, S.; Di Censo, D.; Comte, P.; Liska, P.; Viscardi, G.; Quagliotto, P.; De Angelis, F.; Ito, S.; & Grätzel, M. (2006). Synthesis, Characterization, and DFT-TDDFT Computational Study of a Ruthenium Complex Containing a Functionalized Tetradentate Ligand, *Inorg. Chem.* Vol. 45, pp. 4642-4653.

Bessho, T.; Yoneda, E.; Yum, J.; Guglielmi, M.; Tavernelli, I.; Imai, H.; Rothlisberger, U.; Nazeeruddin, M. K. & Grätzel, M. (2009). New Paradigm in Molecular Engineering of Sensitizers for Solar Cell Applications, *J. Am. Chem. Soc.*, Vol. *131*, pp. 5930–5934.

Bonhôte, P.; Moser J. E.; Humphry-Baker, R.; Vlachopoulos, N.; Zakeeruddin, S. M.; Walder, L. & Grätzel, M. (1999). Long-Lived Photoinduced Charge Separation and Redox-Type Photochromism on Mesoporous Oxide Films Sensitized by Molecular Dyads, *J. Am. Chem. Soc.*, Vol. 121, pp. 1324-1336.

Buchachenko, A.L. (1976). Chemical nuclear polarization. *Russian Chemical Review*. Vol. 45, pp. 375-392.

Chang, C.-J.; Yang, C.-H.; Chen, K.; Chi, Y.; Shu, C.-F.; Ho, M.-L.; Yeh, Y.-S. & Chou, P.-T. (2007). Color tuning associated with heteroleptic cyclometalated Iu(III) complexes; influence of the ancillary ligand. *Dalton Transactions*. Pp. 1881-1890.

Chen, C.; Chen, J.; Wu, S.; Li, J.; Wu, C.; & Ho, K.; (2008). Multifunctionalized Ruthenium-Based Supersensitizers for Highly Efficient Dye-Sensitized Solar Cells, *Angew. Chem. Int. Ed.*, Vol. 47, pp. 7342-7345.

Chen, C.; Wu, S.; Li, J.; Wu, C.; Chen, J.; & Ho, K.; (2007). A New Route to Enhance the Light-Harvesting Capability of Ruthenium Complexes for Dye-Sensitized Solar Cells, *Adv. Mater.* Vol. *19*, pp. *3888–3891*

Chen, L.; You, H.; Yang, C.; Lyu, Y.Y.; Chang, S.; Kwon, O.; Han, E.; Kim, H.; Kim, M.; Lee, H.J. & Das R.R. (2007). Novel, highly efficient blue-emitting heteroleptic iridium(III) complexes based on fluorinated 1,3,4-oxadiazole: tuning to blue by dithiolate ancillary ligands. *Chemical. Communications*, (13) pp.1352-1354.

Cheng, G.; Li, F.; Duan, Y.; Feng, J.; Liu, S.; Qiu, S.; Lin, D.; Ma, Y. & Li, S.T. (2003). White organic light-emitting devices using a phosphorescent sensitizer. *Applied Physics Letters*, Vol. 82, pp. 4224-4226.

Chou, P.T. & Chi, Y. (2007). Phosphorescent dyes for organic light-emitting diodes. *Chemistry - A European Journal*, Vol. 13(2) pp. 380-395.

Cundari, T.R. & Stevens, W.J. (1993). Effective core potential basis sets. *J. Chem. Physics*, Vol. 98, pp. 5555-5565.

De Angelis F., Fantacci S., Evans N., et al. (2007). Controlling phosphorescence color and quantum yields in cationic iridium complexes: a combined experimental and theoretical study. *Inorganic Chemistry*, 46(15) p5989-6001.

Deaton, J.C.; Young, R.H.; Lenhard, J.R.; Rajeswaran, M. & Huo, S. (2010). Photophysical Properties of the Series fac- and mer-(1-Phenylisoquinolinato-N((sect))C(2'))(x)(2-phenylpyridinato-N((sect))C(2'))(3-x)Iridium(III) (x = 1-3). *Inorganic Chemistry*, Vol. 49(20) pp. 9151-9161.

Dedeian K, Shi J, Forsythe E, et al. (2007). Blue phosphorescence from mixed cyano-isocyanide cyclometalated iridium(III) complexes. *Inorganic Chemistry*, Vol. 46(5) pp. 1603-1611.

Forrest, S.R. (2004). The path to ubiquitous and low-cost organic electronic appliances on plastic. *Nature*, Vol. 428, pp. 911.

Gao, F.; Cheng, Y.; Yu, Q.; Liu, S.; Shi, D.; Li, Y. & Wang P. (2008). Ruthenium Sensitizers for High Performance Dye-Sensitized Solar Cells, *Inorg. Chem.* Vol. 48 (6) pp. 2664-2669.

Gao, F.; Wang, Y.; Shi, D.; Zhang, J.; Wang, M.; Jing, X.; Humphry-Baker, R.; Wang, P.; Zakeeruddin, S. M. & Grätzel, M.; (2008). Enhance the Optical Absorptivity of Nanocrystalline TiO_2 Film with High Molar Extinction Coefficient Ruthenium Sensitizers for High Performance Dye-Sensitized Solar Cells, *J. Am. Chem. Soc.*, Vol. 130, pp. 10720-10728.

Haque, S. A.; Handa, S.; Peter, K.; Palomares, E.; Thelakkat, M. & Durrant, J. R. (2005). Supermolecular Control of Charge Transfer in Dye-Sensitized Nanocrystalline TiO2 Films: Towards a Quantitative Structure–Function Relationship, *Angew. Chem. Int. Ed.*, Vol. 44, pp. 5740-5744.

Hayashi, H. & Sakaguchi, Y. (2005). Magnetic field effects and CIDEP due to the d-type triplet mechanism in intra-molecular reactions. *Journal of Photochemistry and Photobiology, C*, Vol. 6, pp. 25-36.

Hirata, N.; Lagref, J.-J.; Palomares, E. J.; Durrant, J. R.; Nazeeruddin, Md. K.; Grätzel, M. & Di Censo, D. (2004). Supramolecular Control of Charge-Transfer Dynamics on Dye-sensitized Nanocrystalline TiO2 Films, *Chem. Eur. J.*, Vol. 10, pp. 595-602.

Hofbeck, T. & Yersin, H. (2010). The triplet state of fac-Ir(ppy)$_3$. *Inorganic Chemistry*, Vol. 49(12) pp. 9290-9299.

Jansson, E.; Minaev, B.; Schrader, S. & Ågren, H. (2007). Time-dependent density functional calculations of phosphorescence parameters for fac-tris(2-phenylpyridine) iridium. *Chemical Physics*, Vol. 333, pp. 157-167.

Jin, Z.; Masuda, H.; Yamanaka, N.; Minami, M.; Nakamura, T. & Nishikitani, Y. (2009). Efficient Electron Transfer Ruthenium Sensitizers for Dye-Sensitized Solar Cells, *J. Phys. Chem. C* Vol. *113*, pp. 2618–2623.

Karthikeyan, C. S.; Wietasch, H. & Thelakkat, M. (2007). Highly Efficient Solid-State Dye-Sensitized TiO2 Solar Cells Using Donor-Antenna Dyes Capable of Multistep Charge-Transfer Cascades, *Adv. Mater.*, Vol. 19, pp. 1091.

Koseki, S.; Schmidt, M.W. & Gordon, M.S. (1998). Effective nuclear charges for the first-through third-row transition metal elements in spin-orbit calculations. *Journal of Physical Chemistry, A*, Vol. 102, pp. 10430-10435.

Koseki, S.; Fedorov, D.G.; Schmidt, M.W. & Gordon, M.S. (2001). Spin-orbit splittings in the through third-row transition elements: comparison of effective nuclear charge and full Breit-Pauli calculations. *Journal of Physical Chemistry, A*, Vol. 105, pp. 8262-8268.

Kuang, D.; Ito, S.; Wenger, B.; Klein, C.; Moser, J.; Humphry-Baker, R.; Zakeeruddin, S. M. & Grätzel, M. (2006). High Molar Extinction Coefficient Heteroleptic Ruthenium Complexes for Thin Film Dye-Sensitized Solar Cells, *J. Am. Chem. Soc.*, Vol. 128, pp. 4146-4154.

Köhler, A. & Bässler, H. (2009). Triplet states in organic semiconductors. *Material Scienece and Engineering R*, Vol. 66, pp. 71-109.

Köhler, A.; Wittmann, H.F.; Friend, R.H.; Khan, M.S. & Lewis, J. (1994). Organic solar cell based on triplet excitons. *Synthetic Metals*, Vol. 67, pp. 245-248.

Köhler, A.; Wittmann, H.F.; Friend, R.H.; Khan, M.S. & Lewis, J. (1996). Organic solar cell based on triplet excitons in a bulk heterojunction. *Synthetic Metals*, Vol. 77, pp. 147-150.

Ladouceur, S.; Fortin, D. & Zysman-Colman, E. (2010). Role of Substitution on the Photophysical Properties of 5,5'-Diaryl-2,2'-bipyridine (bpy*) in [Ir(ppy)(2)(bpy*)]PF(6) Complexes: A Combined Experimental and Theoretical Study. *Inorganic Chemistry*, 49(12) p5625-5641.

Lee S.C., Seo J.H., Kim Y.K. & Kim Y.S. (2009). Studies of efficient heteroleptic Ir(III) complexes containing tpy and dfppy ligands for phosphorescent organic light-emitting devices. *Journal Nanoscience Nanotechnology*,Vol. 9(12), pp. 7094-7098.

Li, Y.; Cao, L.; Ning, Z.; Huang, Z.; Cao, Y. & Tian, H. (2007). Soluble porphyrin-bisindolylmaleimides dyad and pentamer as saturated red luminescent materials. *Tetrahedron Letters*, Vol. 48, pp. 975-978.

Li, X.; Zhang Q.; Tu, Y.; Ågren, H. & Tian, H. (2010). Modulation of iridium(III) phosphorescence via photochromic ligands: a density functional theory study. *Phys. Chem. Chem. Phys.* Vol. 12(41) pp. 13730-1376.

Li, X.; Minaev, B.; Ågren, H. & Tian, H. (2011). Theoretical study of phosphorescence of iridium complexes with fluorine-substituted phenylpyridine ligands. *Eur. J. Inorg. Chem.* DOI: 10.1002/ejic.201100084.

Lindgren, M.; Minaev, B.; Glimsdal, E.; Vestberg, R.; Westlund, R. & Malmstrom, E. (2007). Electronic states and phosphorescence of dendron functionalized platinum(II) acetylides. *Journal of Luminescence*, Vol. 124, pp. 302-310.

Liu T., Zhang H.X. & Shu X. (2007). Theoretical studies on structures and spectroscopic properties of a series of novel mixed-ligand Ir(III) complexes [Ir(Mebib)(ppy)X]. *Dalton Transactions*. pp.1922-1928.

Liu, T.; Zhang, H.X. & Xia, B.H. (2007a). Theoretical studies on structures and spectroscopic properties of a series of novel cationic [trans-(C/N)$_2$Ir(PH$_3$)$_2$]$^+$ (C/N = ppy, bzq, ppz, dfppy). *J. Phys. Chem. A*, Vol. 111(35) pp. 8724-8730.

Liu, Z.; Nie, D.; Bian, Z.; Chen, F.; Lou, B.; Bian, J. & Huang, C. (2008). Photophysical properties of heteroleptic iridium complexes containing carbazole-functionalized beta-diketonates. *ChemPhysChem*, 2008, 9(4) p634-640.

Minaev, B.F. & Terpugova, A.F. (1969). Spin-orbit interaction in charge-transfer complexes. *Journal of Soviet Physics*, No. 10, pp. 30-36.

Minaev, B.F. (1978). Spin-orbit interaction in molecules and mechanism of the external magnetic field on luminescence. *Optics and Spectroscopy*, Vol. 44, No. 2, pp. 256-260.

Minaev, B.F. (1972). Spin-orbit interaction in doublet states of molecules. *Optics and Spectroscopy*, Vol. 32, No. 1, pp. 22-27.

Minaev, B.; Minaeva, V.; & Ågren, H. (2009). Theoretical Study of the Cyclometalated Iridium(III) Complexes Used as Chromophores for Organic Light-Emitting Diodes. *J. Phys. Chem. A*. Vol. 113, pp. 726-735.

Minaev, B.; Ågren, H. & De Angelis, F. (2009a). Theoretical design of phosphorescence parameters for organic electro-luminescence devices based on iridium complexes. *Chemical Physics*, Vol. 358, pp. 245-257.

Minaev, B.; Jansson, E. & Lindgren, M. (2006). Application of density functional theory for studies of excited states and phosphorescence of platinum(II) acetylides. *J. Chem. Physics*, Vol. 125, pp. 094306-094313.

Minaev, B. & Ågren, H. (2005). Theoretical DFT study of phosphorescence from porphyrins. *Chem. Physics*, Vol. 315, pp. 215-239.

Minaev, B. & Ågren, H. (1999). Spin uncoupling in molecular hydrogen activation by platinum clusters. *J. Molecular Catalysis, A: Chemical*, Vol. 149, pp. 179-195.

Minaev, B.; Wang, Y.H.; Wang, C.K.; Luo, Y. & Ågren, H. (2005). Density functional study of vibronic structure of the first absorption Qx band in free-base porphin. *Spectrochimica Acta, A*. Vol. 65, pp. 308-323.

Minaev, B.F.; Jansson E.; Ågren, H. & Schrader, S. (2006). Theoretical study of phosphorescence in dye doped light emitting diods. *J. Chem. Physics*, Vol. 125, No. 23, pp. 234704.

Minaev, B.F.; Minaeva, V.O.; Baryshnikov, G.V.; Girtu, M. & Ågren, H. (2009b). Theoretical study of vibration spectra of sensitizing dyes for photoelectrical converters based on ruthenium (II) and iridium (III) complexes *Rus. J. Appl. Chem.* Vol. 82, pp.. 1211–1221.

Nazeeruddin, Md. K.; Kay, A.; Rodicio, I.; Humpbry-Baker, R.; Miiller, E.; Liska, P.; Vlachopoulos, N. & Grätzel, M. (1993). Conversion of light to electricity by cis-X2bis(2,2'-bipyridyl-4,4'-dicarboxylate)ruthenium(II) charge-transfer sensitizers (X = Cl^-, Br^-, I^-, CN^-, and SCN^-) on nanocrystalline titanium dioxide electrodes, *J. Am. Chem. Soc.*, Vol. 115, pp. 6382-6390.

Nazeeruddin, M.K.; Klein, C.; Grätzel, M.; Zuppiroli, L. & Berner, D. (2009). Molecular engineering of iridium complexes and their application in OLED. In: *Highly Efficient OLED with Phosphorescent Materials*. Yersin, H. ed.Wiley-VCH Verlag GmbH & Co. KGaA, Weinheim.

Nazeeruddin, M. K.; Péchy, P.; Renouard, T.; Zakeeruddin, S. M.; Humphry-Baker, R.; Comte, P.; Liska, P.; Cevey, L.; Costa, E.; Shklover, V.; Spiccia, L.; Deacon, G. B.; Bignozzi, C. A. & Grätzel, M. (2001). Engineering of Efficient Panchromatic Sensitizers for Nanocrystalline TiO_2-Based Solar Cells, *J. Am. Chem. Soc.* Vol. 123, pp. 1613-1624.

Ning, Z.; Chen, Z.; Zhang, Q.; Yan, Y.; Qian, S.; Cao, Y. & Tian, H. (2007). *Aggregation-induced emission (AIE)-active starburst triarylamine fluorophores as potential non-doped red emitter for organic light-emitting diodes and Cl_2 gas chemodosimeter. Adv. Funct. Mater.* Vol. 17, pp. 3799-3805.

Ning, Z.; Fu, Y. & Tian, H. (2010) Improvement of dye-sensitized solar cells: what we know and what we need to know. *Energy Environ. Sci.*, Vol. 3, pp. 1170-1181.

Ning, Z. & Tian, H. (2009) Triarylamine: a promising core unit for efficient photovoltaic materials, *Chem. Commun.*, Vol. 37, pp. 5483-5495.

Ning, Z.; Zhang, Q.; Wu, W. & Tian, H. (2009a) Novel iridium complex with carboxyl pyridyl ligand for dye-sensitized solar cells: High fluorescence intensity, high electron injection efficiency? *J. Organomet. Chem.*, Vol. 694, pp. 2705-2711.

Ning, Z.; Zhou, Y.; Zhang, Q.; Zhang, J. & Tian, H. (2007a). *Bisindolylmaleimide derivatives as non-doped red organic light-emitting materials. J. Photochem. Photobio. A: Chemistry*, 192, pp. 8-13.

Nozaki, K. (2006). Theoretical study of the triplet state of fac-Ir(ppy)$_3$. *J. Chin. Chemical Society*, Vol. 53, pp. 101-112.

O'Regan, B. C. & Durrant, J. R. (2009) Kinetic and Energetic Paradigms for Dye-Sensitized Solar Cells: Moving from the Ideal to the Real, *Acc. Chem. Res.*, Vol. 42, pp. 1799-1808.

O'Regan, B. & Grätzel, M. (1991) A low-cost, high-efficiency solar cell based on dye-sensitized colloidal TiO_2 films, *Nature*, Vol. 353, pp. 737-740.

O'Regan, B. C.; Walley, K.; Juozapavicius, M.; Anderson, A.; Matar, F.; Ghaddar, T.; Zakeeruddin, S. M.; Klein, C. & Durrant, J. R. (2009a) Structure/Function Relationships in Dyes for Solar Energy Conversion: A Two-Atom Change in Dye Structure and the Mechanism for Its Effect on Cell Voltage, *J. Am. Chem. Soc.*, Vol. 131, pp. 3541-3548.

Pope, M.; Kallmann, H.P. & Maganate, P. (1963). *J. Chem. Physics*, Vol. 38, pp. 2042-2050.

Pope, M. & Swenberg, C.E. (1999). Electronic Processes in Organic Crystals and Polymers. *Oxford University Press*, Oxford.

Rausch, A.F.; Thompson, M.E. & Yersin, H. (2009). Blue light emitting Ir(III) compounds for OLEDs - new insights into ancillary ligand effects on the emitting triplet state. *J. Phys. Chem. A.*, Vol. 113(20) pp. 5927-5932.

Rausch, A.F.; Homeier, H.H. & Yersin, H. (2010). Organometalic Pt(II) and Ir(III) triplet emitters for OLED applications. *Topics Orgnometal Chemistry*, Vol. 29, pp. 193-235.

Reynal, A.; Forneli, A.; Martinez-Ferrero, E.; Sánchez-Díaz, A.; Vidal-Ferran, A.; O'Regan, B. C. & Palomares, E. (2008) Interfacial Charge Recombination Between e^-–TiO_2 and the I^-/I^{3-} Electrolyte in Ruthenium Heteroleptic Complexes: Dye Molecular Structure–Open Circuit Voltage Relationship, *J. Am. Chem. Soc.* Vol. 130, pp. 13558-13567.

Salikhov, K.M.; Molin, Y.N.; Sagdeev, R.A. & Buchachenko, A.L. (1984). *Spin Polarization and Magnetic Effects in Radical Reactions*, Elsevier, Amsterdam.

Schmidt-Mende, L.; Kroeze, J. E.; Durrant, J. R.; Nazeeruddin, Md. K. & Grätzel, M. (2005) Effect of Hydrocarbon Chain Length of Amphiphilic Ruthenium Dyes on Solid-State Dye-Sensitized Photovoltaics, *Nano. Lett.* Vol. 5, pp. 1315-1320.

Serebrennikov, Y.A. & Minaev, B.F. (1987). Magnetic field effects due to spin-orbit coupling in transient intermediates. *Chemical Physics*, Vol. 114, pp. 359-367.

Snaith, H. J.; Moule, A. J.; Klein, C.; Meerholz, K.; Friend, R. H. & Grätzel, M. (2007) Efficiency Enhancements in Solid-State Hybrid Solar Cells via Reduced Charge Recombination and Increased Light Capture, *Nano. Lett.* Vol. 7, pp. 3372-3376.

Takizawa S.Y., Nishida J., Tsuzuki T., Tokito S. & Yamashita Y. (2007). Phosphorescent iridium complexes based on 2-phenylimidazo[1,2-a]pyridine ligands: tuning of emission color toward the blue region and application to polymer light-emitting devices. *Inorganic Chemistry*, Vol. 46(10) pp. 4308-4319.

Tan, W.; Zhang, Q.; Zhang J. & Tian, H. (2009). Near-Infrared Photochromic Diarylethene Iridium (III) Complex. *Org. Lett.*, Vol. 11, pp. 161–164.

Tang, C.W. & VanSylke, S.A. (1987). Organic electroluminescent diodes. *Applied Phys. Letters*, Vol. 51, No. 11, pp. 913-915.

Vahtras, O.; Lobods, O.; Minaev, B.: Ågren, H. & Ruud, K. (2002). Ab initio calculations of zero-field splitting parameters. *Chemical Physics*, Vol. 279, pp. 133-142.

Volpi G.; Garino C.; Salassa L.; Fiedler J.; Hardcastle K.I.; Gobetto R. & Nervi C. (2009). Cationic heteroleptic cyclometalated iridium complexes with 1-pyridylimidazo[1,5-alpha]pyridine ligands: exploitation of an efficient intersystem crossing. *Chem. Eur. J.* Vol. 15, pp. 6415-6427.

Wong, W.Y. (2008). Metallopolyyne polymers as new functional materials for photovoltaic and solar cell applications. *Macromolecular Chemistry and Physics*, Vol. 209, pp. 14-24.

Wu, Q.-X.; Shi, L.-L.; Zhao, S.-S.; Wu, S.-X.; Liao, Y & Su, Z.-H. (2010). Theoretical studies on photophysical properties of the 2-phenylpyridine iridium (III) complex and its derivatives. *Chem. Journal Chinese Universities*, Vol. 31, pp. 777-781.

Xie, J.; Ning, Z. & Tian, H. (2005). A soluble 5-carbazolium-8-hydroxyquinoline Al(III) complex as a dipolar luminescent material. *Tetrahedron Letters*, Vol. 46, pp. 8559-8562.

Xu, B. & Yan, B. (2007). Photophysical properties of novel lanthanide (Tb^{3+}, Dy^{3+}, Eu^{3+}) complexes with long chain para-carboxyphenol ester p-L-benzoate (L=dodecanoyloxy, myristoyloxy, palmitoyloxy and stearoyloxy). *Spectrochim. Acta A: Mol. Biomol. Spectrosc.* Vol. 66(2) pp. 236-242.

Yang L.; Okuda F.; Kobayashi K.; Nozaki K.; Tanabe Y.; Ishii Y. & Haga M.A. (2008). Syntheses and phosphorescent properties of blue emissive iridium complexes with tridentate pyrazolyl ligands. *Inorg. Chem.* Vol. 47(16) p. 7154-7165.

Yersin, H. & Finkenzeller, W.J. (2008). Triplet emitters for OLED: Basic properties. In: *Highly Efficient OLED with Phosphorescent Materials.* Yersin, H. ed.Wiley-VCH Verlag GmbH & Co. KGaA, Weinheim.

Zapata F.; Caballero A. & Espinosa A. (2009). A redox-fluorescent molecular switch based on a heterobimetallic Ir(III) complex with a ferrocenyl azaheterocycle as ancillary ligand. *Dalton Transactions*, pp. 3900-3902.

Zeng, X.; Tavasli, M. & Perepichka, I.F. (2008). Cationic bis-cyclometallated iridium(III) phenanthroline complexes with pendant fluorenyl substituents: synthesis, redox, photophysical properties and light-emitting cells. *Chemistry – A European Journal*, Vol. 14, pp. 933-943.

Ågren, H.; Vahtras, O. & Minaev, B. (1996). Response theory and calculations of spin-orbit coupling phenomena in molecules. *Advances Quantum Chemistry*, Vol. 27, pp. 71-162.

4

High Efficiency Red Phosphorescent Organic Light-Emitting Diodes with Simple Structure

Ramchandra Pode[1] and Jang Hyuk Kwon[2]
[1]Department of Physics
[2]Department of Information Display
Kyung Hee University
Korea

1. Introduction

After the first report of electroluminescence in anthracene organic materials in monolayer devices in 1963 by Pope et al. (Pope et al., 1963) and by Helfrich and Schneider in 1965 (Helfrich & Schneider, 1965), this phenomenon remained of pure academic interest for the next two decades owing to the difficulty of growing large-size single crystals and the requirement of a very high voltage (~ 1000 V) to produce the luminance. The evolution of OLED devices is summarized in Fig. 1. Tang and his group demonstrated that the poor performance of the monolayer early device was dramatically improved in two layers device by the addition of a hole transport layer (HTL) with the thin amorphous film stacking in the device structure (VanSlyke & Tang, 1985; Tang et al., 1988). Organic electroluminescent devices having improved power conversion efficiencies by doping the emitting layer were also realized around the same time by the Kodak group. Subsequently, heterostructure configurations to improve the device performance were implemented by inserting several layers like buffer layer between anode and hole transport layer (HTL) (VanSlyke et al., 1996; Shirota et al., 1994; Deng et al., 1999) electron transport layer (ETL), hole blocking layer (HBL) (Adamovich et al., 2003) or interlayer between cathode and ETL (Hung et al., 1997; Kido and Lizumi, 1998) in the device structure. Such multilayer device structure often enhances the drive voltages of OLEDs. Usually, the operating voltage for higher brightness was much higher than the thermodynamic limit which is 2.4 eV for a green device. Chemical doping with either electron donors (for electron transport materials) or electron acceptors (for hole transport materials) can significantly reduce the voltage drop across these films. These devices with either HTL or ETL doped layer show improved performance; but the operating voltages were still rather higher than the thermodynamic limit. Subsequently, Leo and his group proposed the concept of p-type doped HTL and n-type doped ETL (J. Huang et al., 2002). These p-i-n structure devices show high luminance and efficiency at extremely low operating voltages. Indeed all these devices have multilayer structure with high current- and power-efficiencies, but thin emitting layer. Nevertheless, narrow thickness of emitting layer in p-i-n OLEDs and complex design architecture of phosphorescent OLEDs are not desirable from the manufacturing perspective.

In recent years, white phosphorescent OLEDs (PHOLEDs) have received a great deal of attention owing to their potential use in high performance and brightness displays, solid state lighting, and back lighting for Liquid Crystal Displays. White emission can be achieved by mixing three primary colors (red, green, and blue) (D'Andrade et al., 2004; Holmes et al. 2003) or two complementary colors from different emitters (Li et al., 2003; J. Liu et al., 2006; Al Attar et al., 2005). Issues of undesired chromaticity as well as poor batch-to-batch reproducibility resulting in low image quality displays in three colors mixing white OLEDs, are minimized in two colors mixing involving an orange emitter complemented with a blue emitter to produce a white light using a combination of fluorescent/phosphorescent or phosphorescent/phosphorescent emitters in doped OLEDs. Consequently, the demand for the efficient true red bright color for multiple color display and lighting purposes has been significantly enhanced. Indeed, interest in employing red emitters in combination with blue emitters to achieve a white light emission with the simpler OLED architecture is spurred in recent days (Li et al., 2003; J. Liu et al., 2006; Al Attar et al., 2005; Seo et al., 2007; Ho et al., 2008a, 2008b; Chen et al., 2008; Shoustikov et al., 1997).

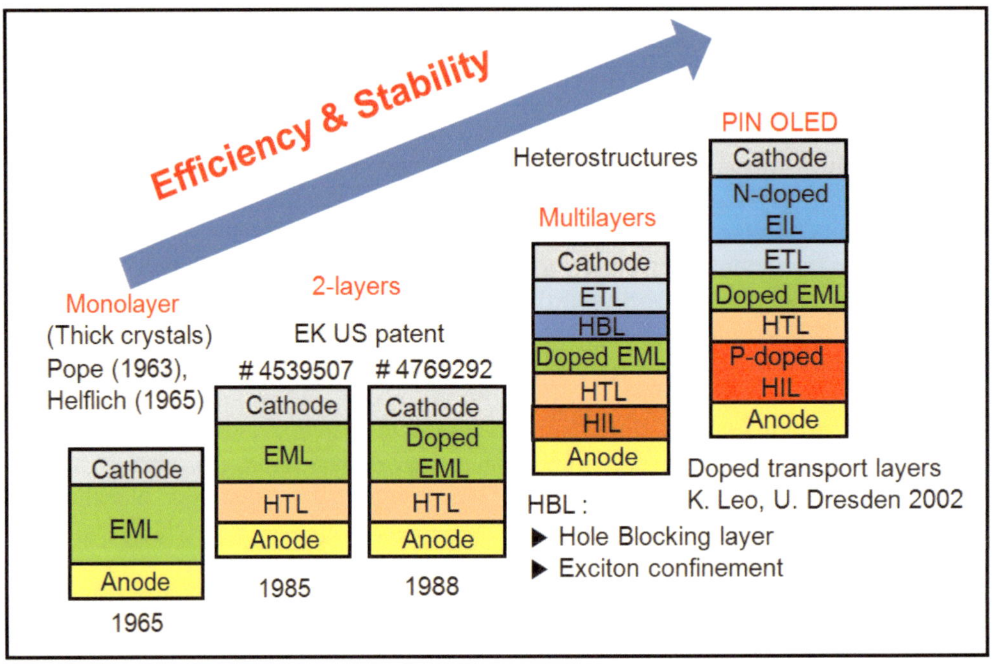

Fig. 1. Evolution of OLED devices (HIL: hole injection layer, HTL: hole transport layer, EML: emissive layer, HBL: hole blocking layer, ETL: electron transport layer)

In this chapter, we discuss efficient red phosphorescent organic light-emitting diodes implemented using multiple quantum well structure, two layers, single layer structures, and ideal host and guest system configurations. The importance of the topic is discussed in this section. The current status of phosphorescent red OLEDs, multiple quantum well, two layers, and single layer configurations for red PHOLEDs are discussed in sections 2, 3, 4, and 5, respectively. Ideal host and guest system for the optimum performance of the red PHOLEDs is presented in section 6. Finally, the conclusion of the present study is illustrated in the section 7 of this chapter.

2. Phosphorescent OLED devices

In recent years, phosphorescent organic light-emitting devices (PHOLEDs) are acquiring the mainstream position in the field of organic displays owing to their potential use in high brightness applications. Schematic of phosphorescence OLEDs and emission mechanism are displaced in Fig. 2. An upper limit on the external quantum efficiency of 5 % in fluorescent small molecule organic devices has been overcome in PHOLEDs by harvesting the singlet and triplet excitons to emission of photons (Baldo et al., 1998; Adachi et al., 2001a). A PHOLED with an internal quantum efficiency of nearly 100% has been demonstrated due to the harvest of both singlet and triplet excitons, leading to devices with high efficiencies (Adachi et al., 2001b; Ikai et al., 2001; Fukase and Kido, 2002; Williams et al., 2007). To achieve the high quantum efficiency in phosphorescent OLEDs, the excited energy of the phosphorescent emitter has to be confined within the emitter itself using wide-energy-gap host materials and carrier-transporting materials, which have higher triplet excited energy levels than that of the emitter and multilayer architecture comprising electron/hole injection and transport layers as shown in Fig. 2. Such multilayer structure often enhances the drive voltages of PHOLEDs. The turn-on voltage of conventional PHOLEDs is relatively high about 1 ~ 2 V compared to that of fluorescent OLEDs as the device designed has multilayer structures for good charge balance and excitons confinement within an emitting material layer (EML), limiting their use in display industries (Wakimoto et al., 1997; Endo et al., 2002).

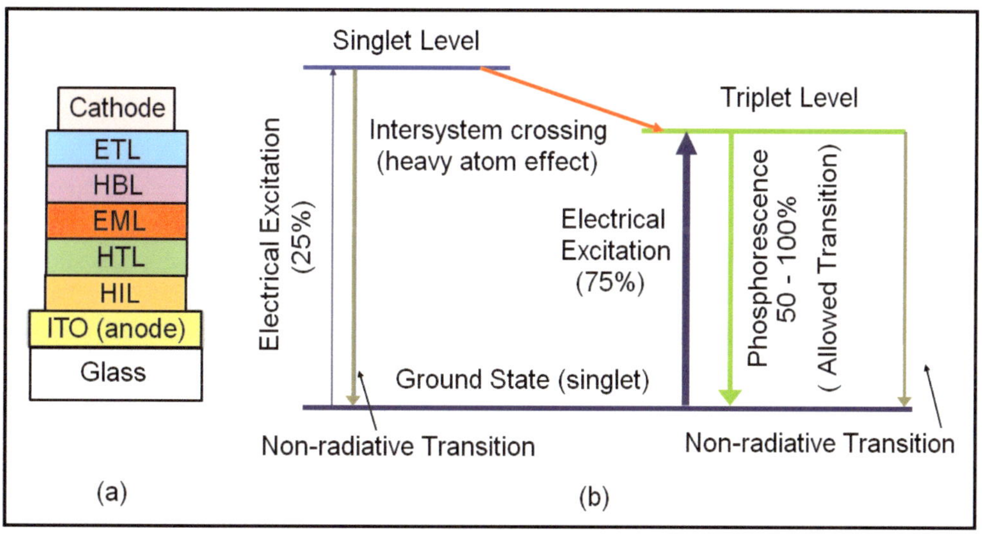

Fig. 2. (a) Schematic of small molecule based phosphorescence OLED, (b) Phosphorescence emission mechanism in phosphorescent OLEDs.

Usually, wide energy gap 4,4′-bis(N-carbazolyl)-1,1′-biphenyl (CBP) is used as a host material for red (~ 2.0 eV) or green (~ 2.3 - 2.4 eV) phosphorescent guests. Iridium (III) and platinum (II) phosphorescent emitters are widely used as triplet dopants molecule. Various red emitting Ir(III) phosphorescent complexes are summarized in Table 1 (Lamansky et al., 2001; Tsuboyama et al., 2003; Duan et al., 2003; H.-K. Kim et al., 2007; Ohmori et al., 2007; J. Huang et al., 2007; Tsuzuki and Tokito, 2008; Mi et al., 2009; T.-C. Lee et al., 2009; Pode et al., 2010; K.-K. Kim et al., 2010; Tsujimoto et al., 2010). The wide band gap host and narrow band gap (E_g)

guest red light emitting system has a significant difference in HOMO (highest occupied molecular orbital) and/or LUMO (lowest unoccupied molecular orbital) levels between the guest and host materials. Thus, the guest molecules are thought to act as *deep traps for electrons and holes in the emitting layer*, causing an increase in the drive voltage of the PHOLED. Further, the dopant concentration in such a host-guest system is usually as high as about 6 ~ 10 percent by weight (wt%) because injected charges move through dopant molecules in the emitting layer. Therefore, self-quenching or triplet-triplet annihilation by dopant molecules is an inevitable problem in host-guest systems with high doping concentrations. Table 2 shows the material performance of red emitting small molecule and polymer PHOLEDs. Table 3 illustrates the suppliers of various materials used in PHOLEDs fabrication.

Sr. No.	Ir complex	Soluble in	Emission wavelength (nm)	Ref.
1)	$Ir(btp)_2(acac)$	2-methyltetrahydrofur-an	612 (PL)	Lamansky et al., 2001
2)	$Ir(piq)_3$	toluene	620	Tsuboyama et al., 2003
3)	$Ir(DBQ)_2(acac)$ $Ir(MDQ)_2(acac)$	CH_2Cl_2	618 (PL) 608 (PL)	Duan et al., 2003
4)	$Ir(piq)_3$	1,2-dichlorobenzene	620	H.-K. Kim et al., 2007
5)	$Ir(piq)_3$	1,2-dichlorobenzene	630	Ohmori et al., 2007
6)	$Ir(C8piq)_3$ $Ir(4F5mpiq)_3$ [F = Fluorine, M = methyl	p-xylene	621 608	J. Huang et al., 2007
7)	$Ir(C4\text{-}piq)_3$	1,2-dichlorobenzene	619 or 617	Tsuzuki and Tokito, 2008
8)	$Ir(BPPa)_3$	CH_2Cl_2	625 (PL)	Mi et al., 2009
9)	$(piq)_2Ir(PO)$ $(nazo)_2Ir(PO)$ $(piq)Ir(PO)_2$ $(nazo)Ir(PO)_2$	CH_2Cl_2	652 (PL) 657 (PL) 591, 620 (PL) 690 (PL)	T.-C. Lee et al., 2009
10	i) $(Et\text{-}Cvz\text{-}PhQ)_2Ir(pic)$ ii) $(EO\text{-}Cvz\text{-}PhQ)_2Ir(picN\text{-}O)$ iii) $(EO\text{-}Cvz\text{-}PhQ)_2Ir(pic)$	1,2-dichlorobenzene	600	Pode et al., 2010
11	$(Ir(phq)_2acac)$ $Ir(piq)_2acac$	1,2-dichlorobenzene	596 and 597	K.-K. Kim et al., 2010
12	$Ir(dbfiq)_2(bdbp)$	Toluene Device	640 (PL) 636 to 642	Tsujimoto et al., 2010

PL : Photoluminescence

Table 1. Red emitting phosphorescent Ir complexes

Device	Lifetime (t50) (h)	Efficacy (cd/A)	Source
Small molecule (Ph) @1000 cd/m^2	120 - 500K	22 - 28	Universal Display
Polymer @1000 cd/m^2	200 - 350K	11 - 31	CDT

Table 2. Red materials performance, 2009

Light Emitting Hosts and Dopants	Injectors/Transporters
• Cambridge Display Technology - Polymers • DuPont - Solution Based Phosphorescent small Molecule • Idemitsu Kosan - Fluorescent and Phosphorescent Small Molecule • Merck - Polymers, Small Molecule • Universal Display - Phosphorescent Small Molecule • Dow Chemical - Fluorescent and Phosphorescent Small Molecule • Sunfine Chem - Fluorescent and Phosphorescent Small Molecule • LG Chemical - Fluorescent Small Molecule	• DS Himetal Dow Chemical • H.C. Starck Group • LG Chemical • Cheil Industries Inc. • Toray • Merck • Nippon Steel Chemical Co., Ltd. • Nissan Chemical Industries • Novaled - P/N Doping • Plextronics • BASF

Table 3. Organic Materials Suppliers

Although PHOLEDs are becoming increasingly important for high brightness displays and lighting applications, there are several issues which need to be addressed sooner or later such as:

- Complex architecture (Multilayer Structure)
- High driving voltage
- Low power efficiency
- Interfacial barrier and charge built-up at interfaces
- Poor performance at driving current densities exceeding 1 mA/cm^2
- Doping concentration about 6 ~ 10 wt%
- Cost competitiveness.

Earlier, Kawamura et al. had reported that the phosphorescence photoluminescence quantum efficiency of $Ir(ppy)_3$ could be decreased by ~5% with an increasing in doping concentration from 2 to 6% (Kawamura et al., 2006). Consequently, the selection of suitable host candidates is a critical issue in fabricating high efficiency PHOLEDs. More recently in order to address device performance and manufacturing constraints, an ideal host-guest system to produce a high efficiency phosphorescent device using a narrow band gap fluorescent host to prevent the hole/or electron trapping has been presented (Jeon et al., 2008a, 2008b; Jeon et al., 2009; Pode et al., 2009). A class of narrow band gap fluorescent material utilizing beryllium complexes as host and ETL for efficient red phosphorescent devices has been proposed. Characteristics of narrow band-gap phosphorescent hosts are: (1) *Small energy band gap,* (2) *Small energy gap between singlet state and triplet state, and* (3) *Good electron transport characteristic.* Simple structure red PHOLED, using narrow band gap fluorescent host materials has demonstrated a *high device performance* and *low manufacturing cost.*

3. Multiple quantum well structure

3.1 Introduction

Organic light emitting diodes (OLEDs) have attracted considerable attention because of their potential applicability to flat-panel displays (FPDs) (Sheats et al., 1996; Shen et al., 1997; Friend et al., 1999), backlighting, and candidates for the next generation lighting (Destruel et al., 1999; D'Andrade and Forrest, 2004), owing to wide viewing angle, low driving voltage, thin, light-weight, and possibly also flexible displays. Indispensable requirement for these applications is the high efficiency of OLEDs devices. In order to achieve the high efficiency in OLEDs, various approaches such as use of highly efficient (high luminescence quantum efficiency) organic materials, insertion of the excition blocking layer and/or hole and electron blocking layers, and optimization of the doping concentration of OLEDs to reduce self-quenching have been reported (Baldo et al., 1998; Bulovic et al., 1998; Baldo et al., 1999). Among these approaches, especially exciton confinement approach by introducing a carrier and/or exciton blocking layer(s) is the most effective and mainly used until now.

Quantum confinement approach using a multiple quantum well (MQW) structure or multi-quantum barrier is widely used in inorganic LED as it leads to a higher efficiency compared to the double hetero- structure or single quantum well (QW) structure. While in OLEDs, only few reports about the MQW structure with good carrier confinement ability were presented till to date. Qiu et al. (Qiu et al., 2002a; 2002b) and Huang et al. (J. Huang et al., 2000) have reported the organic MQW structure by using copper phthalocyanine (CuPc) and N,N'-bis(1-naphthyl)-N,N'-diphenyl-1,1'-biphenyl-4,4'-diamine (NPB) or rubrene. In these articles, the MQW effect has been reported in the fluorescent devices, wherein real device efficiency is not so high besides the poor emission color stability. Recently, the triplet quantum well structure has been reported by Kim et al. using a 4,4'-bis(N-carbazolyl)-1,10-biphenyl (CBP) and PH1 host (S. H. Kim et al., 2007). Since $Ir(ppy)_3$ was doped in all quantum well layers, charge carriers couldn't be effectively confined in this device as carriers move via dopant molecules. Consequently, stable high efficiency results in such a MQW structure couldn't be realized.

In this section, we report the real MQW device structure having various triplet quantum well devices from a single to five quantum wells. Owing to confinement of the triplet energy at the emitting layers in the fabricated MQW device, the highest phosphorescent efficiency is obtained among reported tris(1-phenylisoquinoline)iridium ($Ir(piq)_3$) dopant OLEDs (H. Kim et al., 2008) with a very good color emission stability. The MQW structure is realized using a wide band-gap hole and electron transporting layers, narrow band-gap host and dopant materials, and charge control layers (CCL). Bis(10-hydroxybenzo[h] quinolinato)beryllium complex ($Bebq_2$) and bis[2-(2-hydroxyphenyl)-pyridine] beryllium ($Bepp_2$) are used as a narrow band-gap host material and a CCL material, respectively.

3.2 Experimental

Figure 3(a) shows the configuration of fabricated red PHOLEDs, having a MQW structure. Here, n consists of [$Bebq_2$:$Ir(piq)_3$/ CCL] (red electroluminescence (R-EL) unit) varying from

1 to 5. The 40 nm thick 4, 4′, 4″-tri(N-carbazolyl)triphenylamine (TCTA) hole transport layer (HTL) doped with WO_3 (doping concentration 30 %) is deposited on an indium tin oxide (ITO) coated glass substrate. To prevent the non-radiative quenching of triplet excitons generated at the heavily doped HTL, an electron blocking buffer layer of 12 nm thick TCTA was deposited. Subsequently, emissive layers (EMLs) with quantum-well structures were deposited. The emission layer structures of PHOLEDs were increased by adding the R-EL unit. In order to confine and control a hole and electron in the EML, the CCL layer with a thickness of 5 nm was deposited inside of EML. Later, $Bepp_2$ hole and exciton blocking buffer layer was deposited and followed by a 10 % Cs_2CO_3-doped $Bepp_2$ electron transport layer (ETL). The triplet energies of TCTA, $Bepp_2$, $Bebq_2$, and $Ir(piq)_3$ are 2.7, 2.7, 2.2, and 2.0 eV, respectively (Jeon et al., 2009; S. Y. Kim et al., 2009; Tsuboi et al., 2009). As triplet energies of charge carrier layers and CCL are higher than the host molecule (Fig. 3(b)), all triplet energies are confined in the emitting layers. Finally, Al cathode was deposited in another deposition chamber without breaking the vacuum. Deposition rate of Al was 5~10 Å/sec. The devices were fabricated on ITO coated glass with a sheet resistance of 20 Ω/ϒ. The substrates were cleaned with acetone and isopropyl alcohol sequentially, rinsed in de-ionized water, and then treated in UV-ozone immediately before loading into the high vacuum chamber (~ 2×10^{-7} Torr). The current density-voltage (J–V) and luminance–voltage (L–V) data of red PHOLEDs were measured by Keithley 2635 A and Minolta CS-1000A, respectively. The red PHOLED emitting area was 2 mm^2 for all the samples studied in the present work.

3.3 Results & discussion

In order to select the best CCL, we fabricated red PHOLEDs (n=2) with different CCL materials (CBP; device B, TCTA; device C, $Bepp_2$; device D) at the fixed CCL thickness of 5 nm. Device A was made without any CCL. In J-V-L results (not reproduced here), all three devices were measured until 10,000 cd/m^2 brightness value. The driving voltages (at 1000 cd/m^2) of these devices A, B, C and D are 3.8, 5.6, 6.0 and 4.2 V, respectively. As expected, the driving voltage increases by inserting the CCL. The external quantum efficiency (EQE) characteristics are shown in Fig. 4(a). At a given constant luminance of 1000 cd/m^2, the EQE values are 10.8, 5.1, 5.2, and 13.8 % for the devices A, B, C, and D, respectively. The EQE of the device D with $Bepp_2$ CCL is significantly higher than those of devices A~C. The HOMO energy levels of $Ir(piq)_3$, $Bebq_2$, CBP, TCTA and $Bepp_2$ were at 5.1 eV, 5.5 eV, 5.9 eV, 5.8 eV, and 5.7 eV, respectively. While the LUMO energy levels of $Ir(piq)_3$, $Bebq_2$, CBP, TCTA and $Bepp_2$ were at 3.1 eV, 2.8 eV, 2.6 eV, 2.4 eV, and 2.6 eV respectively. With the TCTA and CBP CCL layers devices, the deep HOMO and high LUMO levels of TCTA and CBP block the movement of holes and electrons at the $Bebq_2$:$Ir(piq)_3$/CCL interface. Therefore, holes and electrons cannot be easily transported through the CCL, resulting in the rise of driving voltage and efficiency decrease. However, the suitable HOMO and LUMO energy levels in $Bepp_2$ CCL can control the carrier movement at ease. As a result, $Bepp_2$ CCL partially confines holes and electrons at the first EML and some of holes and electrons arrive at the second EML after transporting through the $Bepp_2$ CCL. The CCL thickness is varied to optimize device characteristics from 3~10nm. The 5nm thickness of $Bepp_2$ CCL shows reasonable efficiency and voltage increase values.

In our double QW devices, hole barriers by CCLs are probably a dominant factor to control the current flow as hole barriers between HOMO levels of dopant and CCL are relatively high compared with electron barriers. Usually electrons easily overcome its barriers.

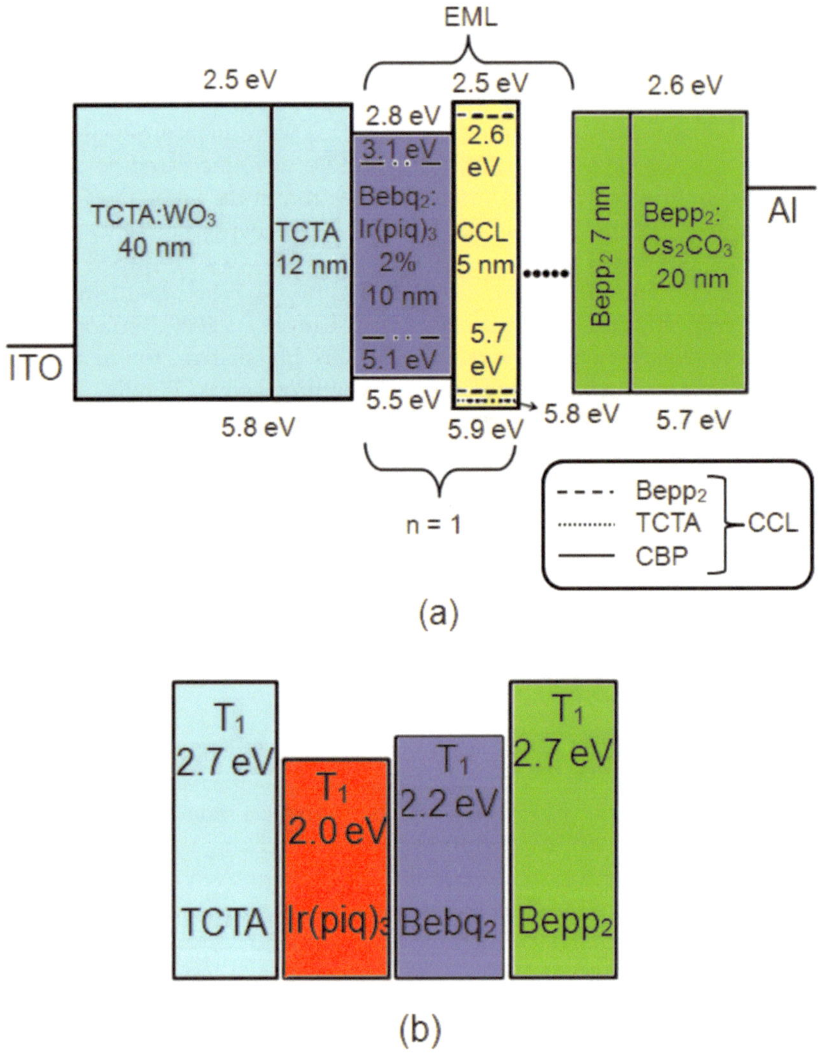

Fig. 3. (a) Energy band diagrams of fabricated red PHOLEDs with multiple quantum well structures. (b)Triplet energies of materials used in the present study.

In order to estimate the carrier confinement percentage in each EMLs with different CCL in double QW structure, the current density values are compared with the hole barriers obtained between HOMO levels of each CCL and $Ir(piq)_3$. According to the thermionic emission barrier model (Hong et al., 2005), ln(J) has a good linear relationship with the potential barrier (Φ).

Figure 4(b) shows a good agreement between Δln(J) and $\Delta\Phi$ at 5V, indicating that the hole barrier is the main factor to determine the current flow in our devices. Here Δln(J) was calculated from the current density differences of single QW and double QW devices at 5 V and $\Delta\Phi$ is the HOMO energy levels difference between the dopant and CCL. Almost similar behaviors are noticed at various voltages. Due to lower hole barrier with $Bepp_2$ compared with other CCL materials, the current flow is much easier with no hindrance. Hole carriers can over-flow in the $Bepp_2$ CCL, creating more excitons in the double QW structure. Therefore, the device D with the $Bepp_2$ CCL improves the recombination efficiency of the electron-hole pairs.

Indeed, the efficiency with two quantum well device structure with $Bepp_2$ CCL is significantly improved. Further to investigate the influence of the quantum wells on the device performance, if any, we have fabricated PHOLEDs with MQW from 1 to 5 wells.

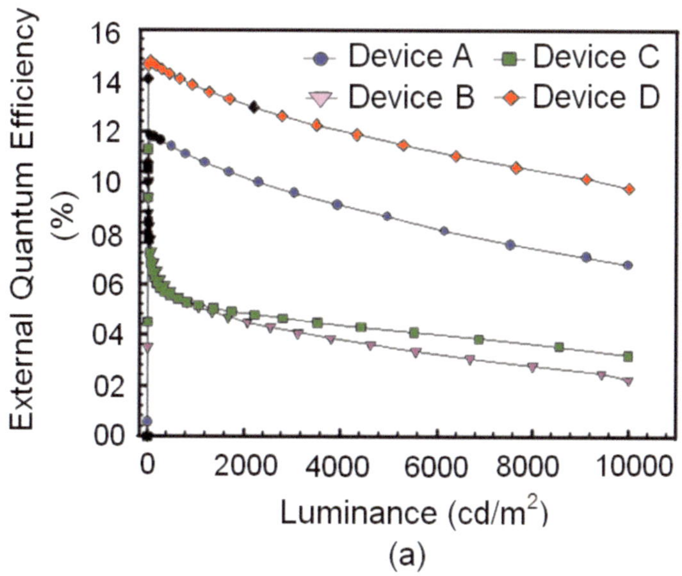

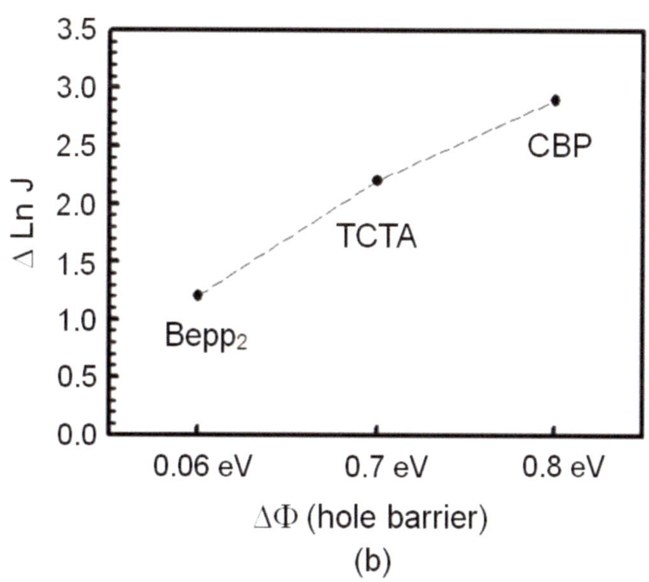

Fig. 4. (a) EQE characteristics of fabricated red PHOLEDs with and without CCL. (b) The current density difference between single quantum well device and double quantum well devices with different CCL

Figure 5 shows the J-V-L characteristics of fabricated red PHOLEDs with the increasing number of R-EL units from 1 to 5. The turn on voltages of MQW red PHOLEDs are 2.4 V for n=1 (device A), 2.5 V for n=2 (device D), 2.6 V for n=3, 2.8 V for n=4, and 3.2 V for n=5, respectively. The driving voltage to reach 1000 cd/m² is 3.8 V for n=1, 4.2 V for n=2, 4.8 V for n=3, 6.0 V for the n=4, and 7.4 V for n=5. The operating voltages in the MQW structure were increased by increasing the number of R-EL units because any addition of QW units offers additional resistance to the conduction of current. From Δln (J) data between single QW and double QW, we have calculated that 29% hole carriers can go the second EML through a $Bepp_2$ CCL.

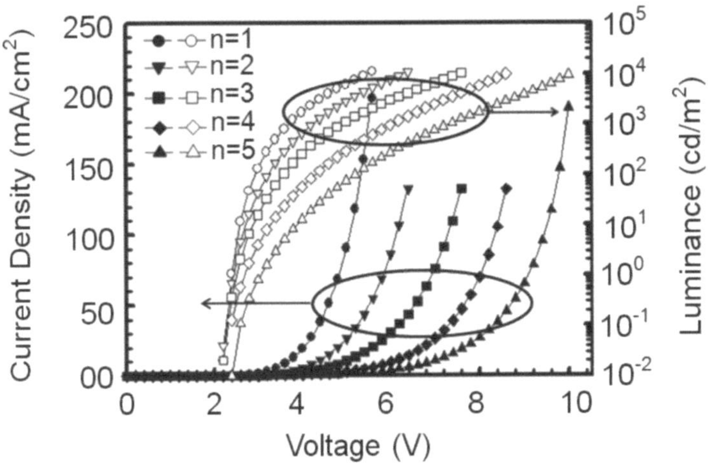

Fig. 5. J-V-L characteristics of fabricated red PHOLEDs with increasing R-EL unit from 1 to 5.

Figure 6 shows the maximum EQE characteristics of fabricated five red PHOLEDs while various electrical parameters of these devices are summarized in Table 4. The maximum EQE characteristics are 11.8 % for n=1, 14.8 % for n=2, 13.6 % for n=3, 12.8 % for n=4, and 8.6 % for n=5, respectively. *The best EL performances are obtained with n=2 among the five red PHOLEDs.* The over-flowing ratio of hole carriers with repeating additional QW and $Bepp_2$ CCL are shown in the inset of Fig. 6. From the J-V characteristics as displayed in Fig. 5, the over-flowing ratio of hole is estimated as [J(n=2)/J(n=1)] x 100% at 5 V (i.e. (90.40 mA/cm² / 26.39 mA/cm²) x 100% = 29%). Only 29% of hole carriers can reach to the second EML through a $Bepp_2$ CCL. The simple calculation results for n=3 and 4 were obtained by assuming 29% hole carrier overflow result for n=2. Real experimental data obtained from the J-V characteristics at 5V well agree with the calculated results, indicating our carrier overflow assumption is reasonable. The excitons can be confined upto 71% in the first QW existing adjacent to the TCTA buffer layer and 21% excitons in to the next second QW. The most excitons can be confined in first and second QWs. Therefore, the best EL performances seem to be obtained with n=2. By increasing the number of quantum wells to n=3 and n=4, the efficiency drop is not significant (over 12%) because electrons can reach to first and second QWs due to the negligible barrier to electron transport. However, the driving voltage is enhanced with increasing the number of QW structures and eventually 5 QW structure does not work properly. In our MQW devices, all devices show excellent color stability with the same CIE coordinate as (0.66, 0.33) as shown Table 4. Our results reveal that the MQW structure improves the external quantum efficiency with no change in the CIE coordinate of red emitting PHOLEDs.

3.4 Conclusions

In summary, the maximum external quantum efficiency of 14.8 % with a two quantum well device structure is obtained, which is the highest value among the reported Ir(piq)$_3$ dopant red phosphorescent OLEDs.

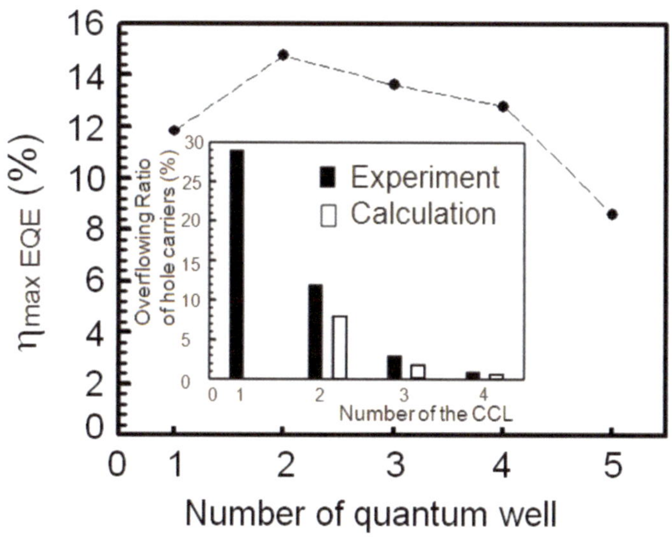

Fig. 6. Maximum EQE characteristics of fabricated five red PHOLEDs with increasing R-EL unit from 1 to 5. Inset: Overflowing ratios of hole carriers with increasing R-EL units.

Parameters	n=1	n=2	n=3	n=4	n=5
Turn on voltage (@ 1cd/m^2)	2.4 V	2.5 V	2.6 V	2.8 V	3.2 V
Operating voltage (@ 1000 cd/m^2)	3.8 V	4.2 V	4.8 V	6.0 V	7.4 V
Maximum current efficiency	9.9 cd/A	12.4 cd/A	11.5 cd/A	0.8 cd/A	7.2 cd/A
Maximum External Quantum Efficiency	11.8 %	14.8 %	13.6 %	12.8 %	8.6 %
CIE (x, y) (@ 1000 cd/m2)	0.66, 0.33	0.66, 0.33	0.66, 0.33	0.66, 0.33	0.66, 0.33

Table 4. Summary of performances of multiple quantum well red PHOLEDs in this study

4. Two layers structure

4.1 Introduction

Performance and efficiencies of red PHOLEDs devices have been improved in recent days, particularly in p-i-n type OLEDs (J. Huang et al., 2002; Pfeiffer et al., 2002). Good charge balance in emitting layers and low barrier to charge carriers injection in p-i-n devices demonstrate a low operating voltage and high efficiency. CBP is the most widely used host material in red and green emitting PHOLEDs (Chin et al., 2005; Tsuzuki, and Tokito , 2007). Other host materials such as 4,4′,4″-tris(N-carbazolyl)-triphenylamine (TCTA), 3-phenyl-4-(1′-naphthyl)-5-phenyl-1,2,4-triazole (TAZ), 1,3,5-tris(N-phenylbenzimidizol-2yl)benzene (TPBI) and aluminum (III) bis(2-methyl-8-quinolinato)-4-phenylphenolate (BAlq) for PHOLEDs are also commercially available and used as a matter of convenience for many guest–host applications (Zhou et al., 2003; Che et al., 2006; J. H. Kim et al., 2003). HOMO , LUMO and triplet energy of these host materials are listed in Table 5. The HOMO and LUMO energy levels of bis(10-hydroxybenzo [h] quinolinato)beryllium complex ($Bebq_2$) are reported at 5.5 and 2.8 eV, respectively (S. W. Liu et al. 2004). High luminance in OLEDs with $Bebq_2$ as an emitter was reported by Hamada et al. (Hamada et al., 1993). Since $Bebq_2$ and beryllium complexes have very good electron transporting characteristics with high electron mobility of ~10^{-4} cm^2/Vs (Y. Liu et al., 2001; Vanslyke et al., 1991; J.-H. Lee et al., 2005) and narrow band gap, $Bebq_2$ can make a suitable candidate for the host of red emitting PHOLEDs (Jeon et al., 2008a, 2008b; Jeon et al., 2009; Pode et al., 2009).

Compounds	HOMO (eV)	LUMO (eV)	Reported Triplet Energy (eV)	Calculated Triplet Energy (eV)
CBP	5.8	2.5	2.6	2.8
TCTA	5.9	2.7	2.8	2.7
TPBI	6.3	2.8	…..	2.8
BAlq	5.9	3.0	2.2	2.6
TAZ	6.6	2.6	….	3.3
$Bebq_2$	5.5	2.8	….	2.5

Table 5. HOMO, LUMO and triplet energy levels of some fluorescent host materials for PHOLEDs

In this section, we report a narrow band gap electron transporting host material, $Bebq_2$, for red light-emitting PHOLEDs. The triplet energy of $Bebq_2$ host was estimated using density functional theory (DFT). Simple bi-layered PHOLEDs, tris(1-phenylisoquinoline)iridium ($Ir(piq)_3$) doped in $Bebq_2$ host, were fabricated and studied.

4.2 Experimental

Beryllium compound has been reported to have a strong fluorescence characteristic. Although long-lived phosphorescence, caused by spin-forbidden decay from the first triplet state (T_1), is a ubiquitous property of organic molecules, no report about the estimation of triplet energy of $Bebq_2$ host and its role on the device performance has been available to date. Therefore to estimate the triplet state energy, the phosphorescent spectrum of $Bebq_2$ was investigated at low temperature. However, no signature of phosphorescent peak in $Bebq_2$ complex is observed at 77 K. It only exhibits a strong fluorescence emission at 466 nm. Therefore, the molecular simulation method was employed to deduce the triplet energy of $Bebq_2$.

The triplet energy state, estimated by molecular modeling and DFT using DMoL3 program (version 4.2), was found to be about 3.0 eV (Park et al., 2008). Usually in phosphorescent host materials, singlet and triplet exchange energy value is about 0.5 eV. However, $Bebq_2$ host shows a very small exchange energy value of 0.2 eV and is a signature of strong electron-electron correlation. Triplet phosphorescent dopants such as $Ir(piq)_3$ and bis(2-phenylquinoline)(acetylacetonate)iridium ($Ir(phq)_2acac$), used in red light-emitting PHOLEDs, have triplet energy states (actual LUMO level) at 2.8 and 3.1 eV, respectively (Chin et al., 2005; T.-H. Kim et al., 2006). This triplet energy of dopant is very close to the triplet energy of the $Bebq_2$ host material, thus facilitating the electron movement in emitting layer.
Furthermore, corroboration of triplet energy state of $Bebq_2$ host and possible energy transfer from the host to dopant were confirmed by fluorescent and phosphorescent quenching experiments using iridium dopants and $Bebq_2$ host in tetrahydrofuran solution. The $Bebq_2$ fluorescence peak is efficiently quenched by $Ir(piq)_3$ dopant, transferring all its singlet energy directly to the dopant triplet state. As a consequence, we conclude that the triplet energy level of $Bebq_2$ is lower than that of the $Ir(piq)_3$ dopant and exchange energy of host material between singlet and triplet must be very small. However, the reported triplet energy value of ($Ir(piq)_3$) in Ref 63 is 2.8 eV which is lower than that of the $Bebq_2$ host (3.0 eV). So, the LUMO energy level (i.e. triplet state) of $Ir(piq)_3$ dopant was confirmed by the optical band-gap and cyclic voltametry measurements and was found to be 3.1 eV. Both host and dopant molecules seem to have almost same value of triplet energy.

4.3 Results & discussion

Figure 7 shows the structures of three red PHOLEDs devices fabricated for the present study. Devices A and B have a conventional multilayer structure containing hole and electron transport and injection, and hole blocking layers with CBP and $Bebq_2$ host materials, respectively. Device A with a CBP host material is used as a control device, while the fabricated device C with $Bebq_2$ host has a simple bilayered structure. Red phosphorescent OLEDs were fabricated as follows:
Devices A & B: ITO / α-NPB (40 nm) / HOST : Dopant (10 wt%, 30 nm) / Balq (5 nm) / Alq_3 (20 nm) / LiF(0.5 nm) / Al(100 nm), and Device C: ITO/α-NPB (40 nm) / HOST : Dopant (10 wt%, 50 nm)/ LiF(0.5 nm) / Al(100 nm)

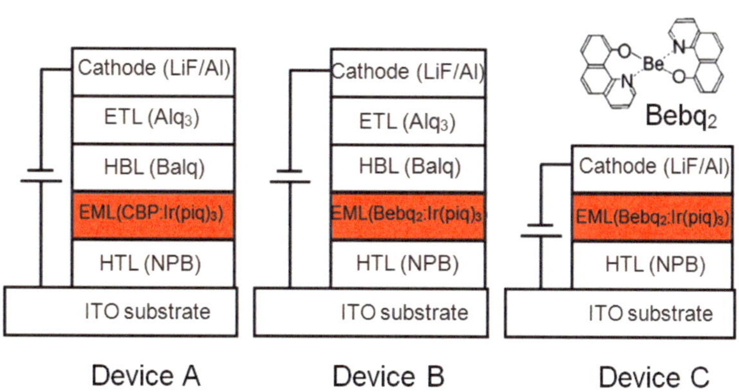

Fig. 7. Structures of fabricated three PHOLEDs: device A - CBP: $Ir(piq)_3$, device B - $Bebq_2$: $Ir(piq)_3$, device C - $Bebq_2$: $Ir(piq)_3$ without HBL and ETL.

Figure 8 shows the I-V-L characteristics of fabricated red phosphorescent devices. At a given constant voltage of 5 V, current density values of 0.82, 2.83, and 18.99 mA/cm^2 in the fabricated devices A, B, and C are noticed as displayed in Fig. 8, respectively. The driving voltage for the device A to reach 1000 cd/m^2 is 8.8 V, 6.8 V for the device B, and 4.5 V for the device C. A low turn-on voltage of 4.5 V in device C with a simple bi-layered structure compared to control device A with CBP host (8.8 V), is observed. The resistance to current conduction in bilayered device C is significantly reduced. As the HOMO energy of $Bebq_2$ host is at 5.5 eV, holes injected from the hole transport layer (HTL) trap directly at the

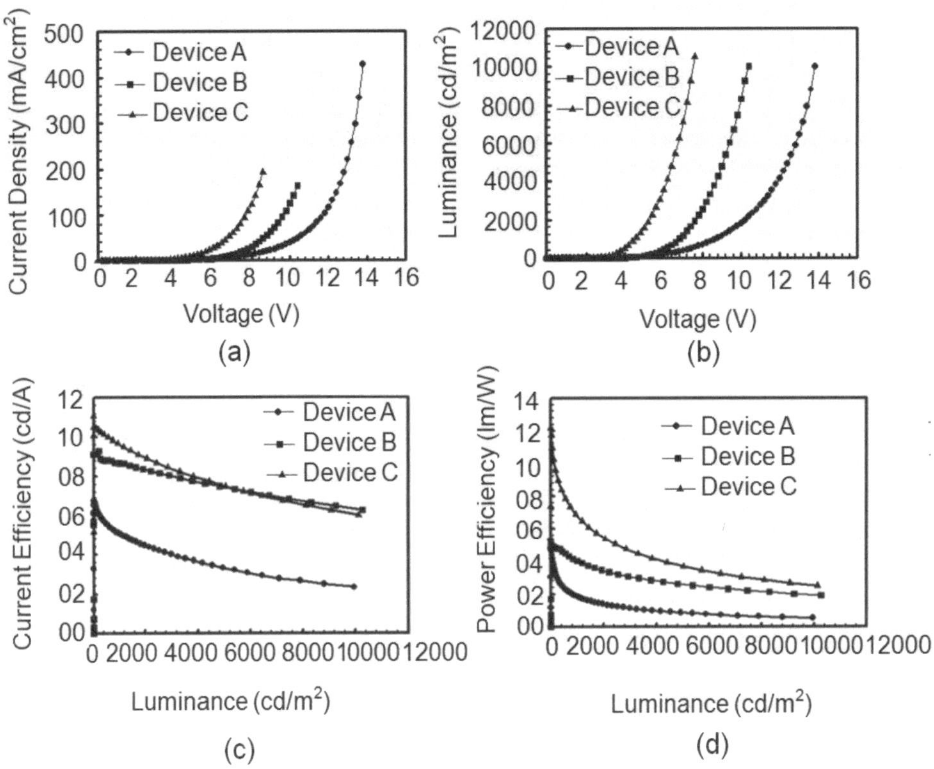

Fig. 8. I-V-L characteristics of fabricated three PHOLEDs ; (a) current-voltage characteristics (b) luminance-voltage characteristics (c) current efficiency-luminance (d) power efficiency luminance.

HOMO level (5.1 eV) of dopant. Barrier to hole injection in the device C is almost negligible. Also, electrons injected from the cathode move freely in the emitting layer as the LUMO (triplet) of dopant and triplet of host are at the same energy and finally captured at the trapped hole sites giving rise to phosphorescent emission. Multilayer structure as displayed in devices A and B introduces heterobarriers to electron and hole injection into emitting layers, thus enhancing the turn-on voltages, although some reduction of driving voltage in device B due to narrow band gap $Bebq_2$ host materials is noticed. Moreover in CBP based PHOLEDs, severe charge trapping at NPB interface has been reported by several researchers [63, 64]. Figure 9 shows the energy band diagram of device C. The current and power efficiency characteristics of fabricated devices are shown in Fig. 8 (c) & (d). At a given constant luminance of 1000 cd/m^2, the current and power efficiencies are 9.66 cd/A and 6.90 lm/W for the device C, 8.67 cd/A and 4.00 lm/W for the device B, and 5.05 cd/A and 1.80 lm/W for the device A, respectively. These values of device C are improved by a factor of 1.9 and 3.8 times compared with those of device A, respectively. In device reliability tests, device A and C show very different behaviors. Device A shows about 120 h lifetime at 1000 nit, while lifetime of 150-160 h is noticed in $Bebq_2$ device. Relatively small initial decrease of brightness value and gradual decay curve is observed in device C, which indicates $Bebq_2$ device reliability is relatively very good. However, material stability of $Bebq_2$ seems not to be good. Figure 10 shows the electroluminescence spectra at a brightness of 1000 cd/m^2 of different fabricated phosphorescent red-emitting devices. Clean red light emission at 632 nm observed in device C indicates the complete energy transfer from a novel narrow band gap $Bebq_2$ host material to $Ir(piq)_3$ dopant. The CIE coordinate of three red devices show the same coordinate as CIE (0.67, 0.33).

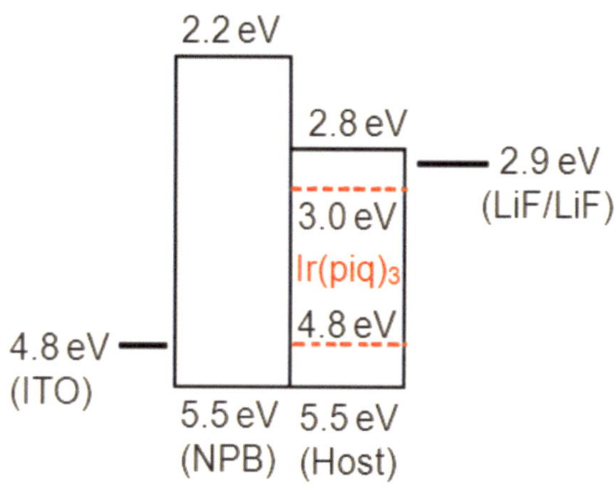

Fig. 9. Energy diagram of red organic bi-layered PHOLEDs with $Bebq_2$:$Ir(piq)_3$ host (Device C).

Anyway, interesting and intriguing results on the performance of bi-layered device C have been obtained. The LUMO level of $Bebq_2$ material (2.8 eV), very close to LUMO values of Balq, Alq_3 and LiF cathode, offers almost no barrier to electron injection between the emitting layer and LiF cathode. Furthermore, excellent electron transporting property of $Bebq_2$ material favors to mobility of electrons which provides a good charge balance in the emitting layer. HOMO levels of $Bebq_2$ host and NPB hole transport layer in the fabricated device C are very close while LUMO energy levels of host and dopant are almost same. Therefore, the emission process in PHOLEDs device C via electron trapping at LUMO and hole trapping at HOMO seems to be minimized, giving to low driving voltage value. In this device C, the emission of red light may be originated from the direct electron capturing from the host and recombining at holes trapped at the HOMO of the dopant in the emitting layer. Indeed, the hole trapping in bilayered device C is not a serious issue. To investigate the influence of recombination zone position on the emission and hole trapping, three PHOLEDs with emitting zone at X = 0, 10, and 20 nm from the HTL/EML interface were fabricated and studied as displayed in Table 6 and Fig. 11. Results reveal excellent emission of red light in all devices, except some contribution to the emission from the $Bebq_2$ host material in devices with X = 10 and 20 nm. These results demonstrate that the emission zone in simple bilayered

PHOLEDs is very broad and hole trapping is not so severe. The EL emission spectra of devices D, E, and F are shown in Fig. 12(a) and CIE coordinates in Fig. 12(b).

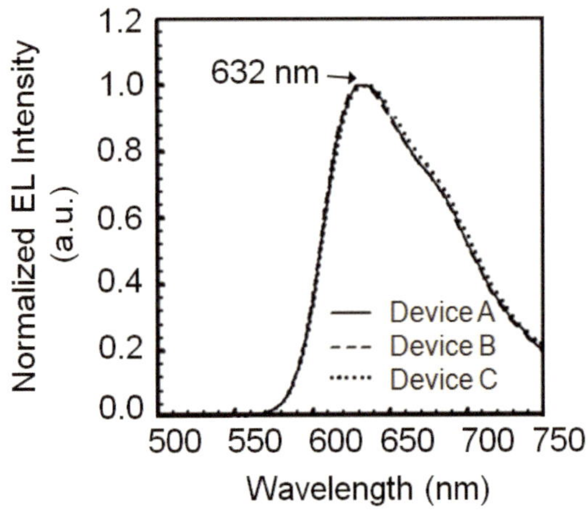

Fig. 10. Normalized electroluminescent spectra of devices A, B, and C at the luminance of 1000 cd/m².

Thickness (Å)	Device D	Device E	Device F
X (nm)	0	10	20
$Bebq_2$:$Ir(piq)_3$	100	100	100
$Bebq_2$	400	300	200

Table 6. Recombination zone position in Device C from the HTL/EML interface

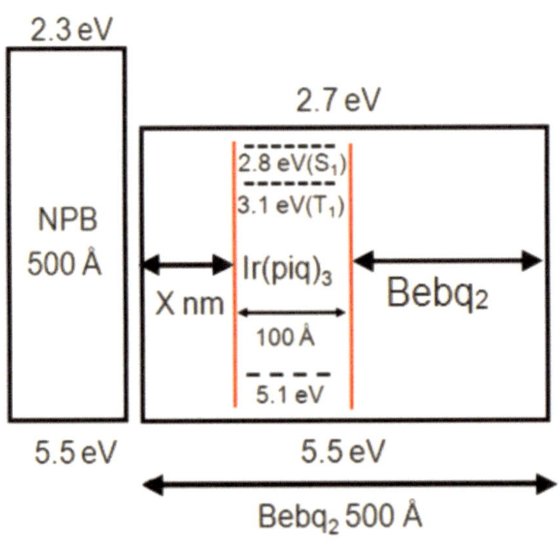

Fig. 11. Recombination zone position in Device C

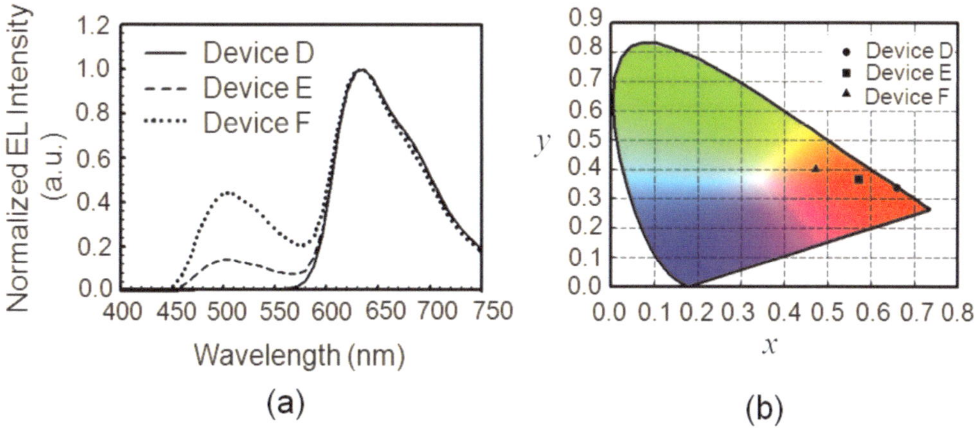

Fig. 12. (a) EL emission spectra, and (b) CIE coordinates of devices D, E, and F.

4.2 Conclusions

A narrow band-gap host material, $Bebq_2$, for red PHOLEDs with a very small exchange energy value of 0.2 eV between singlet and triplet states has been demonstrated. It shows almost no barrier to injection of charge carriers and charge trapping issue in PHOLEDs is minimized. High current and power efficiency values of 9.66 cd/A and 6.90 lm/W in bi-layered simple structure PHOLEDs are obtained, respectively. The operating voltage of bi-layered PHOLEDs at a luminance of 1000 cd/m^2 was 4.5 V. In conclusion, simple bilayerd red emitting device with $Bebq_2$ host could be a promising way to achieve efficient, economical, and ease manufacturing process, important for display and lighting production.

5. Single layer structure

5.1 Introduction

Organic light emitting devices (OLEDs) have made significant stride (Pfeiffer et al., 2002) and the technology has already been commercialized to mobile flat panel display applications. Thermal evaporation technique and complicated fabrication process consisting of multiple layers for charge carriers balancing and exciton confinement (Baldo and Forrest, 2002; Coushi et al., 2004; Tanaka et al., 2007) are employed in highly efficient phosphorescent OLEDs. In order to overcome such complex device architecture, many good approaches are enduring until now. High efficiency devices with pure organic bilayered OLEDs have been reported by several researchers (Jeon et al., 2008b; Pode et al., 2009; Park et al., 2008; Meyer et al., 2007; Z. W. Liu et al., 2009). Furthermore, bilayered devices consisting of an organic single layer with a buffer layer on the electrode have also been reported without any significant improvement of the device performances (Q. Huang et al., 2002; Gao et al., 2003; Wang et al., 2006; Tse et al., 2007). However, truly organic single layered approach is almost rare. To date, only an exclusive article on the red emitting PHOLED single layer device with a tris[1-phenylisoqunolinato-C2,N]iridium (III) ($Ir(piq)_3$) (21 wt%) doped in TPBi (100 nm) with low values of current and power efficiencies under 3.7 cd/A and 3.2 lm/W at 1 cd/m^2 have been reported, respectively (Z. Liu et al., 2009).

In this section, we have presented efficient and simple red PHOLEDs with only single organic layer using thermal evaporation technique. The key to the simplification is the direct injection of holes and electrons into the mixed host materials through electrodes. In conventional OLEDs, usually the Fermi energy gap between cathode (~ 2.9 eV) and surface treated anode (~ 5.1 eV) is about 2.0~2.2 eV which is close to the red light emission energy (1.9 ~2.0 eV). As a consequence, red devices do not at all require any charge injection and transporter layer if the host material has proper HOMO and LUMO energy levels. However, such host materials are very rare. The most suitable option to address such issues is to employ the mixed host system to adequately match the energy levels between emitting host and electrodes. Mixed host system of electron and hole transporting materials to inject electrons and holes from electrodes into the organic layer without any barrier has been studied, respectively and employed for the charge balance. Thus, hole type host materials are required to have HOMO energy levels at 5.1~5.4 eV to match with the Fermi energy of surface treated ITO (5.1 eV). While 2.8~3.0 eV LUMO energy levels of electron transporting host materials are necessary to match the Fermi level of cathode. 4,4′,4″-Tris(N-3-methylphenyl-N-phenyl-amino)triphenylamine(m-MTDATA) and N,N′-diphenyl-N,N′-bis(1,1′-biphenyl)-4,4′-diamine (α-NPB) were used as the hole transporting host materials. Bis(10-hydroxybenzo[h]quinolinato)beryllium ($Bebq_2$) with 2.8 eV LUMO energy was used as the electron transport host material and $Ir(piq)_3$ was employed as a red phosphorescent guest.

5.2 Experimental

m-MTDATA and α-NPB as hole transporting host materials, $Bebq_2$ as an electron transporting host material, and $Ir(piq)_3$ as a red dopant were obtained from Gracel Corporation. Details of the fabrication process have been discussed section 3. The emitting area of PHOLED was 2 mm^2 for all the samples studied in the present work.

5.3 Results & discussion

Figure 13 shows the energy band-diagram of the single layer red PHOLEDs used in the present work. For the evaluation of single layer with different mixed host systems, the following devices were fabricated:

Device A: ITO/m-MTDATA:$Bebq_2$: $Ir(piq)_3$ [1~4 wt%, 100 nm]/LiF (0.5 nm)/Al (100 nm), and Device B: ITO/α-NPB:$Bebq_2$: $Ir(piq)_3$ [1~4 wt%, 100 nm]/LiF (0.5 nm)/Al (100 nm).

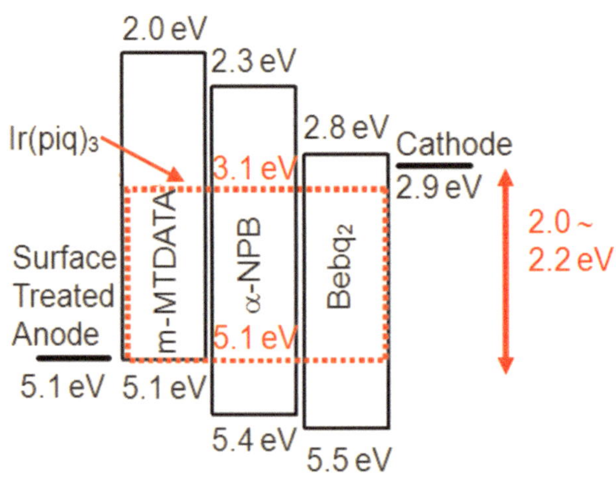

Fig. 13. Energy band-diagram of the single layer red PHOLEDs.

The ratio of the hole and electron transporting hosts was fixed to 1:1. The doping concentrations were varied from 1% to 4% to optimize the device performance. Table 7 shows the performance of red PHOLEDs devices comprising a single emitting layer. The current and power efficiencies values of 7.44 cd/A, and 3.43 lm/W at 1000 cd/m^2 brightness value are reported in 4wt% doped device A, respectively. The driving voltage (to reach 1000 cd/m^2) is 6.9 V. Very similar device performances are obtained in 2 wt% doped device A. The optimum doping condition for Device A seems to be 4 wt% as the highest efficiency is observed at an acceptable brightness value (1000 cd/m^2). Whereas, the driving voltage, current and power efficiencies values of 5.4 V, 9.02 cd/A, and 5.25 lm/W at brightness value of 1000 cd/m^2 are reported in device B with 1 wt% of optimum doping condition, respectively. Maximum current efficiency values for devices A and B were appeared in 4 and 1 wt% of $Ir(piq)_3$ doped mixed hosts, respectively. The color coordinates are (0.66, 0.33) or (0.67, 0.32) for all devices. Even in 1% doped device, a good red emission color is observed.

	Device A			Device B		
	1% Doping	2% Doping	4% Doping	1% Doping	2% Doping	4% Doping
Turn on voltage (@ 1cd/m^2)	2.5 V	2.4 V	2.3 V	2.4 V	2.4 V	2.4V
Operating voltage (@ 1000 cd/m^2)	7.2 V	7.1 V	6.9 V	5.4 V	5.4 V	5.3 V
Maximum current and power efficiency	8.12 cd/A 7.84 lm/W	8.19 cd/A 9.86 lm/W	8.04 cd/A 10.96 lm/W	9.44 cd/A 10.62 lm/W	8.36 cd/A 9.82 lm/W	7.04 cd/A 8.11 lm/W
current and power efficiency (@ 1000 cd/m^2)	7.28 cd/A 3.18 lm/W	7.34 cd/A 3.29 lm/W	7.44 cd/A 3.43 lm/W	9.02 cd/A 5.25 lm/W	8.26 cd/A 4.80 lm/W	7.04 cd/A 4.10 lm/W
CIE (x, y) (@ 1000 cd/m^2)	(0.66, 0.33)	(0.67, 0.32)	(0.67, 0.32)	(0.66, 0.33)	(0.67, 0.32)	(0.67, 0.32)

Table 7. Device performances of various single red devices with different doping concentration

The results of device B (1wt %) is significantly superior to $Ir(piq)_3$ doped multi-layer red PHOLEDs [73]. Device B shows that the doping concentration in PHOLEDs can be reduced until 1~2% range with higher efficiency provided HOMO-HOMO and LUMO-LUMO differences between host and dopant molecules are within ~0.3 eV and emission zone is within 50nm. Device B displays exactly similar behavior although the HOMO-HOMO gap is relatively higher as compared to that in device A. However, unlike device B, similar device properties in device A regardless of doping condition from 1 to 4% are obtained. The self quenching by dopants seems to be not so serious in this device A. This indicates that the emission zone of device A is very broad and the charge balance is also relatively poor. The efficiency of device A is low compared to device B, but 4% doped condition in device A has a little better charge balance.

The J-V-L curve and efficiency characteristics of devices A and B are shown in Fig. 14. The best efficiency yields of 9.44 cd/A (EQE 14.6%) and 10.62 lm/W are noticed in the device B as shown in Fig 14(b). As seen from the results of Fig. 14(a), the driving voltage in device A with m-MTDATA:$Bebq_2$:$Ir(piq)_3$ [4 wt%] is 6.9 V at the brightness of 1,000 cd/m^2. The device B with α-NPB:$Bebq_2$:$Ir(piq)_3$ [1 wt%] shows a driving voltage of 5.4 V at 1000 cd/m^2.

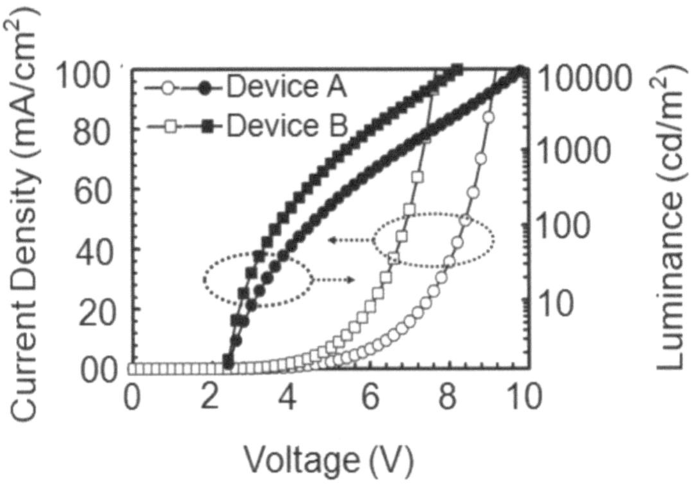

(a)

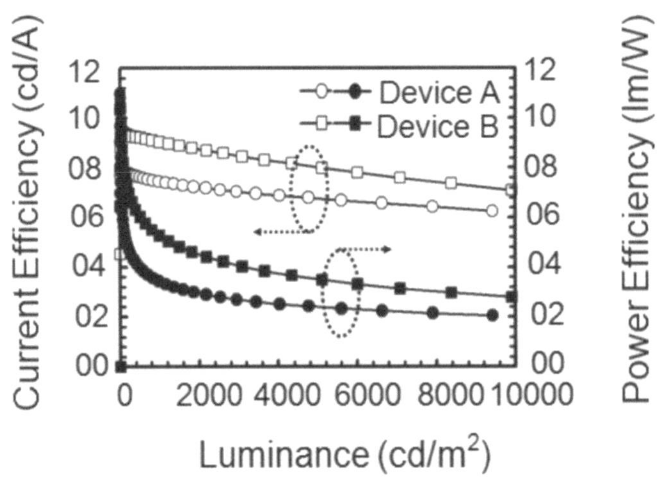

(b)

Fig. 14. Current density (J)-Voltage(V)-Luminance (L) and Efficiency characteristics of single layer red PHOLEDs. (a) J-V-L characteristics, (b) L vs. current and power efficiencies characteristics. Device A(4%) and Device B(1%) fully doped.

In m-MTDATA, no barrier for hole injection from the surface treated ITO (5.1 eV) to the HOMO (5.1 eV) of the m-MTDATA exists. Further, this energy level matches with the HOMO (5.1 eV) of the $Ir(piq)_3$. While, electrons injected from the cathode move freely on the LUMO energy of $Bebq_2$. In case of the device B, the HOMO energy in the α-NPB material at 5.4 eV as against 5.1 eV in the surface treated ITO (HOMO difference ~ 0.3 eV) offers some barrier to the hole injection into the emitting layer. While electrons injected from cathode move freely over the LUMO energy of $Bebq_2$. *To understand the injection barrier situation in m-MTDATA and α-NPB, J-V of hole only devices were investigated.* An ideal Ohmic contact (Giebeler et al., 1998) at ITO and m-MTDATA interface was reported. Whereas, the NPB hole only device had reported to have the injection limited current behavior. When a buffer layer like PEDOT:PSS (poly(3,4-ethylenedioxythiophene)-poly(4-stylenesurfonate) or C60 was introduced at ITO interface, the Ohmic characteristic was observed in this device (Tse et al., 2006; Koo et al., 2008). Form these previously reported results, a high value of driving voltage in the α-NPB mixed device B due to the high barrier to hole injection into the emitting layer was expected. However in reality, the device B with α-NPB hole transporting host shows a lower driving voltage implying a low resistance to the current flow. Here, devices A and B were realized using two different hole transporting host materials having different charge carriers transport abilities, particularly the hole mobility. α-NPB has an ambipolar transporting ability with the hole mobility faster than that of m-MTDATA (S. W. Liu et al., 2007). Thus, mobilities of hole carriers in these mixed host single layer systems rather than hole injection barrier at the ITO/mixed host interface seems to be crucial in deciding the driving voltage. In order to elucidate the conduction and emission processes in single layer devices, we have fabricated following several devices and investigated.

We have made devices C and D without $Ir(piq)_3$ dopant and results were compared with those of devices A and B, respectively. Fig. 15 shows J-V characteristics of devices A,B,C,D. Results on bi-layered ITO/α-NPB (40 nm) / $Bebq_2$: $Ir(piq)_3$ (10 wt%, 50 nm) /LiF (0.5 nm) /Al(100 nm) red emitting PHOLEDs [73], reproduced here for better comparison, show a low driving voltage value of 4.5 V to reach a luminance of 1000 cd/m^2. As displayed in Fig. 15, both devices C and D (undoped) show J-V characteristics similar to $Ir(piq)_3$ doped devices A and B, respectively. Furthermore in our devices A and B, hole and electron injection barriers by dopant molecules are negligible due to no barrier at ITO and cathode interfaces, respectively. Doping in the device may affect carrier mobility due to carrier trapping by dopant molecules. Usually, J-V characteristics of PHOLEDs are changed significantly by adding dopant molecules when HOMO-HOMO and LUMO-LUMO differences between host and dopant molecules are high over 0.3 eV. In device C and D, these energy differences are within 0.3 eV. In this case, the J-V characteristic does not change because trapped charges in dopant molecules easily overcome to host energy level by thermal energy. Described results demonstrate that the conduction of current in a hole and electron transporting mixed host layer is almost independent of (i) the charge trapping at dopant molecules and (ii) hole injection barrier at the ITO/mixed host interface. Further, all mixed single layer devices offer a high resistance to current flow than bi-layered red device with hetero junction (see Fig. 15). The interesting and intriguing results on J-V in mixed host single layer devices may be explained on the basis of existing knowledge on carrier mobilities in organic materials. α-NPB exhibits an ambipolar transporting ability with electron and hole mobility values of 9×10^{-4} and 6×10^{-4} cm^2/Vs, respectively (S. W. Liu et al., 2007), while the hole mobility value in m-MTDATA is 3×10^{-5} cm^2/Vs. Earlier, it was shown that the charge transport behaviors in mixed thin films of α-NPB and Alq_3 are sensitive to (i) compositional fraction, and (ii) charge carriers mobilities of neat compounds (S. W. Liu et al., 2007). The 1:1 mixed layer of α-NPB and Alq_3 appeared to give lower charge carrier mobility of 10^{-2}~10^{-3} order than neat films (S. W. Liu et al., 2007). As a consequence, the fast current flow in the device B despite the large hole injection barrier is attributed to the high hole mobility value and ambipolar nature of α-NPB. Higher driving voltage of single layer devices compared to the bilayer device is also well understood by the decrease in carrier mobility in the mixed host system.

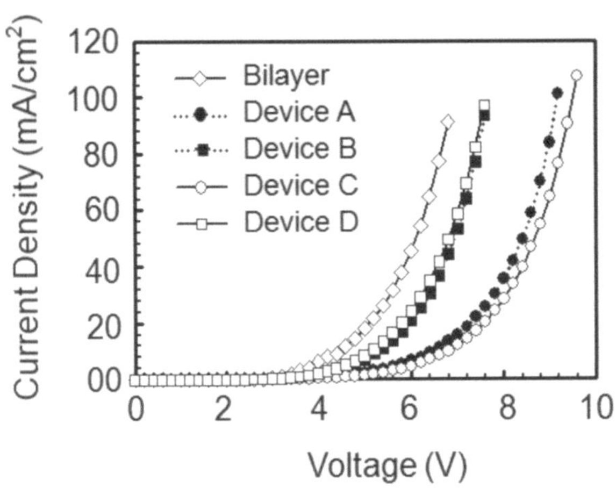

Bilayered device: ITO/α-NPB (40 nm) / $Bebq_2$: $Ir(piq)_3$ (10 wt%, 50 nm) /LiF (0.5 nm)/Al(100 nm); Device A: ITO/m-MTDATA:$Bebq_2$: $Ir(piq)_3$ [4 wt%, 100 nm]/LiF (0.5 nm)/Al (100 nm); Device B: ITO/α-NPB:$Bebq_2$: $Ir(piq)_3$ [1 wt%, 100 nm]/LiF (0.5 nm)/Al (100 nm); Device C: ITO/m-MTDATA:$Bebq_2$ [100 nm]/LiF (0.5 nm)/Al (100 nm); Device D: ITO/α-NPB:$Bebq_2$ [100 nm]/LiF (0.5 nm)/Al (100 nm)

Fig. 15. J-V characteristics of bi-layered and A~D red emitting PHOLEDs devices.

Since the charge transport behaviors in mixed hosts are sensitive to the composition and intrinsic mobilities in neat films, the location of the recombination region may be important to understand the device efficiency. To investigate the recombination zone position, we have evaluated three devices with doped emissive layer located at different positions as:

1. Device A-(L) : ITO/m-MTDATA:$Bebq_2$:$Ir(piq)_3$ [4 wt%, 30 nm]/m-MTDATA:$Bebq_2$ [70 nm]/LiF (0.5 nm)/Al (100 nm);
2. Device A-(C) : ITO/m-MTDATA:$Bebq_2$ [35 nm]/m-MTDATA:$Bebq_2$:$Ir(piq)_3$ [4 wt%, 30 nm]/m-MTDATA:$Bebq_2$ [35 nm]/LiF (0.5 nm)/Al (100 nm);
3. Device A-(R) : ITO/m-MTDATA:$Bebq_2$ [30 nm]/m-MTDATA:$Bebq_2$:$Ir(piq)_3$ [4 wt%, 70 nm]/LiF (0.5 nm)/Al (100 nm).

Similarly, Devices B-(L), (C) and (R) were fabricated using α-NPB instead of m-MTDATA and 1 wt% of $Ir(piq)_3$. The doping region was fixed to 30 nm in all devices. The anode side doped devices show the best current efficiency performance as displayed in Fig. 16 (Devices A-(L) and B-(L)), indicating that the recombination zone is around the ITO/mixed host interface. Further, the emission efficiency performance deteriorates as the doped region is moved toward the cathode side. High current efficiency in α-NPB/$Bebq_2$ mixed host system is the consequences of the better charge balance in the recombination zone. Figure 17 shows electroluminescence (EL) spectra dependence on the emission zone location in doped and undoped devices. Broad and clean EL peak at 620 nm in undoped mixed m-MTDATA/$Bebq_2$ host organic device C is due to exciplex emissions. While the strong and asymmetric EL emission peak at 620 in devices A- (L) to A- (R) due to emissions of exciplex and $Ir(piq)_3$ red phosphorescent dopant are noticed. In these devices, exciplexes are formed as the energy difference between HOMO of m-MTDATA and LUMO of $Bebq_2$ is about 2.3 eV. Whereas in case of fully doped (device B) and undoped (device D) α-NPB/$Bebq_2$ mixed devices, clean peaks at 510 and 620 nm due to strong emission of $Bebq_2$ and $Ir(piq)_3$ dopant are appeared, respectively. Upon moving the doped region toward the anode side, EL spectra show both emission peaks at 510 and 620 nm due to $Bebq_2$ host and $Ir(piq)_3$ dopant, respectively, but with the reduced intensity of 510 nm emission peak of $Bebq_2$. The electron charge carriers are transported over the LUMO of $Bebq_2$ through the doped region and reach the anode side, resulting in the emission due to $Bebq_2$ host.

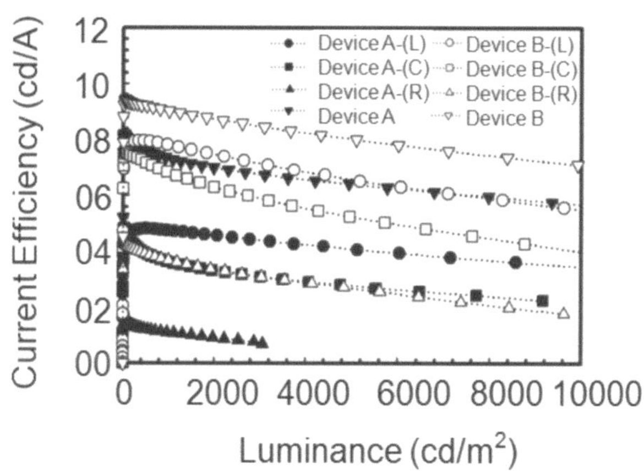

Fig. 16. Luminance-current Efficiency characteristics of various single layer devices fabricated with different locations of doped regions. Device A – Fully doped, Device B- Fully doped.

Device A: ITO/m-MTDATA:$Bebq_2$: $Ir(piq)_3$ [4 wt%, 100 nm]/LiF (0.5 nm)/Al (100 nm) – Fully doped; Device C: undoped mixed m-MTDATA/$Bebq_2$ organic host device
Device B: ITO/α-NPB:$Bebq_2$: $Ir(piq)_3$ [1 wt%, 100 nm]/LiF (0.5 nm)/Al (100 nm)- Fully doped; Device D: undoped mixed α-NPB:$Bebq_2$ organic host device
Although holes are easily injected into the m-MTDATA/$Bebq_2$ organic layer (device A), they are slowly transported due to low hole mobility in m-MTDATA which is further reduced in the mixed host system. While transport behavior in α-NPB/$Bebq_2$ mixed host system is relatively better due to the high hole mobility in α-NPB. Whereas, electrons in both doped devices A and B are transported freely over the LUMO of the $Bebq_2$. These results corroborate that the recombination zone in devices A and B are located between the anode and the center of the emitting layer.

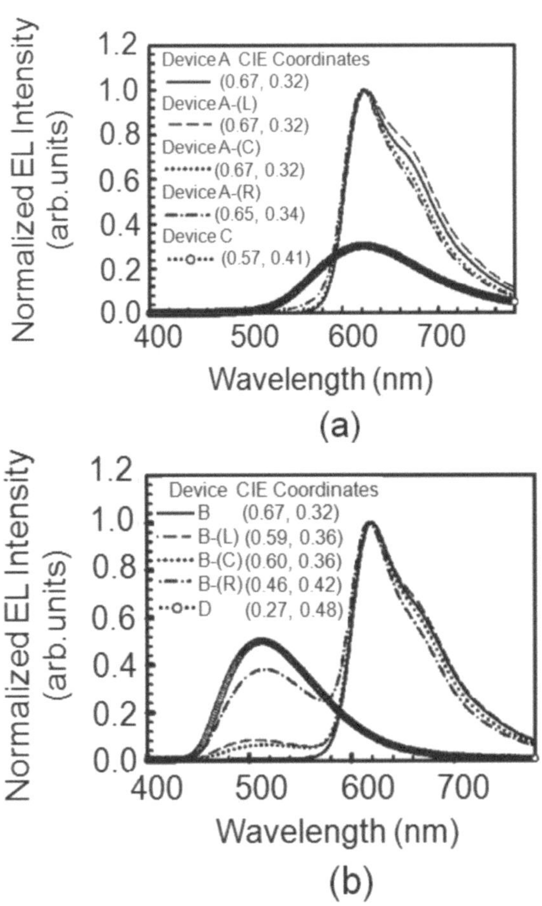

Fig. 17. Electroluminescence (EL) spectra of various single layer devices fabricated with different locations of doped regions at the brightness of 1000 cd/m^2.

5.4 Conclusions

In conclusion, we have demonstrated high efficiency red PHOLEDs comprising only single emitting layer. The key to the simplification is the direct injection of holes and electrons into the mixed host materials through electrodes. The driving voltage of 5.4 V to reach the 1000 cd/m^2 and maximum current and power efficiency values of 9.44 cd/A and 10.62 lm/W, respectively, in the α-NPB/ $Bebq_2$ mixed single layer structure PHOLEDs with the $Ir(piq)_3$ dopant as low as 1 wt% are obtained. We found that carrier mobility is significantly important parameter to simplify the device architecture. The obtained characteristics of red PHOLEDs pave the way to simplify the device structure with reasonable reduction in the manufacturing cost of passive and active matrix OLEDs.

6. Ideal host and guest system

6.1 Introduction

In phosphorescent devices, theoretically 100% internal quantum efficiency (IQE) is achieved by harvesting both singlet and triplet excitons generated by electrical injection which is four times that of fluorescent organic light-emitting devices (OLEDs) (Gong et al., 2002; Tsuzuki et al., 2003; Adachi et al., 2000). Förster and/or Dexter energy transfer processes (Tanaka and Tokito, 2008) between host and guest molecules play an important role in confining the triplet energy excitons in the phosphorescent guest. This determines the triplet state emission efficiency in PHOLEDs. Förster energy transfer (Forster, 1959) is a long range process (up to ~ 10 nm) due to dipole-dipole coupling of donor host and acceptor guest molecules, while Dexter energy transfer (Dexter, 1953) is a short range process (typically ~ 1 to 3 nm) which requires overlapping of the molecular orbital of adjacent molecules (intermolecular electron exchange).

The phosphorescence emission in the conventional host-guest phosphorescent system occurs either with Förster transfer from the excited triplet S_1 state of the host to the excited triplet S_1 state of the guest and Dexter transfer from the excited triplet T_1 state of the host to the excited triplet T_1 state of the guest or direct exciton formation on the phosphorescent guest molecules, resulting in a reasonable good efficiency. However, emission mechanism in phosphorescent OLEDs whether due to charge trapping by guest molecules and/or energy transfer from the host to the guest, is not clearly understood. Till date, several researchers have reported that the charge trapping at guest molecules is the main cause for the emission of PHOLEDs.

Amongst well-known iridium (III) and platinum (II) phosphorescent emitters, Iridium (III) complexes have been shown to be the most efficient triplet dopants employed in highly efficient PHOLEDs (Adachi et al., 2001b; Baldo et al., 1999). Usually, wide energy gap 4,4′-bis(N-carbazolyl)-1,1′-biphenyl (CBP) is used as a host material for red (~ 2.0 eV) or green (~ 2.3 - 2.4 eV) phosphorescent guests [63, 64]. Such a wide energy gap host has the advantage of high T_1 energy of 2.6 eV (Baldo & Forrest, 2000) or 2.55 eV (Tanaka et al., 2004) and long triplet lifetime > 1 *s* (Baldo & Forrest, 2000), while the optical band gap value (E_g) is 3.1 eV (Baldo et al., 1999).

Fig. 18(a) shows both the energy level diagram of *fac*-tris(2-phenyl-pyridinato)iridium(III) ($Ir(ppy)_3$) green and the tris(1-phenylisoquinoline)iridium ($Ir(piq)_3$) red phosphorescent complexes used in doping the CPB host. However, the wide band gap host and narrow band gap (E_g) guest system often causes an increase in driving voltage due to the difference in HOMO and/or LUMO levels between the guest and host materials (Tsuzuki & Tokito, 2007). Thus, the guest molecules are thought to act as deep trapping centers for electrons and holes in the emitting layer, causing an increase in the drive voltage of the PHOLED (Gong et al., 2003). The dopant concentration in such a host-guest system is usually as high as about 6 ~ 10 percent by weight (wt%) because injected charges move through dopant molecules in the emitting layer. Therefore, self-quenching or triplet-triplet annihilation by dopant molecules is an inevitable problem in host-guest systems with high doping concentrations. Earlier, Kawamura et al. had reported that the phosphorescence photoluminescence quantum efficiency of $Ir(ppy)_3$ could be decreased by ~5% with an increasing in doping concentration from 2 to 6% (Kawamura et al., 2005). Consequently, the selection of suitable host candidates is a critical issue in fabricating high efficiency PHOLEDs.

In this section, the minimized charge trapped host-dopant system is investigated by using a narrow band-gap fluorescent host material in order to address device performance and manufacturing constraints. Here, we report an ideal host-guest system that requires only 1% guest doping condition for good energy transfer and provides ideal quantum efficiency in PHOLEDs. We also report that strong fluorescent host materials function very well in phosphorescent OLEDs due to efficient Förster energy transfers from the host singlet state to the guest singlet and triplet mixing state which appears to be the key mechanism.

6.2 Experimental

N,N′-di(4-(N,N′-diphenyl-amino)phenyl)-N,N′-diphenylbenzidine (DNTPD) as a hole transporting layer, CBP and bis(10-hydroxybenzo [h] quinolinato)beryllium complex (Bebq2) as host materials, bis(2-phenylquinoline)(acetylacetonate)iridium (Ir(phq)2acac), tris(1-phenylisoquinoline)iridium (Ir(piq)3) as red dopants, aluminum (III) bis(2-methyl-8-quinolinato)-4-phenylphenolate (BAlq) as a hole blocking layer and Tris-(8-hydroxyquinoline)aluminum (Alq_3) as an electron transporting layer were purchased from Gracel and Chemipro Corporation and were used. The fabricated devices are characterized as described in the section 3. The OLED area was 2 mm^2 for all the samples studied in this work.

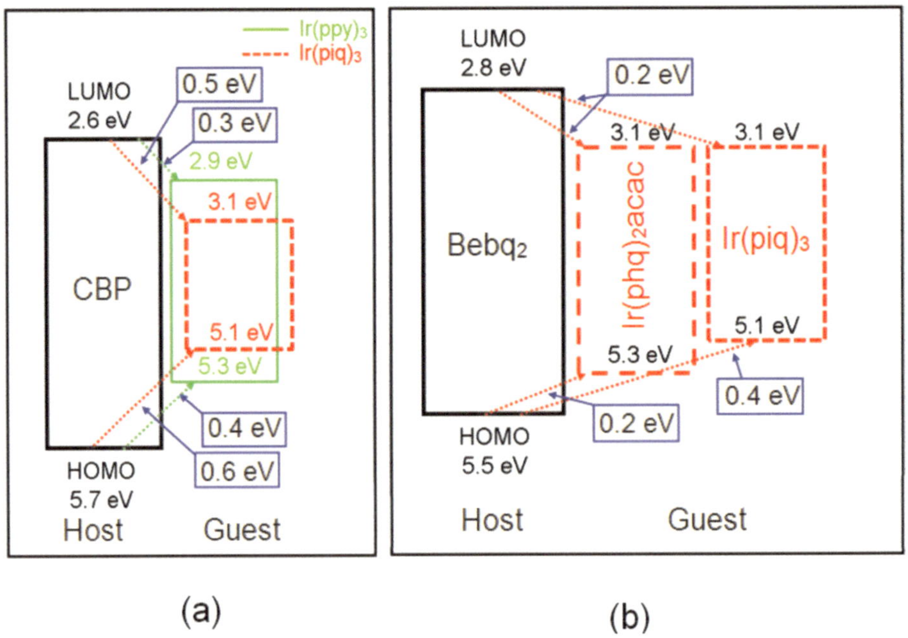

Fig. 18. (a) Energy level diagram of the $Ir(ppy)_3$ green and $Ir(piq)_3$ red phosphorescent complex doped by the CPB host. (b) Energy level diagram of the $Bebq_2$ fluorescent host and ($Ir(phq)_2acac$) and $Ir(piq)_3$ red phosphorescent dopant materials.

6.3 Results & discussion

Fig. 18(b) shows an energy band diagram of the fluorescent host and orange-red phosphorescent dopant materials used in the device fabrication. The simple bilayer PHOLED comprises a DNTPD hole transport layer (HTL), a $Bebq_2$ narrow band gap fluorescent host and an electron transport layer (ETL) plus $Ir(phq)_2acac$ dopant. In the present investigation, the fabricated PHOLED was:

ITO/DNTPD (40 nm) / $Bebq_2$: $Ir(phq)_2acac$ (50 nm, 1%)/ LiF(0.5 nm) / Al(100 nm).

Fig. 19 (a) & (b) and Table 8 (Device B) illustrate the electrical performance of the fabricated phosphorescent device. A luminance of 1000 cd/m^2 was obtained with a driving voltage of 3.7 V, and current and power efficiency values of 20.53 cd/A and 23.14 lm/W, respectively.

Furthermore, the maximum current and power efficiencies were 26.53 cd/A and 29.58 lm/W, respectively. The external quantum efficiency (EQE) value of 21% in the fabricated PHOLED slightly exceeded the theoretical limit of about 20% derived from simple classical optics. Moreover, this can be further improved by optimizing the output coupling. These remarkable results brought some pleasant surprises.

$Ir(phq)_2acac$ concentration (wt%)	Device A (0.5)	Device B (1.0)	Device C (1.5)	Device D (2.0)
Turn-on voltage (V) (at 1 cd/m²)	2.1	2.1	2.1	2.1
Operating voltage (V) (1000 cd/m²)	3.7	3.7	3.6	3.6
Efficiency (at 1000 cd/m²) Current (cd/A) Power (lm/W)	 20.96 18.29	 20.53 23.14	 22.61 19.73	 21.45 18.72
Maximum Efficiency Current (cd/A) Power (lm/W)	 21.25 24.62	 26.53 29.58	 23.46 29.94	 22.73 27.94
CIE (x,y) (1000 cd/m²)	(0.61,0.38)	(0.62,0.37)	(0.62,0.37)	(0.62,0.37)
EQE (%)(maximum)	16.6	21.0	18.9	18.6

Table 8. Key parameters from $Bebq_2$:$Ir(phq)_2acac$ (0.5 - 2 wt%) orange-red emitting ITO/DNTPD (40nm) / $Bebq_2$: $Ir(phq)_2acac$ (50 nm, 0.5 to 2%)/ LiF(0.5 nm) /Al(100 nm) PHOLED devices.

Indeed, because of the extraordinarily low doping concentration (~ 1%) by contrast with most phosphorescent devices (6 ~ 10 wt%), the enhancement of the performance of $Bebq_2$:$Ir(phq)_2acac$ PHOLEDs was never expected. In order to investigate the origin for the enhanced performance, we fabricated several PHOLEDs by varying the doping concentration from 0.5 to 2% in the host-guest system. Current and luminance as a function of voltage are presented in Fig. 19(a), while current and power efficiencies as a function of luminance are presented in Fig. 19(b). This data provides evidence for: (1) complete energy transfer from the fluorescent host to phosphorescent guest, except at extremely low doping concentrations (~0.5%); (2) no significant difference between measured I-V characteristics for identical devices but with different dopant concentrations lying between 0.5 and 2 wt%;

and, (3) the quenching of both luminance, and current and power efficiencies with higher doping concentrations (~ 2 wt%). A summary of the key electrical and optical parameters (Table 8) reveals the excellent device performance for doping concentration as low as 0.5 – 2%, in contrast with conventional PHOLEDs which require a guest concentration typically in the range of 6 to 10 wt%. Therefore, a highly efficient simple bilayer PHOLED structure with a $Ir(phq)_2acac$ guest doping concentration as low as 1% in the narrow band gap $Bebq_2$ fluorescent host is demonstrated here. Previously, $(ppy)_2Ir(acac):Ir(piq)_3$ (0.3 - 1wt %) red (Tsuzuki & Tokito, 2007) and $CBP:Ir(phq)_2acac$ (6 wt%) orange-red PHOLEDs (Kwong et al. 2002) demonstrated an EQE of 9.2% with a power efficiency of 11.0 lm/W and a power efficiency of 17.6 cd/A with an EQE of 10.3% at 600 nit, respectively.

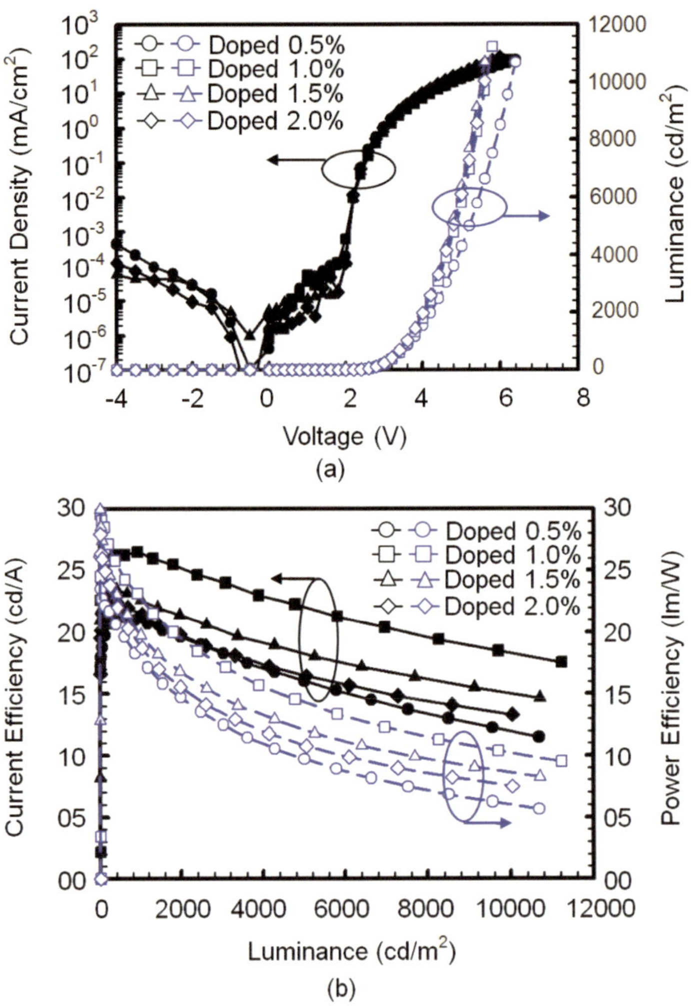

Fig. 19. (a) A J-V-L plot and (b) Current and power efficiencies as a function of luminance from red PHOLEDs doped with different concentrations (0.5 – 2 %) of $Ir(phq)_2acac$.

Figure 20 provides an evidence of energy transfer from the $Bebq_2$ fluorescent host to the $Ir(phq)_2acac$ phosphorescent guest by comparing the electroluminescence (EL) spectra of

PHOLEDs as a function of $Ir(phq)_2acac$ doping concentration from 0.5 to 2%. The strong red light emission peak at 605 nm for all EL curves at 1000 nit is attributed to the phosphorescence of $Ir(phq)_2acac$. The Commission Internationale de l'Eclairage (CIE) color emission coordinates are (0.61, 0.38), (0.62, 0.37), (0.62, 0.37), (0.62, 0.37) for doping concentrations of 0.5, 1.0, 1.5, and 2.0 wt% of $Ir(phq)_2acac$, respectively (as seen in Fig. 21). A slight emission at 500 nm due to the $Bebq_2$ host plus a dominant doping peak at 605 nm when the doping concentration is extremely low (~ 0.5%), suggests an incomplete energy transfer from the $Bebq_2$ host to the $Ir(phq)_2acac$ guest. Furthermore, it indicates that the recombination of injected holes and electrons occurs at host molecule sites and then the excited energy is rapidly transferred from the host to the guest. The presence of a clean EL peak (no emissions at 500 nm) in other devices with doping concentrations of $Ir(phq)_2acac$ > 0.5% indicates a complete energy transfer from the host to the guest.

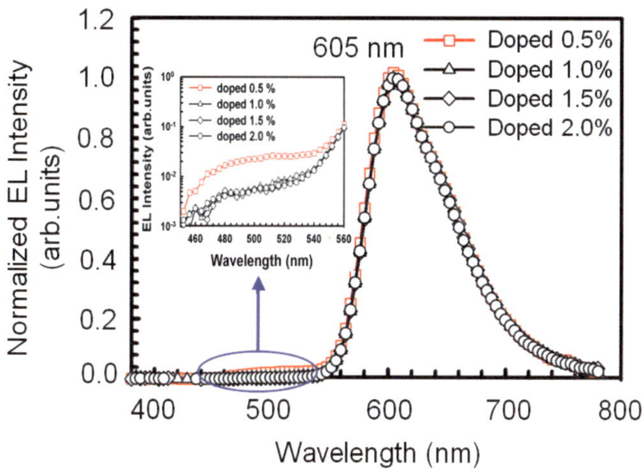

Fig. 20. EL spectra as a function of dopant concentration: $Ir(phq)_2acac$ of an ITO/DNTPD (40 nm) / $Bebq_2$: $Ir(phq)_2acac$ (50 nm, 0.5 to 2%)/ LiF(0.5 nm) / Al(100 nm) PHOLEDs at 1000 cd/m^2.

To understand the phosphorescence emission mechanism more precisely in the $Bebq_2$:$Ir(phq)_2acac$ host-guest system, we fabricated a series of PHOLEDs and studied. At first, we used the well known wide band gap CBP host material instead of $Bebq_2$ and fabricated the multilayer devices with a structure: NPB (40nm) / CBP:$Ir(piq)_3$ (30nm, 10%) / BAlq (5nm) / Alq_3 (20nm) / LiF (0.5nm) / Al (100nm) (Device A) and NPB (40nm) / CBP:$Ir(phq)_2acac$ (30nm, 10%) / BAlq (5nm) / Alq_3 (20nm) / LiF (0.5nm) / Al (100nm) (Device B). Table 9 displays the electrical performance of the fabricated phosphorescent devices. At a luminance of 1000 cd/m^2 the resultant operating voltage was 7.1 V with current and power efficiencies of 14.41 cd/A and 6.28 lm/W, respectively, and an EQE of 11.5%. Furthermore, the maximum current and power efficiency values were 14.43 cd/A and 8.99 lm/W, respectively. Obviously, the two fold increase in driving voltage is a consequence of the trapping of injected holes and electrons at deep $Ir(phq)_2acac$ molecules in the CBP:$Ir(phq)_2acac$ system. Direct charge trapping on the $Ir(phq)_2acac$ guest molecules seems to be the key mechanism for phosphorescence emission in this host-guest system.

Later, bilayer PHOLED device was fabricated using $Ir(piq)_3$ red emitting phosphorescent doping instead of $Ir(phq)_2acac$ and a $Bebq_2$ host. The fabricated devices were: DNTPD

(40nm) / $Bebq_2$:Ir(piq)$_3$ (50 nm, 4~10 wt%) / LiF (0.5 nm) / Al (100 nm). Current density-Voltage-Luminance and current and power efficiencies as a function of luminance plots are shown in Fig. 22. Electrical performances of the fabricated phosphorescent devices are illustrated in Table 10. A weak emission peak at 500 nm in the EL spectra due to the $Bebq_2$ host arises at a doping concentration of 4 wt% (significantly high by comparison with an Ir(phq)$_2$acac doping concentration ~ 0.5 wt%), accompanied by a strong peak at 620 nm (CIE coordinates x = 0.67 and y = 0.32) due to an Ir(piq)$_3$ doping molecule (Fig. 23). At luminance of 1000 cd/m^2, the corresponding operating voltage, current and power efficiencies were 3.5V, 8.41 cd/A and 7.34 lm/W, respectively. Furthermore, the maximum current and power efficiency values were 9.38 cd/A and 11.72 lm/W, respectively. Increasing the Ir(piq)$_3$ concentration to 6 wt% suppresses the $Bebq_2$ host emission and results in a clean EL red emitting peak at 620 nm due to the Ir(piq)$_3$ doping molecules. However, the device performance deteriorates with increasing doping concentration due to a self quenching process as seen in Table 10.

Parameters	Device A CBP:Ir(piq)$_3$	Device B CBP:Ir(phq)$_2$acac
Turn-on voltage (at 1 cd/m^2)	3.3 V	3.1 V
Operating voltage (1000 cd/m^2)	7.2 V	7.1 V
Efficiency (1000 cd/m^2)	5.71 cd/A 2.49 lm/W	14.41 cd/A 6.28 lm/W
Efficiency (Maximum)	6.47 cd/A 4.45 lm/W	14.43 cd/A 8.99 lm/W
CIE (x,y) (1000 cd/m^2)	(0.66, 0.33)	(0.61, 0.38)
Quantum efficiency (maximum)	11.2 %	11.5 %
Roll off (1000 nt vs 10000 nt)	48 %	86 %

Table 9. Electrical performance of the multilayer CBP:Ir(piq)$_3$ (Device A) and CBP:Ir(phq)$_2$acac (Device B) fabricated phosphorescent devices

The primary mechanism for the phosphorescence emission in the $Bebq_2$:$Ir(piq)_3$ host-guest system (Fig. 18 (b)) appears to be due to the energetically favorable electron transport and hole trapping at deep trapping centers in $Ir(piq)_3$ molecules. Thus, an appropriate selection of the host and phosphorescent dopant materials plays a significant role in determining the emission mechanism on phosphorescent devices. These results on phosphorescent emission in $Bebq_2$:$Ir(phq)_2$acac host-guest systems are very interesting and intriguing. The mechanism of phosphorescence emission is not believed to be due to the direct charge trapping in $Ir(phq)_2$acac phosphorescent guest molecules. Attempts have been made here to explain these results on the basis of the existing knowledge of Förster and Dexter energy transfer processes in host-guest systems.

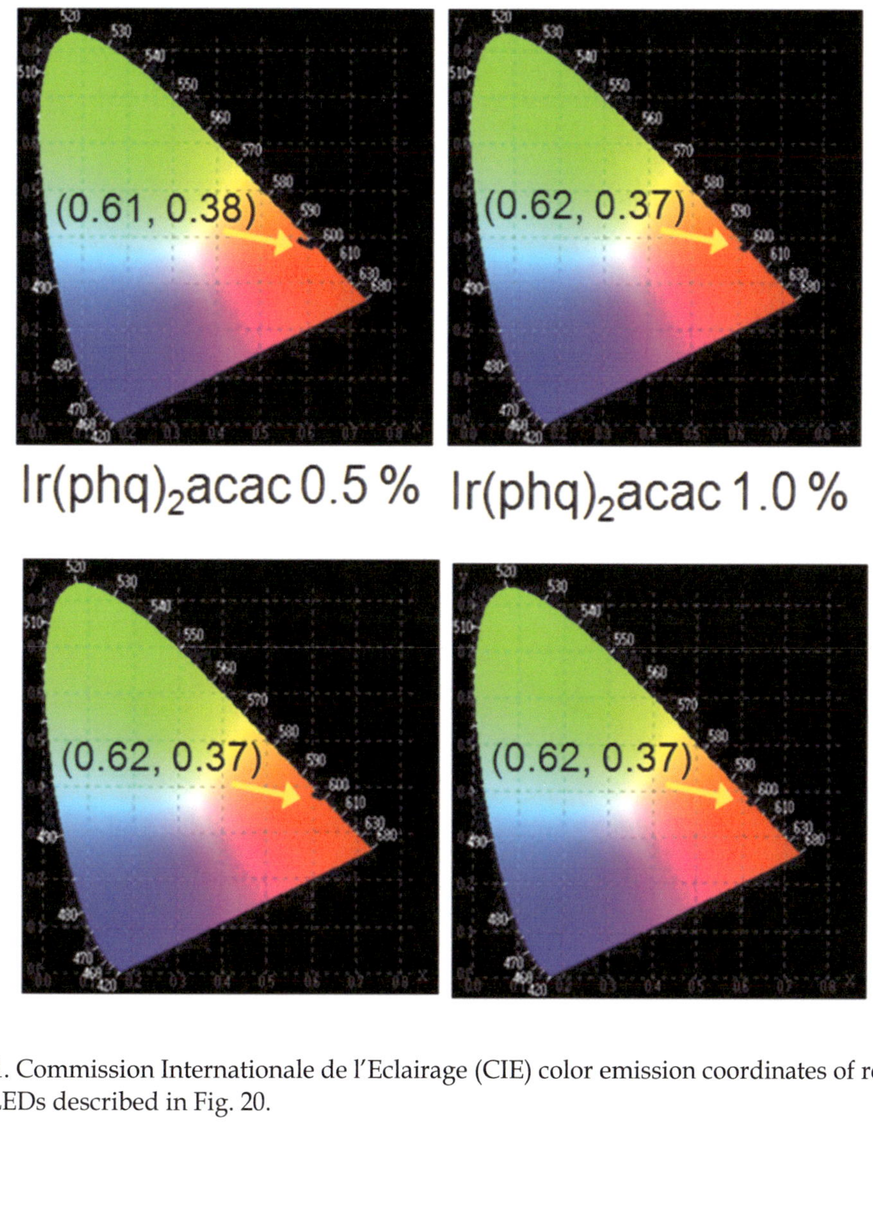

Fig. 21. Commission Internationale de l'Eclairage (CIE) color emission coordinates of red PHOLEDs described in Fig. 20.

The $Bebq_2$ host material produces a strong fluorescence emission but no phosphorescence emission signature even at 77 K. The efficient use of $Bebq_2$ host in the described phosphorescent devices is an extraordinary phenomenon since strong fluorescent host materials are believed to provide poor phosphorescent performance. Therefore, we suspect efficient Förster energy transfer between the host singlet and the metal-to-ligand charge-transfer (MLCT) state of the iridium (III) metal complex. Earlier, Förster energy transfer in phosphorescent OLEDs was postulated by Gong *et al.* and Ramos-Ortiz. *et al.* in solid photoluminescence studies (Gong et al., 2003; Ramos-Ortiz et al., 2002). To investigate Förster energy transfer from the $Bebq_2$ host to the $Ir(phq)_2acac$, time resolved spectroscopy and a Stern-Volmer plot in THF solution measurements techniques were employed. The

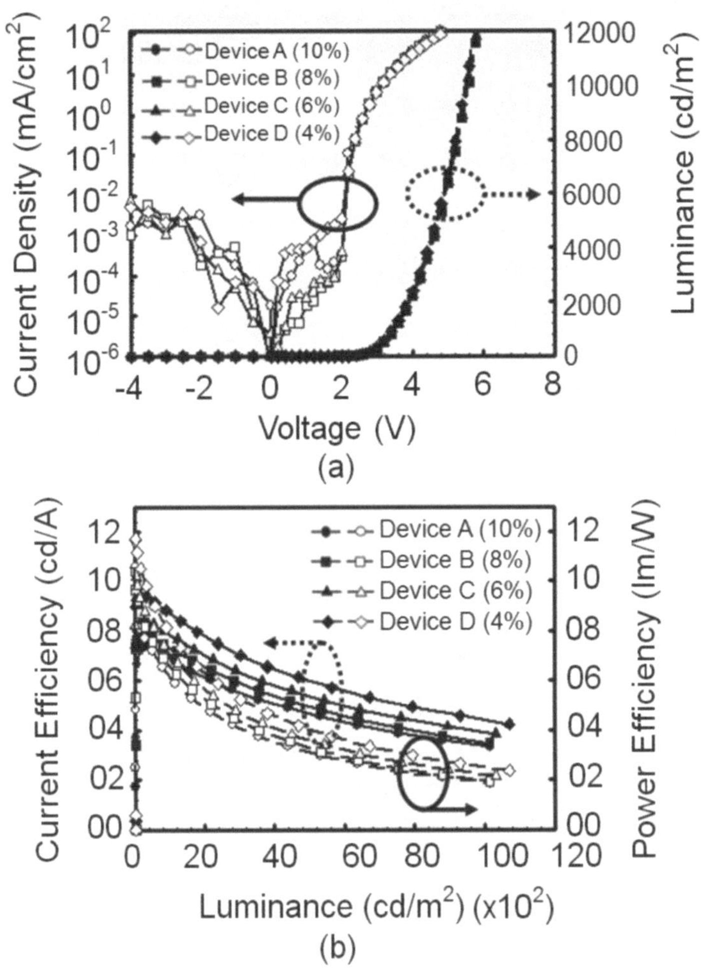

Fig. 22. Current density-Voltage-Luminance and current and power efficiencies as a function of luminance plots of DNTPD (40nm) / $Bebq_2$:$Ir(piq)_3$ (50 nm, 4~10 wt%) / LiF (0.5 nm) / Al (100 nm) red PHOLEDs

singlet fluorescent lifetime of $Bebq_2$ is 5.0 *ns*. From the slope of the linear Stern-Volmer plot (Fig. 24), the calculated energy transfer rate is $k_q = 8\times10^{12}$ $sec^{-1}M^{-1}$, indicating that the energy transfer from the excited singlet state of the host to the dopant triplet occurs quantitatively and ideally. Furthermore, the strong spin orbital coupling induced by the transition metal ion indicates that a narrow energy gap exists between the ^{1}MLCT and ^{3}MLCT states (~ 0.3eV) and opens a pathway for efficient energy transfer from the singlet to the emitting triplet states. Therefore, two channels for Förster energy transfer from the host singlet to the ^{1}MLCT and ^{3}MLCT states of the iridium complex are available as shown in Fig. 25. Overlapping of the host emission and dopant absorption spectra substantiates the hypothesis of efficient Förster energy transfer from the host singlet to the guest emitting triplet states via two channels (Fig. 26). Furthermore, the strong fluorescent quantum efficiency of 0.39 in the host ($Bebq_2$) in solution, obtained using a relative quantum yield measurement, favors Förster energy transfer.

Parameters	Device A (10%)	Device B (8%)	Device C (6%)	Device D (4%)
Turn-on voltage (at 1 cd/m^2)	2.1 V	2.1 V	2.1 V	2.1 V
Operating voltage (1000 cd/m^2)	3.5 V	3.5 V	3.5 V	3.5 V
Efficiency (1000 cd/m^2)	6.78 cd/A 5.92 lm/W	7.18 cd/A 6.26 lm/W	7.65 cd/A 6.68 lm/W	8.41 cd/A 7.34 lm/W
Efficiency (Maximum)	7.38 cd/A 8.10 lm/W	7.82 cd/A 10.40 lm/W	8.37 cd/A 10.67 lm/W	9.38 cd/A 11.72 lm/W
CIE (x,y) (1000 cd/m^2)	(0.67,0.32)	(0.67,0.32)	(0.67,0.32)	(0.67,0.32)
Quantum efficiency (maximum)	11.4 %	13.0 %	14.4 %	16.3 %
Roll off (1000 nt vs 10000 nt)	48 %	48 %	50 %	47 %

Table 10. Electrical performances of the fabricated DNTPD (40nm) / $Bebq_2$:Ir(piq)$_3$ (50 nm, 4~10 wt%) / LiF (0.5 nm) / Al (100 nm) red phosphorescent devices

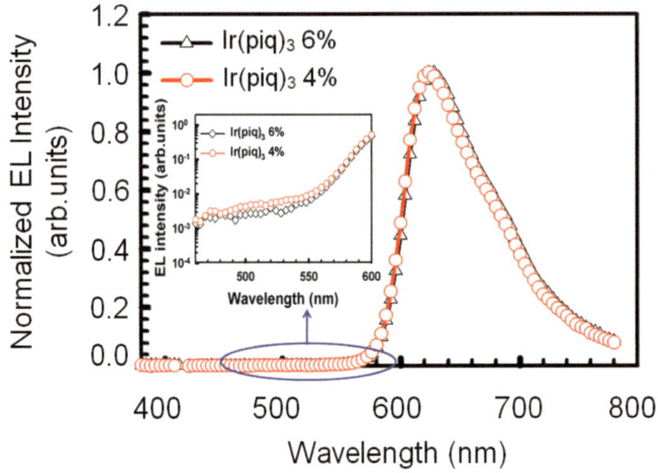

Fig. 23. EL spectra of DNTPD (40nm) / $Bebq_2$:$Ir(piq)_3$ (50 nm, 4 and 6 wt%) / LiF (0.5 nm) / Al (100 nm) PHOLEDs at 8000 cd/m^2.

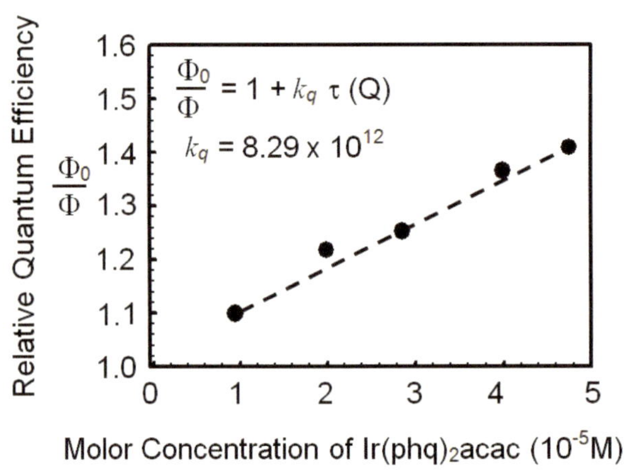

Fig. 24. Stern-Volmer plot showing the effect of $Bebq_2$ fluorescence quenching by ($Ir(phq)_2acac$) dopant.

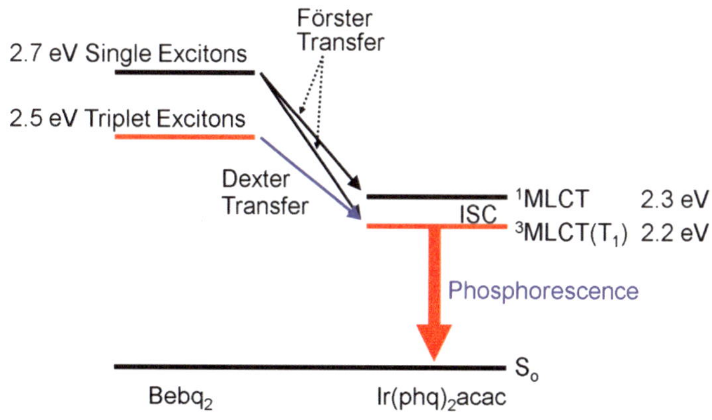

Fig. 25. Förster and Dexter energy transfer mechanism in the $Bebq_2$:Ir(phq)$_2$acac system.

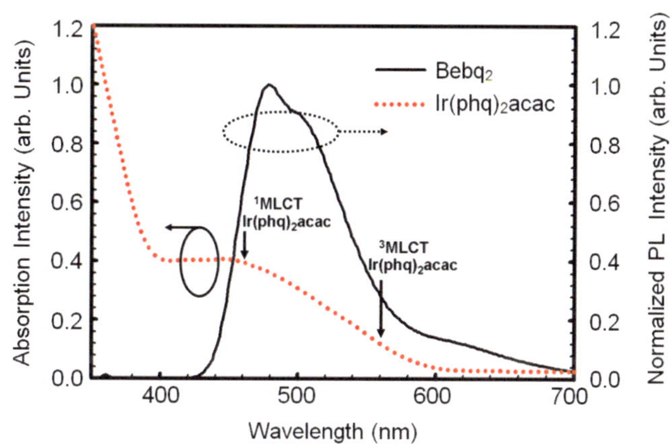

Fig. 26. Spectral overlapping of the photoluminescence spectrum from $Bebq_2$ and the absorption spectrum from Ir(phq)$_2$acac.

The Förster radius (R_0), critical distance for the concentration quenching, was estimated as 1.3 *nm* (similar to a previously reported value) using the following equation:

$$R_0^6 = \frac{9000(\ln 10)k^2\Phi_{PL}}{128\pi^5 N_A n^4}\int_0^\infty F_D(\lambda)\varepsilon_A(\lambda)\lambda^4 d\lambda$$

where k^2 = orientation factor, Φ_{PL} = photoluminescence quantum efficiency, N_A = Avogadro's number, n = refractive index, $\int_0^\infty F_D(\lambda)\varepsilon_A(\lambda)$ = spectral overlap integral between donor photoluminescence ($F_D(\lambda)$), and $\varepsilon_A(\lambda)$ = acceptor absorption, and λ = wavelength.
The triplet exciton energy transfer from the $Bebq_2$ host to ^{3}MLCT is governed by Dexter energy transfer. The rate constant of Dexter energy transfer (Kawamura et al., 2006) is

$$k_{ET} = K\, J \exp(-2R_{DA}/L)$$

where *K* is related to the specific orbital interaction, *J* is a spectral overlap integral, and R_{DA} is donor-acceptor separation relative to their van der Waals radii, L. The ideal Dexter radius is the distance between the host and dopant molecule diameter considering overlapping molecular orbitals in adjacent molecules (Turro, 1991a).

Using the equation: R = [(molecular density in film) × mol% of the dopant in a film]$^{-1/3}$ as reported by Kawamura et al. (Kawamura et al., 2006) yields an average distance of about 58.5Å and 44.9Å between $Ir(phq)_2acac$ molecules for doping concentrations of 0.5%, 1.0%, respectively (Fig.27). By considering the host (13.6Å) and dopant (13.7Å) diameters calculated using a molecular modeling program (DMOL3) (Delley, 2000) and van der Waals interaction distance (usually very small ~ within 2Å) for the ideal Dexter energy transfer condition, the estimated distance between the host and guest molecules is about 15.6Å (i.e. the half diameters of the host and dopant molecules are, 13.7/2Å plus 13.6Å/2, including a 2Å van der Waals interaction distance). In solid state films, the minimum doping concentration is desirable to prevent triplet quenching processes.

Considering that two host molecules are located between two dopant molecules, an ideal Dexter condition (Fig. 27(b)), all host molecules are adjacent to a dopant molecule and dopant molecules are well separated in the host matrix. In such a host and guest molecule arrangement, the separation between two dopant molecules appears to be about 46.9Å. However, when the doping concentration is 0.5%, the estimated separation between $Ir(phq)_2acac$ molecules is 58.5Å, suggesting that more than two host molecules are located between two dopant molecules (Fig. 27(b). If the doping concentration is increased to 1%, the separation between dopant molecules is less than 46.9Å, which results in an efficient energy transfer from the $Bebq_2$ host to the emitting triplet state of $Ir(phq)_2acac$ via Dexter processes. The triplets generated due to the $Bebq_2$ host molecules diffuse to an average distance of only 15.6Å, with a doping concentration of 1 wt%, until they are harvested by $Ir(phq)_2acac$ (Fig. 27(b). If the doping concentration is increased to greater than 1%, the device performance deteriorates due to quenching processes of triplet excitons caused by two closely separated dopant molecules.

The influence of the Förster quenching process in the described system is not serious at all doping concentrations (Fig. 27(a). The singlet exciton energies generated by charge injection in host molecules can be transferred to triplet states of dopant molecules by the efficient Förster energy transfer. In our system, the rate constant of Förster energy transfer is much faster than that of the intersystem crossing of host singlet states. Typical intersystem crossing rate constants fall within the range ~10^6-10^8 sec^{-1} (Turro, 1991b). Thus, strong fluorescent materials, such as $Bebq_2$, are excellent as host materials in PHOLEDs.

6.4 Conclusions

In conclusion, we have demonstrated here an ideal host-guest energy transfer and quantum efficiency conditions with a dopant concentration of approximately 1% in phosphorescent OLEDs. We also report that strong fluorescent host materials function very well in phosphorescent OLEDs. The operating mechanism for the phosphorescence emission is twofold: Firstly, an efficient Förster energy transfer process from the host singlet exciton to the ^{1}MLCT and ^{3}MLCT states of the guest. And, secondly, a Dexter energy transfer process from the host triplet exciton to the ^{3}MLCT state of the guest. The extremely low doping concept for highly efficient PHOLEDs has potential uses in future display and lighting applications.

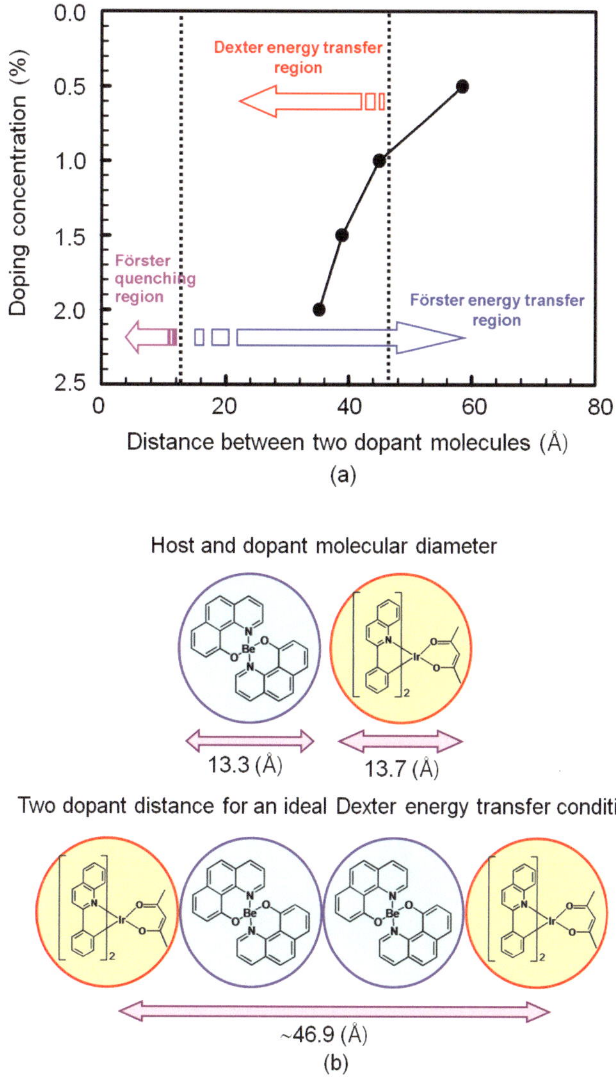

Fig. 27. (a) Förster and Dexter energy transfer conditions as a function of doping concentration and distance between two dopants. (b) The molecular diameters of the host and dopant and the dopant-dopant distance for an ideal Dexter energy transfer.

7. Conclusions

Simple structure red PHOLED, using **narrow band gap fluorescent host materials have been demonstrated**, having a:
(1) Simple structure, (2) Low driving voltage, (3) High efficiency (lm/W), (4) No charge trapping at phosphorescent guest molecules, (5) Low doping concentration, and (6) Low manufacturing cost.
These results are summarized in Table 11. Various triplet quantum well devices from a single to five quantum wells are realized using a wide band-gap hole and an electron transporting layers, $Bebq_2$ narrow band-gap host and $Ir(piq)_3$ red dopant materials, and

$Bepp_2$ charge control layers (CCL). Triplet energies in fabricated MQW devices are confined at the emitting layers. The maximum external quantum efficiency of 14.8 % with a two quantum well device structure is obtained, which is the highest value among the reported $Ir(piq)_3$ dopant red phosphorescent OLEDs. The described MQW device concept can be very useful to future OLED display and lighting applications.

Parameters	[1]MQW p-i-n device n=2	[2]Bilayer device	[3]Single Layer 4% Doping	[4]Single Layer 1% Doping	[5]Ideal host-Guest Device (1%)
Turn on voltage (@ 1cd/m^2)	2.5 V		2.3 V	2.4 V	2.1
Operating voltage (@ 1000 cd/m^2)	4.2 V	4.5V	6.9 V	5.4 V	3.7
Current (cd/A) & Power (lm/W) efficiencies @ 1000 cd/m^2		9.66, 6.90	7.44, 3.43	9.02, 5.25	20.53 23.14
Maximum Current (cd/A) & Power (lm/W) Efficiencies	12.4		8.04, 10.96	9.44, 10.62	26.53, 29.58
CIE (x, y) (@ 1000 cd/m^2)	(0.66, 0.33)	(0.67, 0.33)	(0.67, 0.32)	(0.66, 0.33)	(0.62, 0.37)
EQE (%) (maximum)	14.8				21.0

[1]ITO/TCTA:WO_3(30%,40nm)/TCTA(12nm)/**$Bebq_2$:Ir(2%,10nm)/$Bepp_2$(CCL,5nm)(n=2)** /$Bepp_2$ (7 nm)/$Bepp_2$:Cs_2CO_3 (10%, 20 nm)/ Al(100 nm)
[2]ITO/α-NPB (40 nm) / **$Bebq_2$: $Ir(piq)_3$ (10 wt%, 50 nm)**/ LiF(0.5 nm) / Al(100 nm
[3]ITO/**m-MTDATA:$Bebq_2$: $Ir(piq)_3$ (4 wt%, 100 nm)**/LiF (0.5 nm)/Al (100 nm)
[4]ITO/**α-NPB:$Bebq_2$: $Ir(piq)_3$ (1 wt%, 100 nm)**/LiF (0.5 nm)/Al (100 nm).
[5]ITO/DNTPD (40nm) / **$Bebq_2$: $Ir(phq)_2acac$ (50 nm, 1%)**/ LiF(0.5 nm) /Al(100 nm)

Table 11. Summary of performances of red PHOLEDs in this study

Bi-layered simple structure red PHOLED demonstrates high current and power efficiency values of 9.66 cd/A and 6.90 lm/W, respectively. The operating voltage of bi-layered PHOLEDs at a luminance of 1000 cd/m^2 was 4.5 V. A simple bilayerd red emitting device with $Bebq_2$ host could be a promising way to achieve efficient, economical, and ease manufacturing process, important for display and lighting production.

We have also demonstrated high efficiency red PHOLEDs comprising only single emitting layer. The driving voltage of 5.4 V to reach the 1000 cd/m^2 and maximum current and power efficiency values of 9.44 cd/A and 10.62 lm/W, respectively, in the α-NPB/ $Bebq_2$ mixed single layer structure PHOLEDs with the $Ir(piq)_3$ dopant as low as 1 wt% are obtained. The obtained results could be useful to simplify the device structure with a reasonable reduction in the manufacturing cost of passive and active matrix OLEDs.
An ideal host-guest system displays efficient energy transfer and quantum efficiency conditions with a dopant concentration of approximately 1% in phosphorescent OLEDs. A luminance of 1000 cd/m^2 was obtained with a driving voltage of 3.7 V, and current and power efficiencies of 20.53 cd/A and 23.14 lm/W (maximum current efficiency: 26.53 cd/A and maximum power efficiency: 29.58 lm/W), respectively. The external quantum efficiency (EQE) value of 21% in the fabricated PHOLED, slightly exceeded the theoretical limit of about 20% derived from simple classical optics, is recorded.
In summary, narrow band gap fluorescent host materials with the extremely low doping concept for highly efficient PHOLEDs has potential uses in future display and lighting applications.

8. Acknowledgement

Authors are thankful to Professor Yup Kim, Physics Department (Kyung Hee University, Seoul) for his encouragement during the span of this work.

9. References

Al Attar, H.A., Monkman, A.P., Tavasli, M., Bettington, S. & Bryce, M.R. (2005). White polymeric light-emitting diode based on a fluorine polymer/Ir complex blend system, *Appl. Phys. Lett.* 86: 121101-1 - 121101-3.

Adachi, C., Baldo, M.A., Forrest, S.R. & Thompson, M.E. (2000). High-efficiency organic electrophosphorescent devices with tris(2-phenylpyridine)iridium doped into electron-transporting materials, *App. Phys. Lett.* 77: 904-906.

Adachi, C., Baldo, M.A., Forrest, S.R., Lamansky, S., Thompson, M.E. & Kwong, R.C. (2001a). High-efficiency red electrophosphorescence devices, *Appl. Phys. Lett.* 78: 1622-1624.

Adachi, C., Baldo, M. A., Thompson, M. E. & Forrest, S. R. (2001b). Nearly 100% internal phosphorescence efficiency in an organic light-emitting device, *J. Appl. Phys.* 90: 5048-5051.

Adamovich, V.I., Cordero, S.R., Djurovich, P.I., Tamayo, A., Thompson, M.E., D'Andrade, B. W. & Forrest, S.R. (2003). New charge-carrier blocking materials for high efficiency OLEDs, *Org. Electron.* 4 (2-3): 77-87.

Baldo, M.A., O'Brien, D.F., You, Y., Shoustikov, A., Sibley, S., Thompson, M.E. & Forrest, S.R. (1998). Highly efficient phosphorescent emission from organic electroluminescent devices, *Nature (London)* 395: 151-154.

Baldo, M. A., Lamansky, S., Burrows, P. E., Thompson, M. E. & Forrest, S. R. (1999). Very high-efficiency green organic light-emitting devices based on electrophosphorescence, *Appl. Phys. Lett.* 75: 4-6.

Baldo, M. A. & Forrest, S. R. (2000). Transient analysis of organic electrophosphorescence: I Transient analysis of triplet energy transfer, *Phy. Rev. B* 62, 10958-10966.
Bulovic, V., Khalfin, V. B., Gu, G. & Burrows, P. E. (1998). Weak microcavity effects in organic light-emitting devices, *Phys. Rev. B* 58: 3730-3740.
Che, G., Su, Z., Li, W., Chu, B., Li, M., Hu, Z. & Zhang, Z. (2006). Highly efficient and color-tuning electrophosphorescent devices based on Cu^{I} complex, *Appl. Phys. Lett.* 89: 103511-1 - 103511-3.
Chen, P., Xue, Q., Xie, W., Duan, Y., Xie, G., Hou, J., Liu, S., Zhang, L. & Li, B. (2008). Color-stable and efficient stacked white organic light-emitting devices comprising blue fluorescent and orange phosphorescent emissive units, *Appl. Phys. Lett.* 93: 153508-1 - 153508-3.
Chin, B. D., Suh, M. C., Kim, M. H., Lee, S. T., Kim, H. D. & Chung, H. K. (2005). Carrier trapping and efficient recombination of electrophosphorescent device with stepwise doping profile, *Appl. Phys. Lett.* 86: 133505-1 - 133505-3.
Coushi, K., Kwon, R., Brown, J. J., Sasabe, H. & Adachi, C. (2004). Triplet exciton confinement and unconfinement by adjacent hole-transport layers, *J. Appl. Phys.* 95: 7798-7802.
D'Andrade, B.W., Holmes, R.J. & Forrest, S.R. (2004). Efficient organic electrophosphorescent white-light-emitting device with a triple doped emissive layer, *Adv. Mater.* 16: 624-628.
D'Andrade, B. W. & Forrest, S. R. (2004). White organic light-emitting devices for solid- state lighting, *Adv. Mater.* (Weinheim, Ger.) 16: 1585-1595.
Delley, B. (2000). From molecules to solids with the DMoL3 approach, *J. Chem. Phys.* 113: 7756-7764.
Deng, Z. B., Ding, X. M., Lee, S.T. & Gambling, W.A. (1999). Enhanced brightness and efficiency in organic electroluminescent devices using SiO_2 buffer layers, *Appl. Phys. Lett.* 74: 2227-2279.
Destruel, P., Jolinat, P., Clergereaux, R. & Farenc, J. (1999). Pressure dependence of electrical and optical characteristics of Alq_3 based organic electroluminescent diodes, *J. Appl. Phys.* 85: 397-400.
Dexter, D.L. (1953). A Theory of Sensitized Luminescence in Solids, *J. Chem. Phys.* 21: 836-850.
Duan, J.-P., Sun, P.-P. & Cheng, C.-H. (2003). New iridium complexes as highly efficient orange-red emitters in organic light-emitting diodes, *Adv. Mater.* 15, 224-228.
Endo, J., Matsumoto, T. & Kido, J. (2002). Organic Electroluminescent Devices Having Metal Complexes as Cathode Interface Layer, *Jpn. J. Appl. Phys.* 41: L800-L803.
Forster, T. (1959). Transfer mechanisms of electronic excitation, *Discuss. Faraday Soc.* 27: 7-17.
Friend, R. H., Gymer, R. W., Holmes, A. B., Burroughes, J. H., Marks, R. N., Taliani, C., Bradley, D. D. C., Dos Santos, D. A., Bredas, J. L., Logdlund, M. & Salaneck, W. R. (1999). Electroluminescence in conjugated polymers, *Nature (London)* 397, 121-128.
Fukase, A. & Kido, J. (2002). High efficiency organic electroluminescent devices using Iridium complex emitter and arylamine-containing polymer buffer layer, *Polym. Adv. Technol.* 13: 601-604.

Gao, Y., Wang, L., Zhang, D., Duan, L., Dong, G. & Qiu, Y. (2003). Bright single-active layer small-molecular organic light-emitting diodes with a polytetrafluoroethylene barrier, *Appl. Phys. Lett.* 82: 155-157.

Giebeler, C., Antoniadis, H., Bradley, D. D. C. & Shirota, Y. (1998). Space-charge-limited charge injection from indium tin oxide into a starburst amine and its implications for organic light-emitting diodes, *Appl. Phys. Lett.* 72: 2448-2450.

Gong, X., Ostrowski, J.C., Moses, D., Bazan, G.C. & Heeger, A.J. (2002). Red electrophosphorescence from polymer doped with iridium complex, *App. Phys. Lett.* 81: 3711-3713.

Gong, X., Ostrowski, J.C., Moses, D., Bazan, G.C. & Heeger, A.J. (2003). Electrophosphorescence from a polymer guest-host system with an Iridium complex as guest: Forster energy transfer and charge trapping, *Adv. Funct. Mater.* 13: 439-444.

Hamada, Y., Sano, T., Fujita, M., Fujii, T., Nishio, Y. & Shibata, K. (1993). High luminance in organic electroluminescent devices with bis (10-hydroxybenzo [h]quinolinato) beryllium as an emitter, *Chem. Lett.* 22: 905-909.

Helfrich, W. & Schneider, W.G. (1965). Recombination Radiation in Anthracene Crystals, *Phys. Rev. Lett.* 14: 229-231.

Ho, C.-L., Wong, W.-Y., Wang, Q., Ma, D., Wang, L. & Lin, Z. (2008a). A Multifunctional Iridium-Carbazolyl Orange Phosphor for High-Performance Two-Element WOLED Exploiting Exciton-Managed Fluorescence/Phosphorescence, *Adv. Funct. Mater.* 18: 928-937.

Ho, C.-L., Lin, M.-F., Wong, W.-Y., Wong, W.K. & Chen, C.H. (2008b). High-efficiency and color-stable white organic light-emitting devices based on sky blue electrofluorescence and orange electrophosphorescence, *Appl. Phys. Lett.* 92: 083301-1 - 083301-3.

Holmes, R.J., D'Andrande, B.W., Ren, X., Li, J., Thompson, M.E. & Forrest, S.R. (2003). Efficient, deep-blue organic electrophosphorescence by guest charge trapping, *Appl. Phys. Lett.* 83: 3818-3820.

Hong, I-H., Lee, M.-W., Koo, Y.-M., Jeong, H., Kim, T.-S. & Song, O.-K. (2005). Effective hole injection of organic light-emitting diodes by introducing buckminsterfullerene on the indium tin oxide anode, *Appl. Phys. Lett.* 87: 063502-1 - 063502-3.

Huang, J., Yang, K., Liu, S. & Jiang, H. (2000). High-brightness organic double-quantum-well electroluminescent devices, *Appl. Phys. Lett.* 77: 1750-1752.

Huang, J., Pfeiffer, M., Werner, A., Blochwitz, J., Leo, K. & Liu, S. (2002). Low-voltage organic electroluminescent devices using *pin* structures, Appl. Phys. Lett. 80, 139-141.

Huang, Q., Cui, J., Yan, H., Veinot, J. G. C. & Marks, T. J. (2002). Small molecule organic light-emitting diodes can exhibit high performance without conventional hole transport layers, *Appl. Phys. Lett.* 81: 3528-3530.

Huang, J., Watanabe, T., Ueno, K. & Yang, Y. (2007). Highly Efficient Red Emission Polymer Phosphorescent Lighting Emitting Diodes based on two novel Ir(piq)3 derivatives, *Adv. Mater.* 19: 739-743.

Hung, L.S., Tang, C.W. & Mason, M.G. (1997). Enhanced electron injection in organic electroluminescence devices using an Al/LiF electrode, *Appl. Phys. Lett.* 70: 151-153.

Ikai, M., Tokito, S., Sakamoto, Y., Suzuki, T. & Taga, Y. (2001). Highly efficient phosphorescence from organic light-emitting devices with an exciton-block layer, *Appl. Phys. Lett.* 79: 156-158.

Jeon, W. S., Park, T. J., Park, J. J., Kim, S. Y., Jang, J., Kwon, J. H. & Pode, R. (2008a). Highly efficient bilayer green phosphorescent organic light emitting devices, *Appl. Phys. Lett.* 92: 113311-1 113311-3.

Jeon, W. S., Park, T. J., Park, J. J., Kim, S. Y., Pode, R., Jang, J. & Kwon, J. H. (2008b). Low roll-off efficiency green phosphorescent organic light-emitting devices with simple double emissive layer structure, *Appl. Phys. Lett.* 93: 063303-1 - 063303-3.

Jeon, W.S., Park, T. J., Kim, S. Y., Pode, R., Jang, J. & Kwon, J.-H. (2009). Ideal host and guest system in phosphorescent OLEDs *Org. Electron.* 10: 240-246.

Kawamura, Y., Goushi, K., Brooks, J., Brown, J., Sasabe, H. & Adachi, C. (2005). 100% phosphorescence quantum efficiency of Ir(III) complexes in organic semiconductor films, *Appl. Phys. Lett.* 86: 071104-1 – 071104-3.

Kawamura, Y., Brooks, J., Brown, J., Sasabe, H. & Adachi, C., (2006). Intermolecular Interaction and a Concentration-Quenching Mechanism of Phosphorescent Ir(III) Complexes in a Solid Film, *Phy. Rev. Lett.* 96: 017404-1 – 017404-4.

Kido, J. & Lizumi, Y. (1998). Fabrication of highly efficient organic electroluminescent devices, *Appl. Phys. Lett.* 73: 2721-2723.

Kim, H.-K., Byun, Y.-H., Das, R. R., Choi, B.-K. & Ahn, P.-S. (2007). Small molecule based and solution processed highly efficient red electrophosphorescent organic light emitting devices, *Appl. Phys. Lett.* 91: 093512-1 – 093512-3.

Kim, J. H., Liu, M. S., Jen, A. K. Y., Carlson, B., Dalton, L. R., Shu, C. F. & Dodda, R. (2003). Bright red-emitting electrophosphorescent device using osmium complex as a triplet emitter, *Appl. Phys. Lett.* 83: 776-778.

Kim, K.-K., Lee, J.-Y., Park, T.-J., Jeon, W.-S., Kennedy, G.P. & Kwon, J.-H. (2010). Small molecule host system for solution-processed red phosphorescent OLEDs, *Synthetic Met.* 160: 631-635.

Kim, S. H., Jang, J. S., Hong, J. M. & Lee, J. Y. (2007). High efficiency phosphorescent organic light emitting diodes using triplet quantum well structure, *Appl. Phys. Lett.* 90: 173501-1 - 173501-3.

Kim, S. Y., Jeon, W. S., Park, T. J., Pode, R., Jang, J. & Kwon, J. H. (2009). Low voltage efficient simple *p-i-n* type electrophosphorescent green organic light-emitting devices, *Appl. Phys. Lett.* 94: 133303-1 – 133303-3.

Kim, T.-H., Lee, H. K., Park, O. O., Chin, B. D., Lee, S.-H. & Kim, J. K. (2006). White-Light-Emitting Diodes Based on Iridium Complexes via Efficient Energy Transfer from a Conjugated Polymer, *Adv. Fun. Mater.* 16: 611-617.

Koo, Y.-M., Choi, S.-J., Chu, T.-Y., Song, O.-K., Shin, W.-J., Lee, J.-Y., Kim, J. C. & Yoon, T.-H. (2008). Ohmic contact probed by dark injection space-charge-limited current measurements, *J. Appl. Phys.* 104: 123707-1 – 123797-4.

Kwong, R.C., Nugent, M.R., Michalski, L., Ngo, T., Rajan, K., Tung, Y.-J., Weaver, M.S., Zhou, T.X., Hack, M., Thompson, M.E., Forrest, S.R. & Brown, J.J. (2002). High operational stability of electrophosphorescent devices, *Appl. Phys. Lett.* 81: 162-164.

Lamansky, S., Djurovich, P., Murphy, D., Abdel-Razzaq, F., Lee, H.-E., Adachi, C. Burrows, P.E., Mui, B., Forrest, S.R. & Thompson, M.E. (2001). Highly Phosphorescent Bis-Cyclometalated Iridium Complexes: Synthesis, Photophysical Characterization, and Use in Organic Light Emitting Diodes, *J. Am. Chem. Soc.* 40: 4304-4312.

Lee, T.-C., Chang, C.-F., Chiu, Y.-C., Chi, Y., Chan, T.-Y., Cheng, Y.-M., Lai, C.-H., Chou, P.-T., Lee, G.-H., Chien, C.-H., Shu, C.-F. & Leonhardt, J. (2009). Syntheses, photophysics, and application of iridium(III) phosphorescent emitters for highly efficient, long-life organic light-emitting diodes, *Chemistry – An Asian Journal* 4: 742-753.

Lee, J.-H., Lin, T.-C., Liao, C.-C. & Yang, F. H. (2005). Study on organic light-emitting device with more balanced charge transport, *Proc. of SPIE* 5632: 220-225.

Li, F., Cheng, G., Zhao, Y., Feng, F. & Liu, S.Y. (2003). White-electrophosphorescence devices based on rhenium complexes, *Appl. Phys. Lett.* 83: 4716-4719.

Liu, J., Zhou, Q., Cheng, Y., Geng, Y., Wang, L., Ma, D., Jing, X. & Wang, F. (2006). White electrolumineseence from a single-polymer system with simultaneous two-color emission, Polyfluorene as the blue host and a 2,1,3-benzothiadiazole derivative as the orange dopant on the main chain, *Adv. Funct.Mater.* 16: 957-965.

Liu, S. W., Huang, C. A., Lee, J. H., Yang, K. H., Chen, C. C. & Chang, Y. (2004). Blue mixed host organic light emitting devices, *Thin Solid Films* 453-454: 312-315.

Liu, Y., Guo, J., Feng, J., Zhang, H., Li, Y. & Wang, Y. (2001). High-performance blue electroluminescent devices based on hydroxyphenyl-pyridine beryllium complex, *Appl. Phys. Lett.* 78: 2300-2302.

Liu, Z. W., Helander, M. G., Wang, Z. B. & Lu, Z. H. (2009). Efficient bilayer phosphorescent organic light-emitting diodes: Direct hole injection into triplet dopants, *Appl. Phys. Lett.* 94, 113305-1 - 113305-3.

Liu, Z., Helander, M. G., Wang, Z. & Lu, Z. (2009). Efficient single layer RGB phosphorescent organic light-emitting diodes, *Organic Electron.*10: 1146-1151.

Liu, S. W., Lee, J. H., Lee, C. C., Chen, C. T. & Wang, J. K. (2007). Charge carrier mobility of mixed-layer organic light-emitting diodes, *Appl. Phys. Lett.* 91: 142106-1 - 142106-3.

Meyer, J., Hamwi, S., Bülow, T., Johannes, H.-H., Riedl, T. & Kowalsky, W. (2007). Highly efficient simplified organic light emitting diodes, *Appl. Phys. Lett.* 91: 113506-1 - 113506-3.

Mi, B. X., Wang, P. F., Gao, Z. Q., Lee, C. S., Lee, S. T., Hong, H. L., Chen, X. M., Wong, M. S., Xia, P. F., Cheah, K. W., Chen, C. H. & Huang, W. (2009). Strong Luminescent Iridium Complexes with C^N=N Structure in Ligands and Their Potential in Efficient and Thermal-stable Phosphorescent OLEDs, *Adv. Mater.* 21: 339-343.

Ohmori, Y., Kajii, H. & Hino, Y. (2007). Organic Light-Emitting Diodes Fabricated by a Solution Process and Their Stress Tolerance, *J. of Display Technology* 3: 238-244.

Park, T.-J., Jeon, W.-S., Park, J.-J., Kim, S.-Y., Lee, Y.-K., Jang, J., Kwon, J.-H. & Pode, R. (2008). Efficient simple structure red phosphorescent organic light emitting devices with narrow band-gap fluorescent host, *Appl. Phys. Lett.* 92: 113308-1 - 113308-3.

Pfeiffer, M., Forrest, S.R., Leo, K. & Thompson, M.E. (2002). Electrophosphorescent p-i-n organic light-emitting devices for very-high- efficiency flat-panel displays, *Adv. Mater.* 14: 1633-1636.

Pode, R., Kim, K.-H., Kwon, J.-H., Lee, S.-J., Shin, I.-A. & Jin, S.-H. (2010). Solution processed efficient orange phosphorescent organic light-emitting device with small molecule host, *J Phys. D: Appl. Phys.* 43: 025101-1 - 025101-5.

Pode, R., Ahn, J.-S., Jeon, W. S., Park, T. J. & Kwon, J.-H. (2009). Efficient red light phosphorescence emission in simple bi-layered structure organic devices with fluorescent host-phosphorescent guest system, *Current Applied Physics* 9, 1151-1154.

Pope, M., Kallmann, H.P. & Magnante, P. (1963). Electroluminescence in Organic Crystals, *The Journal of Chemical Physics* 38: 2042- 2043.

Qiu, Y., Gao, Y., Wei, P. & Wang, L. (2002a). Organic light-emitting diodes with improved hole-electron balance by using copper phthalocyanine/aromatic diamine multiple quantum wells, *Appl. Phys. Lett.* 80: 2628-2630.

Qiu, Y., Gao, Y., Wang, L., Wei, P., Duan, L., Zhang, D. & Dong, G. (2002b). High-efficiency organic light-emitting diodes with tunable light emission by using aromatic diamine/5,6,11,12-tetraphenylnaphthacene multiple quantum wells, *Appl. Phys. Lett.* 81: 3540-3542.

Ramos-Ortiz, G., Oki, Y., Domercq, B. & Kippelen, B. (2002). Förster energy transfer from a fluorescent dye to a phosphorescent dopant: a concentration and intensity study, *Physical Chemistry Chemical Physics (RSC Publishing)* 4: 4109-4114.

Sheats, J. R., Antoniadis, H., Hueschen, M., Leonard, W., Miller, J., Moon, R., Roitman, D. & Stocking, A. (1996). Organic Electroluminescent Devices, *Science* 273: 884-888.

Shirota, Y., Kuwabara, Y., Inada, H., Wakimoto, T. , Nakada, H., Yonemoto, Y., Kawami, S. & Imai, K. (1994). Multilayered organic electroluminescent device using a novel starburst molecule, 4,4',4''-tris(3-methylphenylphenylamino)triphenylamine, as a hole transport material, *Appl. Phys. Lett.* 65: 807-809.

Seo, J.H., Park, J.H., Kim, Y.K., Kim, J.H., Hyung, G.W., Lee, K.H. & Yoon, S.S. (2007). Highly efficient white organic light-emitting diodes using two emitting materials for three primary colors (red, green, and blue), *Appl. Phys. Lett.* 90: 203507-1 - 203507-3.

Shen, Z., Burrows, P. E., Bulovic, V., Forrest, S. R. & Thompson, M. E. (1997). Three-Color, Tunable, Organic Light-Emitting Devices, *Science* 276: 2009-2011.

Shoustikov, A., You, Y., Burrows, P.E., Thompson, M.E. & Forrest, S.R. (1997). Orange and red organic light-emitting devices using aluminum tris(5-hydroxyquinoxaline), *Synth. Met.* 91, 217-221.

Tanaka, D., Sasabe, H., Li, Y.J., Su, S.-J., Takeda, T. & Kido, J. (2007). Ultra High Efficiency Green Organic Light-Emitting Devices, *Jpn. J. Appl. Phys. 2 Lett. (Japan)* 46: L10-L12.

Tanaka, I. & Tokito, S. (2008). Energy-transfer processes between phosphorescent guest and fluorescent host molecules in phosphorescent OLEDs, Edited by H. Yersin, *Highly efficient OLEDs with Phosphorescent materials*, Wiley -VCH verlag GmbH & Co. KGaA, Weinheim.

Tanaka, I., Tabata, Y. & Tokito, S. (2004). Energy-transfer and light-emission mechanism of blue phosphorescent molecules in guest-host systems, *Chem. Phys. Lett.* 400: 86-89.

Tang, C.W., Chen, C. H. & Goswami, R. (1988). Electroluminescent device with modified thin film luminescent zone, US Patent No. 4 769 292.

Tse, S. C., Tsung, K. K. & So, S. K. (2007). Single-layer organic light-emitting diodes using naphthyl diamine, *Appl. Phys. Lett.* 90: 213502-1 - 213502-3.

Tse, S. C., Tsang, S. W. & So, S. K. (2006). Polymeric conducting anode for small organic transporting molecules in dark injection experiments, *J. Appl. Phys.* 100: 063708-1 - 063708-5.

Tsuboi, T., Jeon, W. S. & Kwon, J. H. (2009). Observation of phosphorescence from fluorescent organic material $Bebq_2$ using phosphorescent sensitizer, *Opt. Mater.* 31: 1755-1758.

Tsuboyama, A., Iwawaki, H., Furugori, M., Mukaide, T., Kamatani, J., Igawa, S., Moriyama, T., Miura, S., Takiguchi, T., Okada, S., Hoshino, M. & Ueno, K. (2003). Homoleptic Cyclometalated Iridium Complexes with Highly Efficient Red Phosphorescence and Application to Organic Light-Emitting Diode, *J. Am. Chem. Soc.* 125: 12971-12979.

Tsujimoto, H., Yagi, S., Asuka, H., Inui, Y., Ikawa, S., Maeda, T., Nakazumi, H. & Sakurai, Y. (2010). Pure red electrophosphorescence from polymer light-emitting diodes doped with highly emissive bis-cyclometalated iridium(III) complexes, *J. of Organometallic Chemistry* 695: 1972-1978.

Tsuzuki, T. & Tokito, S. (2008). Highly Efficient Phosphorescent Organic Light-Emitting Diodes Using Alkyl-Substituted Iridium Complexes as a Solution-Processible Host Material, *Appl. Phys. Express* 1: 02185-1 - 02185-3.

Tsuzuki, T. & Tokito, S. (2007). Highly Efficient, Low-Voltage Phosphorescent Organic Light-Emitting Diodes Using an Iridium Complex as the Host Material, *Adv. Mater.* 19: 276-280.

Tsuzuki, T., Shirasawa, N., Suzuki, T. & Tokito, S. (2003). Color tunable organic light-emitting diodes using pentafluorophenyl-substituted iridium complexes, *Adv. Mater.* 15: 1455-1458.

Turro, N. J. (1991a). *Modern Molecular Photochemistry*, University Science Books, USA, pp 305.

Turro, N.J. (1991b). *Modern Molecular Photochemistry*, University Science Books, USA, pp 186.

VanSlyke, S. A. & Tang, C.W. (1985). Organic electroluminescent devices having improved power conversion efficiencies, US Patent 4539507.

VanSlyke, S. A., Chen, C.H. & Tang, C.W. (1996). Organic electroluminescent devices with improved stability, *Appl. Phys. Lett.* 69: 2160-2162.

Vanslyke, S. A., Tang, C. W., O'brien, M. E. & Chen, C. H. (1991). Electroluminescent device with organic electroluminescent medium, US Patent 5061569.

Wakimoto, T., Fukuda, Y., Nagayama, K., Yokoi, A., Nakada, H. & Tsuchida, M. (1997). Organic EL Cells Using Alkaline Metal Compounds as Electron Injection Materials, *IEEE Trans. Electron Devices* 44: 1245-1248.

Wang, H. F., Wang, L. D., Wu, Z. X., Zhang, D. Q., Qiao, J., Qui, Y. & Wang, X. G. (2006). Efficient single-active-layer organic light-emitting diodes with fluoropolymer buffer layers, *Appl. Phys. Lett.* 88: 131113-1 - 131113-3.

Williams, E. L. , Haavisto, K., Li, J. & Jabbour, G. E. (2007). Excimer-Based White Phosphorescent Organic Light-Emitting Diodes with Nearly 100% Internal Quantum Efficiency, *Adv. Mater.* 19: 197-202.

Zhou, X., Nimoth, J. B., Pfeiffer, M., Maennig, B., Drechsel, J., Werner, A. & Leo, K. (2003). Inverted transparent multi-layered vacuum deposited organic light-emitting diodes with electrically doped carrier transport layers and coumarin doped emissive layer, *Synth. Met.* 138: 193-196.

5

Organic Field-Effect Transistors Using Hetero-Layered Structure with OLED Materials

Ken-ichi Nakayama, Yong-Jin Pu, Junji Kido and Masaaki Yokoyama
Yamagata University, Osaka University
Japan

1. Introduction

In recent years, organic transistors have attracted much attention due to their advantages in developing low-cost, flexible, and large-area production. So far, many kinds of organic materials have been reported to achieve high-performance organic field-effect transistors (OFETs). There are two types of organic semiconductors, p-type and n-type, whose majority carriers are holes and electrons, respectively. For logic gates application, both types and similar performance OFETs are required for CMOS application. Pentacene is the most popular material in p-type OFET, and many kinds of polymer materials are also reported (McCulloch et al., 2006). On the other hand, the performance of n-type OFETs is generally inferior to that of p-type (Dimitrakopoulos and Malenfant, 2002). In particular, stability in air is the most serious problem in n-type OFET. Fullerene is the most standard n-type material showing the highest mobility (Singh et al., 2007); however, the device cannot operate in air.

There are two guidelines to achieve high mobility and high stability in n-type OFET. One is to develop a new material having deeper LUMO level. Oxygen and water deteriorate OFET performance by accepting electrons from the semiconductor molecule. Therefore, enough deep LUMO level is an efficient way to avoid effect of oxygen or water. In fact, there have been many materials having deep LUMO levels, for example, perylene bisimide compound, fullerene derivatives, fluorinated compounds, and so on.

The other important point is surface treatment of the insulator. The field-effect mobility of the organic semiconductor is strongly affected by the device fabrication process. Various methods on surface treatments have been reported to improve the carrier mobility. The HMDS treatment is a standard and efficient way to make the surface hydrophobic (Lin et al., 1997; Lim et al., 2005). Organic semiconductor can aggregate with high crystallinity on the hydrophobic surface without influence of the substrate surface. These methods were developed in p-type OFETs; however, they are also efficient to improve the mobility and stability of n-type OFETs. Recently, it has been pointed out that low mobility and instability in air of n-type organic semiconductor is attributed to the surface electron traps of the gate insulator, and if electron traps can be perfectly eliminated, almost organic semiconductors can be operate in n-type mode (Chua et al., 2005). Therefore, it has been believed that the gate insulator surface should be as possible as inert to achieve high mobility and stability in n-type OFETs.

In this chapter, we introduce a new concept of a hetero-layered OFET to improve the performance of OFETs instead of conventional surface treatment methods. The hetero-layered OFET includes an interfacial layer of electronic active organic semiconductor having opposite transport polarity between the insulator and channel layer. For the interfacial layer of n-type OFET, we employed various types of hole transporting material, which are generally used for organic light-emitting diodes (OLEDs). For p-type OFET, electron transporting material was employed.

Such a hetero-layered OFETs composed of p-type and n-type organic semiconductors have been studied for ambipolar organic transistors, which aimed at the simple inverter circuit or organic light-emitting transistors (Rost, 2004; Rost et al., 2004). On the other hand, our proposed hetero-layered OFET employs charge transport material of OLEDs. They generally form amorphous films resulting in no FET operation by themselves.

The proposed hetero-layered OFET showed improvement of the mobility compared to the conventional surface treatment. In addition, we found that the stability in air was drastically improved in n-type OFET by using a hole transporting material having higher HOMO level. We discuss the relationship between the OFET performance and the electronic property of the interfacial layer.

2. Perylene bisimide and hole transporting materials

In this section, we will introduce the results of perylene bisimide (PTCDI-C8H) for the channel layer and the hole transporting material of NPD, TAPC and m-MTDATA for the interfacial layer. Perylene bisimide compounds are promising n-type organic semiconductor having deep LUMO levels and high crystallinity. In particular, PTCDI with long alkyl chains bring about a highly ordered film structure, and very high electron mobility has been reported (Tatemichi et al., 2006). On the other hand, NPD and TAPC having triphenyl amine structure are very standard hole transporting material for OLED devices. They show comparably high hole mobility and good film formation.

Figure 1 shows the hetero-layered structure OFET with top contact and the molecular structures of m-MTDATA and NPD. Organic transistors were fabricated on a heavily doped Si substrate with SiO_2 layer (300 nm) that works as a common gate electrode. The interfacial semiconductor layer of m-MTDATA and NPD (20 nm ~ 30 nm) were deposited by thermal evaporation. For the comparison, the substrates with well-known surface treatment by octadecyltrichlorosilane (OTS) and hexamethyldisilazane (HMDS) were also prepared. Au source and drain electrodes were deposited through a shadow mask. Channel length and width were defined to be 50 μm and 5.5 mm, respectively. The current modulation of OFETs were measured by a semiconductor parameter analyzer in the glove box, where the concentration of oxygen and water were less than 1 ppm. The field-effect mobility, threshold voltage and on/off ratio were estimated from the equation of saturation regime, $I_D=[(WC\mu)/2L](V_G - V_T)^2$, where C is the capacitance per unit area of the gate dielectrics, W is the channel width, L is the channel length, μ is the carrier mobility, and V_T is the threshold voltage.

Figure 2 shows the transfer characteristics of OFETs with an interfacial layer of NPD, those subjected to HMDS surface treatment, and those without any interfacial layer and not subjected to surface treatment (None). In all the devices, the source-drain current (I_D) increased with the positive gate voltage (V_G), which indicates that these OFETs operate only in the n-type mode, and the hole-transporting layer does not acts as a p-type channel layer.

The performances of OFETs with different interfacial layers are summarized in **Table 1**. The optimum thickness of the interfacial layer is also indicated. The mobility was improved with increasing thickness of the hole transporting layer and showed a maximum around 20 nm. The mobility for heterolayered device was estimated assuming the gate capacitance of only SiO_2 because it is difficult to determine the channel interface. The conventional HMDS treatment resulted in an improvement in the mobility from 2.5×10^{-2} cm^2/Vs (None) to 6.9×10^{-2} cm^2/Vs. Interfacial layers composed of NPD and m-MTDATA increased the mobility up to 0.11 and 0.13 cm^2/Vs, respectively.

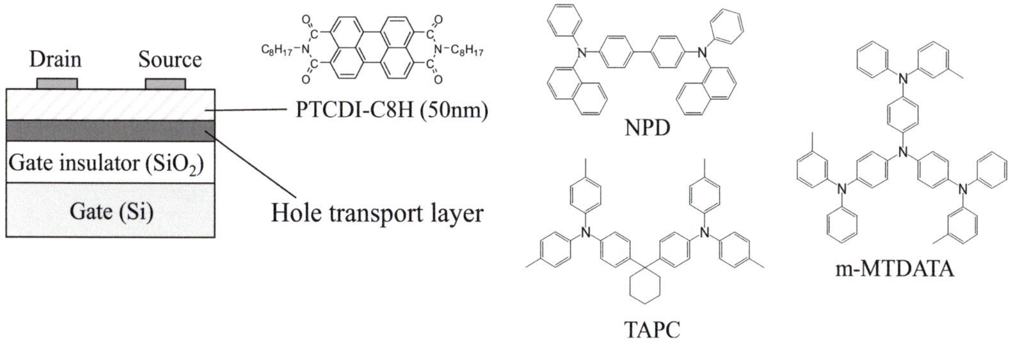

Fig. 1. Device structure of hetero-layered OFET using PTCDI-C8H and hole transporting materials.

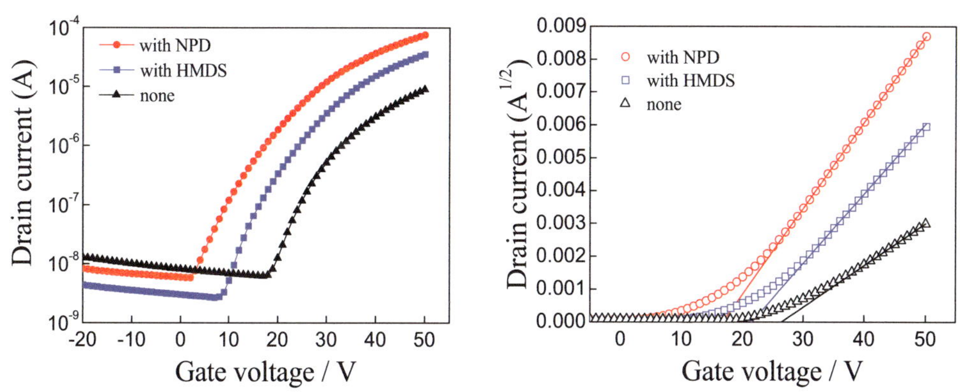

Fig. 2. Transfer characteristics of the n-type OFET with hetero-layered structure and conventional surface treatment.

These results indicate that an electronically active material can be used to fabricate an interfacial layer, and high performance can be achieved without a surface treatment of self-assembly monolayer. We also investigated some other organic materials, n-hexatriacontane that is perfectly inert material, and Alq_3 that is a well-known emissive and electron transporting material. In these cases, the mobility was rather reduced to be 10^{-3} cm^2/Vs. These results enable us to conclude that hole transporting materials are responsible for enhancing mobility.

Mobility can also be improved by modifying the structure of the semiconductor film. X-ray diffraction patterns of PTCDI-C8H films with and without the NPD (10 nm) interfacial layer were measured under the same condition (**Fig. 3**). Patterns of both the films showed a very strong peak at 4.3° corresponding to d = 2.05 nm. This peak is assigned to the long axis of the molecules, which are aligned vertically on the surface. However, in the case of the PTCDI-C8H films with the NPD interfacial layer, the diffraction peaks are rather weak, which is also supported by the fact that the high order peaks become unclear, as shown in the magnified inset of **Fig. 3**. This result indicates that the improvement in mobility caused by the hole-transporting interfacial layer is not attributed to the increase in crystallinity of the PTCDI-C8H film. This interpretation is also supported by contact angle measurements. The contact angle of the interfacial layer was 85.8° for m-MTDATA and 92.5° for NPD. These values are comparable to that of HMDS-treated SiO_2 surface. This fact also indicates that the mobility improvement can be attributed to the electronic effect of hole transporting layer.

Surface	Mobility ($cm^2 V^{-1} s^{-1}$)	Threshold (V)	On/off ratio
Bare	0.0351	20.4	5.18×10^3
HMDS	0.0690	21.5	1.32×10^4
n-hexatriacontane (15 nm)	0.0407	11.8	3.00×10^3
TAPC (20 nm)	0.0713	15.5	1.28×10^4
NPD (10 nm)	0.110	16.7	1.17×10^4
m-MTDATA (20 nm)	0.127	25.3	1.33×10^4
Alq_3 (15 nm)	0.00686	12.4	-

Table 1. Performances of PTCDI-C8H OFETs with different interfacial layers between the gate insulator and the channel layer.

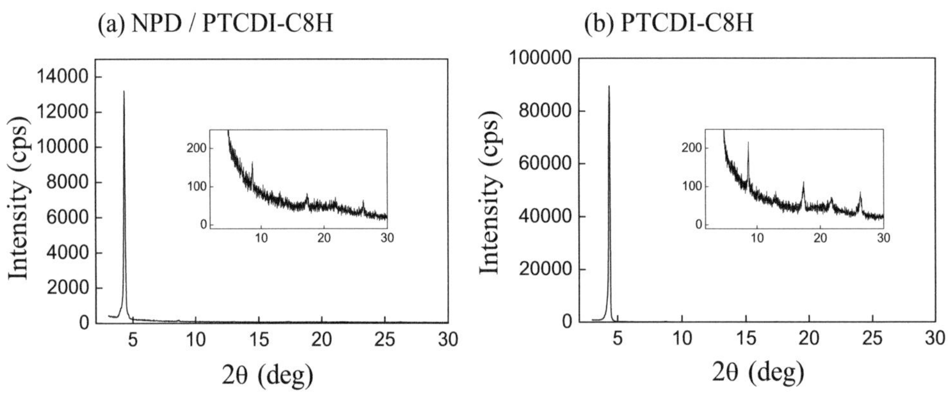

Fig. 3. X-ray diffraction patterns of (a) NPD (10 nm)/PTCDI-C8H (50 nm) film and (b) PTCDI-C8H (50 nm) film deposited on the Si/SiO_2 substrate.

Energy levels (highest occupied molecular orbital (HOMO) and LUMO levels) of the organic semiconductors used in this study are shown in **Fig. 4**. The n-type organic semiconductor,

PTCDI-C8H, has a deep LUMO level of 4.6 eV. On the other hand, hole transport material of NPD and TAPC has higher HOMO level and wide energy gap exceeding 3 eV. Therefore, LUMO level of the interfacial layer is much higher than that of the channel layer.

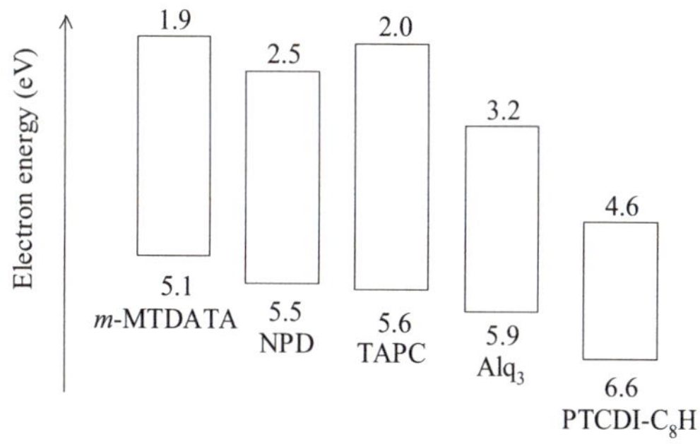

Fig. 4. Energy level diagrams of organic semiconductors used in this study. Upper and lower values indicate the LUMO and HOMO.

These results can be interpreted as following model. Figure 5 shows the schematic energy diagram of the heterolayered OFET. Since the LUMO level of the interfacial layer was considerably higher than that of the n-type semiconductor layer, electrons would not enter the interfacial layer. In addition, the hole-transporting layer did not show p-type FET operation. Therefore, it was concluded that an n-type channel was formed at the interface between the hole-transporting layer and the n-type semiconductor film. The role of the interfacial layer can be basically attributed to the separation of the channel carriers from the surface electron traps, similar to the conventional hydrophobic surface treatment. However, it was noted that the mobility or threshold voltage had a correlation with the HOMO level of the inserted layer. Mobility increased in the order of m-MTDATA > NPD > TAPC, which corresponded to the order of the HOMO levels, i.e., the interfacial layer with a higher HOMO level exhibited better performance. This result suggests that the nature of semiconductor of the interfacial layer affected the electron transportation process at the interfacial channel.

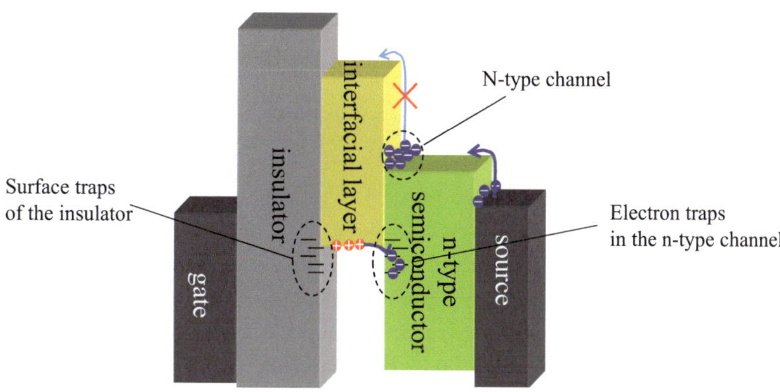

Fig. 5. Schematic relationship of energy levels in the hetero-layered OFET composed of hole transporting layer and n-type organic semiconductors.

This additional effect should be discussed from the viewpoint of electronic interaction between the hole transporting layer and the n-type channel layer. In the single layer device

of PTCDI-C8H, the surface electron traps of SiO_2 can be passivated by inert surface treatment like HMDS. However, there would be many electron traps in the PTCDI-C8H film itself. They cannot be eliminated by surface treatment of the substrates. On the other hand, the hole transporting materials generally have higher HOMO levels, in other words, electron donating character. Therefore, the interfacial layer tends to give electrons toward PTCDI-C8H film at the interface. It may not eliminate the shallow electron traps because the HOMO level of NPD is far from the LUMO level of PTCDI-C8H, but the deep electron traps are expected to be filled in advance by thermally activated charge transfer. As a result, the injected electrons can move smoothly at the interface, resulting in the observed high electron mobility. We conclude that this trap-filling effect is essential of the hetero-layered OFET.
Thus, we proposed the concept of hetero-layered OFETs and ascertained its validity. The performance was improved by insertion of the electronic active material rather than an inert surface treatment. Because the film structure of the deposited PTCDI-C8H was not changed by the surface treatment or interfacial layer, we concluded that this improvement is attributed to electron donating character of the hole transporting layer.

3. C_{60} and hole transporting materials

In this section, the concept of hetero-layered OFET is applied to the combination of C_{60} channel layer and hole transporting material (**Fig. 6**). C_{60} is the most standard material of n-type organic semiconductors and the highest performance in n-type OFET has been reported. The device structure is the same structure with the previous section. For the interfacial layer, typical hole transporting material of NPD and m-MTDATA were used.

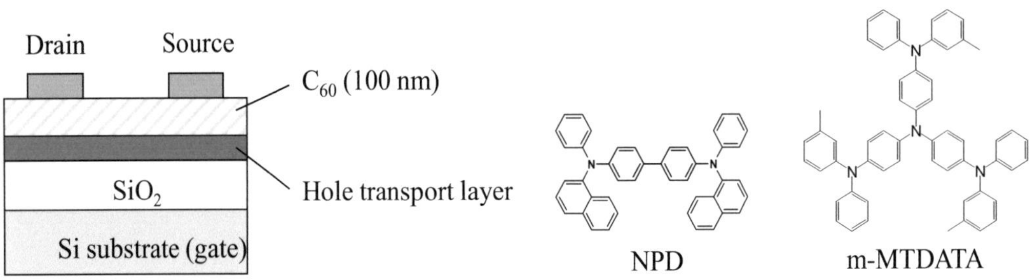

Fig. 6. Device structure of hetero-layered OFET composed of C_{60} and hole transporting materials.

Figure 7 shows the drain current–gate voltage (I_D-V_G) characteristics of the hetero-layered OFETs with m-MTDATA and NPD interfacial layer, and the single layer C_{60} OFETs on OTS-treated, HMDS-treated, and non-treated substrates. Also in this case, the I_D - V_G curves for the hetero-layered devices increased only for positively biased gate voltage with almost no hysteresis for forward and backward sweeps. This means that they did not operate as an ambipolar transistor, and the interfacial layer of the hetero-layered device did not work as a p-type channel layer.
The performance of each device were summarized in **Table 2**. The conventional surface treatment by OTS and HMDS brought about high electron mobility of 0.50 and 0.80 cm^2/Vs, respectively, whereas the normal device on the non-treated substrate showed low mobility of 7.5×10^{-3} cm^2/Vs. However, it should be noted that the hetero-layered device with m-

MTDATA and NPD achieved very high electron mobility of 1.1 and 1.8 cm²/Vs, respectively. These values are the highest value for C_{60} FETs without any surface treatment or substrate heating. Thus, it was revealed that the hetero-layered OFET is generally efficient to improve the performance even in high performance OFETs using C_{60}.

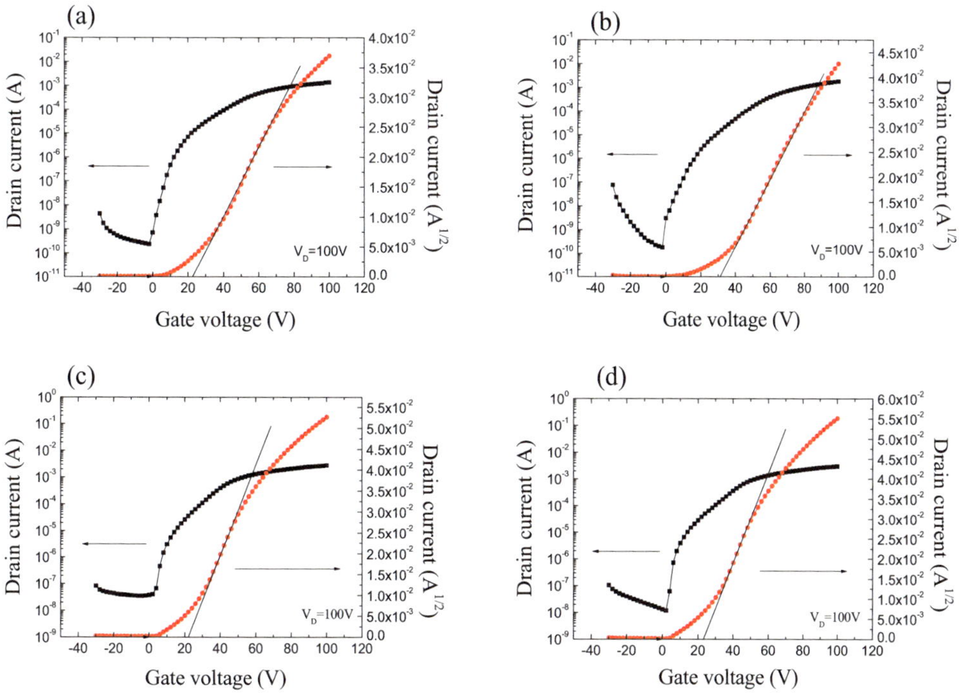

Fig. 7. Transfer characteristics of OFETs devices of C_{60} film on the various kinds of surface, (a) OTS, (b) HMDS, (c) m-MTDATA (20 nm) films, and (d) NPD (30 nm).

Surface	Mobility ($cm^2 V^{-1} s^{-1}$)	Threshold (V)	On/off ratio
SiO_2	0.0075	66	6.6×10^4
OTS	0.50	23	5.8×10^6
HMDS	0.80	30	1.0×10^7
m-MTDATA	1.8	23	7.8×10^4
NPD	1.8	22	2.6×10^5

Table 2. The performances of the OFETs with various interfacial layers.

The mobility improvement can be caused also by change of the film structure. In this section, the film structures were evaluated by using atomic force microscopy (AFM) because

C_{60} films deposited at room temperature generally shows no diffraction peak in XRD measurements. **Figure 8** shows the morphology of the C_{60} films deposited on m-MTDATA and NPD interfacial layer, and those on the non-treated, OTS-treated and HMDS-treated substrates. The deposited film of C_{60} has granular surface with a diameter around 100 nm, and almost no difference was observed for all the films. The root-mean-square (RMS) roughness of the C_{60} films on the m-MTDATA (1.9 nm) and NPD (1.9 nm) are almost the same with those on the non-treated substrate (1.5 nm), OTS-treated substrate (1.5 nm), and HMDS-treated substrate (2.9 nm). These results indicate that the observed improvement of electron mobility was not due to the morphological change of the C_{60} films.

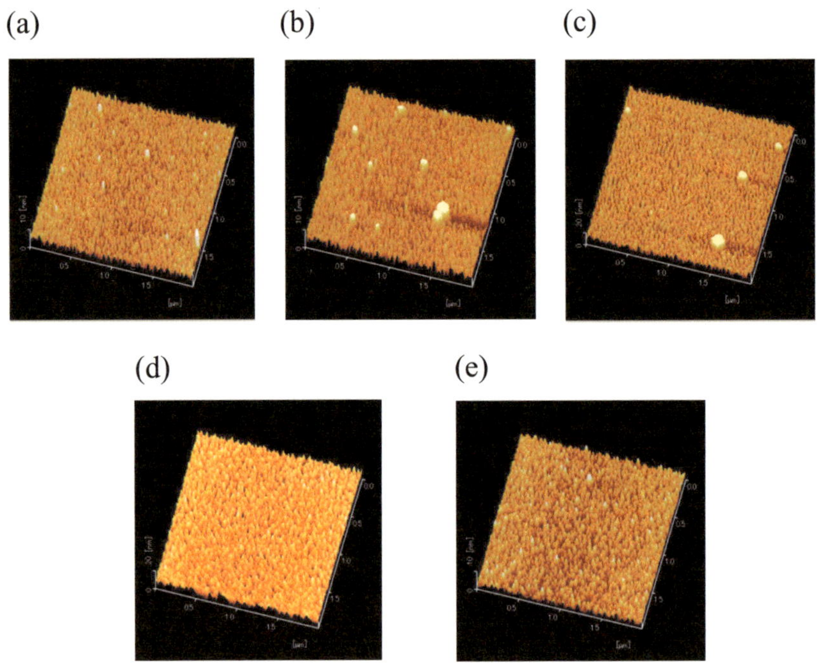

Fig. 8. The AFM images (2×2 μm^2) of C_{60} deposited film (100 nm) on the various kinds of surface, (a) non-treated SiO_2, (b) OTS, (c) HMDS, (d) m-MTDATA (20 nm), and (e) NPD (30 nm).

In the same manner as the previous section, the improvement by the hetero-layered structure of C_{60} OFET is attributed to the electronic effect at the interface. There is a large electron injection barrier from C_{60} (4.20 eV) to m-MTDATA (1.90 eV) or NPD (2.40 eV). Therefore, when the gate is positively biased, injected electrons from the source electrode would accumulate at the interface between C_{60} and hole transporting layer. Also in this case, the primary effect of the interfacial layer would be isolation of channel electrons from the SiO_2 substrate surface having electron traps. This interpretation is supported by the fact that the threshold voltage becomes smaller (negatively shift) than that of non-treated device, which is similar to the effect of OTS and HMDS treatment. However, the hetero-layered OFETs with m-MTDATA showed higher mobility than that with OTS and HMDS treatment. This also means, electron donating character of hole transport layer and electron accepting character of C_{60} would cause partial electron transfer to fill the surface or interfacial traps in the C_{60} film.

These effects also affect the air stability of n-type operation. It is well-known that n-type OFET is very sensitive to oxygen and water and does not work in air. **Figure 9** shows the degradation characteristics of the field-effect mobility of C_{60} OFETs with exposure time to air. The device was placed in a dark box without any sealing under humidity of 30 ~ 40 %. The normal device with a bare surface showed rapid decrease of the field-effect mobility after exposure to air, and almost no operation was observed within 100 hours. The OTS treatment improved the initial performance; however, the degradation in air could not be prevented. On the other hand, the device of the hetero-layered OFETs with hole transporting layer showed much better stability. The device composed of NPD and C_{60} showed the mobility larger than 10^{-2} cm^2/Vs after 1000 hours exposure to air.

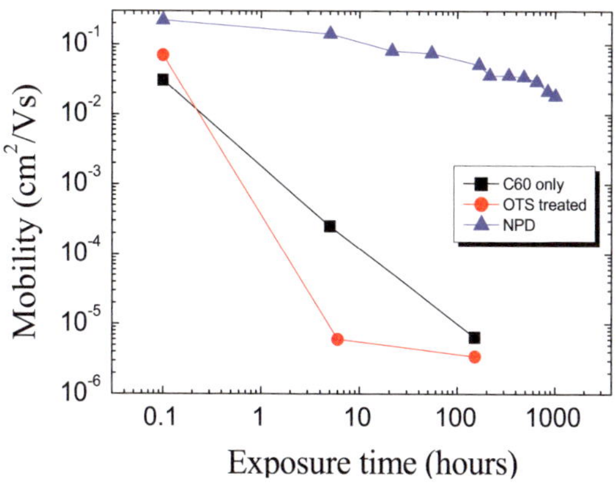

Fig. 9. Degradation characteristics of the field-effect mobility of C_{60} OFETs in air.

Generally, degradation of field-effect mobility in n-type OFET is interpreted as an increase of electron traps caused by oxygen or water at the channel layer or its interface with the insulator. In the same way as the initial performance, hole transporting layer having higher HOMO level gives electrons to C_{60} in the ground state to fill the electron traps. Because the LUMO level of C_{60} is higher than the HOMO level of NPD, this charge transfer is partial and requires thermal activation. Therefore, additional oxygen or water by exposure to air would be compensated by the interfacial layer, resulting in long lifetime under atmospheric condition.

Thus, the concept of the hetero-layered OFET was extended to high performance n-type OFET using C_{60}. Also in this case, the electron mobility was improved by the interfacial layer of NPD. In addition, the stability in air was drastically improved. These results also can be explained by partial electron transfer from the hole transporting layer to n-type channel layer leading to trap filling.

4. Pentacene and electron transporting materials

In this section, we extend the hetero-layered concept to the opposite combination of materials, that is, p-type organic semiconductor and electron transporting interfacial layer. For p-type semiconductor, pentacene was used as the most standard material. For the interfacial layer, many kinds of electron transporting materials were employed as shown in **Figure 10**. Most of them are electron transporting materials for OLED device forming amorphous film. The n-type organic semiconductor like NTCDA and $HAT(CN)_6$ were also investigated. The HOMO and LUMO level of each material is shown in **Figure 11**. The materials are arranged by the LUMO levels representing electron accepting characters to discuss the energetic effects later.

The devices were fabricated in the same way and the device performance was measured in the glove box purged with dry nitrogen gas. The thickness of was 1.0 nm for the interfacial layer and 50 nm for the pentacene film.

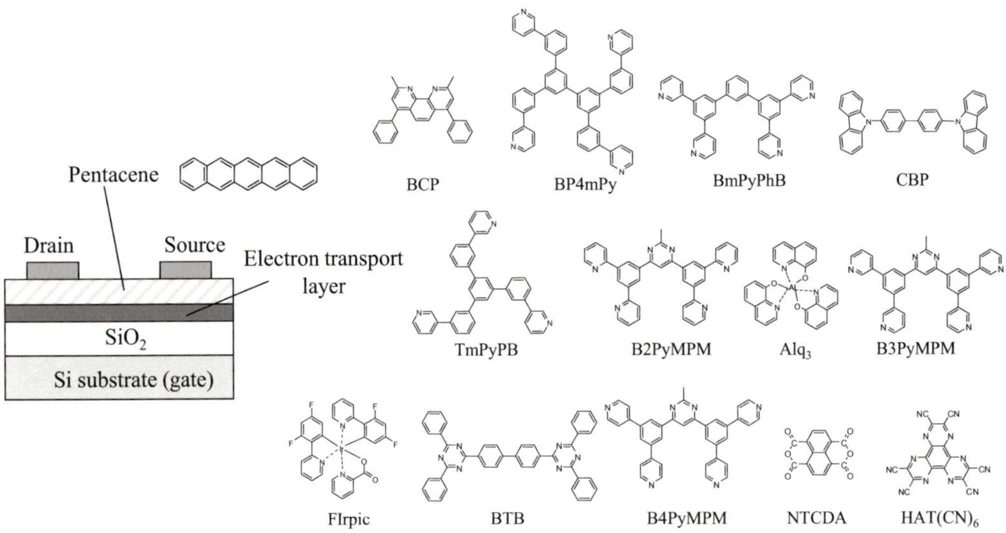

Fig. 10. Device structure of hetero-layered structure OFET using pentacene and electron transporting layer.

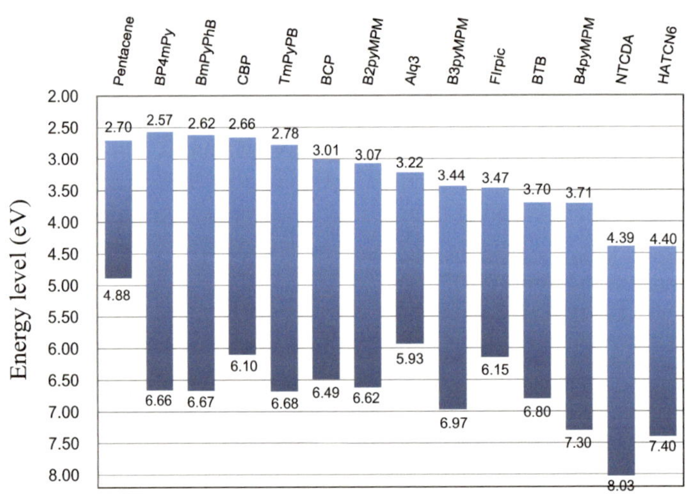

Fig. 11. Energy diagrams of the electron transporting materials used for interfacial layers.

The output curves of the hetero-layered OFETs using each interfacial layer were shown in **Figure 12**. All the devices showed pure p-type operation and no ambipolar operation was observed. The off current was decreased by inserting an interfacial layer of most electron transporting materials. On the other hand, in the case of NTCDA and $HAT(CN)_6$, the off current was increased and on/off ratio became lower.

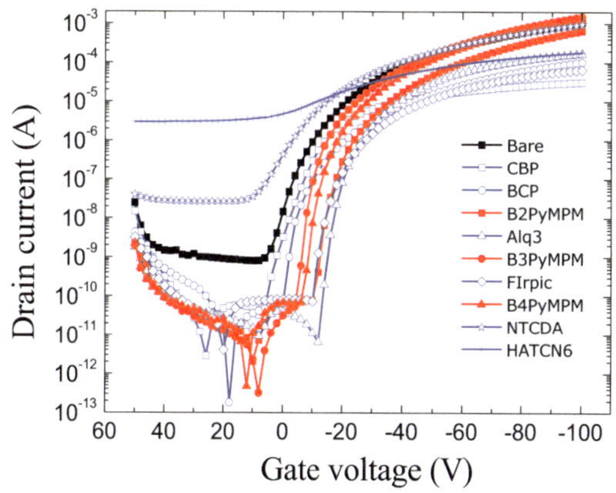

Fig. 12. Transfer curves of the heterolayered OFET composed of p-type pentacene and electron transporting interfacial layers.

Surface	Mobility ($cm^2 V^{-1}s^{-1}$)	Threshold (V)	On/off ratio
Bare	0.282	-21	1.35×10^6
BP4mPy	0.276	-28.7	1.89×10^9
BmPyPhB	0.361	-31.8	6.13×10^{10}
CBP	0.019	-4.7	1.09×10^7
TmPyPB	0.167	-29.1	2.03×10^9
BCP	0.024	-18.2	5.67×10^9
B2PyMPM	0.320	-36.7	3.20×10^9
Alq_3	0.068	-22.8	2.57×10^7
B3PyMPM	0.486	-28.4	4.68×10^9
Firpic	0.019	-18.6	1.61×10^7
B4PyMPM	0.388	-29.1	2.57×10^9
NTCDA	0.193	-11.3	3.61×10^4
$HAT(CN)_6$	0.033	10.7	6.05×10^1
BTB	0.045	-14.8	1.48×10^9

Table 3. FET performance of the hetero-layered OFET composed of p-type pentacene and electron transporting interfacial layers.

The FET performance of each device is summarized in **Table 3**. The field-effect mobility was improved in some cases, and B3PyMPM showed the highest mobility of 0.486 cm^2/Vs among these interfacial materials. The thickness of interfacial layers was 1 nm that is much thinner compared to the HTL/n-type layered OFET. In p-type hetero-layered device, thick interfacial layer ~ 10 nm rather decreased the mobility in many cases. These results imply that role of the interfacial layer is different with n-type hetero-layered devices.

We discussed the correlation between the OFET performance and the LUMO level of the interfacial materials. **Figure 13** shows the correlation between LUMO levels and field-effect mobility, and threshold voltages. From **Fig. 13 (a)**, no correlation with the field-effect mobility was observed. These results imply the mobility is not determined by the electronic property of the interfacial material and charge transfer effect is not concerned. It would be because the hole transport is more stable and less affected by the interfacial traps compared to electron transport. On the other hand, weak correlation with threshold voltage was observed as shown in **Fig. 13 (b)**. It was found that the threshold voltage becomes higher (positive shift) as the LUMO levels of the interfacial layer becomes lower (deeper). These results indicates that the charge transfer from pentacene to the electron transporting material promotes hole accumulation in the pentacene film to the gate voltage application. In the case of NTCDA and $HAT(CN)_6$, their electron accepting character is so strong that hole doping occurred and off current was increased.

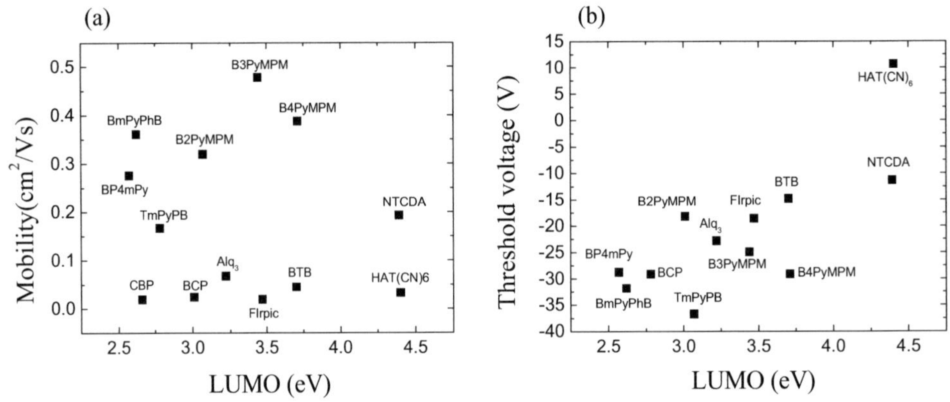

Fig. 13. Correlation between LUMO levels of interfacial electron transporting materials and (a) field-effect mobility, (b) threshold voltages.

The observed mobility can be explained by the film structure rather than energetic properties. **Figure 14** shows the atomic force microscope (AFM) surface images of the thin pentacene film deposited on the interfacial layer. In the device showing high mobility, large and rigid granular domains were observed, for example, in the case of B3PyMPM, B4PyMPM, B2PyMPM, and BmPyPhB. On the other hand, small grains or amorphous-like surface were observed, for example, in the case of Alq_3, FIrpic, and so on. From these results, we concluded that the mobility in the p-type hetero-layered OFETs composed of pentacene and electron transporting material is determined by the structural effects rather than the energetic effects.

The material group of B3PyMPM, B4PyMPM, B2PyMPM, and BmPyPhB showed large grains and higher mobility. These molecules were developed for electron transporting materials of OLED devices and very high performance was achieved (Tanaka et al., 2007; Sasabe et al., 2008). However, their LUMO levels are distributed from 3.71 eV (B4PyMPM) to 2.62 eV (BmPyPhB). Therefore it is difficult to group these four materials by energetic properties, and electron accepting character leading to charge transfer seems to be not concerned. One plausible explanation is the molecular arrangement of the interfacial layer. These molecules include nitrogen atoms in the benzene ring. Since nitrogen atom has higher electron affinity that carbon atom, the nitrogen part become negatively charged. On the other hand, SiO_2 surface without inert surface treatment has OH (hydroxyl) groups and its proton becomes positively charged. Therefore the nitrogen atoms in these molecules are attracted to the weak positive charge and the molecule would lie flat on the surface. This effect can be interpreted as electrostatic interaction between two point charges (nitrogen δ^- and hydrogen δ^+), rather than interaction between dipole moments between the molecule and hydroxyl group. In order to maximize Coulomb stabilization, all the nitrogen atoms should touch the surface, resulting in flat arrangement of the molecule on the surface. Consequently, very smooth and flat surface is achieved and pentacene film is expected to form high crystalline film with large grains.

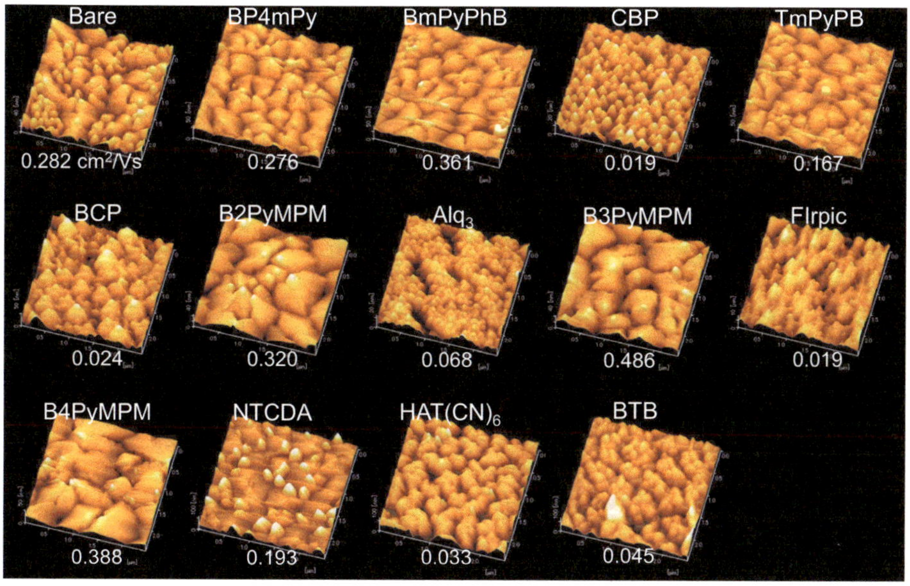

Fig. 14. AFM surface images of the pentacene film deposited on various interfacial layers.

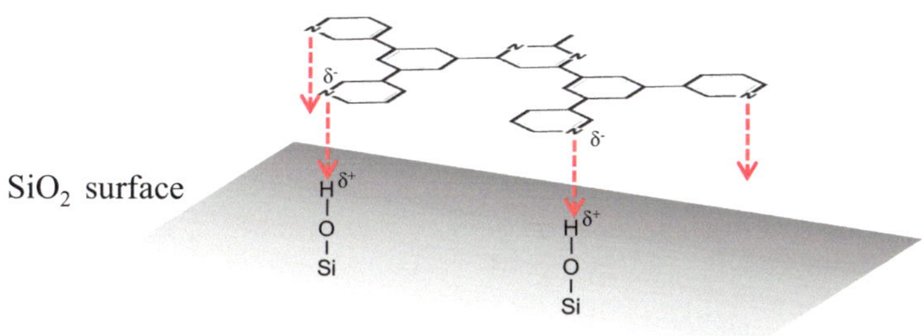

Fig. 15. Molecular arrangement model of B3PyMPM on the SiO_2 surface.

Thus, the concept of heterolayered OFET was extended to p-type pentacene OFET with electron transporting interfacial layer. The performance was slightly improved, but it was mainly attributed to the effect of film structures. It was suggested that the electron transport material including nitrogen atoms forms a preferable underlayer to improve the crystallinity of the pentacene film on it.

5. Conclusion

In this chapter, we introduced a concept of heterolayered OFET composed of the channel organic semiconductor layer and the interfacial organic semiconductor having opposite polarity. In the HTL/n-type devices, the initial performance and stability in air was significantly improved. This effect can be attributed to electron transfer from HTL to n-type semiconductor at the interface, resulting in filling interfacial traps in advance. In the ETL/p-type devices, the performance was slightly improved, but that was mainly attributed to structural effect of film formation.

The hetero-layered OFET is very simple method. The device can be fabricated only by subsequent evaporation of two materials. It does not need self-assembly monolayer treatment taking a long time. Furthermore, it can be expected to solve the most serious problems in n-type OFET of mobility and stability in air. Our results suggest that air stable OFET without designing a new material having deep LUMO level. We expect a novel science and engineering for this "in-plane" carrier transport at the interface subjected to electrostatic gradient.

6. Acknowledgement

This study was partially supported by the New Energy and Industrial Technology Development Organization (NEDO), Precursory Research for Embryonic Science and Technology (PRESTO) program of the Japan Science and Technology agency (JST), and Grant-in-Aid for Scientific Research in Japan.

7. References

Chua, L. L., Zaumseil, J., Chang, J. F., Ou, E. C. W., Ho, P. K. H., Sirringhaus, H., and Friend, R. H., (2005) General observation of n-type field-effect behaviour in organic semiconductors, *Nature* Vol.434, No.7030, pp. 194-199, 193.

Dimitrakopoulos, C. D. and Malenfant, P. R. L., (2002) Organic thin film transistors for large area electronics, *Adv. Mater.* Vol.14, No.2, pp. 99-+, 510.

Lim, S. C., Kim, S. H., Lee, J. H., Kim, M. K., Kim, D. J., and Zyung, T., (2005) Surface-treatment effects on organic thin-film transistors, *Synth. Met.* Vol.148, No.1, pp. 75-79, 22.

Lin, Y. Y., Gundlach, D. J., Nelson, S. F., and Jackson, T. N., (1997) Stacked pentacene layer organic thin-film transistors with improved characteristics, *IEEE Electron Device Lett.* Vol.18, No.12, pp. 606-608, 189.

McCulloch, I., Heeney, M., Bailey, C., Genevicius, K., Macdonald, I., Shkunov, M., Sparrowe, D., Tierney, S., Wagner, R., Zhang, W., Chabinyc, M. L., Kline, R. J., McGehee, M. D., and Toney, M. F., (2006) Liquid-crystalline semiconducting polymers with high charge-carrier mobility, *Nat Mater* Vol.5, No.4, pp. 328-333, 1476-1122.

Rost, C., (2004) Ambipolar organic field-effect transistor based on an organic heterostructure, Vol.95, No.10, pp. 5782, 00218979.

Rost, C., Karg, S., Riess, W., Loi, M. A., Murgia, M., and Muccini, M., (2004) Ambipolar light-emitting organic field-effect transistor, Vol.85, No.9, pp. 1613, 00036951.

Sasabe, H., Chiba, T., Su, S. J., Pu, Y. J., Nakayama, K., and Kido, J., (2008) 2-Phenylpyrimidine skeleton-based electron-transport materials for extremely efficient green organic light-emitting devices, *Chem Commun (Camb)* No.44, pp. 5821-5823, 1359-7345.

Singh, T. B., Sariciftci, N. S., Yang, H., Yang, L., Plochberger, B., and Sitter, H., (2007) Correlation of crystalline and structural properties of C[sub 60] thin films grown at various temperature with charge carrier mobility, Vol.90, No.21, pp. 213512, 00036951.

Tanaka, D., Sasabe, H., Li, Y.-J., Su, S.-J., Takeda, T., and Kido, J., (2007) Ultra High Efficiency Green Organic Light-Emitting Devices, Vol.46, No.1, pp. L10-L12, 0021-4922 1347-4065.

Tatemichi, S., Ichikawa, M., Koyama, T., and Taniguchi, Y., (2006) High mobility n-type thin-film transistors based on N,N '-ditridecyl perylene diimide with thermal treatments, *Appl. Phys. Lett.* Vol.89, No.11, pp. 21.

6

Organic Light Emitting Diodes Based on Novel Zn and Al Complexes

Petia Klimentova Petrova, Reni Lyubomirova Tomova
and Rumiana Toteva Stoycheva-Topalova
Institute of Optical Materials and Technologies "Acad. J. Malinowski"
Bulgarian Academy of Sciences
Up to 1 July 2010 Central Laboratory of Photoprocesses "Acad. J.Malinowski"
Bulgaria

1. Introduction

Organic light emitting diodes (OLEDs) have gained great interest in the last years due to their potential for future flat panel display and solid state lighting applications. OLEDs are a novel and very attractive class of solid-state light sources, which generate a diffuse, non-glaring illumination with high color rendering. Compared to the other major lighting technologies in the market - incandescent, fluorescent, high intensity discharge (HID) lamps, LED and electroluminescent, OLED technology has the potential of achieving substantial energy and CO_2 savings, without compromising color rendering or switching speed. The unique features of OLED lighting are inspired the imagination of designers who are exploring various OLED applications: windows, curtains, automotive light, decorative lighting and wall papers. The OLED technology is now being commercialized as a multi-billion dollar market. OLEDs are already used in small displays in cellular phones, car stereos, digital cameras, etc. The rapidly growing market for OLED displays and lighting is driving research in both advanced materials and improved manufacturing processes. In spite of the spectacular results achieved, there are still many problems concerning the efficiency, stability and lifetime of OLEDs, materials selection and optimization, encapsulation, uniformity over large areas, manufacturing cost, colour saturation, etc. to be solved.

OLED represents a quite complicated system of many very thin layers of various materials situated between electrode layers (one of which is transparent); this system emits light when placed under electric potential. The type of material used as the light emitter determines the specific characteristics of such devices.

Two types of OLEDs are developed - on the bases of "small" molecules (SM-OLED) (Tang & VanSlyke, 1987) and conjugated polymers (PLED) (Burroughes et al., 1990), oligomers, etc. Potential emitters for SM-OLED are metal complexes from the lanthanide and platinum groups as well as complexes of Al, Zn, Cd, Cu, Be, B with carefully selected ligands from the group of heterocyclic compounds like as hydroxyquinoline, benzoxazole, benzothiazole, triarylamines, etc. (Petrova & Tomova, 2009). The first generation of efficient devices, pioneered by Tang and Van Slyke from Eastman Kodak (1987), was based on fluorescent materials. In this case, the emission of light is the result of the recombination of singlet excitons, but the internal quantum efficiency is limited to 25%. The second generation uses phosphorescent materials where all excitons can be converted into emissive triplet state through efficient intersystem crossing (Baldo et al.,1998). Such materials are up to four times more efficient than fluorescent materials.

An important aspect to improve OLEDs performances is suitable selection of materials for functional OLED layers. In this work we have presented our successful decisions for all functional layers – hole transporting, electron transporting, buffer, hole blocking, electroluminescent in the structures of OLEDs. The new examined electroluminescent Zn and Al complexes were synthesized in the Laboratory of Dyes Synthesis at the Department of Applied Organic Chemistry, Faculty of Chemistry, Sofia University "St. Kl. Ohridski".

2. OLED structure

The simplest OLED structure is a single layer device architecture, where the organic emitter is deposited between two electrodes and acts as emitter and as charge transport material (holes and electrons) at the same time. If a forward bias voltage is applied to the electrodes of an OLED device as depicted in Fig.1a, electrons from the cathode and holes from the anode are injected into the organic semiconductor. The oppositely charged carriers move towards each other across the organic semiconductor, encountere, recombine to form excitons and some of them decay radiatively. The efficiency of an OLED is determined by the number of charge carriers that are injected and the number of holes and electrons that actually recombine during emission of light. In order to improve the device efficiency, the multi layer OLED architecture was introduced (Fig.1b).

Nowadays devices may have a total of 7-9 layers of active materials: an anode; anode buffer layer (ABL), hole injecting layer (HIL) or electron blocking layer (EBL); hole transporting layer (HTL); emissive layer (EML); electron transporting layer (ETL) or hole blocking layer (HBL), electron injecting layer (EIL); cathode buffer layer (CBL), a cathode and a protective barrier layer (Tomova et al., 2007). Inserting of these layers facilitates charge carrier injection by reducing the respective injection barriers; enhances the recombination of electrons and holes in the emissive layer (due to accumulation of charges in the EL); shifts the recombination area towards the middle of the device and thus prevents the quenching of the excitons at the electrodes.

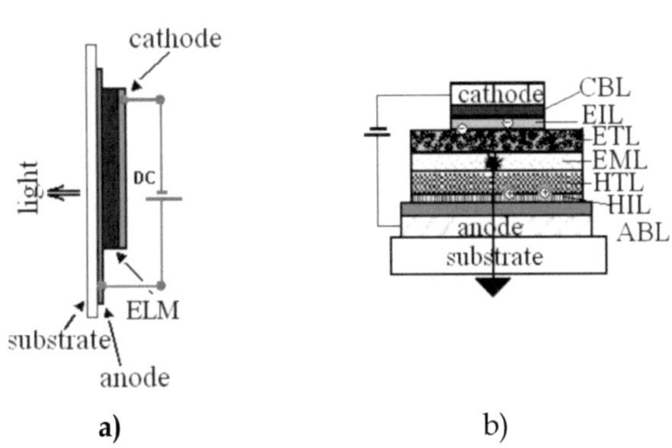

Fig. 1. Structure of: a) monolayer OLED; b) multilayer OLED.

The bilayer OLED consisting of hole transporting layer and emissive layer of different electroluminescent "small" moleculear materials is a basic structure in our investigations. They were prepared by thermal evaporation in vacuum better than 10^{-4} Pa at rates 2-5 A/s on commercial polyethylene terephtalate (PET) flexible substrate, coated with transparent anode of In_2O_3:SnO_2 (ITO - 40 Ω/sq). As cathode was used Al electrode, thermal evaporated in the same vacuum cycle.
We studied the morphology, photoluminescence (PL), electroluminescence (EL) and the performance of the devices measuring the current-voltage (I/V), luminescence-voltage (L/V) and electroluminescence-voltage (EL/V) characteristics. The electroluminescent efficiency (η_L) was calculated by equation (1) and used for quantifying the properties of the OLEDs.

$$\eta_L = L / I \quad (1)$$

(where *L* is the luminescence (in cd/m^2) and *I* is the current density (in A/m^2) and used for quantifying the properties of the OLEDs.
All measurements were carried out with unpackaged devices with area of 1cm^2, at room temperature and ambient conditions.

3. Hole transporting and buffer layers

The operating mechanisms of OLEDs involve injection of electrons and holes into the organic emitter layers from the electrodes. During recombination, electrons and holes generate molecular excitons (Kido et al., 1998), which result in the emission of light from the emitter layer. Therefore the effective recombination of electrons and holes affects on the electroluminescence efficiency of organic light-emitting diodes. That's why, it is important to balance the number of holes and electrons in EL devices. The mobility of holes in OLED materials used as the hole transport layer (HTL) is some orders of magnitude greater than that of the electrons in the ETL (Zheng et al., 2005). The recombination zone is shifted towards the cathode, which usually leads to a non-radiative loss of energy (Rothberg et al., 1996) and decreasing of an OLED efficiency (Sheats et al., 1996). For that reason, by reducing the mobility of holes in HTL or promoting electron injection into ETL can improve the balance of carriers in OLED. The reducing of holes mobility can be achieved via inserting a proper buffer layer between anode and hole transporting layer. On the other hand introducing of a buffer layer improve the ITO morphology such as inhomogenity or protrusions, impede the diffusion of indium into the organic layer during device operation, which is correlated with the decay of a device's performance (Schlatmann et al., 1996).
The ITO/organic interface morphology play a key role to stable operation and efficiency of the device. For that reaseon, a lot of work has been devoted to the anode buffer layers (ABLs) between ITO and the organic material. The introduced buffer layers mainly can be divided in inorganic and organic compounds. Among the reported inorganic anodic buffer layers good inorganic insulators such as transparent metal oxides Pr_2O_3, Y_2O_3, Tb_4O_7, ZnO (Xu et al., 2001), Al_2O_3 (Li et al., 1997; Xu et al., 2001), SiO_2 (Deng et al., 1999; Xu et al., 2001), silicon nitride Si_3N_4 (Jiang et al., 2000; Xu et al., 2001), carbon nitride a-C:N (Reyes et al., 2004), transition metal oxides, also V_2O_5, (Wu et al., 2007; Guo et al., 2005), MoOx (You et al., 2007; Jiang et al., 2007), WO_3, (Jiang et al., 2007; Meyer et al., 2007), CuOx, (Hu et al, 2002; Xu et al., 2001), NiO, (Chan et al., 2004; Im et al., 2007) and Ta_2O_5 (Lu & Yokoyama, 2003), have

attracted much attention due to their capability to lower the hole-injection barrier and improve the interface morphology.
As the organic buffer layers a variety of materials as copper phthalocyanine (Van Slyke et al., 1996; Shi & Tang, 1997; Tadayyon et al., 2004), α-Septithiophene (Park et al., 2002), Langmuir-Blodjett films of polymethylmethacrylate (Kim et al., 1996), polytetrafluoroethylene (Gao et al., 2003), fluoropolymers (Wang et al., 2006), fluorene-based poly(iminoarylene)s (Jung et al., 2002), conductive polymer such as polythiophene (Arai et al., 1997), poly(3,4-ethylenedioxythiophene) (Carter et al., 1997; Berntsen et al., 1998), and polyaniline (Krag et al., 1996) etc. have been tested.
We explored the effect of p-isopropenylcalix[8]arenestyrene copolymer (iPrCS and polycarbonate (PC), as buffer layers in OLED, and the incorporation of TPD with PVK as hole transporting layer.

3.1 p-Isopropenylcalix[8]arenestyrene copolymer (iPrCS)

The calixarenes are a class of bowl-shaped cyclo-oligomeres obtained via phenol-formaldehyde condensation with a defined upper and lower rim, and a cavity. This speciality enable them to act as host molecules due to their cavities, and allow utilized them as chemical sensors, extractants for radioactive waste processing, materials for non-linear optics, bio-active compounds.

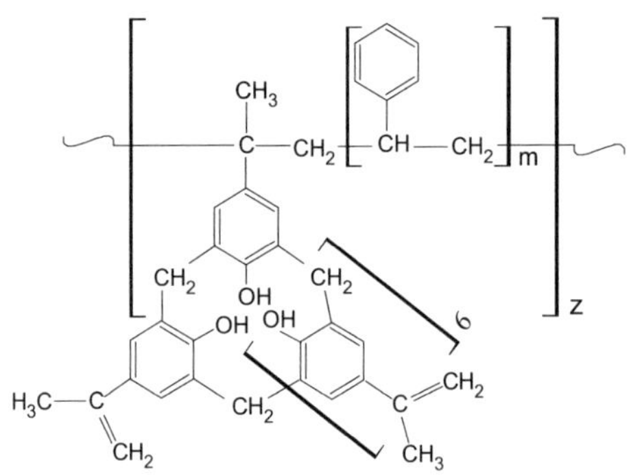

Fig. 2. Chemical structure of the *p*-Isopropenylcalix[8]arenestyrene copolymer used in device fabrication as buffer layer.

In this work, we offer the *p*-Isopropenylcalix[8]arenestyrene copolymer (iPrCS) as a novel anode buffer layer (ABL) for the fabrication of OLED with improved efficiency and life time. The *p*-Isopropenylcalix[8]arenestyrene copolymer (Fig.2) employed (Petrova et al., 2010) for this study was for the first time synthesized according to described procedure (Miloshev & Petrova, 2006), in University of Chemical Technology and Metallurgy, Sofia. Until now the calix[4]arene compounds were used only for design of electroluminescent complexes - for ex. a calix[4]arene $[Al\ I]^{3+}$ complex (Legnani et al., 2004), lanthanide complexes with calix[4]arene derivatives (Wei et al., 2007).
Two types of devices were investigated: ITO/ABL/TPD/Alq_3/Al, and ITO/TPD/Alq_3/Al as a reference structure. The buffer layer (δ = 10 - 16 nm) of iPrCS was deposited on PET/ITO substrates by spin-coating from 0.1 - 0.3% solution in THF at 2000 rpm. N, N'-bis(3-methylphenyl)-N, N'-diphenylbenzidine (TPD) (δ = 30 nm) as hole transporting and tris(8-hydroxyquinoline) aluminum (Alq_3, δ = 50, 75 nm) as electroluminesent and electron

transporting layer were used. TPD, Alq_3 and the Al cathode (δ = 120 nm) were deposited via thermal evaporation in a vacuum better than 10^{-4} Pa at rates of 2-5A/s.
Figure 3 presents typical nonlinear current/voltage (Fig.3a), luminescence/voltage (Fig.3b) and efficiency characteristics (Fig.3c) of ITO/iPrCS $_{(10\text{-}16nm)}$/TPD $_{(30nm)}$/Alq_3 $_{(50nm)}$/Al and ITO/TPD $_{(30nm)}$/Alq_3 $_{(50nm)}$/Al as a reference structure. It was shown that the turn-on voltage slightly decreased with increasing of the thickness of iPrCS, while the luminescence and efficiency of devices increased and reached maximum values at a thickness of 13 nm. The efficiency of 2 cd/A at 13 nm iPrCS was nearly 80% higher than those of 1.2 cd/A of the reference structure.

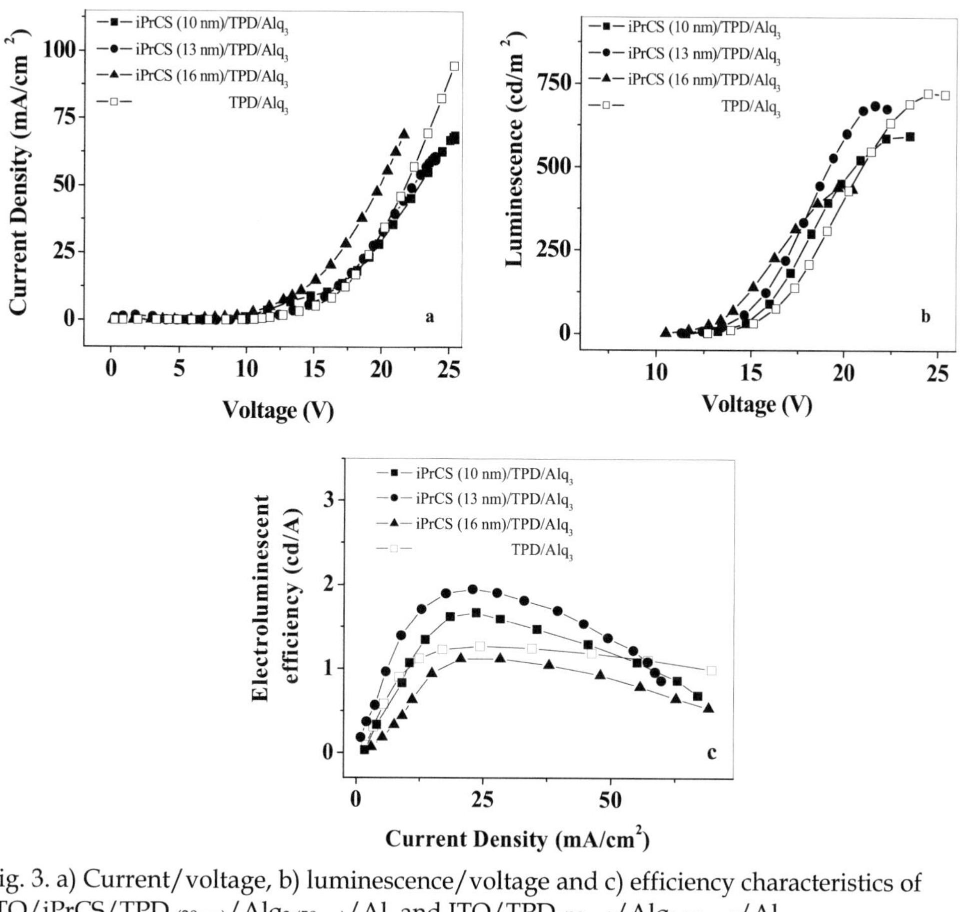

Fig. 3. a) Current/voltage, b) luminescence/voltage and c) efficiency characteristics of ITO/iPrCS/TPD $_{(30nm)}$/Alq_3 $_{(50nm)}$/Al, and ITO/TPD $_{(30nm)}$/Alq_3 $_{(50\ nm)}$/Al.

Effect of the two thicknesses of Alq_3 on the performance of the devices with 13 nm film iPrCS is presented in Fig.4. It was established that the luminescence (Fig.4b) and efficiency (Fig.4c) of the devices with iPrCS were higher compared to the reference structures and that the device with 75 nm emissive layer of Alq_3 showed the best characteristics. The efficiency of 3.04 cd/A at the current density of 20 mA/cm^2 of the device with iPrCS is similar to those of 3.4 cd/A at the same current density, reported by Okamoto for the structure ITO/CF_x/NPB $_{(60\ nm)}$/Alq_3 $_{(60nm)}$/LiF/Al (Okamoto et al., 2006).

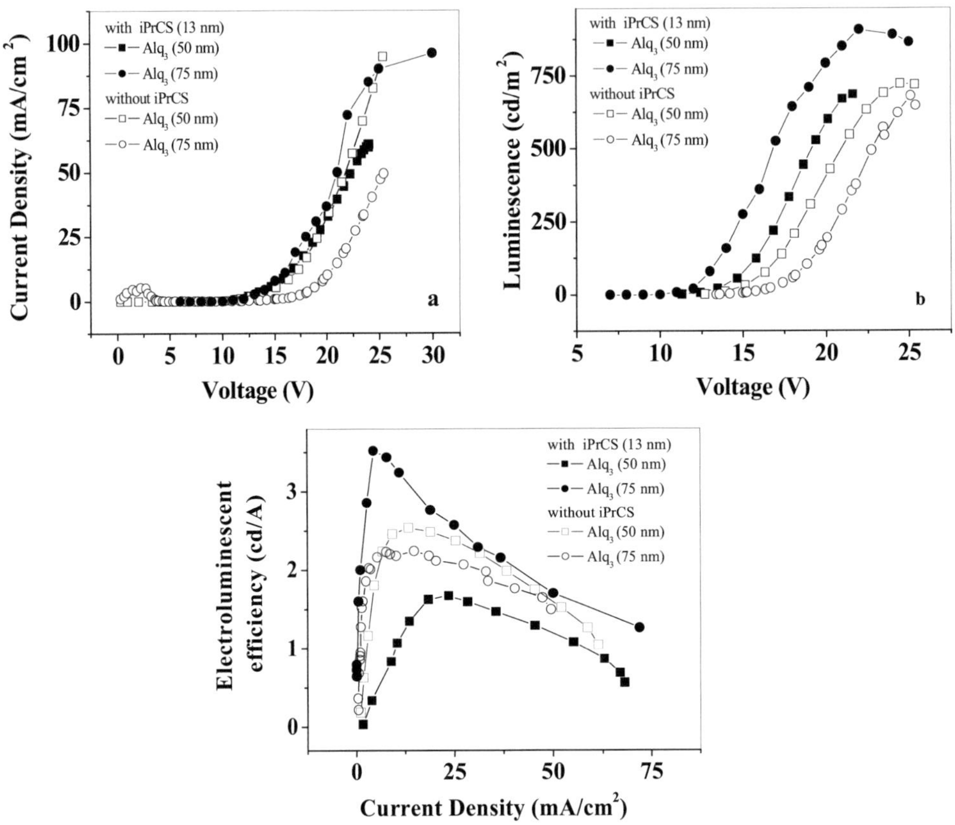

Fig. 4. a) Current/voltage, b) luminescence/voltage and c) efficiency characteristics of ITO/TPD $_{(30nm)}$/Alq_3 $_{(50, 75nm)}$/Al with and without iPrCS

It could be supposed that notwithstanding the iPrCS is an insulator, it seems to enhance the hole injection thus improving a hole-electron balance in OLED and makes the tunneling injection in OLED.

2.2 Polycarbonate (PC)

The PC is a rigid, transparent and amorphous material with high Tg 140-155 ^{o}C. It possesses excellent dielectric and optical characteristics. The possibility of usage of PC as buffer layer in OLED with ITO/PC/TPD/Alq_3/Al structure was investigated. The PC layers with thicknesses of 9, 12 and 17 nm were deposited via spin-coating from 0.1%, 0.2% and 0.3% dichlorethane solutions. The basic characteristics of OLED structure with different thickness of PC buffer layer are presented in Fig.5. It was found that inserting of 9 nm buffer layer in OLED devices decreased the turn on voltage from 12.5 to 8 V, and increased the current density from 10 to 24 mA/cm^2 and the luminescence from 220 to 650 cd/m^2 (at 17.5 V) compared to the reference structure. Further increasing of the thickness of PC buffer layer decreases the current density and the luminescence, and shift the turn on voltage toward higher values (Fig.5b), as was established with iPrCS.

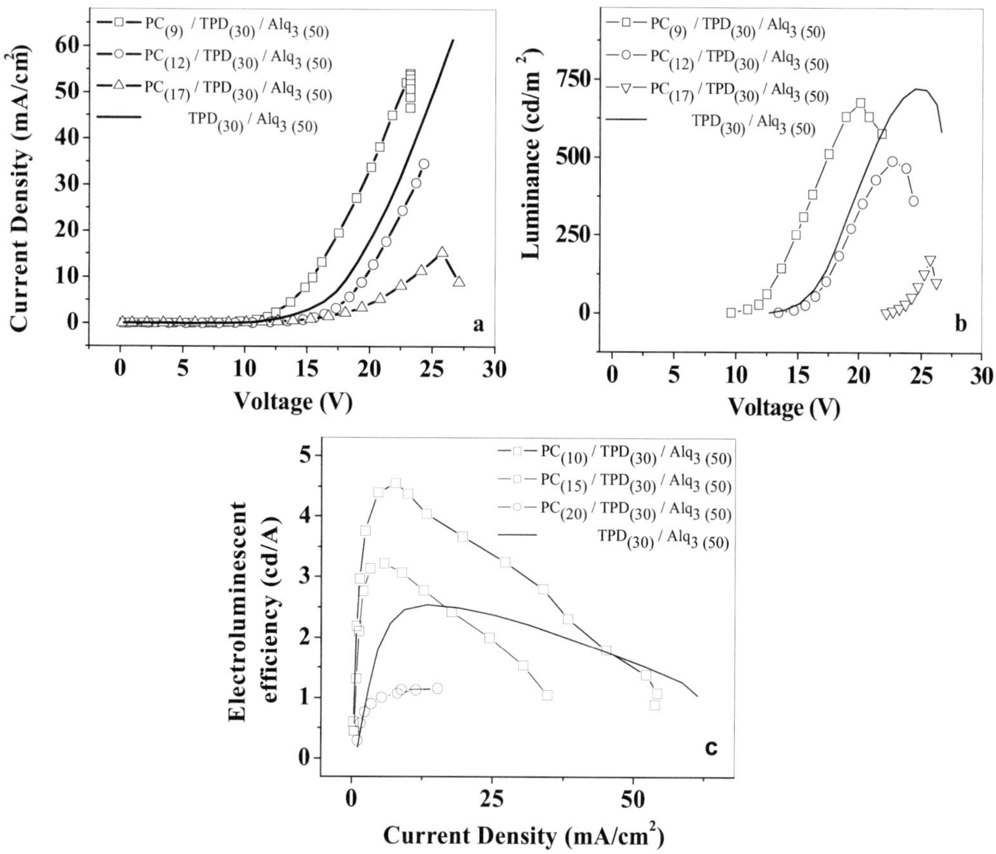

Fig. 5. Current/voltage (5a), luminescence/voltage (5b) and efficiency (5c) for inset in legends structures.

The best characteristics - the lowest turn-on voltage, the highest luminescence and the highest efficiency showed OLED with 9 nm PC buffer layer. It should be noted that the efficiency of the device with 9 nm buffer layer is more than 2x higher than that of the reference device. Similar improvement of characteristics of device with 1 nm Teflon buffer layer was observed by Qiu et al. (2002). They supposed that the Teflon layer acts as a stable fence to impede indium diffusion from ITO electrode into the TPD layer and thus enhances the device stability.

It could be supposed that the improvement of EL performance of devices with buffer layers of iPrCS and PC has just the same genesis. Although these compounds are insulators, they seem to enhance the hole injection from anode by tunneling. Thus improving a hole-electron balances in OLED.

We also made attempts to use the PC and iPrCS polymers as a matrix for TPD. In this cases the turn on voltages of the devices with composite buffer layers were lower than that with only PC and iPrCS buffer layers, but unfortunatly the luminescence of the devices were significantly reduced and unsatisfactory. The last one makes the application of PC and iPrCS polymers irrelevant as matrix of TPD for OLEDs.

On the results obtained could be concluded that iPrCS and polycarbonate can be successfully use as buffer layers for obtaining of OLED with good performance.
Further devices with the typical hole transporting layers poly(9-vinylcarbazole) (PVK) and N, N'-bis(3-methylphenyl)-N, N'-diphenylbenzidine (TPD) were studied. That's why we investigated the influence of single layer of PVK, TPD, PVK as a buffer layer with respect to TPD and composite layer of PVK:TPD on the performance of the device structure ITO/HTL/Alq_3/Al. The HTL (31 nm) of PVK and PVK:TPD composite films (10wt% TPD relatively PVK in 0.75% dichloroethane solutions) were deposited by spin-coating.

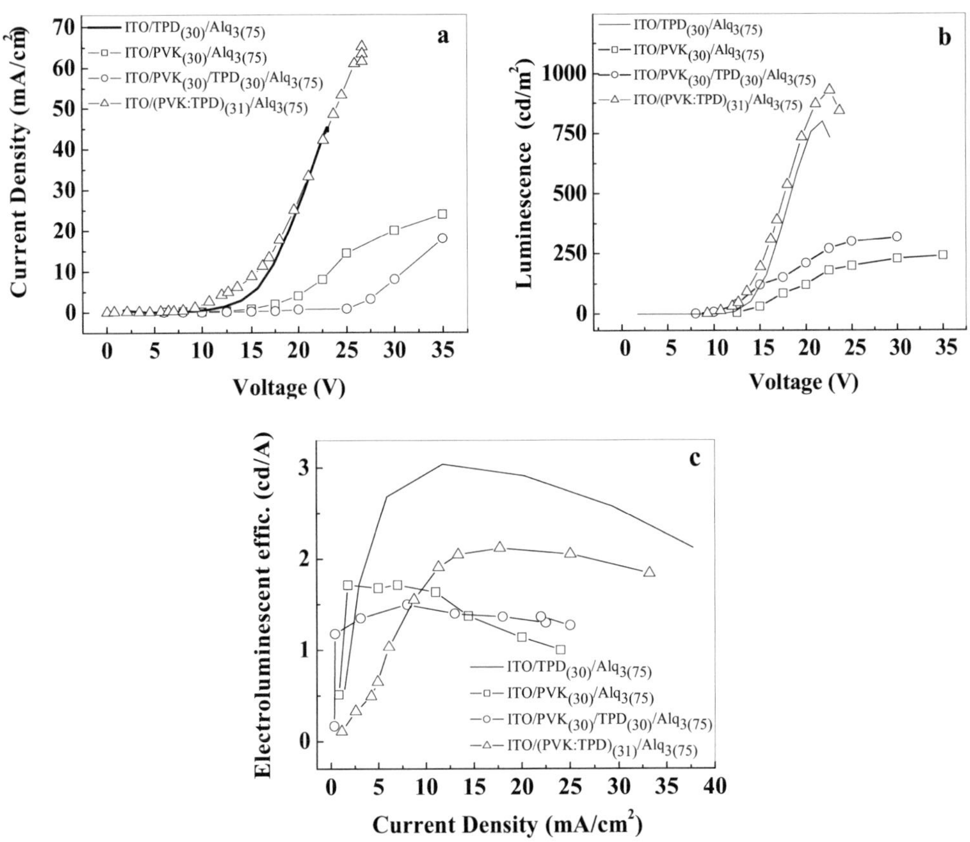

Fig. 6. a) Current/voltage, b) luminescence/voltage and c) efficiency characteristics of devices shown in set.

The optimal I/V, L/V and efficiency characteristics of the devices ITO/PVK/Alq_3/Al, ITO/PVK/TPD/Alq_3/Al, ITO/(PVK:TPD)/Alq_3/Al and ITO/TPD/Alq_3/Al as reference are presented in Fig.6. It is seen that the I/V and L/V curves for ITO/(PVK:TPD)/Alq_3/Al and ITO/TPD/Alq_3/Al structures are almost identical. But it was established that due to the well known trend of TPD thin films to crystallization, the lifetime of the reference device with TPD only is many times shorter than that with composite layer of PVK:TPD. The device structure with only PVK and ITO/PVK/TPD/Alq_3/Al, showed a decrease in the current density, luminescence and efficiency compared to the reference device. Obviously, the use of PVK as HTL, or as a buffer layer in respect of TPD HTL in OLEDs is not felicitous, because impedes the charge transfer.

It could be stressed that the devices with PVK:TPD composite layer demonstrates the best characteristics. The involving of TPD in PVK matrix improves the current density, luminescence and luminescent efficiency, reduces the turn-on voltage and increases the lifetime compared to the others devices.

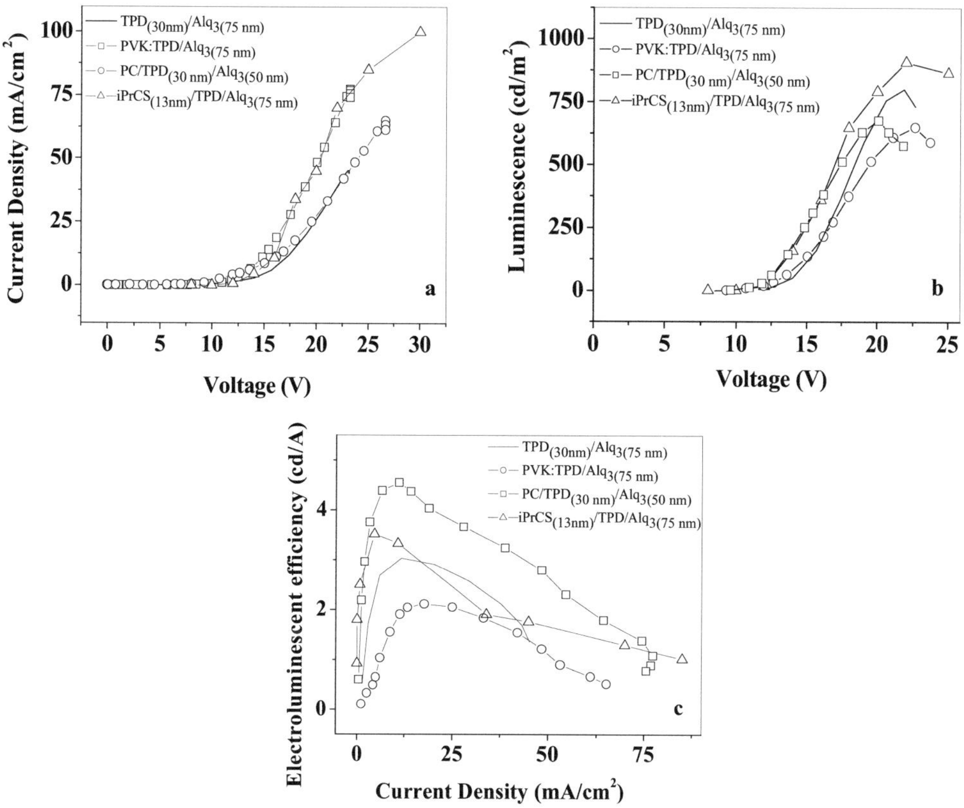

Fig. 7. a) Current/voltage, b) luminescence/voltage and c) efficiency characteristics of devices ITO/iPrCS/TPD/Alq_3/Al, ITO/PC/TPD/Alq_3/Al, ITO/(PVK:TPD)/Alq_3/Al and ITO/TPD/Alq_3/Al shown in set.

The best results obtained for four type devices with different buffer and hole transporting layers are presented in Fig.7. It is clearly seen that inserting of buffer layer caused decreasing of turn on voltage and increasing of current densities, luminescence and efficiency. The best electroluminescence of 570 cd/m^2 at 17.5 V belonged to the device with iPrCS, followed by devices with PC, TPD and PVK:TPD, respectively with 510, 380 and 350 cd/m^2. At the same time the best efficiency of 3.3 cd/A at 37 mA/cm^2 exhibited device with PC followed by devices with TPD (2.17 cd/A), iPrCS (1.88 cd/A) and PVK:TPD (1.73 cd/A). A comparison of the OLED characteristics for the four devices clearly indicates that the device performance is greatly improved when the ITO surface was covered by polymeric film.

Besides that the efficiency of the devices with composite PVK:TPD layer is not so high, this HTL is most perspective due to the synergistic effect from properties of both components. The incorporation of TPD with PVK offers an attractive route to combine the advantiges of easy spin-coating formability of PVK with the better hole transporting properties of TPD. The composite PVK:TPD layers is very reproducible, simplify the obtaining of experimental samples and by reason of that it was used in our basic structure for the study of different electroluminescent compounds as emitting layer in OLEDs.

The efficiency of the OLED is a complexed problem, and depends not only on the energy levels of functional layers of the devices, but also on the interfaces between inorganic electrodes/organic layers. We demonstrate that the thin polymeric films enable to facilitate the transport of carriers and to improve the adhesion and morphology between ITO, and "small" molecular organic layer.

2.3 Effect of morphology

The ITO is common known as an excellent electrode, but its morphology can has an affect on the organic layers evaporated on ITO substrate, where the small spikes in the ITO surface can lead to local crystallization of HTL and EL causing a bright white-spot that may increase the leakage and instability of the device.

The surface morphology of the hole transporting and buffer layers were studied by scaning electron microscopy (SEM) and atom force microscopy (AFM).

SEM micrographs of vaccum deposited TPD and spin-coating composite PVK:TPD hole transporting films on PET/ITO substrates are presented in Fig.8 and Fig.9.

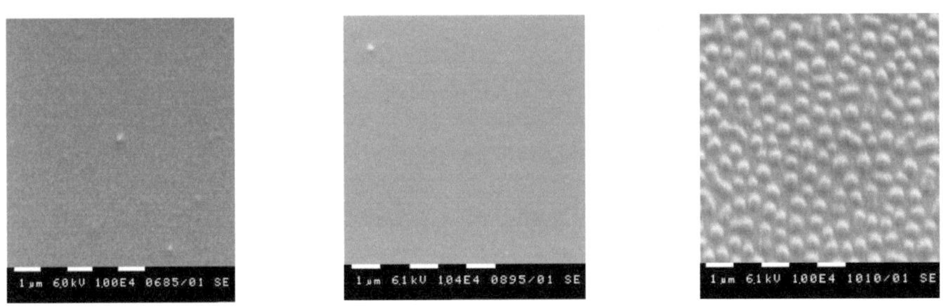

a) bare ITO b) ITO/TPD - as deposited c) ITO/TPD after one day

Fig. 8. SEM images of: a) bare ITO on PET substrate; b) as deposited, and c) after one day vacuum deposited 30nm TPD layer on ITO/PET

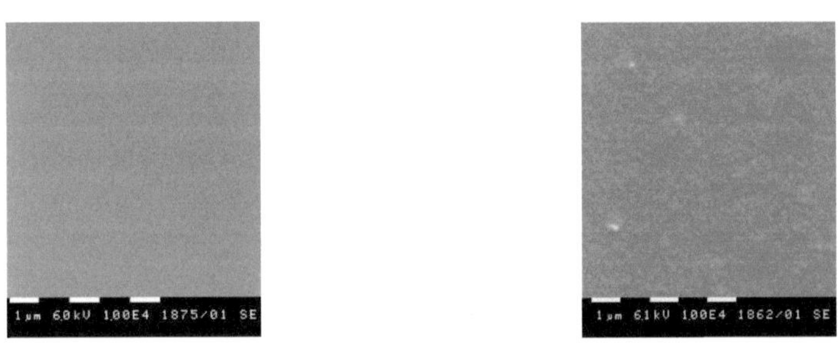

a) ITO/PVK:TPD - as deposited b) ITO/PVK:TPD after one day

Fig. 9. SEM images of composite PVK:TPD spin-coating deposited layer on ITO/PET

The surface morphology of the developed by us composite films of PVK:TPD (Fig.9.) is very smooth and homogeneous, without any defects and cracks, thus creating a suitable conditions for the condensation of the next electroluminesent layer. The similar is the surface morphology of the vacuum as-deposited TPD films on bare ITO (Fig.9b.), but after 1 day storage at ambient temperature, TPD formed an islands structure with bubbles, which is a prerequisite for recrystallization and oxidation (Fig.8c.). At the same time the surface morphology of PVK:TPD, layers does not show any changes after 1 day storage (Fig.9b.) - better stability of devices with composite PVK:TPD hole transporting layer could be expected.
The results of AFM investigations are presented in Fig.10. It is shown that surface of the commercial ITO coated PET substrates is with uniform roughness with some imperfections. The evaporated TPD layer onto this ITO surface makes a granular structure (Fig.10. a, b).
The introducing polymer buffer layers covered the ITO pinholes, spikes and other defects, thus leveling its surface (Fig.10. c, e, and g). The amorphous and very smooth surface of spin-coated polymer thin films creates more suitable conditions for vacuum deposition of TPD thin films compared to the bare ITO. As far as TPD layers deposited onto studied buffer coatings are concerned, a quite even granular structure is observed (Fig.10. d, f, h).

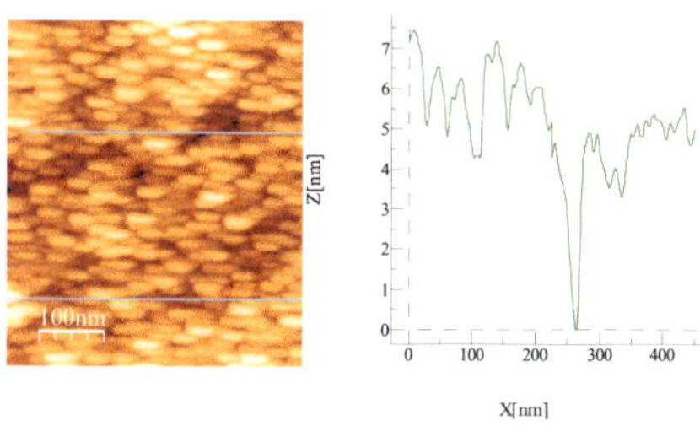

Fig. 10. a) bare ITO surface onto PET substrate.

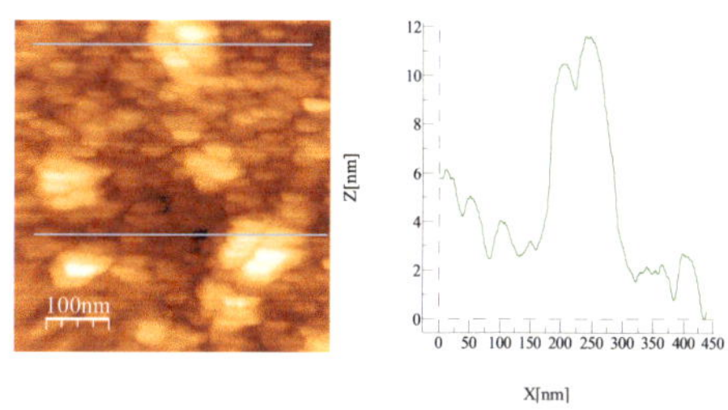

b) ITO/TPD surface

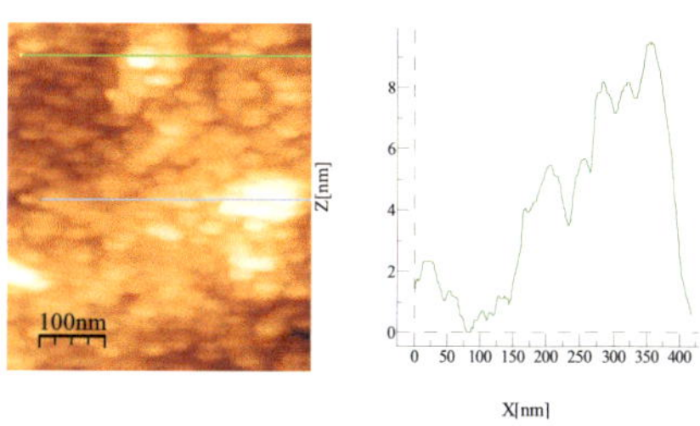

Fig. 10. c) ITO/ iPrCS surface

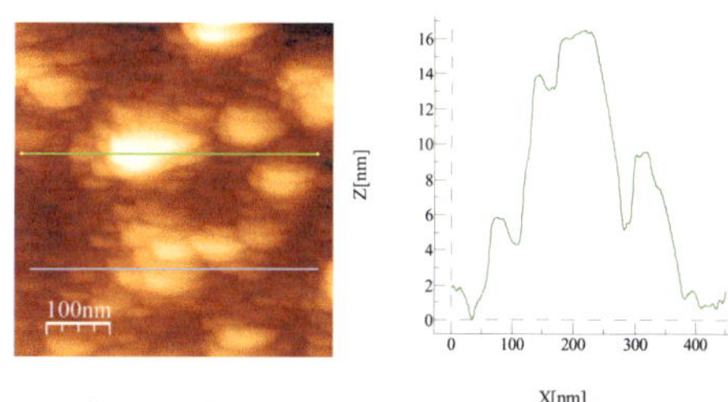

d) ITO/iPrCS/TPD surface

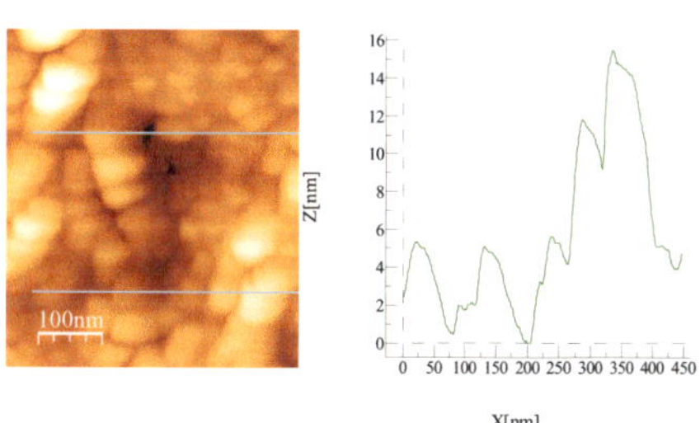

Fig. 10. e) ITO/ PC surface

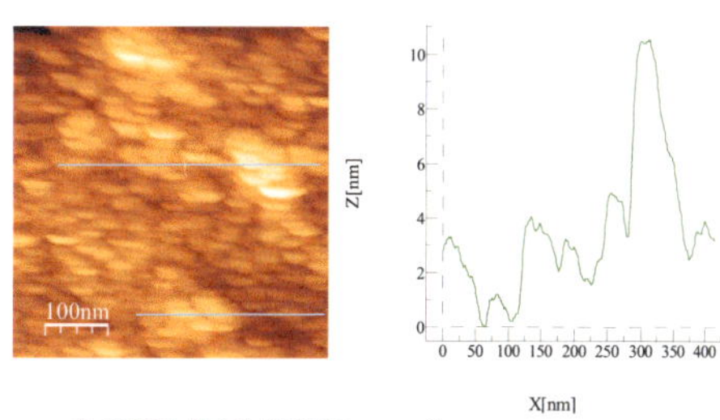

f) ITO/PC/TPD surface

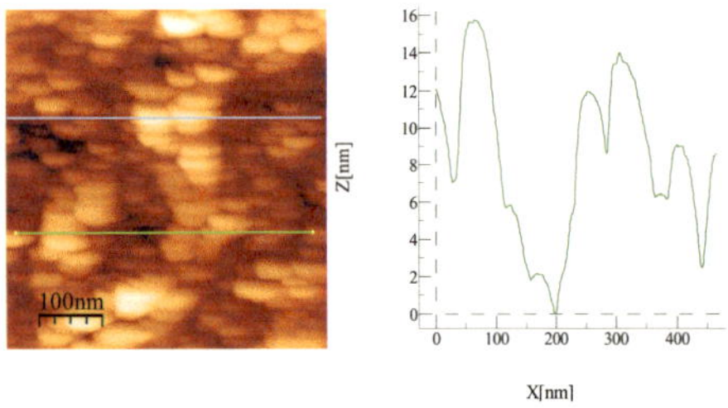

Fig. 10. g) ITO/ PVK surface

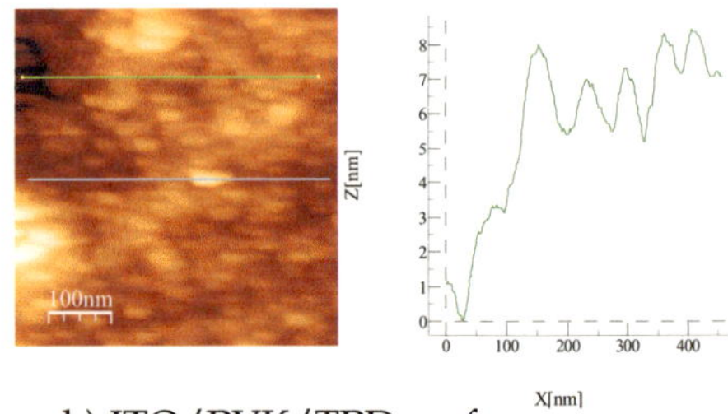

h) ITO/PVK/TPD surface

Fig. 10. AFM images and cross-section profiles of the surfaces of a) bare ITO, b) ITO/TPD,c) ITO/iPrCS surface, d) ITO/iPrCS/TPD surface, e) ITO/PC surface, f) ITO/PC/TPD surface, g) ITO/PVK surface, h) ITO/PVK/TPD surface

Unlike the fast recrystalization of TPD layer deposited on bare ITO, the amorphorous and homogeneous surface of TPD films deposited on the buffer-coated ITO was very stable. The results obtained show that the polymer modifies successfully the film morphology, thus preventing the recrystallization of hole transporting layer (TPD) and following emissive layer. These results definitely have an effect on the current density and luminance characteristics of the devices. Probably, the higher *Tg* of the polymers than that of the TPD, improve the durability of HTL on Joule heat, which arises in OLED operations, thus enable the better performance of OLED.

3. Novel Zn complexes

Many organic materials have been synthesized and extended efforts have been made to obtain high performance electroluminescent devices. In spite of the impressive achievements of the last decade, the problem of searching for the new effective luminescent materials with different emission colours is still topical. Metal-chelate compounds are known to yield broad light emission and seem to provide design freedom needed in controlling photo-physical processes in such devices. Among these materials, Zn complexes have been especially important because of the simplicity in synthesis procedures and wide spectral response. Extensive research work is going on in various laboratories to synthesize new Zn complexes containing new ligands to produce a number of novel luminescent Zn complexes as emitters and electron transporters (Sapochak et al, 2001, 2002; Hamada et al, 1996; Sano et al, 2000; Kim et al, 2007; Rai et al, 2008). Zinc(II) bis[2-(2-hydroxyphenyl) benzothiazole] ($Zn(BTz)_2$) has been studied as an effective white light emissive and electron transporting material in OLED. Hamada et al. (1996) reported that the device with single-emitting layer of $Zn(BTz)_2$ showed a greenish white emission. Later on an efficient white-light-emitting device were developed with electroluminescent layers of $Zn(BTz)_2$ doped with red fluorescent dye of 4-dicyanomethylene-2-methyl-6-[2-(2,3,6,7,-tetrahydro-1H,5H-benzo[i,j]quinolizin-8-yl)vinyl]-4H-pyran (DCM2) (Lim et al, 2002) or rubrene (Zheng et al, 2005; Wu et al, 2005). Recently Zhu et al. (2007) fabricated white OLED with $Zn(BTz)_2$ only as emitter. The obtained white emission is composed of two parts: one is 470 nm, which originates from exciton emission in $Zn(BTz)_2$, the other is 580 nm, which originates from exciplexes formation at the interface of TPD/$Zn(BTz)_2$.

We investigated the new Zn complexes Zinc(II) [2-(2-hydroxyphenyl)benzothiazole] acetylacetonate (AcacZnBTz) and Zinc(II) bis[2-(2-hydroxynaphtyl)benzothiazole)

(Zn(NBTz)$_2$), and known Zinc(II) bis[2-(2-hydroxyphenyl)benzothiazole] (Zn(BTz)$_2$) (Tomova et al, 2008), and Zinc(II) bis(8-hydroxyquinoline) (Znq$_2$) (Fig.11), synthesized by prof. Deligeorgiev as electroluminescent and electron transporting compounds. The basic OLED structure was PET/ITO/(PVK:TPD)/EML/Al.

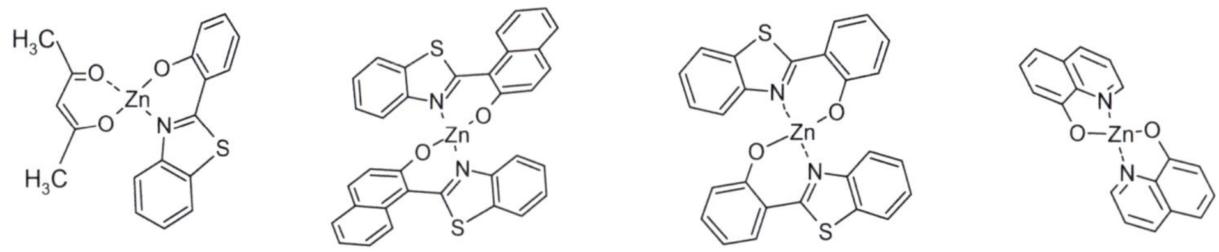

AcacZnBTz Zn(NBTz)$_2$ Zn(BTz)$_2$ Znq$_2$

Fig. 11. The chemical structures of used Zn complexes

The absorption and fluorescent (PL) spectra of the complexes were taken using the Spectro-fluorimeter Perkin Elmer MPF 44 are presented in Fig.12.

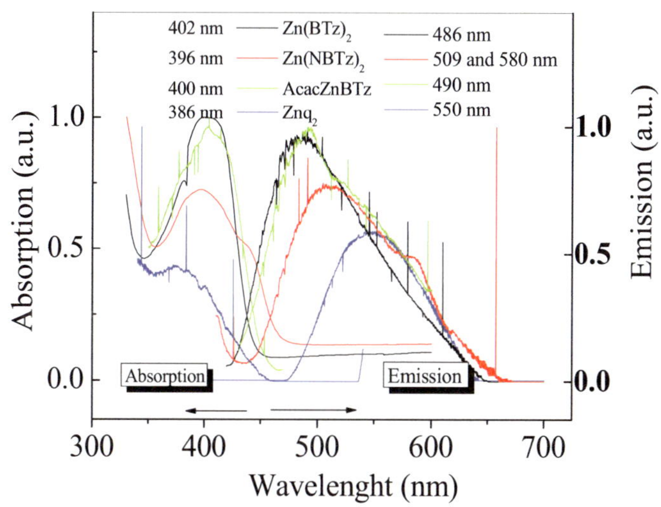

Fig. 12. Absorption and PL emission spectra of 100 nm films of Zn complexes evaporated on glass substrate

The PL peak wavelength of Znq$_2$ is at 550 nm, of Zn(BTz)$_2$ at 486 nm, of AcacZnBTz at 490 nm. Zn(NBTz)$_2$ shows peak at 509 nm and shoulder at 580 nm. The data obtained for PL peaks of Znq$_2$ and Zn(BTz)$_2$ are very closed to the results reported by Shukla & Kumar (2010) for Znq$_2$ (540 nm) and by Qureshi et al. (2005) for Zn(BTz)$_2$ (485 nm).
The electroluminescent (EL) spectra of devices PET/ITO/(PVK:TPD) $_{(31\ nm)}$/EML $_{(75\ nm)}$/Al, obtained at different voltages by Ocean Optics HR2000+ spectrometer are shown in Fig.13.
It was established that the EL spectra of the complexes with benzthiazole ligand were very similar and exhibited a green electroluminescence around 525 nm. Besides the EL spectra of all four compounds were red shifted, about 10 nm for Znq$_2$ and 25 - 30 nm of benzthiazole complexes, compared to their corresponding PL spectra. Take into account the fact that the exciton disassociates easily under the excitation of electric field than the light, red shifting of

EL spectra were quite understandable (Wu et al, 2005). The highest EL intensity showed the devices with AcacZnBTz followed by those with $Zn(BTz)_2$, Znq_2, and $Zn(NBTz)_2$.

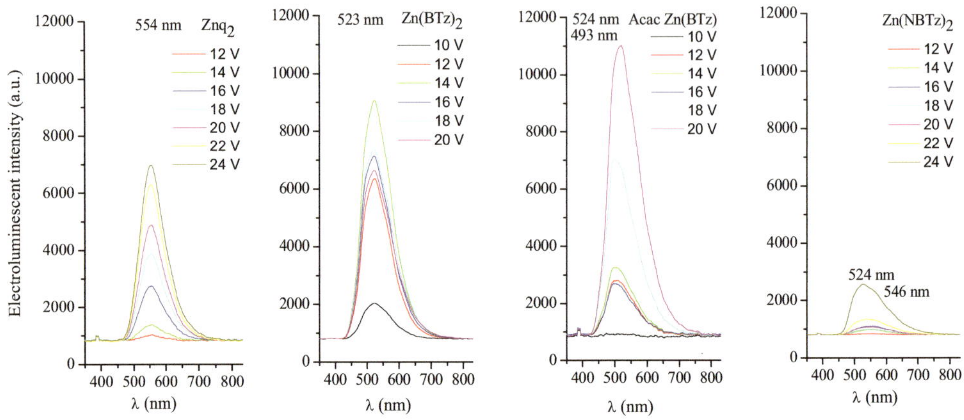

Fig. 13. Electroluminescent spectra of OLEDs with different Zn complexes

The EL peak wavelength of the devices with Znq_2 and $Zn(BTz)_2$ is the same during the device operation independantly on the working voltage, while EL peak of the devices with AcacZnBTz moves from 493 to 524 nm with increasing the working voltage. Our results were quite different from these obtained by Wu et al. (2005), who showed almost identical EL and PL for $Zn(BTz)_2$, and Qureshi et al. (2005) who founded broader EL than PL spectrum AFM images of top surfaces of devices with EML of different Zn complexes are presented in Fig.14.

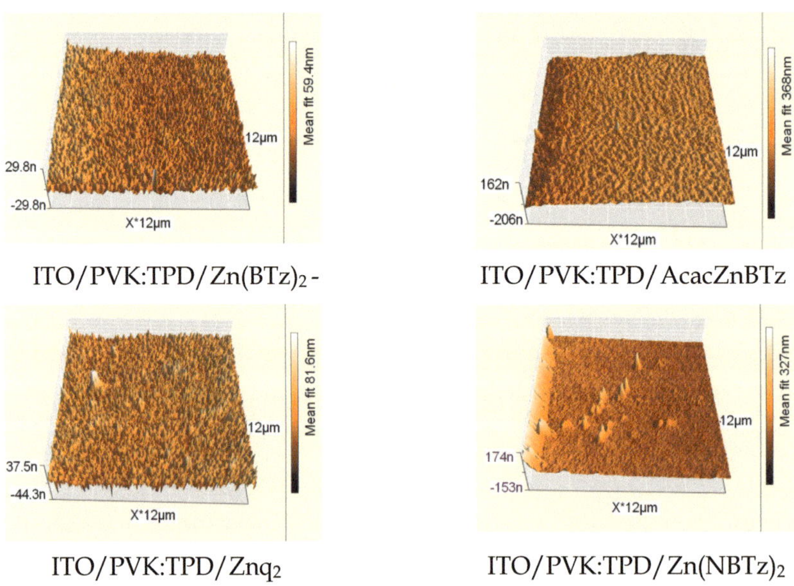

Fig. 14. AFM images of top surfaces of devices with EML of different Zn complexes, performed by "EasyScan 2" produced by "Nanosurf" (Switzerland) on area of 12.5 x 12.5 µm, at measurement mode "scan forward" and Scan mode from down to up.

The AFM images show that evaporated Znq_2 and $Zn(BTz)_2$ compounds, on PET/ITO/PVK:TPD structure, formed similar fine-textured surfaces with root mean square (RMS) roughness respectively 6.88 nm and 4.64 nm. The AcacZnBTz layer made soft outline ridge surface with RMS roughness 20.06 nm.

All three complexes formed smooth and even surfaces requisite for the good performance of OLED on their base. Maybe due to the molecular structure specific of the $Zn(NBTz)_2$ the film obtained from it is very flat (RMS roughness 22.82 nm), but with some acicular formations over 150 nm on some areas. Namely these formations are a precondition for the worse EL performance of OLED with electroluminescent layer of $Zn(NBTz)_2$.

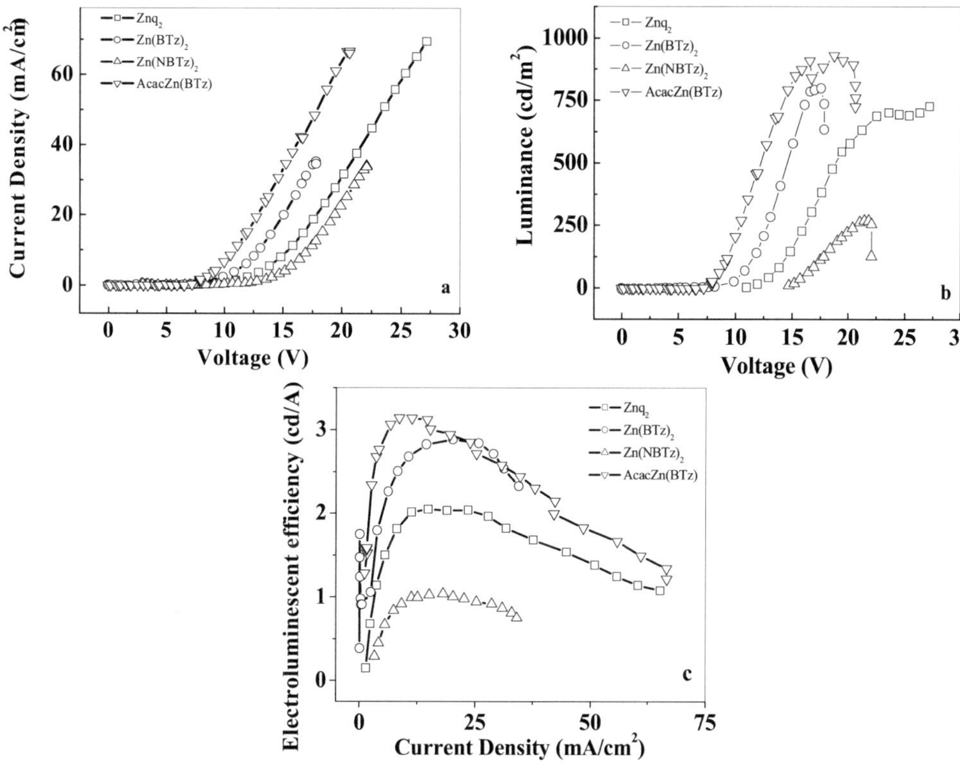

Fig. 15. a) Current/voltage and b) luminescence/voltage characteristics and c) electro-luminescent efficiency for devices with different EML $_{(75nm)}$ and HTL of (PVK:TPD) $_{(31nm)}$

Fig.15. presents the current/voltage, luminance/voltage and efficiency characteristics of four type identical devices with different EML. It was established that the current densities and the luminescence decreased and the turn-on voltage of devices increased in following sequence AcacZnBTz, $Zn(BTz)_2$, Znq_2, $Zn(NBTz)_2$. Luminescence of the device with AcacZnBTz at 15 V DC was nearly 1.5 and 3 times higher than those by $Zn(BTz)_2$ and Znq_2, respectively (Fig.15b). At the same time the electroluminescent efficiencies of the devices with AcacZnBTz and $Zn(BTz)_2$ were nearly the same (around 3 cd/A) and 1. 5 and 3 times higher than that of devices with Znq_2 and $Zn(NBTz)_2$ (Fig.15c).

For OLEDs with similar structures Sano et al. (2000) reported efficiency 1.39 cd/A at luminance 100 cd/m^2 for ITO/TPD/$Zn(BTz)_2$/Mg:In device, Zheng et al. (2005) - 4.05 cd/A

for doped with rubrene $Zn(BTz)_2$ white device at maximum luminescence 4048 cd/m^2 [10] and Rai et al.. (2008) - 1.34 cd/A for ITO/NPD/Zn(Bpy)q/Al.
The results presented in this chapter show that the studied Zn complexes with the exception of $Zn(NBTz)_2$ can be successfully used as emitters and electron transporting layers for OLED. It could be stressed that the efficiency of the devices with $Zn(BTz)_2$ is 2.9 cd/A at luminance 250 cd/m^2 – one of the best reported up to now in the literature for the devices with similar structure. Besides that the devices with new Zn complexes are not optimized, its characteristics are quite promising, especially for AcacZnBTz – the highest luminance and the efficiency 3 cd/A in the range of 10 – 30 mA/cm^2.

4. Aluminum bis(8-hydroxyquinoline)acetylacetonate (Alq_2Acac) complex

Since Tang and VanSlyke (1987) had developed the first organic light-emitting diode (OLED), Aluminum tris(8-hydroxyquinoline) (Alq_3) has been one of the most successful organic materials ever used as the emitting, electron-transport and host material layer in OLEDs. Numerous derivatives on Alq_3 structure were prepared and their optical and semiconductor properties were tested. Alq complex BAlq (bis(2-methyl-8-quinolinate)4-phenyl-phenolate) was first introduced by Kodak group as a blue-emitting material and mostly used as hole blocking layer (Kwong et al., 2002) and as a blue emitter (Kwong et al., 2005; Iwama et al., 2006; Yu et al., 2007). Hopkins and coworkers (1996) have also obtained blue shifted emission from Alq_3 derivate via introduction of the strong electron withdrawing $-SO_2NR_2$ group at C-5 of the 8-hydroxyquinoline ligand.
Azenbacher group investigated the role of 5-(arylethynyl)- (Pohl & Anzenbacher, 2005), 5-(aryl)- (Pohl et al., 2004; Montes et al., 2004, 2006; Pérez-Bolívar et al., 2006), and two C4-aryl- (Pérez-Bolívar et al., 2010) substituents on the quinolinolate rings, in Alq_3 derivatives and their effect on the photophysical properties and electroluminescence. Many methyl-substituted derivatives $nMeq_3Al$ (Kwong et al., 2005; Sapochak et al., 2001; Kido & Iizimi, 1998), phenyl-substituted $Alpq_3$ (Tokito et al., 2000), soluble 5-substituted-Alq_3 derivates (Mishra et al., 2005), aluminum complexes such as Alq_2OR (OR=aryloxy or alkoxy ligand) (Lim et al., 2006), have been developed and have been demonstrated to be useful emissive materials or/and hole blocking/electron transporting materials. Ma et al. (2003) have synthesized a new material dinuclear Aluminum 8-hydroxy-quinoline complex ($DAlq_3$) with two time higher electron mobility than that in Alq_3.
Omar et al. (2009) synthesized and investigated new aluminum tris(8-hydroxyquinoline) derivatives, having nitrogen functionalities at position-4 of the quinolate ligand, acting as efficient emitters with higher luminance and external quantum efficiency than the parent Alq_3 in an identical OLEDs. (The PL and EL emission wavelengths of the new Al complexes can be tunes according to the electronic properties of the substituents at position-4). Bingshe Xu et al. (2008) reported about a mixed-liquand 8-hydroxyquinoline aluminium complex with higher electron mobility and electroluminescent efficiency compared with Alq_3.
Here we presented a new Al complex, aiming the development of OLED with improved performance. The novel mixed-ligand Aluminum bis(8-hydroxyquinoline)acetylacetonate (Alq_2Acac) complex (Fig.16.) was synthesized and it performance as electroluminescent and electron transporting layer for OLED was studied and compared with that of the parent Alq_3 (Petrova et al., 2009).
To investigate the efficiency of the new Al complex as emitter, the devices ITO/HTL/EML/Al with EML layers of Alq_2Acac or commercial Alq_3 were fabricated.

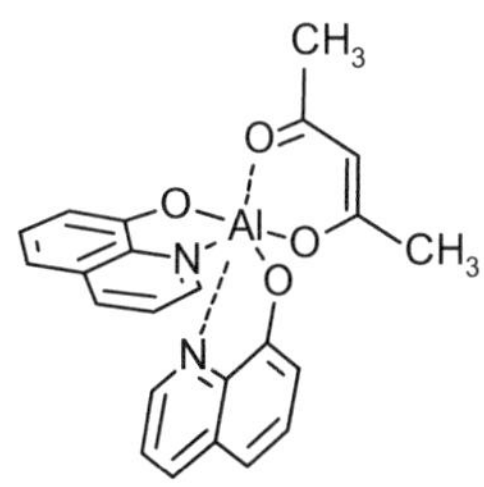

Fig. 16. Structure of Aluminum bis(8-hydroxyquinoline)acetylacetonate (Alq_2Acac)

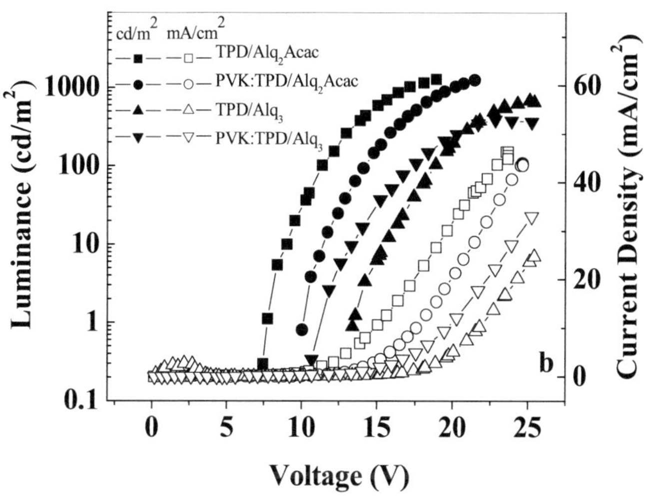

Fig. 17. I/V and L/V characteristics for devices with different $HTL_{(31\ nm)}$ and $EL_{(75\ nm)}$

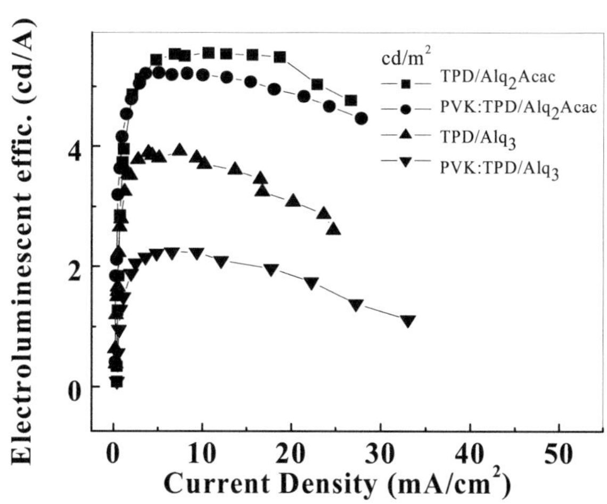

Fig. 18. Electroluminescent efficiency for devices with different $HTL_{(31\ nm)}$ and $EL_{(75\ nm)}$

The current density-voltage and luminescence-voltage characteristics of the studied devices are shown in Fig.17. The I/V curves of the devices with Alq_2Acac were located in lower voltage region compared to the devices with Alq_3. The luminescence of the devices with Alq_2Acac is 2 times higher compared to the similar devices with Alq_3 (Fig.17). The turn-on voltage of the devices with Alq_2Acac is lower compared to those with Alq_3 especially in the case with TPD hole transporting layer – nearly 2 times.
Bingshe Xu et. al. (2008) reported the electron mobilities in Alq_2Acac can be determined to be $2.7–4.4x10^{-6}$ $cm^2/V.s$ at electric fields ranging between $1.42x10^6$ and $2.40x10^6$ V/cm, which is higher than those in Alq_3 published in the literature (Huang et al., 2005; Brütting et al., 2001).
It could be stressed that the efficiency of the devices with Alq_2Acac are nearly 50 % higher compared to those with Alq_3 with HTL of TPD and about 2 times higher with HTL of PVK:TPD (Fig.18).
The ionization potential (Ip) and the electron affinity (Ea) of Alq_2Acac and Alq_3 were determined by cyclic voltammetry of 0.001 M solutions of compounds in $C_2H_4Cl_2$ in presence of 0.1 M tetra-n-butylammonium hexafluorophosphate as supporting electrolyte.

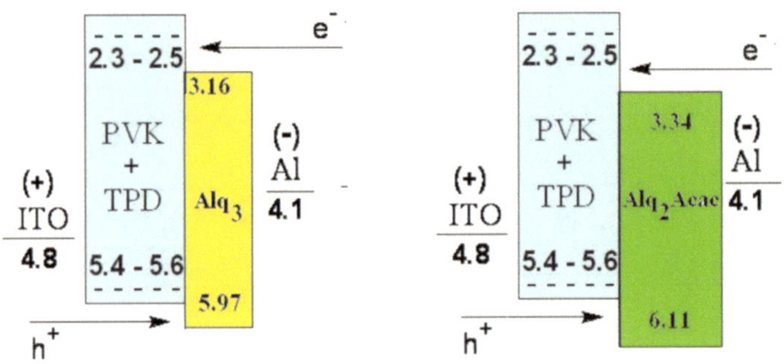

Fig. 19. The energy band diagram of investigated OLEDs

They were: Ip: Alq_2Acac 6.11 eV and Alq_3 5.97 eV; Ea: Alq_2Acac 3.34 eV and Alq_3 3.16 eV.
The band gaps were nearly equal (Eg = 2.77 eV for Alq_2Acac and 2.81 eV for Alq_3), that is in agreement with the values of 2.64 eV obtained by extrapolation of UV-Vis spectrums to absorption edges (Fig.20). As can be seen in Fig.19, both barriers for electrons and for holes are higher at Alq_2Acac compared with these of Alq_3, which explains the better efficiency of devices with Alq_2Acac.
Our results for the devices ITO/TPD $_{(30nm)}$/Alq_2Acac (or Alq_3) $_{(75nm)}$/Al are 5.6 cd/A for Alq_2Acac and 3.9 cd/A for Alq_3. Alq_2Acac based devices performed higher current density and emission efficiency. It indicates that the electron transport of Alq_2Acac is better than that of Alq_3 after the electron injection from the cathode to the electron transport layer, which is in a good agreement with the actual measurement of mobility. Probably, the molecular mixed-liqand structure of Alq_2Acac promoted higher electroluminescence efficiency and led to subsequent increase of the device performance.

4.1 Luminescence studies

The absorption and the fluorescent emission spectra of thin layers Alq_2Acac and Alq_3 are nearly identical see Fig.20. Both complexes emit green light with maximum at 520nm.

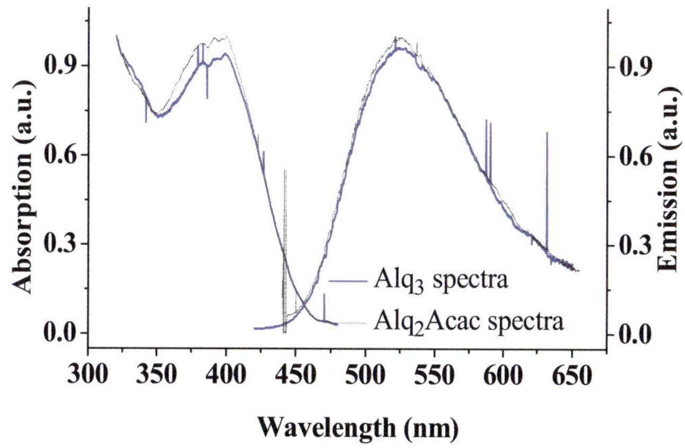

Fig. 20. Absorption and fluorescent emission spectra of Alq_2Acac and Alq_3 100 nm layers deposited on glass substrates.

The EL spectra observed at different voltages from the two studied devices ITO/PVK:TPD/Alq_2Acac and ITO/PVK:TPD/Alq_3 were shown in Fig.21.

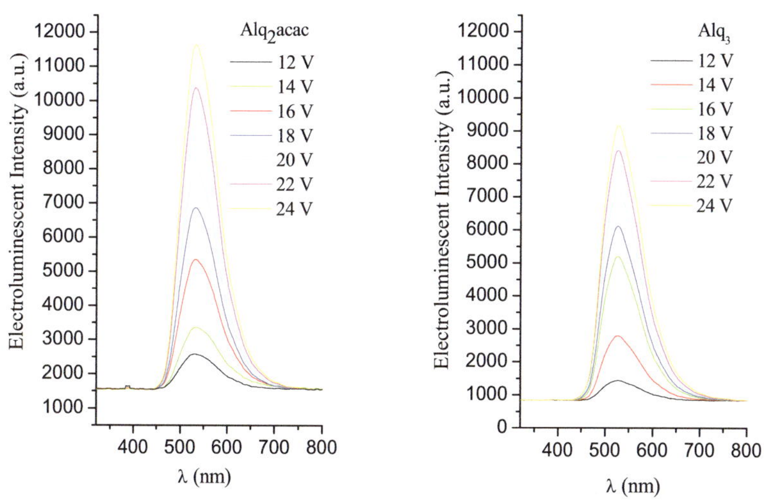

Fig. 21. EL spectra at different voltages of OLED structures ITO/PVK:TPD/Alq_2Acac and ITO/PVK:TPD/Alq_3.

The electroluminescence of both devices was similar to the fluorescence. It was established that the Alq_2Acac emission peak was located at 531 nm, quite close to that of Alq_3 at 525 nm, respectively. As far as concerned to the intensity of the peaks, those of the devices with Alq_2Acac emitter layers are nearly 2 times higher than that with Alq_3 at the identical experiments. It could be take note of that EL spectra of the two devices (Fig.21a and Fig.21b) are nearly identical like the PL spectra of the corresponding Alq_2Acac and Alq_3 thin solid films (Fig.20). It is possible, the included in Al complex acetylacetonate ligand does not participate in the π -- π * transition of quinolinolato ligands responsible for light emission.

Obviously the replacement of a quinolinolato ligand with an acetylacetonate ligand couldn't tune the emission colour but increase the efficiency of the devices.

4.2 Morphology

The performance of OLEDs is greatly influenced by the morphology of organic thin layers. This is due to the important role that morphology of the active organic thin films play in the phenomena that led to light emission. As was mention above, strong recrystalization of the TPD layers after 1 day storage at ambient temperature was established. At the some time composite PVK:TPD films remain stable - very smooth and homogeneous without any defects and cracks.

In this part of the work, the surface roughness of organic thin films was investigated via White Light Interferometer (WLI) MicroXAM S/N 8038. The surface relief profiles of the hole transporting layers of TPD and composite films of PVK:TPD were presented in Fig.22, while the surface profiles of the next electroluminesent layers of Alq_2Acac deposited onto corresponding HTL were presented in Fig.23. The root mean squire (RMS) roughness observed of the different samples are: 2.00±0.15 nm for TPD, 1.65±0.14 nm for PVK:TPD, 2.45±0.13 nm for TPD/Alq_2Acac, and 2.05±0.17 nm for PVK:TPD/Alq_2Acac. The RMS of PVK:TPD/Alq_3 determined from the surface profile shown in Fig.24 is 2.20±0.22. Both electroluminescent layers of Alq_3 and Alq_2Acac deposited on the composite films PVK:TPD show flat and amorphous surfaces which is a prerequisite for good performance of devices.

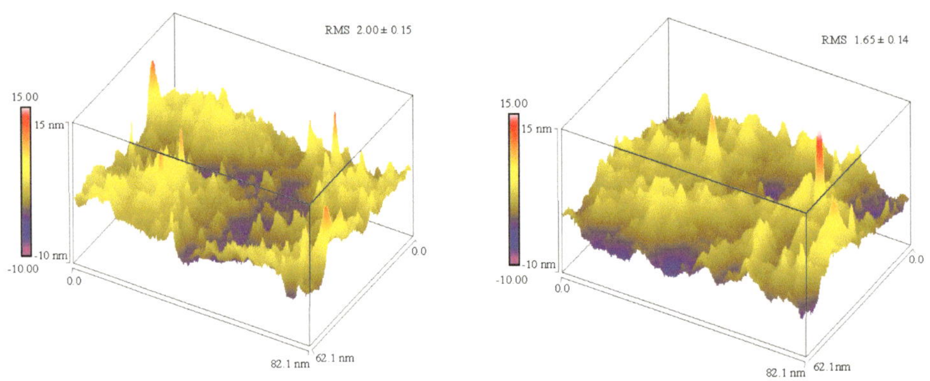

a) PET/ITO/TPD b) PET/ITO/PVK:TPD

Fig. 22. WLI of: OLED HTL layers of TPD (as deposited), and composite PVK:TPD

The similar is the surface morphology of the vacuum as deposited TPD films. Quantitative values indicate that the flexible acetylacetonate moieties help the formatting of more uniform and planarizing molecular film. The PVK:TPD, PVK:TPD/Alq_3 and PVK:TPD/Alq_2Acac layers does not show any changes after 1 day storage - better stability of devices with composite PVK:TPD hole transporting layer could be expected.

In conclusion, must to give prominence that the molecular structure of Alq_2Acac not only promoted the formation of very quantitative thin films, contributing to the high device efficiency, as well as the replacement of quinolinolato ligand with acetylacetonate ligand couldn't tune the emission colour. Alq_3 is still one of the widely-used fundamental materials as emitter and electron transporting layer in OLED due to its excellent thermal stability, high fluorescence efficiency and relatively good electron mobility. The results obtained show that the change of one 8-hydroxyquinoline ligand with acetylacetonate ligand in the novel complex improves substantially the performance of OLED. Besides that the devices with new Al complex are not optimized, its characteristics are quite promising.

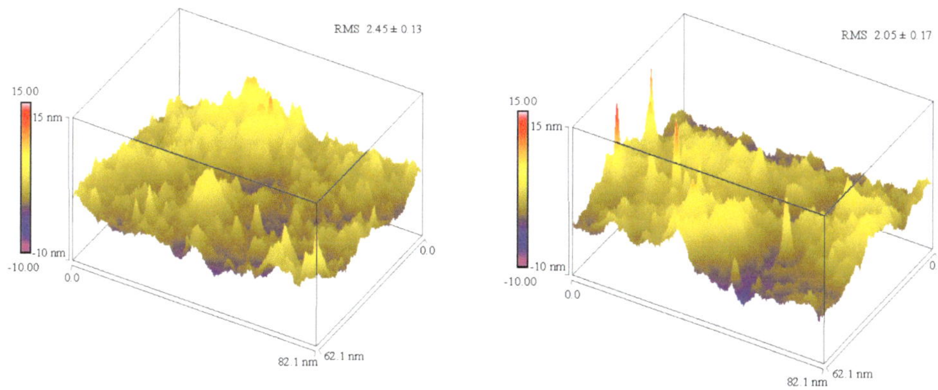

a) PET/ITO/TPD/Alq_2Acac b) PET/ITO/PVK:TPD/Alq_2Acac

Fig. 23. WLI surface of electroluminescent Alq_2Acac layer deposited onto HTLs.

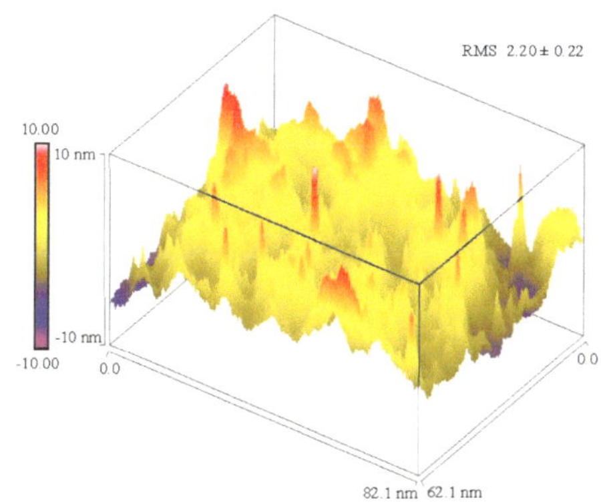

Fig. 24. WLI surface of electroluminescent Alq_3 layer deposited on PET/ITO/PVK:TPD HTL.

5. Bathocuproine as hole-blocking layer

In conventional PVK:TPD/Alq_3/Al OLEDs, the mobility of holes in PVK:TPD is much larger than that of the electrons in Alq_3. Also, the injection barrier of the anode/PVK:TPD interface is lower than that of the cathode/Alq_3 interface, resulting in the imbalance of holes and electrons in the emitting zone (Brütting at al., 2001; Mück et al., 2000). Therefore, it is necessary to confine the redundant holes in the emitting layer in order to increase the efficiency. Many effective methods have been reported to reduce the hole mobility and improve the balance of holes and electrons in the emitting layer (Troadec et al., 2002; Masumoto & Mori, 2008; Mori et al., 2008; Kim et al., 2005).

2,9-Dimethyl-4,7-diphenyl-1,10-phenanthroline (bathocuproine, BCP) was used in OLED and organic photovoltaic cell because of its multiple role as hole blocking (Adamovich et al., 2003; Kim et al., 2008; Tomova et al., 2008, 2010), exciton-blocking layer (Zhang et al., 2005; Tripathi et al., 2008; Mori & Kato, 2007; Wu et al., 2003), electron transporting and buffer layer (Wang et al., 2006), or in combination with NPB in $(NPB/BCP)_n$ (n-number of layers) as hole-trapping layer (Shi et al., 2006).

In this work we present our results concerning the role of bathocuproine as hole blocking layer in OLED structure: ITO/HTL/EML/HBL/ETL/M. HTL of composite PVK:TPD was spin-coated layer, and HBL, EML and ETL -were thermal evaporated films of BCP and Alq_3. The absorption and emission photoluminescence spectrums of evaporated layers of BCP, Alq_3 and BCP/Alq_3, measured by Spectrofluorimeter Perkin Elmer MPF 44 are presented in Fig.25.

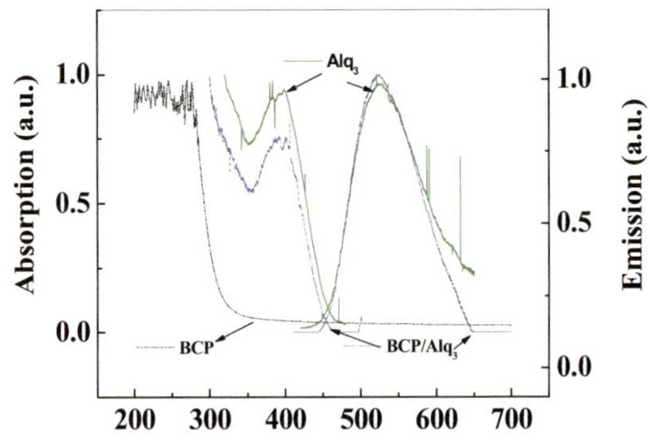

Fig. 25. Absorption and PL emission spectra of thin evaporated films (100 nm) of BCP, Alq3 and BCP/Alq_3

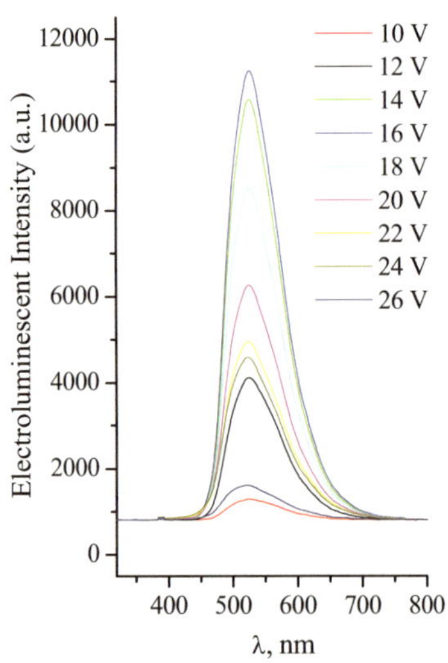

Fig. 26. EL spectra of the ITO/PVK:TPD $_{(31nm)}$/Alq_3 $_{(40\ nm)}$/BCP $_{(1\ nm)}$/Alq_3 $_{(15\ nm)}$ device at different voltages, were measured on Ocean Optics HR2000+ spectrometer

It is seen that the PL spectra of Alq_3 and BCP/Alq_3 layers are nearly identical with PL peak wavelength at 520 nm. Therefore in these wave length ranges BCP neither absorbs nor radiates and observed fluorescent emission originates from Alq_3 only. Besides, the EL spectra (Fig.26) observed from the ITO/PVK:TPD/$Alq_{3\ (40)}$/BCP $_{(1)}$/$Alq_{3\ (15)}$ device at different voltages are quite similar to the fluorescent spectrum of the corresponding Alq_3 film. The emission peaks were located at 525 nm and are at the same position as the peak of Alq_3 based structure (presented earlier in this chapter), which also is an evidence that BCP do not participate in the light emission.

Fig.27 presents the current density (Fig.27a) and luminance (Fig.27b) versus voltage, and current and power efficiency versus luminance (Fig.27c) characteristics of the devices ITO/PVK:TPD/$Alq_{3\ (40\ nm)}$/BCP $_{(x\ nm)}$/$Alq_{3\ (15\ nm)}$/Al, where x is 0; 1; 5 and 15 nm. The I-V curves (Fig.27a) show that insertion of BCP layer decreases the current density and shifts the threshold voltage from 11V to 17V for devices without and with 15 nm BCP. The luminescence initially increases from 750cd/m^2 to 1100cd/m^2 for device with 1 nm BCP and then decreases with increasing the thickness of BCP (Fig.27b).

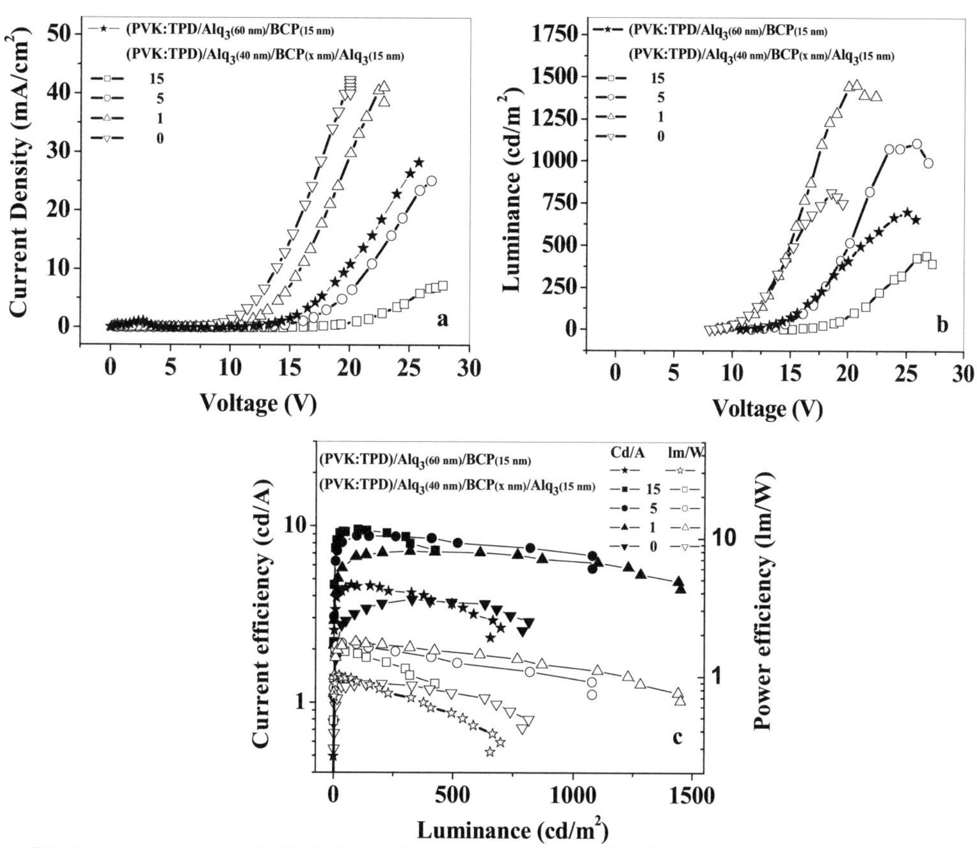

Fig. 27. Current-voltage (I-V) (2a), luminescence-voltage (L-V) (2b), current and power efficiency (2c) curves for devices shown in set

Fig.28 presents the driving voltage and the efficiency at luminance 100 and 200 cd/m^2 in dependence on the thickness of BCP. It is seen that despite of the higher voltage of the

devices with BCP their current efficiency significantly increase from 3.7 to 9.6 cd/A, and power efficiency increase from 0.87 to 1.46 lm/W (at 100 cd/m^2) for devices without and with 15 nm BCP layer. The beneficial influence of BCP is not only related to the magnitude of the efficiency but also to the broader luminance range which could be seen in Fig.27c.
The best characteristics - the lower threshold and working voltage, the highest luminescence and 2 times increased efficiency from 3.7 to 7.1 cd/A and from 0.87 to 1.75 lm/W at 100 cd/m^2 demonstrates device with 1 nm BCP layer compared to the device without BCP.

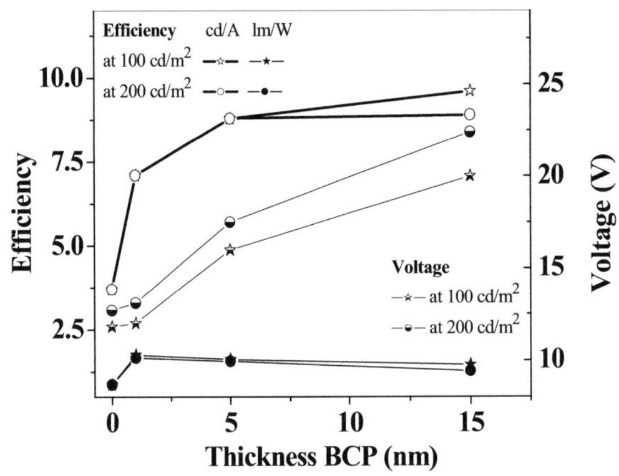

Fig. 28. Current and power efficiency, and voltage v/s thickness of BCP

At first sight increasing of the luminance for device with very thin BCP layer looks strangely, but it can be explain with the island structure of thin layer. On one hand the islands are enough great to confine the holes (due to the high hole barrier from 0.7 eV at the EML/HBL (Fig.29) thus improving the recombination at the EML/BCP interface, but on the other hand they aren't enough dense to cause materially decreasing of the electric field yet.

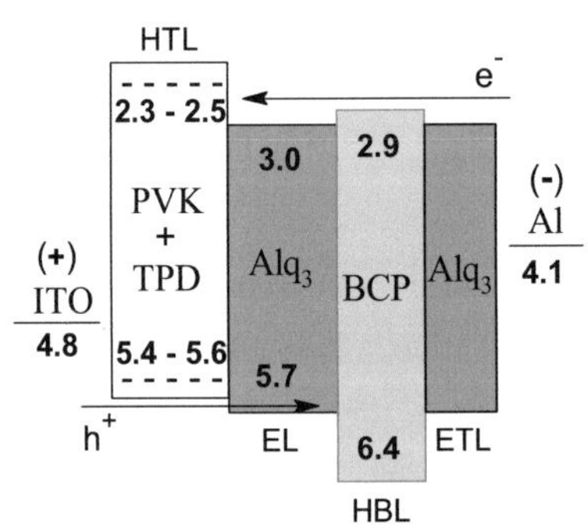

Fig. 29. The energy band diagram of the investigated OLEDs.

Khalifa et al. (2004) estimated that the diffusion length of holes in BCP probably lies in the 15-20 nm range. Thicker BCP layers lead to a decrease of luminance which could be attributed to a decrease of electron density arriving at the Alq_3/BCP interface and thus to a degradation of the carrier balance. The comparison of the results obtained for devices with BCP/Alq_3 and BCP only, at equal thicknesses of BCP, shows more than 2 times higher current efficiency and 70-100% higher power efficiency at approximately equal luminance for the devices with Alq_3 as ETL. Obviously this is due to the higher with 0.1 eV LUMO level of BCP than that of Alq_3, which make the electron injection from Al to BCP more difficult.

The crystallization of organic films in OLEDs is one factor reducing the device performance. It is well known that, because of molecular migration, an evaporated BCP film is immediately crystallized after deposition (Masumoto & Mori, 2008). That's why we investigated the surface relief profiles and roughness, of consequently deposited layers of the device with best performance PVK:TPD/Alq_3 $_{(40nm)}$/BCP $_{(1nm)}$/Alq_3 $_{(15nm)}$. The WLI images presented in Fig.30 show flat and amorphous layers structure with nearly the same roughness - RMS 2.20 ± 0.22 nm for PVK:TPD/Alq_3 $_{(40nm)}$, RMS 2.16 ± 0.34 nm for PVK:TPD/Alq_3 $_{(40nm)}$/BCP $_{(1nm)}$ and RMS 1.89 ± 0.22 nm for PVK:TPD/Alq_3 $_{(40nm)}$/BCP $_{(1nm)}$/Alq_3 $_{(15nm)}$. It is a prerequisite for good performance.

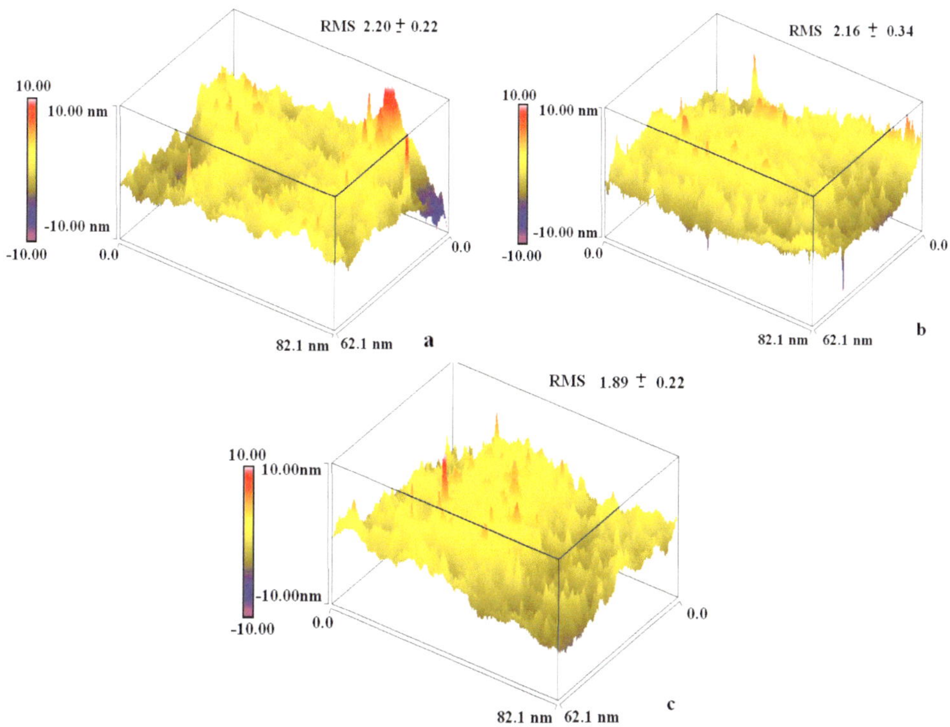

Fig. 30. Surface relief profiles, obtained on White Light Interferometer MicroXAM S/N 8038 of: a) PVK:TPD $_{(31\ nm)}$/Alq_3 $_{(40\ nm)}$; b) PVK:TPD/Alq_3 $_{(40\ nm)}$/BCP $_{(1\ nm)}$; c) PVK:TPD $_{(31\ nm)}$/Alq_3 $_{(40\ nm)}$/BCP $_{(1\ nm)}$/Alq_3 $_{(15\ nm)}$

Becker et al. (2007) in their research of the influence of thermal annealing of BCP and Alq_3 (used as exciton blocking layers) on characteristics of organic photovoltaics were found that BCP is more susceptible to heat and oxygen than Alq_3. They observed that visible on a macroscopic level crystals of BCP were appeared at temperature of 100°C while Alq_3 even at 300°C, in open air, formed only minor pinpoint crystals. In our case probably that is the reason causing the better performance of devices containing additional ETL of Alq_3. OLED with improved efficiency 4.3 cd/A more than twice that of the undoped OLED (1.8 cd/A) by doping BCP into Alq_3 as an ETL and HBL was fabricated by Wu et al. (2003). Our results for efficiencies from 7.2 up to 9.6 cd/A in dependence on the thickness of BCP are quite promising.

In conclusion it can be say that BCP with its deep HOMO level (6.4 eV) is a good hole blocking layer. The influence of HBL in confining the carriers and excitons was clearly evidenced by a strong increase of the device efficiency. BCP offers possibilities to optimize the architecture of the OLED thus improving significantly the performance of the devices. The better performance of the devices with BCP could be attributed to the improved hole-electron balance.

6. Conclusions

On the basis of synthesized novel Zn and Al complexes, successfully chosen functional layers and improved architecture experimental OLEDs with very good characteristics are developed. The results reveal a new approach to the design and preparation of high-performance luminescence materials for the development of full-color flexible displays, new class energy-saving solid state light sources.

7. References

Adamovich V., Cordero S., Djurovich P., Tamayo A., Thompson M., Andrade B., and Forrest S. (2003), New charge-carrier blocking materials for high efficiency OLEDs, *Organic Electronics*, Vol. 4, No 2-3, pp. 77-87, ISSN: 1566-1199.

Arai M., Nakaya K., Onitsuka O., Inoue T., Codama M., Tanaka M. and Tanabe H. (1997), Passive matrix display of organic LEDs, *Synth.Met.*, Vol. 91, No 1-3, pp. 21-25, ISSN: 0379-6779.

Baldo M., O'Brien D., You Y., Shoustikov A., Sibley S., Thomson M., and Forrest S. (1998), Highly efficient phosphorescent emission from organic electroluminescent devices, *Nature*, Vol. 395, No 6698, pp. 151-154, ISSN : 0028-0836

Becker K. (2007), Exciton Blocking Layers in Organic Photovoltaics, Cornell Center for Materials Research., http://matdl.org/repository/view/matdl:753.

Berntsen A., Croonen Y., Liedenbaum C., Schoo H., Visser R, Vieggaar J., and Van de Weijer P. (1998), Stability of polymer LEDs, *Optical Materials*, Vol. 9, No 1-4, pp. 125-133, ISSN: 0925-3467.

Brütting W., Berleb S., and Mückl A. (2001), Device physics of organic light-emitting diodes based on molecular materials, *Org. Electron.*,Vol. 2, No 1, pp. 1-36, ISSN: 1566-1199.

Burroughes J., Bradley D., Brown A., Marks R., Mackey K., Friend R., Burns P., Holmes A. (1990), Light-emitting diodes based on conjugated polymers, *Nature,* Vol. 347, No 6293, pp. 539-541, ISSN : 0028-0836

Carter S., Angelopoulos M., Karg S., Brock P, and Scot J. (1997), Polymeric anodes for improved polymer light-emitting diode performance, *Appl. Phys. Lett.*, Vol. 70, No 16, pp.2067-2070, ISSN: 0021-8979.

Chan I. and Hong F. (2004), Improved performance of the single-layer and double-layer organic light emitting diodes by nickel oxide coated indium tin oxide anode, *Thin Solid Films*, Vol. 450, No 2, pp. 304- 311, ISSN: 0040-6090.

Deng Z., Ding X., Lee S., and Gambling W. (1999), Enhanced brightness and efficiency in organic electroluminescent devices using SiO_2 buffer layers, *Appl. Phys. Lett.*, Vol. 74, No 15, pp. 2227-2230, ISSN: 0021-8979.

Gao Y., Wang L.,.Zhang D, Duan L., Dong G., and Qiu Y. (2003),, Bright single-active layer small-molecular organic light-emitting diodes with a polytetrafluoroethylene barrier, *Appl. Phys. Lett.*, Vol. 82, No 2, pp. 155-158, ISSN: 0021-8979.

Guo F.and Ma D. (2005), White organic light-emitting diodes based on tandem structures, *Appl. Phys. Lett.*, Vol. 87, No 17, 173510, ISSN: 0021-8979.

Hamada Y., Sano T., Fujii H., Nishio Y., Takahashi H. and Shibata K. (1996), White-Light-Emitting Material for Organic Electroluminescent Devices, *Japanese Journal of Applied Physics*, Vol. 35, No 10B, pp. L1339-L1341, ISSN: 0021-4922.

Hopkins T., Meerholz K., Shaheen S., Anderson M., Schmidt A., Kippelen B., Padias A., Hall H.., Peyghambarian N., and Armstrong R. (1996), Substituted Aluminum and Zinc Quinolates with Blue-Shifted Absorbance/Luminescence Bands: Synthesis and Spectroscopic, Photoluminescence, and Electroluminescence Characterization, *Chem. Mater.*, Vol. 8, No 2, pp 344–351, ISSN: 0897-4756

Hu W. and Matsumura M. (2002), Organic single-layer electroluminescent devices fabricated on CuO_x-coated indium tin oxide substrate, Appl. Phys. Lett. Vol. 81, No 5, 806, ISSN: 0021-8979.

Huang C. H., Li F., and Huang W. (2005), Introduction to Organic Light-Emitting Materials and Devices, Fudan University Press, Shanghai.

Im H., Choo D., Kim T., Kim J., Seo J., and Kim Y. (2007), Highly efficient organic light-emitting diodes fabricated utilizing nickel-oxide buffer layers between the anodes and the hole transport layers, , *Thin Solid Films*, Vol. 515, No 12, pp. 5099-5102, ISSN: 0040-6090.

Iwama Y., Itoh T., Mori T., and Mizutani T. (2006), Electroluminescent properties of organic light-emitting diodes using BAlq and Alq_3 co-evaporation layer, *Thin Solid Films* Vol. 499, No 1-2, pp. 364-368, ISSN: 0040-6090.

Jiang H., Zhou Y., Ooi B., Chen Y., Wee T.,. Lam Y, Huang J., and Liu S. (2000), Improvement of organic light-emitting diodes performance by the insertion of a Si_3N_4 layer, *Thin Solid Films*,Vol. 363, No1-2, pp.28-25, ISSN: 0040-6090.

Jiang X, Zhang Z, Cao J., Khan M., Haq K., and Zhu W. (2007), White OLED with high stability and low driving voltage based on a novel buffer layer MoO_x, *J.Phys.D: Applied Physics*, Vol. 40, N 18, 5553, ISSN: 0022-3727

Jung B., Lee J., Chu H., Do L., and Shim H. (2002), Synthesis of Novel Fluorene-Based Poly(iminoarylene)s and Their Application to Buffer Layer in Organic Light-Emitting Diodes, *Macromolecules*, Vol. 35, No 6, pp 2282–2287, ISSN 0024-9297.

Khalifa M., Vaufrey D., and Tardy J. (2004), Opposing influence of hole blocking layer and a doped transport layer on the performance of heterostructure OLEDs, *Organic Electronics*, Vol. 5, No 4, pp. 187-198, ISSN: 1566-1199.

Kido J., and MatsumotoT. (1998), Bright organic electroluminescent devices having a metal-doped electron-injecting layer, *Appl. Phys. Lett.*, Vol. 73, No 20, pp. 2866- 2869, ISSN:0003-6951.

Kido J., and Iizimi Y. (1998), Fabrication of highly efficient organic electroluminescent devices, *Appl. Phys. Lett.*, Vol. 73, No 19, pp. 2721- 2724, ISSN 0003-6951.

Kim Y., Park H., and Kim J. (1996), Enhanced quantum efficiency in polymer electroluminescence devices by inserting a tunneling barrier formed by Langmuir–Blodgett films, *Appl. Phys. Lett.*, Vol. 69, No 5, pp. 599-602, ISSN: 0021-8979.

Kim J., Song M., Seol J., Hwang H., and Park Ch. (2005), Fabrication of Red, Green, and Blue Organic Light-Emitting Diodes Using m-MTDATA as a Common Hole-Injection Layer, *Korean J. Chem. Eng.*, Vol. 22, No 4, pp. 643-647, ISSN: 0256-1115.

Kim D., Kim W., Lee B. and Kwon Y. (2007), White Organic Light-Emitting Diode Using Blue-Light-Emitting $Zn(HPB)_2$ Material, *Japanese Journal of Applied Physics*, Vol. 46, No. 4B, pp. 2749-2753, ISSN: 0021-4922.

Kim D., Kim W., Kim B., Lee B., and Kwon Y. (2008), Characteristics of white OLED using Zn(phen) as a yellowish green emitting layer and BCP as a hole blocking layer, *Colloids and Surfaces A: Physicochem. Eng. Aspects*, Vol 313–314, pp. 320-323, ISSN: 0927-7757.

Krag S., Scott C., Salem R, and Angelopoulos M. (1996), Increased brightness and life time of polymer light-emitting diodes with polyaniline anodes, *Synth. Met.*; Vol. 80, No 2, pp. 111–117, ISSN: 0379-6779

Kwong R., Nugent M., Michalski L., Ngo T., Rajan, K. Tung Y., Weaver M., Zhou Th., Hack M., Thompson M., Forrest S., and Brown J. (2002), High operational stability of electrophosphorescent devices, *Appl. Phys. Lett.*, Vol. 81, No 1, pp.162- 165, ISSN: 0021-8979.

Kwong C., Djurisic A., Choy W., Li D., Xie M., Chan W., Cheah K, Lai P., and Chui P. (2005), Efficiency and stability of different tris(8-hydroxyquinoline) aluminium (Alq_3) derivatives in OLED applications, *Mat. Sci. Eng. B*, Vol., 116, No 1, pp. 75 -81, ISSN: 0921-5107

Legnani C., Reyes R., Cremona M., Bagatin I., and Toma H.. (2004), Tunable blue organic light emitting diode based on aluminum calixarene supramolecular complex, *Appl. Phys. Lett.*, Vol. 85, No 1, pp. 10- 13, ISSN: 0021-8979.

Li F., Tang H., Anderegg J., and Shinar J. (1997), Fabrication and electroluminescence of double-layered organic light-emitting diodes with the Al_2O_3/Al cathode, Appl. Phys. Lett. Vol.. 70, No 10, pp. 1233-1236, ISSN: 0021-8979.

Lim J., Lee N., Ahn Y., Kang G. and Lee C. (2002), White-light-emitting devices based on organic multilayer structure, *Current Applied Physics*, Vol. 2, No 4, pp. 295-298, ISSN: 1567-1739.

Lim J., Jeong C., Lee J., Yeom G., Jeong H., Chai S., Lee I., and Lee W. (2006), Synthesis and characteristics of bis(2,4-dimethyl-8-quinolinolato)(triphenylsilanolato)aluminum (III): A potential hole-blocking material for the organic light-emitting diodes, *J. Organometallic Chemistry*, Vol. 691, No 12, pp. 2701- 2707, ISSN: 0022-328X.

Lu H. and Yokoyama M. (2003), Enhanced emission in organic light-emitting diodes using Ta_2O_5 buffer layers, *Solid-State Electronics*, Vol. 47, No 8, pp. 1409–1412, ISSN: 0038-1101.

Ma D., Wang G., Hu Y., Zhang Y., Wang L., Jing X., Wang F., Lee C., and Lee S. (2003), A dinuclear aluminum 8-hydroxyquinoline complex with high electron mobility for organic light-emitting diodes, *Appl. Phys. Lett.* Vol. 82, No 8, pp. 1296-1299, ISSN: 0021-8979.

Masumoto Y. and Mori T. (2008), Application of organic bathocuproine-based alloy film to organic light-emitting diodes, *Thin Solid Films*, Vol. 516, No 10, pp. 3350-3356, ISSN: 0040-6090

Meyer J., Hamwi S., Bülow T., Johannes H.,. Riedl T, and Kowalsky W. (2007), Highly efficient simplified organic light emitting diodes, *Appl. Phys. Lett,*. Vol. 91, No 11, 113506, ISSN: 0021-8979.

Miloshev S. and Petrova P. (2006), Preparation of copolymers of p-Isopropenylcalix[8]arene and Styrene, *Polymer Bulletin* , Vol. 56, pp. 485-494, ISSN: 01700839.

Mishra A., Periasamy N., Patankar M., and Narasimhan K., (2005), Synthesis and characterisation of soluble aluminium complex dyes based on 5-substituted-8-hydroxyquinoline derivatives for OLED applications, *Dyes and Pigments*, Vol. 66, No 2, pp. 89-97, ISSN: 0143-7208.

Montes V., Li G., Pohl R., Shinar J., Anzenbacher P. (2004), Effective Color Tuning in Organic Light-Emitting Diodes Based on Aluminum Tris(5-aryl-8-hydroxyquinoline) Complexes, *Adv. Mater.*, Vol. 16, No 22, pp. 2001-2003, ISSN: 1521-4095.

Montes V., Li G., Pohl R., Shinar J., Anzenbacher P. (2006), Effective Manipulation of the Electronic Effects and Its Influence on the Emission of 5-Substituted Tris(8-quinolinolate) Aluminum(III) Complexes, *Chemistry - A European Journal*, Vol. 12, No 17, pp. 4523- 4535, ISSN: 1521-3765

Mori T. *and* Masumoto Y. (2006), Effect of Organic Alloy for Suppression of Polycrystallization in BCP Thin Film, *J. Photopolym. Sci. Technol.*, Vo.. 19, No 2, pp. 209-214, ISSN: 0914-9244..

Mori T. *and* Kato K. (2007), Photovoltaic Properties of Organic Thin-Film Solar Cell Using Various Exciton-Diffusion Blocking Materials, *J. Photopolym. Sci. Technol*,. Vol. 20, No. 1, pp. 61-66, ISSN: 0914-9244.

Mori T., Masumoto Y., *and* Itoh T. (2008), Control of Electroluminescence Spectra Using Hole-Blocking Layer for White Organic Light-Emitting Diodes, *J. Photopolym. Sci. Technol.*, Vol. 21, No. 2, pp. 173-180, ISSN: 0914-9244.

Mück A., Berleb S., Brütting W., and Schwoerer M. (2000), Transient electroluminescence measurements on organic heterolayer light emitting diodes, *Synthetic Metals*, Vol. 111-112, No1, pp. 91-94, ISSN: 0379-6779.

Okamoto K., Kanno H., Hamada Y., Takahashi H., and Shibata K. (2006), High efficiency red organic light-emitting devices using tetraphenyldibenzoperiflanthene-doped rubrene as an emitting layer, *Appl. Phys. Lett.*, Vol. 89, No 1, 013502, ISSN: 0021-8979.

Omar W., Haverinen H.,and Horm O. (2009), New Alq_3 derivatives with efficient photoluminescence and electroluminescence properties for organic light-emitting diodes, *Tetrahedron*, Vol. 65, No 47, pp. 9707-9712, ISSN: 0040-4020

Park J., Lee Y., Kwak Y., and Choi J. (2002), Characteristic effects of hole injection on organic electroluminescent devices, *Journal of the Korean Physical Society*, Vol 41, No6, pp. 1050-1053, ISSN: 0374 4884.

Pérez-Bolívar C., Montes V., and Anzenbacher P. (2006), True Blue: Blue-Emitting Aluminum(III) Quinolinolate Complexes, *Inorg. Chem*, .Vol. 45, No 24, pp. 9610-9612, ISSN 0020-1669

Pérez-Bolívar C., Llovera L., E. Lopez S., Anzenbacher P. (2010), 4- vs. 5-phenylquinolinolate aluminum (III) isomers, *Journal of Luminescence*, Vol. 130, No 1145, ISSN: 0022-2313

Petrova P., Tomova R., Stoycheva-Topalova R., Kaloianova S., and Deligeorgiev T. (2009), Novel Al compex as emitter in Organic light emitting diodes, *Optoelectronic and Advanced Materials – Rapid Communications (OAM-RC)*, Vol. 3, No 5, pp. 424-427, ISSN: 1842-6573.

P. Petrova, R. Tomova (2009), Materials used for organic light-emitting diodes - organic electroactive compounds, *Bulgarian Chemical Communications*, Vol. 41, No 3, pp. 211-225, ISSN: 0324-1130.

P. Petrova, R. Tomova Stoycheva-Topalova R., and Miloshev St. (2010), Influence of p-isopropenylcalixarenestyrene copolymer buffer layer over Alq_3 based OLEDs, *Eur. Phys. J. Appl. Phys.*, Vol. 51, 33210, ISSN: 1286-0042.

Pommerehne J., Vestweber H.,. Guss W, Mahrt R., Bässler H., Porsch M.,and Daub J. (1995), Efficient two layer leds on a polymer blend basis, *Adv. Mater.* 7, pp. 551- 554, ISSN: 0935-9648.

Qiu Y., GaoY., Wang L., and Zhang D. (2002), Efficient light emitting diodes with Teflon buffer layer, *Synthetic Metals,* Vol. 130, No 3, pp. 235–237, ISSN: 0379-6779.

Qureshi M., Manohara S., Singh S., and Mahapatra Y. (2005), Electroluminescent properties of dimeric bis(2-(2′-hydroxyl phenyl)benzthiazolate)zinc (II) complex, *Solid State Comm.*, Vol. 133,No 5, pp. 305-309, ISSN: 0038-1098.

Rai V., Srivastava R., Chauhan G., Saxena K., Bhardwaj R., Chuand S., Kamalasan M. (2008), Synthesis and electroluminescence properties of zinc(2,2′ bipyridine)8-hydroxyquinoline, *Mat. Lett.*, Vol. 62, No 17-18, pp. 2561-2563, ISSN: 0167-577X.

Reyes R., Legnani C., Cremona M., Brito H., Britto R, and Achete C. (2004), Amorphous carbon nitride thin films as buffer layer in organic LEDs, *Physica Status Solidi (c)*, Vol.1, No S2 ,pp. S229–S235, ISSN: 1610-1634.

Pohl R. and Anzenbacher P. (2003), Emission Color Tuning in AlQ_3 Complexes with Extended Conjugated Chromophores, *Org. Lett.*,Vol, No 16, pp. 2769- 2772, ISSN 1523-7060.

Pohl R., A. Montes V., Shinar J., and Anzenbacher P. (2004), Red−Green−Blue Emission from Tris(5-aryl-8-quinolinolate)Al(III) Complexes, *J. Org. Chem.*, Vol. 69, No 5, pp. 1723-1725, ISSN: 0022-3263.

Rothberg L. and Lovinger A. (1996), Status of and prospects for organic electroluminescence, *J. Mater. Res.*, Vol.11, No 12, pp. 3174 -3187, ISSN: 0884-2914.

Sano T., Nishio Y., Hamada Y., Takahashi H., Usuki T. and Shibata K. (2000), Design of conjugated molecular materials for optoelectronics, *J. Mater. Chem..,* Vol. 10, No 1, pp. 157-161, ISSN: 0959-9428.

Sapochak L., Padmaperuma A., Washton N., Endrino F., Schmett G., Marshall J.,. Fogarty D, Burrows P. and Forrest S. (2001), Effects of Systematic Methyl Substitution of Metal (III) Tris(*n*-Methyl-8-Quinolinolato) Chelates on Material Properties for Optimum Electroluminescence Device Performance, *J. Am. Chem. Soc.* Vol. 123, No 25, pp. 6300-6307, ISSN 0002-7863.

Sapochak L., Benincasa F., Schofield R., Baker J., Riccio K., Fogarty D., Kohlmann H., Ferris K. and Burrows P. (2002), Electroluminescent Zinc(II) Bis(8-hydroxyquinoline): Structural Effects on Electronic States and Device Performance, *J. Am. Chem. Soc.* Vol. 124, No 21, pp.6119-6125, ISSN 0002-7863.

Schlatmann A., Floet D., Hilberer A., Garten F., Smulders P, Klapwijk T., and Hadziioannou G. (1996), Indium contamination from the indium–tin–oxide electrode in polymer light emitting diodes, *Appl. Phys. Lett.*, Vol. 69, No 12, pp. 1764-1767 ISSN 0003-6951.

Sheats J., Antoniadis H., Hueschen M., Leonard W., Miller J., Moon R., Roitman D., and Stocking A. (1996), Organic electroluminescent devices, *Science*, Vol. 273, No 5277, pp. 884 -888, ISSN: 00368075

Shi J. and Tang C. (1997), Doped organic electroluminescent devices with improved stability, *Appl. Phys. Lett.*, Vol. 70, No 13, pp.1665-1667, ISSN: 0021-8979.

Shi Y., Deng Zh, Xu D., and Xiao J. (2006), Organic light-emitting diodes with improved hole–electron balance and tunable light emission with aromatic diamine/bathocuproine multiple hole-trapping-layer, *Displays*, Vol 27, No 4-5, pp. 166-169, ISSN: 0141-9382.

Shukla V. and Kumar S. (2010), Conversion of a green light emitting zinc-quinolate complex thin film to a stable and highly packed blue emitter film, *Synthetic Metals*, Vol. 160, No 5-6, pp. 450–454, ISSN: 0379-6779.

Tang C. and VanSlyke S. (1987), Organic electroluminescent diodes, *Appl. Phys. Lett.* Vol. 51, No 12, pp. 913-915, ISSN: 0003-6951.

Tadayyon S., Grandin H., Griffiths K., Norton P., Aziz H., and Popovic Z. (2004), CuPc buffer layer role in OLED performance: a study of the interfacial band energies, *Organic Electronics,* Vol. 5, No 4, pp. 157–166, ISSN: 1566-1199.

Tokito S., Noda K., Tanaka H., Taga T., and Tsutsi T., Organic light-emitting diodes using novel metal–chelate complexes, *Syn. Met.*, Vol. 111-112, No 1, pp. 393-397, (2000) ISSN: 0379-6779.

Tomova R., Petrova P., Buroff A., and Stoycheva-Topalova R. (2007), Organic light-emitting diodes (OLEDs) - the basis of next generation light-emitting devices, *Bulgarian Chemical Communications*, Vol. 39, No 4, pp. 247 -259, ISSN: 0324-1130.

Tomova R., Petrova P., Stoycheva-Topalova R, Buroff A., Vassilev A., and, Deligeorgiev T. (2008), Bathocuproine as Hole-blocking and Electron-transporting Layer in Organic Light Emitting Devices, *Nanoscience& Nanotechnology - Proceedings of the 9th Workshop Nanostructured Materials Application and Inovation Transfer,* November 28-30, 2007, Sofia, Bulgaria, eds. E.Balabanova, I.Dragieva, Pub by Prof. Marin DrinovAcademic Publishing House, Sofia, pp. 114-117, ISBN: 978-954-322-286-5.

Tomova R., Petrova P., Kaloianova S., Deligeorgiev T., Stoycheva-Topalova R., and Buroff A. (2008), Organic Light Emitting Devices Based on Novel Zn Complexes, *Proceedings of 14th Int. Workshop on Inorganic and Organic Electroluminescence & 2008 Int. Conference on the Science and Technology of Emissive Displays and Lighting (EL 2008*), Italy 9-12 September 2008, pp. 69-73

Tomova R., Petrova P., and Stoycheva-Topalova R. (2010), Role of bathocuproine as hole-blocking and electron-transporting layer in organic light emitting devices, *Phys. Status Solidi C*, Vol 7, No. 3–4, pp. 992–995, ISSN : 1610-1634

Troadec D., Veriot G. and Moliton A. (2002), Simulation study of the influence of polymer modified anodes on organic LED performance, *Synthetic Metals*, Vol. 127, No 2, pp.165-175, ISSN: 0379-6779.

Tripathi V., Datta D., Samal G., Awasthi A., and Kumar S. (2008), Role of exciton blocking layers in improving efficiency of copper phthalocyanine based organic solar cells, *J. of Non-Crystalline Solids*, Vol. 354, No 19-25, pp. 2901- 2904, ISSN: 0022-3093.

Van Slyke S., Chen C., and Tang C. (1996), Organic electroluminescent devices with improved stability, *Appl. Phys. Lett.*, Vol. 69, No 15, pp. 2160-2163, ISSN: 0021-8979.

Wang H., Wang L., Wu Z., Zhang D., Qiao J., Qiu Y., and Wang X. (2006), Efficient single-active-layer organic light-emitting diodes with fluoropolymer buffer layers, *Appl. Phys. Lett.*, Vol. 88, No 13,131113, ISSN: 0021-8979.

Wang Y., Teng F., Zhou Q., and Wang Y. (2006), Multiple roles of bathocuproine employed as a buffer-layer in organic light-emitting diodes, *Applied Surface Science*, Vol. 252, No 6, pp. 2355-2359, ISSN: 0169-4332.

Wei X., Yang G, Cheng J., Lu Z., and Xie M. (2007), Synthesis of novel light emitting calix[4]arene derivatives and their luminescent properties, *Mat. Chem. Phys.*, Vol. 102, No 2-3, pp. 214-218, ISSN: 0254-0584.

Wu J., Hou J., Cheng Y, Xie Z., and Wang L. (2007), Efficient top-emitting organic light-emitting diodes with a V_2O_5 modified silver anode, *Semicond. Sci. Technol.*, Vol. 22, No 7, pp. 824-828, ISSN: 0268-1242.

Wu X., Hua Y., Wang Z., Zheng J., Feng X. and Sun Y. (2005), White Organic Light-Emitting Devices Based on 2-(2-Hydroxyphenyl) Benzothiazole and Its Chelate Metal Complex, *Chinese Phys. Lett.*, Vol.. 22, No 7, pp. 1797-1799, ISSN: 1741-3540.

Wu Z., Yang H., Duan Y., Xie W., Liu S., and Zhao Y. (2003), Improved efficiency of organic light-emitting devices employing bathocuproine doped in the electron-transporting layer, *Semicond. Sci. Technol.*, Vol. 18, L49–L52, ISSN 0268-1242.

Xu B., Chen L., Liu X., Zhou H., Xu H., Fang X., and Wang Y. (2008), Mixed ligands 8-hydroxyquinoline aluminum complex with high electron mobility for organic light-emitting diodes, *Appl. Phys. Let.*, Vol. 92, No 10, 103305, ISSN: 0021-8979.

Xu X., Yu G., Liu Y.,and Zhu D. (2006), Electrode modification in organic light-emitting diodes, *Displays*, Vol. 27, No 1, pp. 24-34, ISSN: 0141-9382.

You H.,. Dai Y, Zhang Z., and Ma D. (2007), Improved performances of organic light-emitting diodes with metal oxide as anode buffer, *J. Appl. Phys.* Vol. 101, No 2, pp. 026105-026108), ISSN: 0034-6748.

Yu J., Li L., Jiang Y., Ji X., and Wang T. (2007), Luminescent Enhancement of Heterostructure Organic Light-Emitting Devices Based on Aluminum Quinolines, *J. Electr. Sci. Techn. of China*, Vol. 5, No 2, pp. 183-186, ISSN : 1674 862X.

Zhang Y., Cheng G., Zhao Y., Hou J., and Liu Sh. (2005), White organic light-emitting devices based on 4,4'-bis(2,2'-diphenyl vinyl)-1,1'-biphenyl and phosphorescence sensitized 5,6,11,12-tetraphenylnaphthacene, *Appl. Phys. Lett.*, Vol 86, No 1, 011112, ISSN: 0003-6951

Zheng J., Hua Y., Yin S.,. Feng X, Wu X., Sun Y., Li Y., Yang Ch. and Shuai Z. (2005), White organic light-emitting devices using Zn(BTZ)2 doped with Rubrene as emitting layer, *Chinese Science Bulletin*, Vol. 50, No 6, pp. 509-513, ISSN: 1001-6538.

Zheng X., Wu Y., Sun R., Zhu W., Jiang X., Zhang Z., and Xu S., (2005), Efficiency improvement of organic light-emitting diodes using 8-hydroxy-quinolinato lithium as an electron injection layer, *Thin Solid Films* , Vol. 478. No 1-2, pp.252-255, ISSN: 0040-6090.

Zhu F., Hua Y., Yin S., Deng J., Wu K., Niu X., Wu X. and Petty M. (2007), Effect of the thickness of $Zn(BTZ)_2$ emitting layer on the electroluminescent spectra of white organic light-emitting diodes, *J. Luminescence*, Vol. 122-123, pp. 717-719, ISSN: 0022-2313.

Part 2

OLED Processes and Devices

7

Unconventional, Laser Based OLED Material Direct Patterning and Transfer Method

Seung Hwan Ko[1] and Costas P. Grigoropoulos[2]
[1]*Applied Nano Tech and Science Lab, Department of Mechanical Engineering Korea Advanced Institute of Science and Technology, (KAIST), Daejeon*
[2]*Laser Thermal Lab, Department of Mechanical Engineering, University of California Berkeley, California*
[1]*Korea*
[2]*USA*

1. Introduction

Organic light emitting diode (OLED) displays have a number of desirable features such as high contrast and brightness, wide color range, thin structure and light weight, among others.(Hirano et al. 2007) OLED displays have several manufacturing requirements such as large area scalability and an increasing push towards smaller feature sizes, tighter feature shape control, high yield and low cost. However, traditional lithography and thermal evaporation deposition techniques have significant disadvantages, including the need for masks that are typically difficult to make to the required specifications at a reasonable price. The conventional vacuum deposition and photolithographic patterning methods are well developed for inorganic microelectronics. However, organic electronics materials are chemically incompatible with corrosive etchants, resists and developers used in conventional integrated circuit (IC) processing. In practice, conventional IC fabrication processes are subject to limitations, in that they are multi-step, involve high processing temperatures, toxic waste and are therefore expensive. Furthermore, the increasing size of electronic devices such as displays poses great difficulty in adapting standard microfabrication processes, including lithographic patterning. (Zschieschang et al. 2003, Ko et al. 2007)

Therefore, there is a strong need to develop a novel process instead of complex modification of conventional vacuum deposition and photolithography based processes. OLED display manufacturing employs direct write techniques for patterning the various materials. Examples of OLED material direct write technologies include ink jet printing (Hashimoto et al, 2006, Gohda et al. 2002, Lee et al. 2002,, Kobayshi et al. 2002, Shirasaki et al. 2004, Fleuster et al. 2004, Lee et al. 2005, Saafir et al. 2005) screen printing (Shinar et al. 2007, Lee et al. 2009) and laser induced forward transfer (LIFT) (Hirano et al. 2007, Piqué et al. 1999, Suh et al. 2003, Willis et al. 2005, Kyrkis et al. 2006). As described in a recent review on OLED RGB patterning, success of an OLED patterning scheme depends on the material type, device design, pixel array pattern, display format, substrate size, placement accuracy, process TACT-time, and defect density. The type of material and OLED architecture largely determine which type of RGB patterning can be applied. Other factors determining the

viability of the patterning method for active matrix organic light emitting diode (AMOLEDs) depend on the given material set. (Lamansky et al. 2005) Solution processible direct write technologies such as inkjet printing and screen printing are subject to a number of limitations such as the need for solvent removal and contamination into the deposited material. Additionally, the minimum feature size is heavily influenced by the properties of the fluid used to deliver the material of interest and multilayer structure fabrication is difficult. (Kyrkis et al. 2006) Vacuum-processable OLEDs have been patterned mostly by deposition through a shadow mask or fine-metal mask (FMM). (Kang et al. 2003) Deposition can be accomplished either as a conventional physical evaporation or organic vapour phase deposition (OVPD), but FMM-related patterning issues are largely independent of the deposition technique. Remaining FMM patterning issues include difficulty of fabricating high resolution masks for large-area displays, mask lifetime and cleaning, particle contamination, and thermal expansion effects.
In this chapter, unconventional OLED material direct patterning and transfer methods especially laser based forward transfer and patterning approaches will be presented as promising potential alternative to conventional OLED fabrication methods.

Fig. 1. 14 inch OLED display from CDT (Cambridge Display Technology) from CDT ltd.

2. OLED material laser induced forward transfer and patterning techniques

LIFT techniques pattern and transfer materials of interest by laser induced localized thermal evaporation or chemical decomposition of dynamic release layer. This dynamic release layer is the crucial part of the LIFT process and can be (a) a part of material of interest, (b) specially designed light absorbing thin intermediate layers in LITI (laser induced thermal imaging) (Lamansky et al. 2005, Blanchet et al. 2003a, Suh et al. 2003) and LIPS (laser induced pattern-wise sublimation) process (Hirano et al. 2007) or (c) a mixture of active or sensitive material in a UV absorbent matrix in MAPLE DW(matrix assisted pulsed laser evaporation direct writing) (Piqué et al. 1999, Arnold et al. 2007) process. LIFT and several variations have demonstrated deposition of metals, metal oxide films, inorganic dielectric films, ceramics, and polymer and biomaterials. (Arnold et al. 2007, Willis et al. 2005, Kyrkis et al. 2006, Chrisey et al. 2003) Most notable recent advance in LIFT technique is the OLED pixel fabrication using dialkyltriazene polymer as an UV-absorbing and decomposing intermediate sacrificial layer compared with thermal decomposition. (Rardel et al. 2007)

However, most LIFT based techniques in OLED material transfer process still exhibit a number of limitations such as laser selection (wavelength, fluence), resolution, and edge sharpness. Most LIFT based techniques apply ultraviolet (UV) or infrared (IR) laser with relatively high laser fluences (1~10 J/cm^2) to obtain enough pressure for ablative material transfer. UV or IR lasers need complex and expensive laser and optic system. Furthermore, without strict design of light absorbing layer, high power UV or IR lasers have high possibility for organic material damage during the LIFT process because generally organic materials have strong UV and IR absorption bands attributed to electronic and vibrational transitions, respectively. Besides thermal degradation, high laser threshold laser can also induce mechanical cracks on transfer material and problem in edge sharpness. Also resolution was usually limited to 50 to 100 μm.

Ko et. al. reported a nanomaterial enabled laser transfer (NELT) to facilitate the high resolution patterning and transfer of the heat-sensitive OLED material with more versatile laser wavelength selection with one or two order smaller laser energy than conventional LIFT processes. This is characterized by the introduction of an efficient light absorbing, loosely connected nanomaterial layer and the choice of laser wavelength that although is strongly absorbed by the properly engineered nanomaterial, it interacts only weakly with the organic material of interest, leading to effective evaporation and transfer of the material with less damage potential.

2.1 Laser Induced Thermal Imaging [LITI]

Over the last twelve years, we at 3M have developed Laser Induced Thermal Imaging (LITI) as a high resolution, digital patterning technique with a large number of potential applications including the patterning of digital color proofs, plates, and film; LCD color filters, black matrix, and spacers; field emission display (FED) anodes, contrast enhancement filters, and nanoemitters; organic field effect transistor (OFET) fabrication; and OLED emitters, color filters, and color conversion filters. Since 2000, 3M has partnered with Samsung SDI to jointly develop the process for AMOLEDs. (Wolk et al, 2004)

LITI involves the use of a precoated donor film, a large format laser exposure system, and a receptor (e.g. an AMOLED backplane) (Fig. 2). For OLEDs, a stock roll of functional non-transferring layers is prepared and stored. Solvent coating or vapor deposition is used to deposit an ultrathin (e.g. 20-200 nm) layer of red, green, or blue emitting transfer layer(s) to the stock roll shortly before patterning. Patterning of each color is then accomplished by first aligning the receptor (e.g. an AMOLED backplane) to the laser exposure system and then laminating a donor film to the aligned receptor. After the alignment step, the laser system is used to expose the laminated assembly. Exposed regions are released from the donor and adhered to the receptor. The process is repeated from two or more times, depending upon the OLED construction. Alignment is performed only once. (Wolk et al, 2004)

Once a donor is used to pattern OLED materials, it is discarded. Although the transferred area represents less than a third of the coated surface, the exposed donor film now contains a high resolution pattern. Dimensional instability of the film and the physical changes that the film undergoes during the exposure process make it impractical to reuse the exposed donor. (Wolk et al, 2004)
3M's LITI Process is well suited for use in the manufacture of high precision flat panel displays, where high resolution, absolute placement accuracy, and large format imaging are all required. The advantages of the LITI process are significant in situations where the separation of coating and patterning steps resolves a fundamental process. LITI applications include patterning of organic electronic materials for OLEDs and organic transistors, patterning of multilayered OLED stacks, patterning of polarizers or nano-emitters, and the potential of patterning enzymes and other biomaterials. (Wolk et al, 2004)

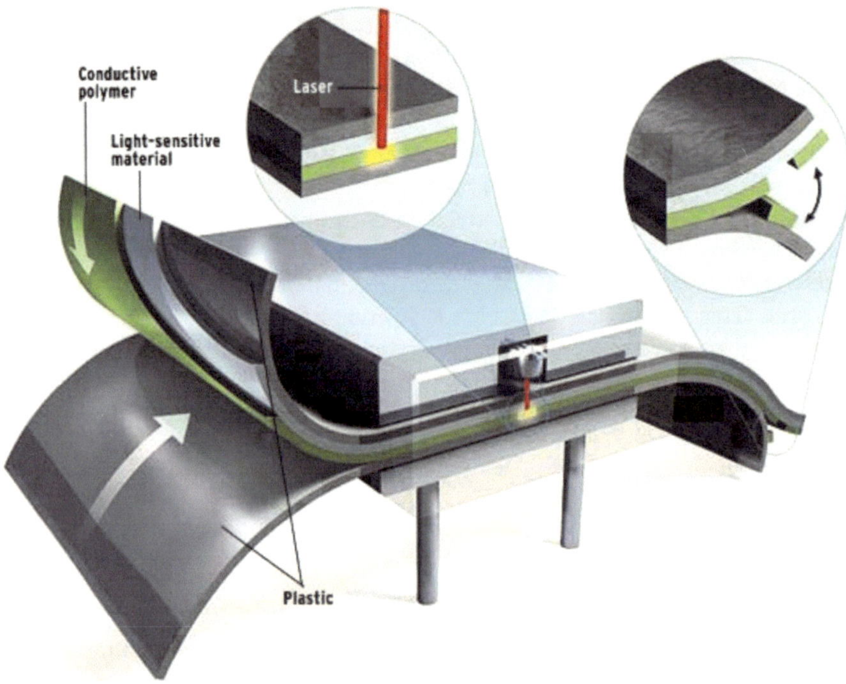

Fig. 2. LITI process schematics. (Blanchet et al. 2003b)

LITI is an emerging technology for high-resolution patterning of materials, including but not exclusive to both solution- and vacuum-processable OLED material sets. (Lamansky et al. 2005) Base steps in the LITI process include deposition of the material to be patterned (transfer material) onto a specially designed donor film, precise optical alignment of a large format laser imaging system to device substrate (receptor) fiducials, lamination of the donor onto the substrate, and patterning of the transfer material onto the substrate by selective exposure of the donor-transfer material-receptor stack to laser radiation. Conversion of laser radiation to heat is achieved in a light-to-heat conversion (LTHC) layer(s), which typically utilizes carbon black as a black body absorber. To generate a patterned RGB OLED display, optical alignment is performed only once, but lamination and exposure have to be performed for at least two colors.

Advantages of LITI over other patterning methods include its applicability to a broad spectrum of OLED material sets, high patterning accuracy (±2-5 μm compared to ±15-20 μm for shadowmasking and ink-jet techniques), ability to pattern multilayer structures in a single step, scalability of the process to large mother glass sizes, and ability to meet TACT time requirements. It is possible that LITI introduces thermal defects in the OLED materials during patterning, but by fine-tuning process conditions, donor structure, and OLED composition, occurrence of such defects can be minimized. The process is also sensitive to particulates and similar contamination of substrate (receptor in LITI terms) and donor surfaces. This puts stringent requirements on the donor, substrate and transfer atmosphere cleanliness.

2.2 Laser Induced Patternwise Sublimation [LIPS]

White OLED with color filter (WOLED+CF) methods and thermal transfer technologies are expected as alternatives to precision mask patterning. Sony demonstrated the WOLED+CF prototype display at SID 2003 (Kashiwabara et al. 2004). However, high power consumption and color impurity are the issues of this method for the TV application. Laser-induced thermal imaging (LITI) (Lee et al. 2004) and radiation-induced sublimation transfer (RIST) (Boroson et al. 2005) have been proposed as thermal transfer technologies. They have some concerns in production process. In the LITI process, contact between the donor film and the emitting area will degrade the device and transfer quality. Though RIST is a sublimation process without the contact, OLED material will be damaged by gases (e.g. O_2, H_2O etc.) released from a polyimide film donor during laser-heating. In addition, they require high precision technique to set flexible film on a large scale glass substrate uniformly without adhesive agents. Imprecise setting of a film donor lowers transfer performance. (Hirano et al 2007)

Sony has proposed a novel laser transfer technology for manufacturing OLED displays. Laser-induced pattern-wise sublimation (LIPS) has been developed to image RGB pixel pattern. OLED materials are precisely patterned from glass donors to a substrate by a scanning laser beam. The LIPS device performance is examined in comparison with conventional evaporated devices. Using this technology, a 27.3-inch active matrix (AM) OLED display has been fabricated. (Hirano et al 2007)

LIPS is a laser thermal transfer process. Two systems has been prepared, as shown in figure 3, in order to develop the LIPS process. One is the laser transfer system composed of alignment equipment, a step-moving (X-axis) laser head and a scanning substrate stage (Y-axis). The radiation source is an 800 nm diode laser. A width of the laser beam is adjusted in accordance with that of the transferred pattern. The other is the vacuum chamber where a glass donor is fixed on a substrate with clamping equipment. (Hirano et al 2007)

Figure 3 also shows the process flow diagrams of LIPS. A glass donor is necessary for each emission layer (EML) to be patterned. Organic material is deposited in a conventional evaporator on a glass donor covered with molybdenum absorption layer. Organic common layers such as a hole injection layer (HIL) and a hole transport layer (HTL) are formed on the glass substrate including a pixel defined layer (PDL) and bottom electrodes, as shown in figure 1(d). The substrate and the glass donor are introduced without exposure to the air and spaced apart in the vacuum chamber. And then the glass donor is put on the substrate and fixed by the clamping equipment. It is moved out of the chamber onto the stage of the laser transfer system in the atmosphere after introducing inert gas into the chamber. The transfer gap between the glass donor and the substrate is precisely controlled all over the substrate by the rigid donor, the PDL and atmospheric pressure. Moreover the PDL prevents the donor from contacting the emitting area on the substrate. After mechanical alignment of the substrate to the laser head, laser beam scans and heats the designated position of the glass donor and organic material is transferred to the substrate by vacuum sublimation. The transferred organic material functions as an EML. The gap atmosphere is kept vacuum by the clamping equipment during the laser transfer. The patterning process is done for each emission layer. Common layers such as an electron transport layer (ETL) and a top electrode are formed on the patterned substrate after removing the donor glass in inert gas. (Hirano et al 2007)

From the viewpoint of productivity, the laser transfer process in the atmosphere can simplify a production system and improve laser positioning accuracy. Multiplying laser beams promise high throughput even for large-scale mother glass. Glass donors can be re-cycled, which saves the production cost. (Hirano et al 2007)

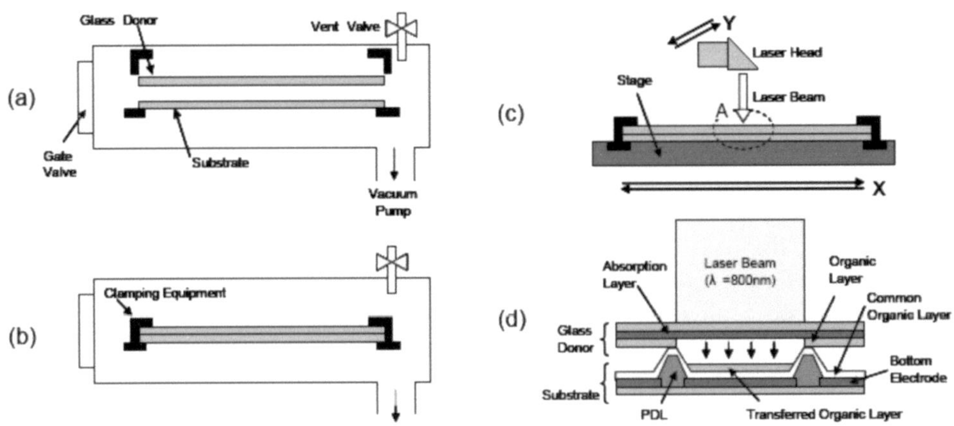

Fig. 3. Schematic diagrams of the LIPS process. (a) Placement of the glass donor and the substrate in the vacuum chamber (b) Setting of the glass donor on the substrate with clamping equipment (c) Placement of the substrate on the laser transfer system (d) An enlarged cross-section diagram of A in figure (c). (Hirano et al 2007)

The gap between a donor sheet and a substrate is critical to transfer accuracy. The advantage of the LIPS process is high precision by use of a glass donor instead of a film donor. The position accuracy is better than 4um. The pattern width variation is within ±2.0um. Using the patterning accuracy, we can realize high aperture ratio more than 60% for large-sized OLED display. The further improvement of patterning accuracy is possible by mechanical adjustment. (Hirano et al 2007)

2.3 Matrix Assisted Pulsed Laser Evaporation – Direct Writing [MAPLE-DW]

MAPLE DW was originally developed as a method to rapidly prototype mesoscopic passive electronic devices such as interconnects, resistors, and capacitors. (Piqué et al. 1999, Chrisey et al. 2000) This technology falls under the category of a "direct-write" approach because, in the same manner as a pen or pencil, it can be used to rapidly form any pattern with the aid of CAD/CAM systems. The schematic of the apparatus is shown in figure 4. The material to be transferred is mixed in a laser-absorbent matrix and coated onto a support, or ribbon, that is transparent to the laser irradiation. A focused laser pulse is directed through the backside of the ribbon so that the laser energy first interacts with the matrix at the ribbon interface. The laser pulse is focused at the matrix-ribbon interface by a UV microscope objective that also serves as an optical guide to determine the area of the matrix to transfer. Layers of matrix near the support interface evaporate due to localized heating from electronic and vibrational excitation. This sublimation releases the remaining material further from the interface by gently and uniformly propelling it away from the quartz support to a substrate positioned 25 μm to several mm away. By removing the ribbon and allowing the laser pulse to interact with the substrate, this approach is also able to micromachine channels and through vias into polymer, semiconductor, and metal surfaces. All micromachining and material transfer can be controlled by computer (CAD/CAM), which enables this tool to rapidly fabricate complex structures without the aid of masks or moulds. When applied to polymers and composites, MAPLE-DW has produced 2-D and 3-D patterns as well as functioning devices. One such device was a chemoresistor fabricated by depositing a polymer/carbon composite (polyepichlorohydrin/graphite mixture) across two electrodes.(Piqué et al. 1999) This device retained function as demonstrated by sensitivity to chemical threats. In addition, polymer thick film (PTF) resistors were fabricated using epoxy-based materials.(Modi et al. 2001) The fabricated PTF resistors spanned four decades of sheet resistances (10 Ω/sq. to 100 k Ω/sq.) and performed consistent to theoretical models for temperature and frequency variance.

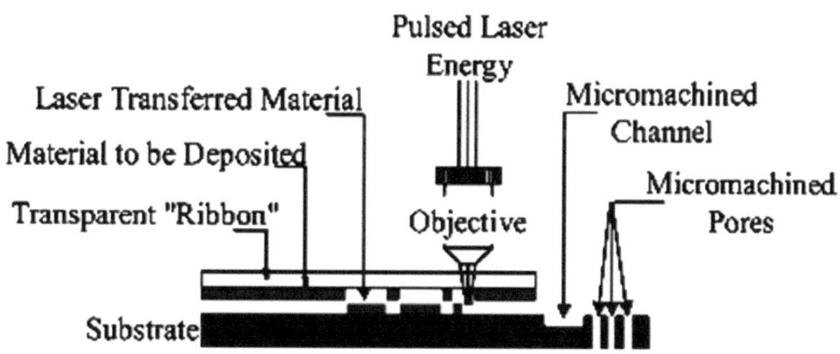

Fig. 4. Schematic diagrams of the MAPLE-DW deposition system. (Chrisey et al 2003)

2.4 Nanomaterial Enabled Laser Transfer [NELT]

Many of the direct write technologies mentioned above are subject to a number of limitations such as the need for solvent removal and contamination into the deposited material for ink jet printing. Additionally, the minimum feature size is heavily influenced by the properties of the fluid used to deliver the material of interest.

The conventional LIFT techniques also exhibit a number of limitations as they involve localized evaporation of either the material of interest or the light-to-heat converting intermediate layer resulting from the laser induced temperature rise. Laser-based techniques have been used successfully to deposit metals, metal oxide films, inorganic dielectric films, and ceramics,(Willis et al. 2005, Kyrkis et al. 2006, Chrisey et al. 2003, Arnold et al. 2007) but have limited success to the deposition of organic materials.

Generally, organic materials have strong ultraviolet (UV) and infrared (IR) absorption bands attributed to electronic and vibrational transitions, respectively as shown in figure 5. Therefore, UV or IR lasers have been typically used for organic material laser transfer by the direct laser absorption in the same organic material or a separate light absorbing organic material. However, organic compounds have high vapor pressures and can be easily damaged by thermal decomposition or degradation induced from photodecomposition by direct UV absorption or thermal decomposition by IR absorption. This is also the case even using an additional light absorbing organic layer. To overcome this problem, a thin metal film layer can be introduced as visible laser light absorber. However, this may cause organic material thermal degradation because the metal film usually has high melting or ablation threshold and exhibits inefficient energy coupling due to high reflectivity.

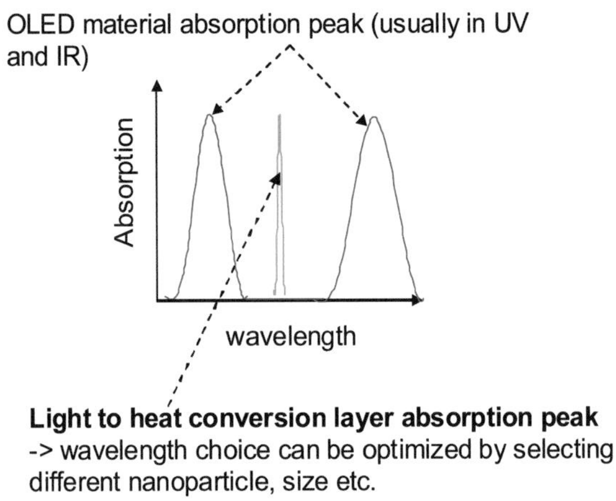

Fig. 5. Light to heat conversion layer engineering for OLED laser transfer

Ko et al. reported a nanomaterial enabled laser transfer (NELT) method to facilitate the patterning and transfer of the heat-sensitive OLED material. This is characterized by the introduction of an efficient light absorbing nanomaterial layer and the choice of laser wavelength that although is strongly absorbed by the properly engineered nanomaterial, it interacts only weakly with the organic material of interest, leading to effective evaporation and transfer of the material with less damage potential. (Ko et al. 2008)

The illustration of the NELT process and the donor multilayer is shown in Figure 6(a). Either a large area homogenized beam or a tightly focused Gaussian Nd:YAG pulsed laser beam (wavelength = 532 nm, pulse width = 5 ns FWHM) were applied normal to the donor substrate to induce the local heating of nanoparticle and the transfer of a target film onto a receptor substrate. The homogenized Nd:YAG laser beam cross section was shaped to a 0.9 × 0.9 mm2 flat top beam profile of good spatial uniformity by a micro-lens laser beam homogenizer while the tightly focused laser beam size had a 2~10 μm (FWHM) $1/e^2$ diameter Gaussian profile. The applied laser fluence was around 0.05~0.15 J/cm^2. The donor was composed of three parts; transparent substrate (glass slide) / laser-to-heat conversion layer (nanomaterials) / target material (OLED material). Self assembled monolayer (SAM) protected silver nanoparticles (Ag NPs) (30~40 nm sized, Figure 6(a) insets) were used as the laser-to-heat conversion nanomaterial layer by spin-coating on the glass substrate to form a 100~200 nm layer. The SAM coating serves to stabilize the Ag NPs and prevent coalescence into a bulk film. (Korge et al. 1998) The optical property measurement by spectroscopic ellipsometry (Figure 6(b)) shows the NP film has strong absorption around 530 nm facilitating highly efficient energy absorption to the NPs laser while the polymer material exhibits little absorption. The target OLED material was tris-(8-hydroxyquinoline)Al (Alq3) layer deposited by a resonant infrared pulsed laser deposition (RIR-PLD) technique to form 200~300 nm layer on top of Ag NP film on a glass slide. The donor substrate was then placed in intimate contact with the receptor substrate in ambient air environment. The receptor substrate was a glass substrate with thin PDMS layer on top as an adhesion promoter.

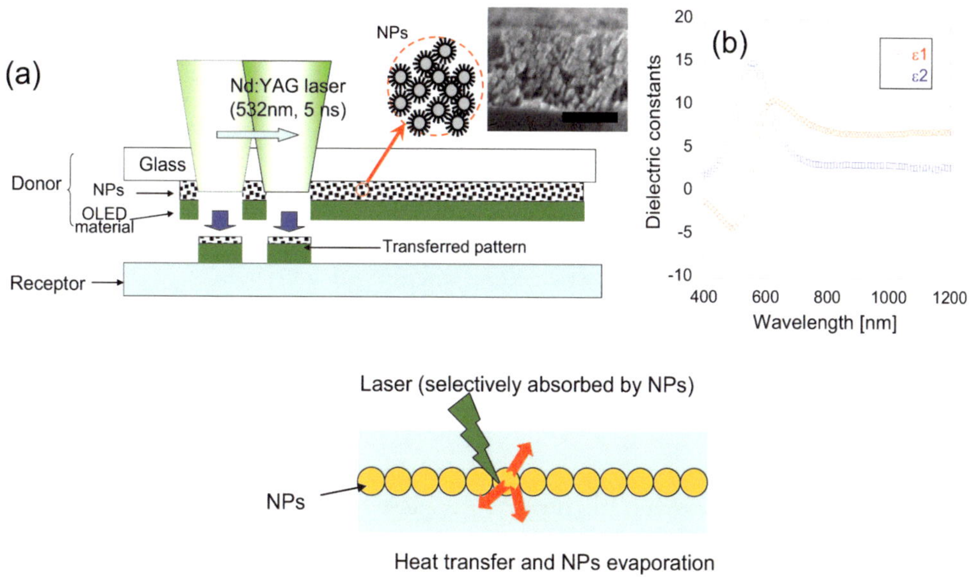

Fig. 6. (a) Schematic illustration of the NELT process and the donor multilayer (Alq3/Ag NPs/glass) structure. The inset pictures are schematics of SAM protected NPs and the SEM picture of the spin coated Ag NPs. Inset scale corresponds 100nm. (b) Ellipsometric measurement of the NPs film optical property. (Ko et al. 2008)

Figure 7(a-c) shows fluorescence pictures of (a) a donor substrate and (b,c) transferred Alq3 patterns on a PDMS/glass substrate after homogenized Nd:YAG laser transfer. The black squares in Figure 7(a) mark donor regions where single laser pulses (0.07 J/cm^2) transferred corresponding Alq3 patterns (0.9×0.9 mm^2 squares with 2 mm pitch) on a receptor substrate (Figure 7(b)). Figure 7(d) shows the donor cross section SEM picture near the laser beam edge and figure 7(e,f) show the corresponding magnified pictures. The micrograph of individual transferred Alq3 pattern (Figure 7(c)) and the SEM picture (Figure 7(g)) indicate that the transfer was very successful with well-defined edges and strong fluorescence without any visible cracks on the Alq3 surface. If the Alq3 layer becomes damaged, the layer will cease to fluoresce or the intensity will be greatly reduced. The thickness of the transferred layer was about 200 ~ 300 nm corresponding to the thickness of the Alq3 layer on the donor substrate structure as well as a small amount of the nanoparticle matrix layer.

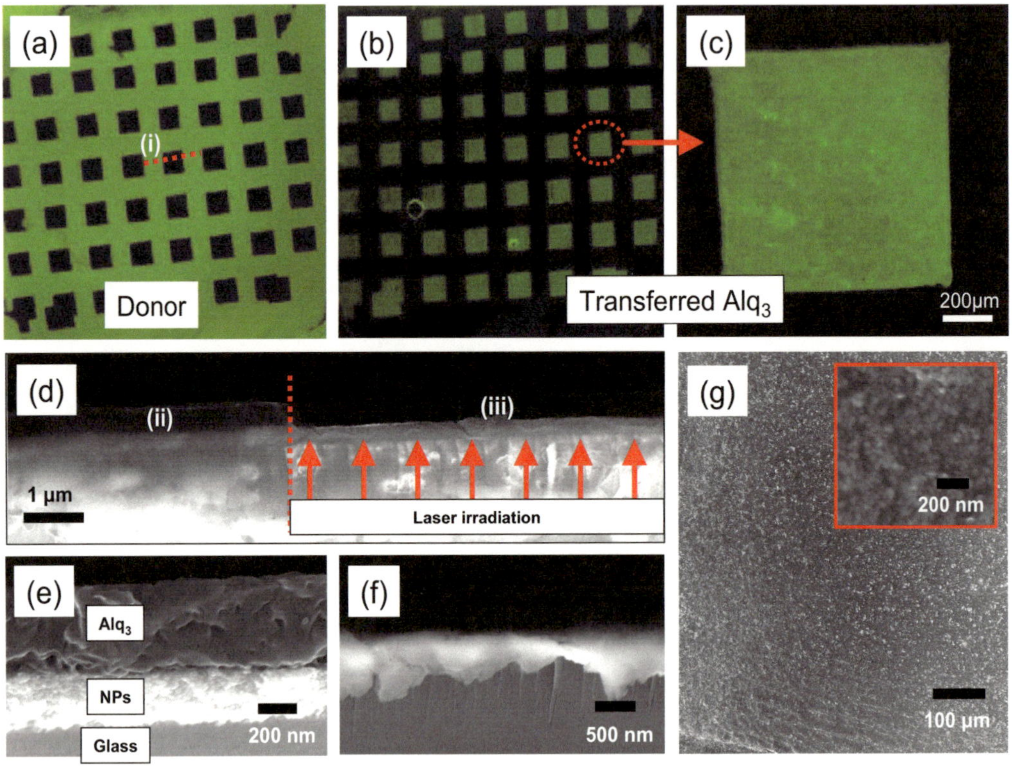

Fig. 7. Fluoresce pictures of (a) a donor substrate and (b) transferred Alq3 patterns (0.9×0.9mm^2 squares with 2 mm pitch) on a PDMS/glass substrate under UV lamp illumination. (c) Magnified fluoresce micrograph of individual transferred Alq3 pattern under UV lamp illumination. (d) SEM cross sectional picture near laser edge along line (i) in (a). Magnified SEM cross sectional pictures of original donor with Alq3/Ag NPs/glass structure before laser transfer and of (f) donor after laser transfer at (ii) and (iii) in (d) respectively. (g) SEM top view picture of transferred Alq3/Ag on PDMS/glass substrate. Inset is the magnified SEM top view picture. (Ko et al. 2008)

Nanomaterials exhibit remarkable properties that may be substantially different from those observed in the bulk counterparts due to the large surface-to-volume ratio, large surface energy, and the confinement of molecules, atoms, and electrons within a small spatial region. Note that the typical laser fluence (0.07-0.12 J/cm^2) used for NELT based on Ag NPs is considerably lower than the reported ablation threshold (Bäuerle et al. 2000) for metal films (1–10 J/cm^2) for nanosecond laser irradiation at visible wavelength.

Proper selection of nanoparticle size, size distribution, as well as the material type may allow highly efficient laser energy coupling via tuning to an absorption peak. Noble metal NPs such as Au and Ag exhibit strong absorption peaks in the visible wavelength region that are typically not observed in the bulk materials due to surface plasmon oscillation modes of conduction electrons in the nanoparticles (Figure 8). Additionally, enhanced electric fields between nanoparticles may contribute to more efficient energy absorption mechanisms. (Ko et al. 2006)

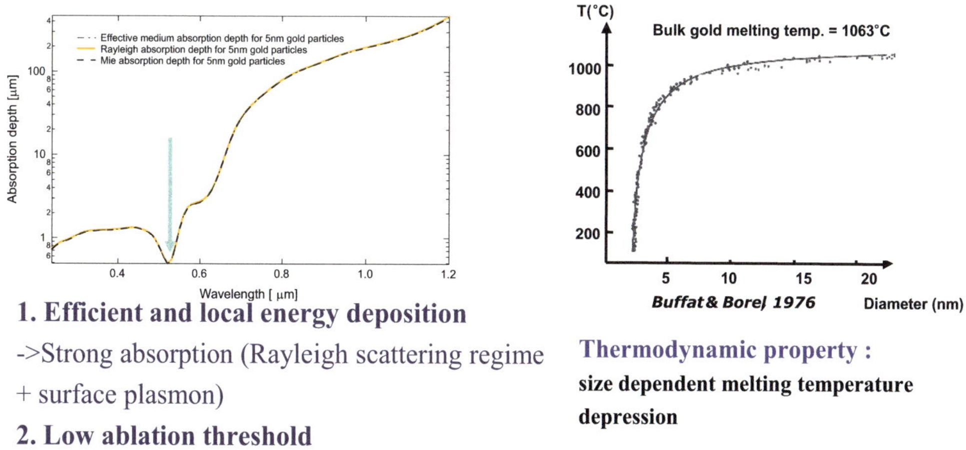

Fig. 8. Optical property and melting temperature depression of metal nanoparticles.

For efficient energy absorption, the deposited energy may be more confined to the laser focal zone due to the reduced thermal diffusion of nanoparticle thin films. This implies that the relatively lower thermal conductivity of a nanoparticle thin film reduces the rate of lateral energy dissipation, thereby facilitating the heating, evaporation and transfer processes. The low thermal conductance of the alkanethiol SAM is due to the vibrational mismatch between the nanoparticle solid core and the surface coating as reported by several researchers. (Wilson et al. 2002, Wang et al. 2006)

The melting temperature depression may also enable ablation driven by the nanomaterial melting and vaporization at much lower laser energies than for bulk materials. (Figure 8) Upon reaching the SAM desorption temperature (typically at about 150~250°C), the SAM coating may be removed. At this stage, molten nanoparticles agglomerate to form larger particles or are expelled by the pressure built up from the volatile species expansion attributed to desorbed SAM, vaporized residual organic solvent and trapped gases. Additionally, due to the presence of the SAM coating, the nanoparticles may be held together by weak physical van der Waals force compared with the strong metal bonding in thin metal film. (Altman et al. 2005) Therefore, the expulsion may be enhanced by the relatively weak bonding between the nanoparticles.

Due to the combined aforementioned effects, NELT results in laser transfer at relatively smaller laser energy with minimal thermal damage to the target OLED material. The Alq3 layer shows strong decrease in fluorescence if the temperature exceeds approximately 300°C. Therefore, the Alq3 layer does not encounter higher temperatures during the laser transfer process.

Figures 9 and 10 illustrate that this process can fabricate Alq3 patterns of arbitrary shape. Figure 9 shows step and pixelized transfer of Alq3 patterns by irradiation of a large area homogenized laser beam (of ~mm size) while figure 4 depicts Alq3 patterning by tightly focused Gaussian laser beam (<10µm). Figure 9 displays photographs of the fluorescing transferred Alq3 patterns on a PDMS/glass substrate under UV lamp illumination. The green letters (a) "UCB", (e) "LTL" and the star shapes correspond to regions where the laser irradiation resulted in the transfer of the Alq3 material from the donor substrate to the receptor substrate. The "UCB" and "LTL" letters were formed from combining the 0.9×0.9

mm2 squares with single laser pulse (0.072 J/cm^2). The lines of stars indicated at (a-i,ii) and (e-i,iii) were patterned using a homemade aluminum mask on a glass slide (placed before the objective lens for scaling down the projection of the mask at 1:10 reduction) using the same pulse energy as for the above letters. Figures 9(b-d), (f-h) show the magnified fluorescence micrograph of individual transferred Alq3 pattern of star shapes (b,f) and letters (c,d,g,h). The bubbles appearing in Figure 9(a,e) are in the PDMS adhesion layer and not a result of the Alq3 laser transfer process. In all cases, the fluorescence of the transferred Alq3 is strong, indicating no damage during the transfer process. Additionally, the edges of the transferred patterns are sharp and well defined, confirming good spatial control of the transfer.

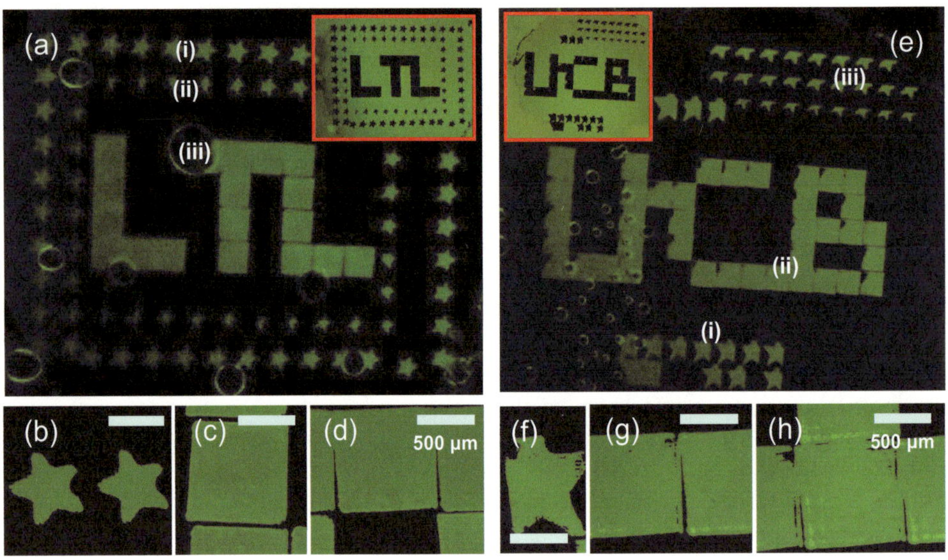

Fig. 9. Step and irradiation of large sized homogemized beam. Fluoresce pictures of transferred Alq3 patterns on a PDMS/glass substrate under UV lamp illumination. (a) (i) star shapes made by homemade mask and (ii~iii) "LTL" letters made by 0.9 × 0.9 mm2 squares. Magnified fluoresce micrograph of individual transferred Alq3 pattern of (b) star shapes, (c~d) letters under UV lamp illumination. (e) (i) star shapes made by homemade mask and (ii~iii) "UCB" letters made by 0.9 × 0.9 mm2 squares. Magnified fluoresce micrograph of individual transferred Alq3 pattern of (f) star shapes, (g~h) letters under UV lamp illumination. Small circles are the bubbles formed in PDMS layer and has nothing to do with current process. (Ko et al. 2008)

Figure 10 shows fluorescence pictures of transferred Alq3 patterns on a PDMS/glass substrate by raster scanning tightly focused Gaussian laser pulses. Alq3 micro dot arrays were transferred by (a,b) 5X, (c) 20X, (d) 50X objective lens focusing. A range of sizes of Alq3 micro dot arrays ((a,b) 20~25 μm (5X), (c) 5 μm (20X), (d) 3 μm (50X)) could be transferred via the objective lens focusing. Alq3 micro dot arrays were smaller than the original focused laser beam size (5X: 10 μm, 20X: 3 μm, 50X: 2 μm). Instead of using single laser irradiation of homogenized large area beam to transfer 1 mm2 square patterns as shown in Figure 7, square patterns of similar size and quality could be achieved by raster scanning a laser beam focused by a 5X objective lens along the arrow direction (Figure 10(d)).

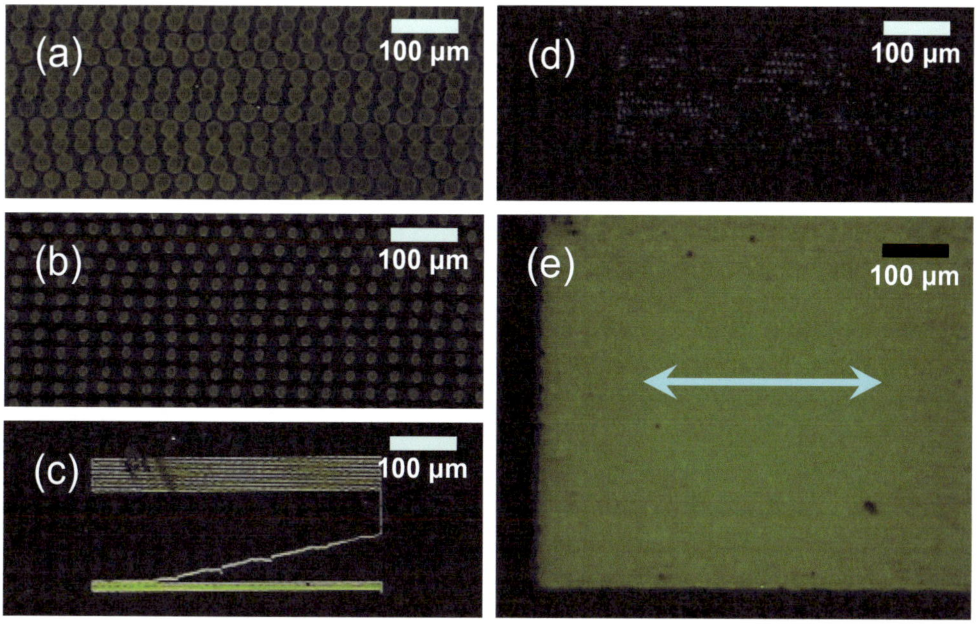

Fig. 10. Raster scanning of focused Gaussian beam. Fluoresce pictures of transferred Alq3 patterns on a PDMS/glass substrate under UV lamp illumination. Alq3 dot arrays transferred by (a,b) 5X, (c) 20X, (d) 50X objective lens focusing. 1 mm^2 square by raster scanning of focused laser with 5X objective lens. Arrow indicates laser spot scanning direction. (Ko et al. 2008)

In summary, the nanomaterial enabled laser transfer (NELT) method was demonstrated to directly pattern and transfer Alq3 through laser absorbing layer engineering. The combined effects of melting temperature depression, lower conductive heat transfer loss, strong absorption of the incident laser beam, and relatively weak bonding between nanoparticles during laser irradiation result in the transfer of patterns with very sharp edges at relatively lower laser energy than commonly used, thus inducing minimal damage to the target OLED material with no evidence of cracks. This technique is not limited to the current specific combination of the Ag NPs with alkanethiol SAM protective layer and the utilized laser wavelength, but can be applied to a broad range of laser wavelengths with proper selection of nanoparticle size and size distribution, as well as the material type. Additionally, NELT may be particularly advantageous for the mass production of temperature sensitive devices.

2.5 Self Assembled Monolayer assisted Nanomaterial Enabled Laser Transfer [SAM-NELT]

NELT process could demonstrate the OLED material pixel transfer. The NPs were used to facilitate the OLED material transfer at low laser energy. Some part of NPs will transferred from donor substrate on OLED material surface. The transferred NPs may be beneficial for some points such that they can be used for metal electrode or may enhance the luminescence by surface plasmon. However, for some application, there is a strong needs for modified NELT process without using NPs. In this section, self assembled monolayer assisted NELT (SAM-NELT) is introduced for variant of NELT process.

Figure 11 illustrates the schematics of NELT process (Figure 11(a)) and the two types of donor multilayer (figure 11(b)). Either a large area homogenized beam or a tightly focused Gaussian Nd:YAG pulsed laser beam (wavelength = 532 nm, pulse width = 5 ns FWHM) were applied normal to the donor substrate which was composed of three parts; transparent substrate (borosilicate glass slide) / laser-to-heat conversion layer (nanomaterials) / target material (OLED material). The applied laser induces the local heating of laser-to-heat conversion layer and the transfer of the OLED material patterns onto the flexible receptor substrate. In this experiment, two types of SAM (self assembled monolayer, Alkanethiol) coated nanomaterials were used as the laser absorbing and dynamic releasing layer: SAM coated Ag nanoparticle thin film (type I) or SAM layer coated Au thin film (type II). SAM protected Ag nanoparticles (30~40 nm sized) were spin-coated on the glass substrate to form a 100~200 nm thin film and SAM coated Au thin film (100nm on 5nm Cr layer) was prepared by e-beam evaporation and subsequent SAM coating. The homogenized Nd:YAG laser beam cross section was shaped to a 0.9 × 0.9 mm^2 flat top beam profile of good spatial uniformity by a micro-lens laser beam homogenizer while the tightly focused laser beam size had a 2~10 μm (FWHM) $1/e^2$ diameter Gaussian profile. The applied laser fluence was around 0.05~0.15 J/cm^2 for type I donor with SAM coated Ag nanoparticle thin film and 0.01~0.05 J/cm^2 for type II donor with SAM coated Au thin film. The optical property measurement by spectroscopic ellipsometry) indicates the nanoparticle thin film has strong absorption around 530 nm facilitating highly efficient Nd:YAG laser absorption to the NPs laser while the polymer material exhibits little absorption. The target OLED material were 1) green fluorescent monomer: tris-(8-hydroxyquinoline)Al (Alq3) layer (200~300 nm) deposited by a resonant infrared pulsed laser deposition (RIR-PLD) technique (Dubb et al, 2001) and 2) blue fluorescent polymer: PFO layer (100~300nm) deposited by spincoating. RIR-PLD was done by Free-Electron Laser (FEL) with the wavelength of 6.67 μm, which corresponds to the resonance peak of Alq3 absorption in IR, 5μs pulse width, and fluence of 1.5 J/cm^2 at 30 Hz for 2-3 minutes. The donor substrate was then placed on top of the receptor substrate with a small gap or intimate contact with the acceptor substrate in ambient air environment. The receptor substrate was a thin PDMS layer.

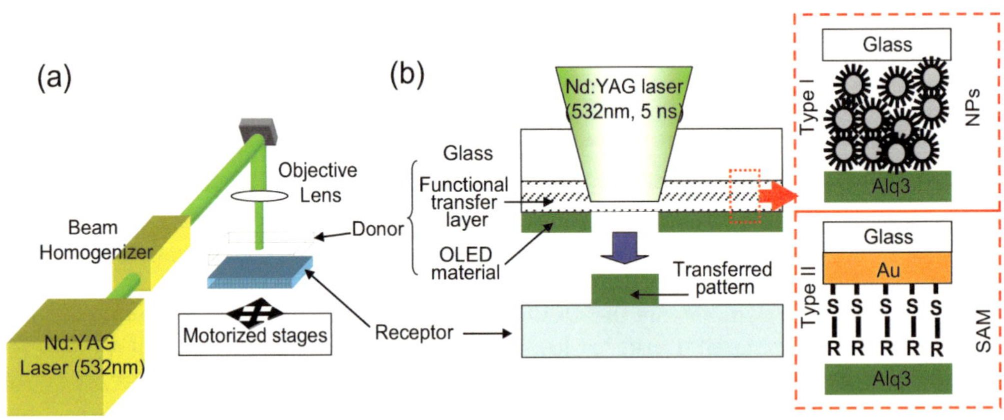

Fig. 11. (a) Schematic diagram of experimental set up for a nanomaterial enabled laser transfer with a homogenized Nd:YAG laser. (b) Magnified schematic illustration of the laser transfer process and the two types of donor multilayer. (Ko et al. 2010)

For certain applications, these nanoparticles transferred with the OLED material can be regarded as the contaminants. This side-effect can be removed by using type II donor. The micrographs of individual transferred Alq3 and PFO patterns from type II donor show successful OLED material transfer without damaging or transferring of Au thin film.
Transferred OLED material pattern can be controlled by laser scanning or laser beam shaping. Alternatively, selective transfer can be induced by pre-patterning of laser absorbing layer. Figure 12 shows the schematics and the fluorescent pictures of transferred Alq3 pattern on the PDMS substrate (Fig. 12(b)) and the donor substrate (Fig. 12(c)). Laser was scanned for large area with constant laser energy, and the transfer happened only for the area with pre-patterned type II donor (SAM coated Au thin film).

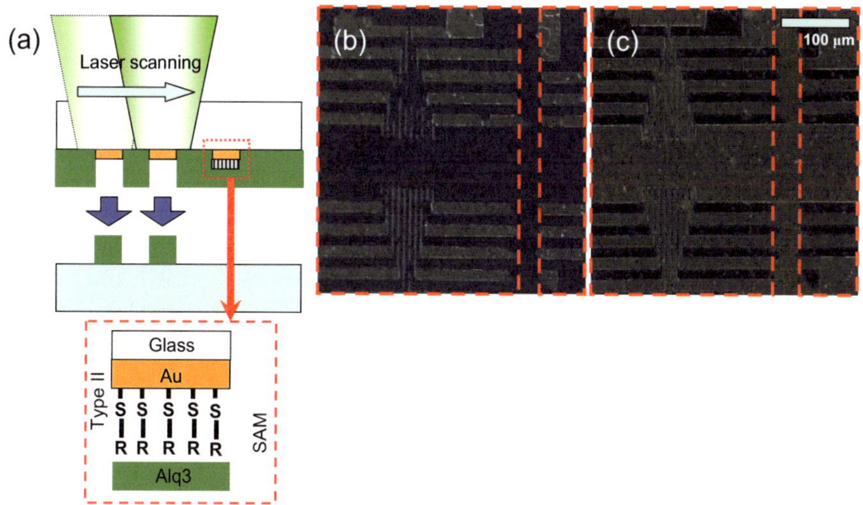

Fig. 12. (a) Schematic diagram of selective NELT process for a pre-patterned type II donor substrate. (b) Fluorescent images of (b) transferred Alq3 patterns on a PDMS film and (c) donor substrate after NELT process. (Ko et al. 2010)

3. Conclusion

Organic light emitting diode (OLED) displays have a number of desirable features such as high contrast and brightness, wide color range, thin structure and light weight, among others and OLED displays have several manufacturing requirements such as large area scalability and an increasing push towards smaller feature sizes, tighter feature shape control, high yield and low cost. However, traditional lithography and thermal evaporation deposition techniques have significant disadvantages, including the need for masks that are typically difficult to make to the required specifications at a reasonable price. Vacuum-processable OLEDs have been patterned mostly by deposition through a shadow mask or fine-metal mask (FMM). Deposition can be accomplished either as a conventional physical evaporation or organic vapour phase deposition (OVPD), but FMM-related patterning issues are largely independent of the deposition technique. Remaining FMM patterning issues include difficulty of fabricating high resolution masks for large-area displays, mask lifetime and cleaning, particle contamination, and thermal expansion effects. Therefore, there is a strong need to develop a novel process instead of complex modification of conventional vacuum deposition and photolithography based processes. OLED display manufacturing employs direct write techniques for patterning the various materials. Examples of OLED material direct write technologies include ink jet printing, screen printing and laser induced forward transfer (LIFT). Solution processible direct write technologies such as inkjet printing and screen printing are subject to a number of limitations such as the need for solvent removal and contamination into the deposited material. Additionally, the minimum feature size is heavily influenced by the properties of the fluid used to deliver the material of interest and multilayer structure fabrication is difficult. In this chapter, unconventional OLED material direct patterning and transfer methods especially laser based forward transfer and patterning approaches were presented as promising potential alternative to conventional OLED fabrication methods.

4. Acknowledgment

The authors would like to thank Dr. H.K. Park of AppliFlex LLC, (CA, USA) for his advices and supports. The authors also would like gratefully acknowledge the financial supports to KAIST by the Industrial Strategic Technology Development Program from the Korea Ministry of Knowledge Economy (Grant No. 10032145), the National Research Foundation of Korea (Grant No. 2010-0003973), the cooperative R&D Program from the Korea Research Council Industrial Science and Technology (Grant No. B551179-10-01-00), and the support to the University of California, Berkeley by the NSF STTR (Grant No. 0930594) through Appliflex LLC, CA, USA.

5. References

Arnold, C.; Serra, P. & Piqué, A. (2007) Laser direct-write techniques for printing of complex materials. *MRS Bull.* Vol.32, No.1, (Jan 2007) pp.23-31, ISSN 0883-7694.

Altman, I.S.; Agranovski, I.E. & Choi, M. (2005) Mechanism of nanoparticle agglomeration during the combustion synthesis. *Appl. Phys. Lett.* Vol.87, No.5, (Aug 2005) 053104 ISSN 0003-6951.

Bao, Z.; Feng, Y.; Dodavalapur, A.; Raju, V.R. & Lovinger, A.J. (1997) High-Performance Plastic Transistors Fabricated by Printing Techniques. *Chem. Mater.* Vol.9, No.6, (Jun 1997) pp.1299-301, ISSN 0897-4756.

Bäuerle, D., Laser Processing and Chemistry (Springer, New York, 2000).

Blanchet, G.B.; Loo, Y.-L.; Rogers, J.A.; Gao F. & Fincher, C.R. (2003a) Large area, high resolution, dry printing of conducting polymers for organic electronics. *Applied Physics Letters* Vol.82, No.3, (November 2002), pp. 463-465, ISSN 0003-6951.

Blanchet, G.B.; Loo, Y.-L.; Rogers, J.A.; Gao F. & Fincher, C.R. (2003b) Large Area Printing of Organic Transistors via a High Throughput Dry Process. *Mat. Res. Soc. Symp. Proc.* pp. 283-287, Vol. 736, Boston, MA, USA, Dec 2-32002.

Boroson, M.; Tutt, L.; Nguyen, K.; Preuss, D.; Culver, M. & Phelan, G. (2005) Non-Contact OLED Color Patterning by Radiation- Induced Sublimation Transfer (RIST). *SID Symposium Digest*, Vol.36, No.1, pp.972-975 (May 2005), ISSN 0003-966X.

Bubb, D.M.; Horwitz, J.S.; Callahan, J.H.; McGill, R.A.; Houser, E.J. & Chrisey, D.B.; Papantonakis, M.R.; Haglund, Jr., R.F.; Galicia M.C. & Vertes, A. (2001) Resonant infrared pulsed-laser deposition of polymer films using a free-electron laser. *J. Vac. Sci. Technol. A* Vol.19, No.5, (Sep 2001) 2698, ISSN 0734-2101.

Chrisey, D.; Piqué, A.; McGill, R.; Horwitz, J.; Ringeisen, B.; Bubb, D. & Wu, P. (2003) Laser deposition of polymer and biomaterial films. *Chem. Rev.* Vol.103, No.2, (Feb 2003) pp.553-576, ISSN 0009-2665.

Chrisey, D. B.; Piqué, A.; Modi, R.; Wu, H.D.; Aueyung, R.C.Y.; Young, H.D. (2000) Direct writing of conformal mesoscopic electronic devices by MAPLE DW. *Appl. Surf. Sci.* Vol.168, No.1-4, (Dec 2000), pp.345-352, ISSN 0169-4332.

Chung, J.; Ko, S.; Bieri, N.R.; Grigoropoulos, C.P.; Poulikakos, D., (2004a) Conductor microstructures by laser curing of printed gold nanoparticle ink. *Appl. Phys. Lett.*, Vol.84, No.5, (Feb 2004), pp. 801-803, ISSN 0003-6951.

Chung, J.; Ko, S.; Grigoropoulos, C.P.; Bieri, N.R.; Dockendorf, C.; Poulikakos, D. (2005) Damage-free low temperature pulsed laser printing of gold nanoinks on polymers, *ASME Journal of Heat Transfer*, Vol.127, No.7, (Jul 2005), pp. 724-732, ISSN 0022-1481.

Chung, J.; Bieri, N.R.; Ko, S.; Grigoropoulos, C.P.; Poulikakos, D. (2004b) In-tandem deposition and sintering of printed gold nanoparticle inks induced by continuous Gaussian laser irradiation, *Applied physics A-Materials science and processing*, Vol.79, No.4-6, (Sep 2004), pp. 1259-1261, ISSN 0947-8396.

Fardel, R.; Nagel, M.; Nüesch, F.; Lippert, T. & Wokaun. A. (2007) Fabrication of organic light-emitting diode pixels by laser-assisted forward transfer. *Appl. Phys. Lett.* Vol.91, No.6, (Aug 2007) 061103 ISSN 0003-6951.

Funamoto, T.; Matsueda, Y.; Yokoyama, O.; Tsuda, A.; Takeshita, H.; Miyashita, S. (2002) A 130-ppi, Full-Color Polymer OLED Display Fabricated Using an Ink-jet Process, *SID Symposium Digest*, Vol.33, No.1, pp.899-901, (May 2002) ISSN 0003-966X.

Fleuster, M.; Klein, M.; Roosmalen, P.V.; Wit, A.D.; Schwab, H. (2004) Mass Manufacturing of Full Color Passive-Matrix and Active-Matrix PLED Displays. *SID Symposium Digest*, Vol.35, No.1, pp. 1276- 1279, (May 2004) ISSN 0003-966X.

Gohda, T.; Kobayashi, Y.; Okano, K.; Inoue, S.; Okamoto, K.; Hashimoto, S.; Yamamoto, E.; Morita, H.; Mitsui, S. & Koden, M. (2006) A 3.6-in. 202-ppi Full-Color AMPLED Display Fabricated by Ink-Jet Method, *SID Symposium Digest*, Vol.37, No.1, pp.1767-1770, (May 2006) ISSN 0003-966X.

Ganier, F.; Hajlaoui, R.; Yasser, A. & Srivastava, P. (1994) All-polymer field effect transistor realized by printing technique. *Science* Vol.265, No.5179, (Sept 1994) pp.1684-86, ISSN 0036-8075.

Hirano, T.; Matsuo, K.; Kohinata, K.; Hanawa, K.; Matsumi, T.; Matsuda, E.; Matsuura, R.; Ishibashi, T.; Yoshida, A. & Sasaoka, T. (2007) Novel laser transfer technology for manufacturing large-sized OLED displays. *SID Symposium Digest*, Vol.38, No.1, pp.1592-1592, (May 2007) ISSN 0003-966X.

Kashiwabara, M.; Hanawa, K.; Asaki, R.; Kobori, I.; Matsuura, R.; Yamada. H.; Yamamoto, T.; Ozawa, A.; Sato, Y.; Terada, S.; Yamada, J.; Sasaoka, T.; Tamura, S. & Urabe, T. (2004) Advanced AM-OLED Display Based on White Emitter with Microcavity Structure. *SID Symposium Digest*, Vol.35, No.1, pp.1017-1019, (May 2004) ISSN 0003-966X.

Ko, S.H.; Pan, H.; Grigoropoulos, C.P.; Luscombe, C.K.; Fréchet, J.M.J.; Poulikakos, D. (2007b) Air stable high resolution organic transistors by selective laser sintering of ink-jet printed metal nanoparticles. *Appl. Phys. Lett.*, Vol.90, No.14, (Apr 2007) 141103 ISSN 0003-6951.

Ko, S.H.; Pan, H.; Grigoropoulos, C.P.; Luscombe, C.K.; Fréchet, J.M.J.; Poulikakos, D. (2007c) All-inkjet-printed flexible electronics fabrication on a polymer substrate by low-temperature high-resolution selective laser sintering of metal nanoparticles. *Nanotechnology*, Vol.18, No.34, (Aug 2007) 345202 ISSN 0957-4484.

Ko, S.H.; Pan, H.; Grigoropoulos, C.P.; Luscombe, C.K.; Fréchet, J.M.J. & Poulikakos, D. (2007) All inkjet printed flexible electronics fabrication on a polymer substrate by low temperature high resolution selective laser sintering of metal nanoparticles. *Nanotechnology* Vol.18, No.34, (August 2007) 345202, ISSN 0957-4484.

Ko, S.H.; Pan, H.; Ryu, S.G.; Misra, N.; Grigoropoulos, C.P. & Park, H.K. (2008) Nanomaterial Enabled Laser Transfer for Organic Light Emitting Material Direct Writing, *Applied Physics Letters* Vol.93, No.15, (Oct 2008) 151110, ISSN 0003-6951.

Ko, S.; Choi, Y.; Hwang, D.J.; Grigoropoulos, C. P.; Chung J. & Poulikakos, D. (2006) Nanosecond laser ablation of gold nanoparticle films. *Appl. Phys. Lett.* Vol.89, No.14, (Oct 2006) 141126, ISSN 0003-6951.

Ko, S.H.; Chung, J.; Pan, H.; Grigoropoulos, C.P.; Poulikakos, D. (2007a) Fabrication of multilayer passive and active electric components on polymer using inkjet printing and low temperature laser processing. *Sensors and Actuators A: Physical*, Vol.134, No.1, (Feb 2007) pp.161-168, ISSN 0924-4247. 17.

Ko, S.H.; Pan, H.; Lee, D.; Grigoropoulos, C.P.; Park, H.K. (2010) Nanoparticle Selective Laser Processing for a Flexible Display Fabrication. *Jpn. J. Appl. Phys.*, Vol.49, No.5, (May 2010) 05EC03 ISSN 1347-4065.

Korgel, B.A.; Fitzmaurice, D. (1998) Self-assembly of silver nanocrystals into two-dimensional nanowire arrays. *Adv. Maters.* Vol.10, No.9, (Jun 1998) pp.661-665 ISSN 0935-9648.

Kobayashi, M.; Hanari, J.; Shibusawa, M.; Sunohara, K.; Ibaraki, N. (2002) A 17-in Full-Color OLED Display by Using Polymer Ink-Jet Technology. *Proc Int Disp Workshops*, pp. 231-234, (2002).

Kyrkis, K.; Andreadaki, A.; Papazoglou, D. & Zergioti, I. (2006) In: *Recent Advances in Laser Processing of Materials*, J. Perrière, E. Millon, and E. Fogarassy (Ed.), 213–241, Elsevier, ISBN 978-0080447278.

Kang, C.H.; Kim, T.S. (2003) United States Patent Appl., US 2003/0221613 A1

Lamansky, S.; Hoffend Jr., T.R.; Le, H.; Jones, V.; Wolk, M.B.; Tolbert, W.A. (2005) Laser Induced Thermal Imaging of Vacuum-Coated OLED Materials, In: *Organic Light-Emitting Materials and Devices IX*, Z.H. Kafafi, P.A. Lane (Ed.), Vol.5937, 593702, SPIE, IBSN 0-8194-5942-9.

Lee, S.; Chin, B.; Kim, Kang, M.T.; Song, M.; Lee, J.; Kim, H.; Chung, H.; Wolk, M.; Bellmann, E.; Baetzold, J.; Lamansky, S.; Savvateev, V.; Hoffend, T.; Staral, J.; Roberts, R.; Li, Y. (2004) A Novel Patterning Method for Full-Color Organic Light- Emitting Devices: Laser Induced Thermal Imaging (LITI). *SID Symposium Digest*, Vol.35, No.1, (May 2004) pp.1008-1011, ISSN 0003-966X.

Lee, S.T.; Lee, J.Y.; Kim, M.H.; Suh, M.C.; Kang, T.M.; Choi, Y.J.; Park, J.Y. (2002) A New Patterning Method for Full-Color Polymer Light-Emitting Devices: Laser Induced Thermal Imaging (LITI). *SID Symposium Digest*, Vol.33, No.1, (May 2002) pp.784-787, ISSN 0003-966X.

Lee, D.; Chung, J.; Rhee, J.S.; Wang, J.P.; Hong, S.M.; Choi, B.R.; Cha, S.W.; Kim, N.D.; Chung, K.; Gregory, H.; Lyon, P.; Creighton, C.; Carter, J.; Hatcher, M. (2005) Ink Jet Printed Full Color Polymer LED Displays. *SID Symposium Digest*, Vol.36, No.1, (May 2005) pp. 527-529, ISSN 0003-966X.

Lee, D.-H.; Choi, J.S.; Chae, H.; Chung C.-H. & Cho, S.M. (2009) Screen-printed white OLED based on polystyrene as a host polymer, *Current Applied Physics*, Vol.9, No.1, (Jan 2009), pp.161-164, ISSN 1567-1739.

Modi, R.; Wu, H. D.; Auyeung, R. C. Y.; Gilmore, C. M.; Chrisey, D.B.J. (2001) Direct writing of polymer thick film resistors using a novel laser transfer technique. *J. Mater. Res.* (2001) Vol.16, No.11, (Nov 2001), pp.3214-3222, ISSN 0884-2914.

Piqué, A.; Chrisey, D.B.; Auyeung, R.C.Y.; Fitz-Gerald, J.; Wu, H.D.; McGill, R.A.; Lakeou, S.; Wu, P.K.; Nguyen, V.; Duignan, M. (1999) A novel laser transfer process for direct writing of electronic and sensor materials. *Appl. Phys. A*, Vol.69, No.7, (1999) S279, ISSN 0947-8396.

Redinger, D.; Molesa, S.; Yin, S.; Farschi, R.; Subramanian, V. (2004) An Ink-Jet-Deposited passive component process for RFID, *IEEE transactions on electron devices*, Vol.51, No.12, (Dec 2004), pp. 1978-1983, ISSN 0018-9383.

Saafir, A.K.; Chung, J.; Joo, I.; Huh, J.; Rhee, J.; Park, S.; Choi, B.; Ko, C.; Koh, B.; Jung, J.; Choi, J.; Kim, N.; Cung, K.; Srdanov, G.; MacPherson, C.; Truong, N.; Stevenson, M.; Johnson, A.; Chen, P.; Cardellino, T.; Pflanzer, R.; Yu, G.; Goenaga, A.; O'Regan, M.; Keys, D. (2005) A 14.1" WXGA Solution Processed OLED Display with a-Si TFT. *SID Symposium Digest*, Vol.36, No.1, (May 2005) pp. 968-971, ISSN 0003-966X.

Shirasaki, T.; Ozaki, T.; Sato, K.; Takei, M.; Toyama, T.; Shimoda, S. & Tano, T. (2004) Full-color Polymer AM-OLED using Ink-jet and a-Si TFT Technologies. *SID Symposium Digest*, Vol.35, No.1, (May 2004) pp. 1516- 1519, ISSN 0003-966X.

Suh, M.C.; Chin, B.D.; Kim, M.; Kang, T.M. & Lee. S.T. (2003) Enhanced luminance of blue light-emitting polymers by blending with hole-transporting materials, *Adv. Maters.*, Vol.15, No. 15, (August 2003), 1254, ISSN 0935-9648.

Shinar, J.; Shinar, R.; Zhou, Z. (2007) Combinatorial fabrication and screening of organic light-emitting device. *Applied Surface Science*, Vol.254, No.3, (Nov 2007), pp.749-756, ISSN 0169-4332.

Wang, R.Y.; Segalman, R.A. & Majumdar, A. (2006) Room temperature thermal conductance of alkanedithiol self-assembled monolayers. *Appl. Phys. Lett.* Vol.89, No.17, (Oct 2006) 173113 ISSN 0003-6951.

Wolk, M.B.; Baetzold, J.; Bellmann, E.; Hoffend Jr, T.R.; Lamansky, S.; Li, Y.; Roberts, R.R.; Savvateev, V.; Staral, J.S.; Tolbert, W.A. (2004) Laser Thermal Patterning of OLED Materials, In: *Organic Light-Emitting Materials and Devices VIII*, Z.H. Kafafi, P.A. Lane (Ed.), Vol.5519, pp. 12-23, IBSN 0-8194-5942-9.

Willis, D.A. & Grosu, V. (2005) Microdroplet deposition by laser-induced forward transfer. *Appl. Phys. Lett.* Vol.86, No.24, (Jun 2005) 244103, ISSN 0003-6951.

Wilson, O.M.; Hu, X.Y.; Cahill, D.G.; Barun, P.V. (2002) Colloidal metal particles as probes of nanoscale thermal transport in fluids. *Physical Review B* Vol.66, No.22, (Dec 2002), 224301 ISSN 1098-0121.

Zschieschang, U.; Klauk, H.; Halik, M.; Schmid G. & C. Dehm (2003). Flexible organic circuits with printed gate electrodes. *Adv. Mater.* Vol.15, No.14, (July 2003), pp. 147, ISSN 0935-9648.

8

Interlayer Processing for Active Matrix Organic Light Emitting Diode (OLED) Displays

Peter Vicca[1,4], Soeren Steudel[1,4], Steve Smout[1,4], Kris Myny[1,4], Jan Genoe[1,3,4], Gerwin G.H. Gelinck[1,4] and Paul Heremans[1,2,4]
[1]*IMEC*
[2]*Katholieke Universiteit Leuven*
[3]*Katholieke Hogeschool Limburg*
[4]*HOLST Centre*
[1,2,3]*Belgium*
[4]*The Netherlands*

1. Introduction

At the moment, the display market is dominated by liquid crystal displays with amorphous silicon (a-Si) thin film transistor (TFT) backplanes processed on glass. Within the last two years, active matrix OLED (AMOLED) displays have rapidly expanded their market share and are poised to break out of their niche application status. The brightness of OLEDs is current controlled and therefore the driving engine behind every pixel is in its most simple implementation comprising 2 transistors and 1 capacitor (2T1C) as in Figure 1. The required current drive for the necessary brightness (> 500 cd/m^2 for cell phone display at > 72 dpi) is however quite challenging for standard a-Si TFTs due to their relatively low mobility (~1 cm^2/Vs).

The simplest implementation of a pixel engine for AMOLED current steering is depicted in Figure 1. As a result, each pixel in the display contains 2 transistors and 1 capacitor. Since OLEDs requires current driving, transistor 2 (T2) is used as a current source, whereby the gate-source voltage determines the current level through the OLED. This gate-source voltage is stored on capacitor C and can be modified by selecting transistor 1 (T1) via the row select line. The voltage on the dataline is then transferred to the capacitor.

It is envisioned that the next generation of displays for handheld device will be flexible or even rollable AMOLED displays realized on foils.

The substrate material for backplanes is typically glass, but could be replaced by a flexible substrate, provided that low temperature materials are employed to create the backplane. A working process for OTFTs in a backplane has been shown by Gelinck *et al.* Truly rollable displays using organic TFTs have been shown recently either using e-ink by Huitema *et al* (PolymerVision) or using OLEDs by Kawashima *et al,* Katsuhara *et al* and by Yagi *et al,* with the organic TFTs possessing a similar performance to those of a-Si. In order to achieve a sufficient light output, the preferred process flow implements top emitting OLEDs on top of the driving engine resulting in an aspect ratio >70%. The choice of the flexible substrate therefore will determine the process flow in a substantial way.

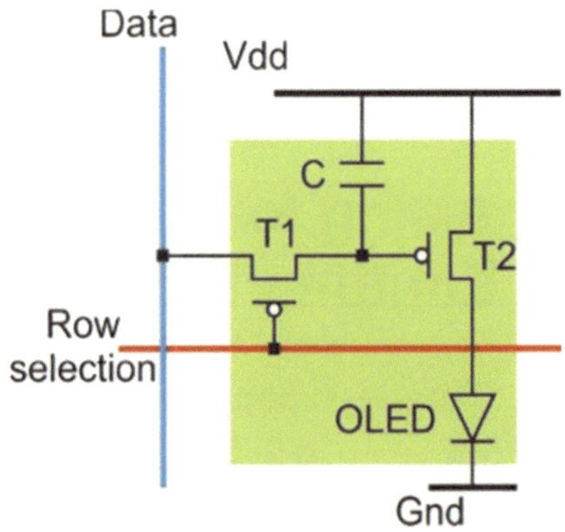

Fig. 1. The schematic layout of a 2T1C pixel driver for AMOLED backplanes

Using a high temperature foil like polyimide (PI) which can sustain temperature of <260 °C allows a process flow very similar to the flow used for a-Si backplanes. On the other hand, if a low-temperature, low-cost foil such as polyethylenetheraphtalate (PET) (<120 °C) or polyethylene naphtalate (PEN) (<160 °C) is used, it will necessitate the development of low temperature materials and processes for the backplanes.
To produce a highly efficient OLED display at the end of the process, the choice for a top emitting OLED on top of the pixel engine as shown in Figure 2 is most favorable to generate a larger emitting surface. In order to electrically isolate the OLED from the pixel engine, an interlayer is required. This implements also the necessity to have a connection between the OTFT and the OLED anode. This connection, or via hole, will have to be generated through the interlayers that protect the OTFT from degrading, and ideally without impacting the performance of the OTFT.

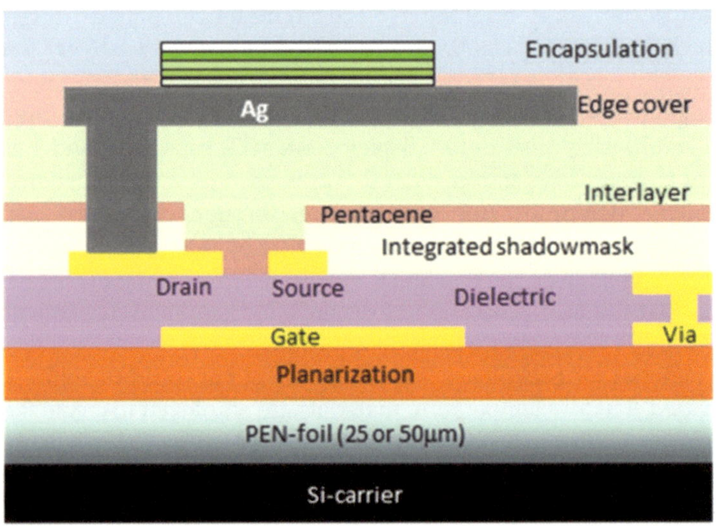

Fig. 2. Schematic cross section of an AM-OLED stack

The OLED pixels (red, green and blue) will be deposited by evaporation, using a shadowmask patterning technique to create different colors. Prior to this deposition, the anode material has to be deposited and patterned on top of the interlayers. Considering solely performance, the most favorable anode material is silver due to its low resistivity and high reflectivity

There are several challenges in realizing top emitting OLEDs on organic TFTs which have a long lifetime because of their limited chemical, physical and environmental stability. In this chapter, we address processing issues for the interlayers and the anode.

Another issue will be the adhesion of materials on top of each other. Organic materials generally have a pronounced polarity which will act to repel or attract other materials. Another drawback of processing on top of OTFTs is their inherent performance sensitivity to a variety of solvents and their degradation in atmospheric conditions.

2. Interlayer with via process

To obtain the flexibility needed to create a rollable display, OTFTs are one of the possible choices. Among OTFTs, one of the most widely studied and used organic semiconductor materials is pentacene, allowing mobilities up to 1 $cm^2/(Vs)$. However, immediately following deposition, oxygen, humidity and solvents will affect the transistor performance adversely. The upper temperature limit the pentacene can sustain is approximately 140 °C; Fukuda *et al* has shown that higher temperatures will result in recrystallization and a decrease of the transistor performance.

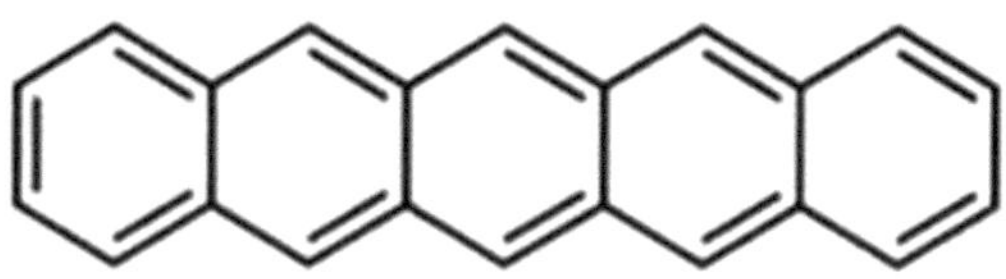

Fig. 3. Pentacene molecule

On top of the MIM stack with OTFTs, an insulating interlayer with via holes needs to be processed on which the reflective metal anode of the OLED is photolithographically processed. This interlayer has to allow a good adhesion of the OLED anode, which is deposited last on top of the interlayer through a shadowmask.

In order to have a very smooth anode, the polymer interlayer needs to have a very low surface roughness, since this will be reproduced in the surface of the anode.

The main requirement of the interlayer, that is, to protect the OTFT from air, suggests the use of a material that has good barrier properties against chemicals, moisture and air, a solvent free deposition technique, low temperature budget and good adhesion. When surveying the options available for such an interlayer, a premium choice would be to use poly(p-xylylene), also known as parylene. The deposition of parylene by chemical vapor deposition (CVD) is known to be pinhole free at thicknesses >600 nm, and has been used in many applications in aerospace, electronics and military for its good barrier properties against water, chemicals and oxygen. These properties therefore would be very useful when it comes to protecting the pentacene from degrading. Also the deposition technique is fully compatible with the semiconductor, since the temperature inside the polymerization chamber does not exceed 30 °C over the entire deposition run. Measurements prove the barrier properties of parylene as depicted in Figure 4. After it has been deposited, the parylene polymer can handle temperatures up to 160 °C. When this temperature is exceeded, the polymer will rather degrade and decompose instead of deform.

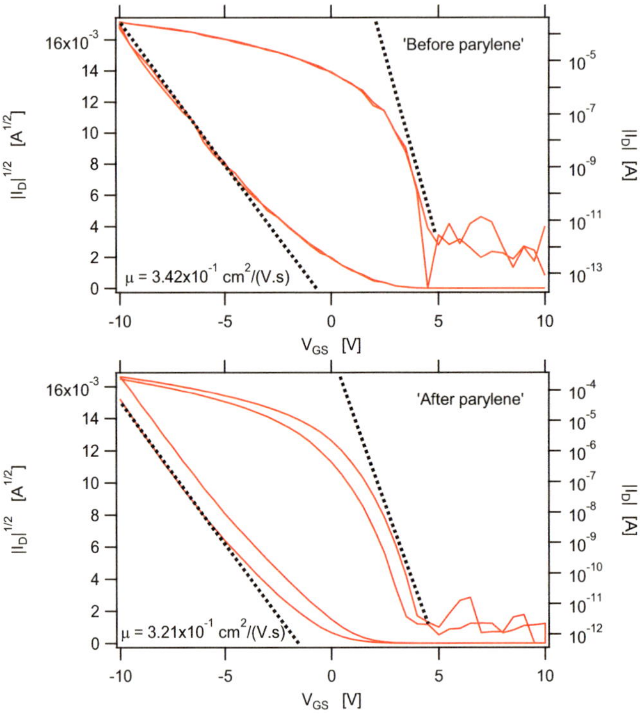

Fig. 4. Transistor transfer curves in saturation measured after before (left) and after (right) parylene deposition. The mobility went down from 0.34 to 0.32 cm²/(Vs) (in both graphs the semiconductor is pentacene)

Parylene itself comes in 3 main derivates; type N, type C and type D, all commercially available. Other types also do exist as commercial products and are chemically modified to be high temperature resistant or have a fluorinated structure. The difference is based on the presence of chlorine atoms on the monomer as depicted below.

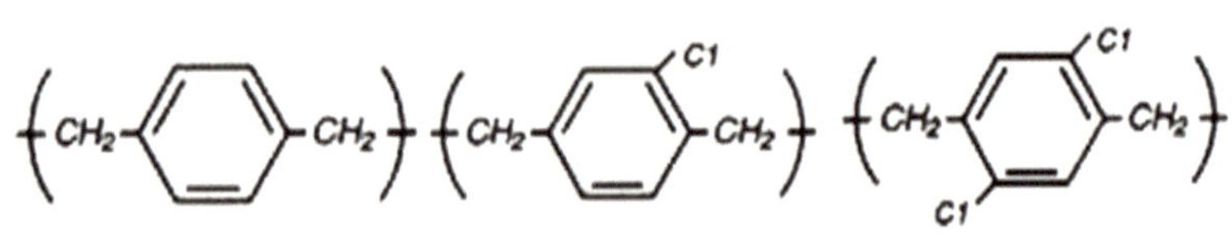

Fig. 5. Different parylene derivatives (left to right) Parylene N, Parylene C and Parylene D

This presence of chlorine atoms on the benzene ring has an influence on the surface energy of the resulting film and therefore on the adhesion of subsequent layers. Previous research has shown that the more chlorine atoms that are bonded on the benzene ring, the worse the adhesion towards silver becomes. This can also be seen in contact angle measurements on the different films and therefore, the choice of the N type is best suited for this application. The maximal thickness that was observed by Vicca *et al*, still adhering sufficiently for further processing, was 2 micrometer.
To ensure that a silver anode adheres well on the parylene layer, a slow deposition rate, < 1,5 Å/s, is required. This will allow sufficient relaxation time of the silver and reduces stress in the metal film. Stress free layers up to 200 nm are possible with this approach.

Patterning of parylene is done by photolithography using a dry etch plasma to define the desired structures. We used oxygen plasma allowing etch rates up to 17 Å/s using a moderate (50 W) etch power in a reactive ion etch (RIE) plasma. The various parameters such as pressure, gas flows, power, etc., are different for each etch chamber and will not be discussed further.
The CVD process will lead to a conformal coating of the sample, meaning that the polycrystalline structure of the pentacene (with root mean square roughness of 10 nm) is projected into the surface of the parylene and thus, in the surface of the anode.
To decrease the resulting surface roughness, a second spin coated layer over the parylene layer will act as a planarization layer as suggested by Yagi *et al*. The requirements for this material are similar to those for the parylene interlayer; solvent and temperature needs to be compatible with the processing steps and should allow a good adhesion of the subsequent OLED anode material.
Considering the requirements (temperature budget <150 °C, crosslinkable and solution processed) there are very few materials available in literature, (e.g. SU8, SC100, photoresist, PVP and PMMA) and after some initial screening experiments, we chose PVP. Having a dielectric constant of around 4.5, this commercially available material has been already used by many research groups as a dielectric in TFT processing, and it has proven to result in smooth surfaces and good adhesion properties. Typically, layer thicknesses around 200 - 400 nm are obtained, covering the parylene film completely without affecting the bending radius of the substrate.

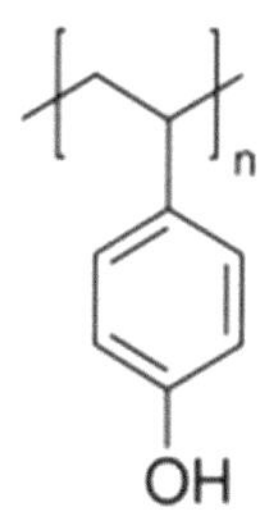

Fig. 6. Monomer of poly-4-vinylphenol

The PVP will not remain on the substrate as a rigid polymer film, but needs crosslinking to form a film that is compatible with chemicals used in further processing. To enable this cross linking, a molecule will bond on the separate PVP polymer chains and create a chemically inert film. The cross linker most commonly used for PVP is poly(melamine-co-fromaldehyde). A process flow is described by Hwang *et al* and reports a thermal initiation of the cross linking reaction at temperatures around 180-200 °C.

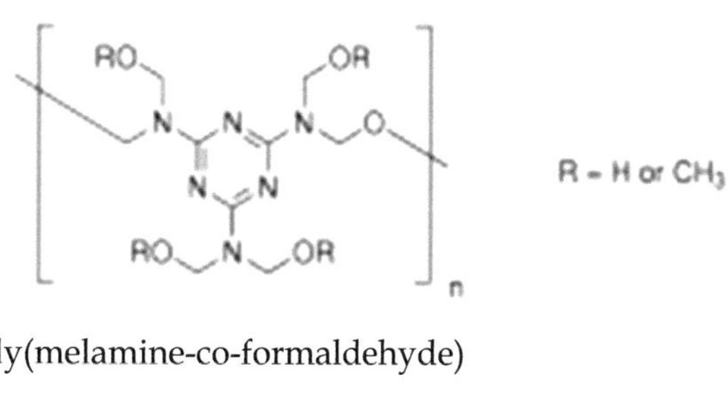

Fig. 7. Crosslinker poly(melamine-co-formaldehyde)

When lowering the cross linking temperature to a more acceptable 145 °C, the cross linking time increased to almost 3 hours. When using this long time for a baking process, it is important to allow the substrate a long cool down period afterwards to prevent stress effects. These stress effects can be seen by bended wafers, which will result in misaligned structures. The use of low temperature cross linkers has been described in literature and can be used for further decreasing of the temperature.
Other ways for decreasing the crosslinking temperature even further were investigated and reported by Vicca *et al* and resulted in the use of another commercially available crosslinking molecule; (hydroxymethyl)benzoguanamine.
Another point that needs to be taken into account is the cross linking reaction itself between the polymer chains and the cross linking molecule. Very often this reaction produces side products such as ions or volatile organic compounds (VOC) or else unused cross linker remains behind. These molecules will very often function like as a charge trap, influencing the transistor characteristics and which has been observed for both crosslinking molecules.
Typical effects that can be seen in the transfer characteristic curves are hysteresis, indicating the presence of charge trapping ions, often left from reaction side products or VOC. This might indicate a need for a longer bake time to evaporate these products out of the film. If the threshold voltage (V_t) shifts in the negative direction, this indicates an excessive amount of cross linker. To solve this effect, a lower percentage of the cross linker should be used. Using these indications, a good recipe for PVP can be optimized.
Once the resist is patterned, dry etching with oxygen plasma (RIE) will remove the PVP accurately. Just like when etching parylene, the discussion on different parameters like flows, power, pressure, etc., will not be applicable here, since this is very dependent on etch chamber architecture.

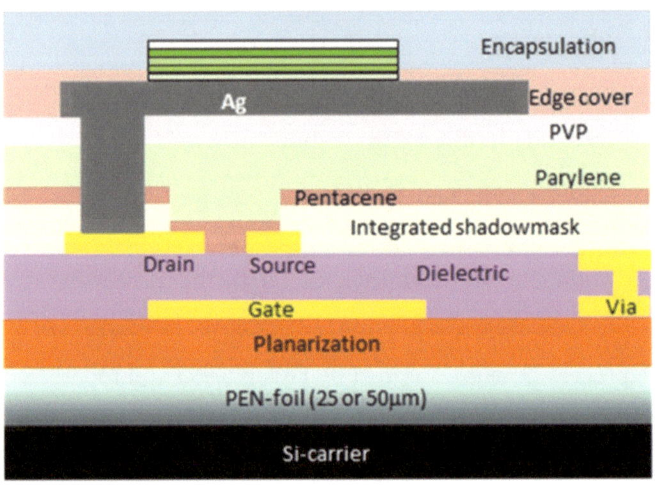

Fig. 8. The full AM-OLED stack using a PVP planarization layer on top of the parylene interlayer

3. Silver anode

To find the optimal material that can be used as an anode material, the different properties of possible candidates must be compared to ensure that the best option is used for the process. Considering that the most advanced OLED stacks have doped- transport/injection layers implemented, the choice of work function is not strongly relevant. To increase the output of the OLED, it would be useful to reflect the light that is emitted towards the backside of the stack. This efficiency increment can be realized by choosing a highly reflective metal that also can be deposited with a sufficient thickness such that it is not transparent. To ensure these thick layers are possible on top of the polymer interlayer, a good degree of adhesion is required to prevent the metal from delaminating.

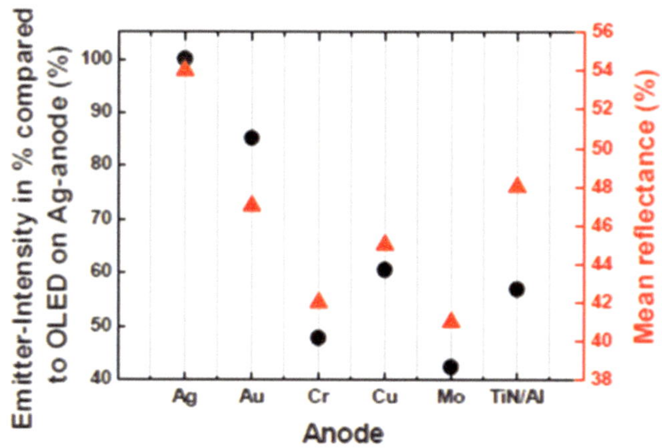

Fig. 9. Comparison of different anode metals for green OLED stacks measured at 525 nm

In Figure 9, a comparison between different metals is made to compare the light intensity with the intensity of the silver, It is clearly shown that the silver always has the highest intensity and also the highest reflectance. This was also calculated for red and blue OLED stacks in Figures 10 and 11, respectively.

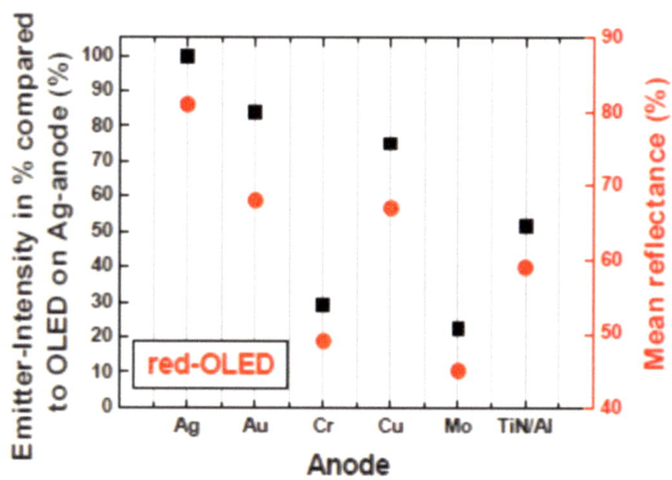

Fig. 10. Comparison of different anode metals for red OLED stacks measured at 630 nm

From a performance point of view, Ag emerges as the optimal choice. In addition, the deposition of the silver anode on PVP or parylene does not require any adhesion layers or adhesion promotors such as HMDS since silver has a good adhesion on the polymer surfaces.

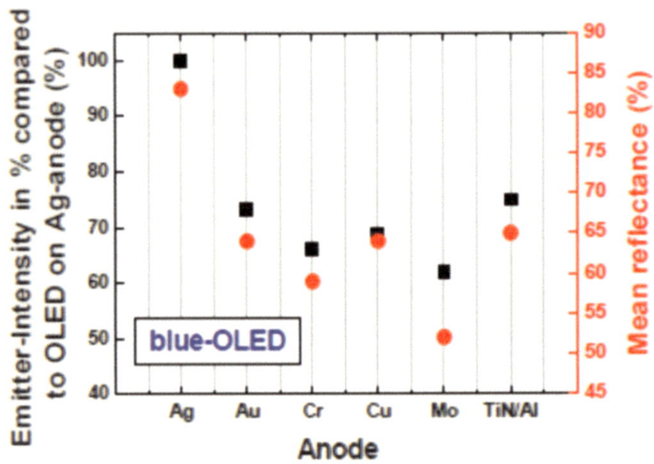

Fig. 11. Comparison of different anode metals for blue OLED stacks measured at 460 nm

3.1 Lift off

Once the silver layer has been deposited, it still needs to be patterned in order to define the separate pixels in the display. For patterning, the main options available are lift off or etching. When using a lift off process, the resist will already be patterned on the wafer prior to the silver deposition and will be dissolved afterwards in a solvent to remove the silver where it has no contact with the polymer interlayer.

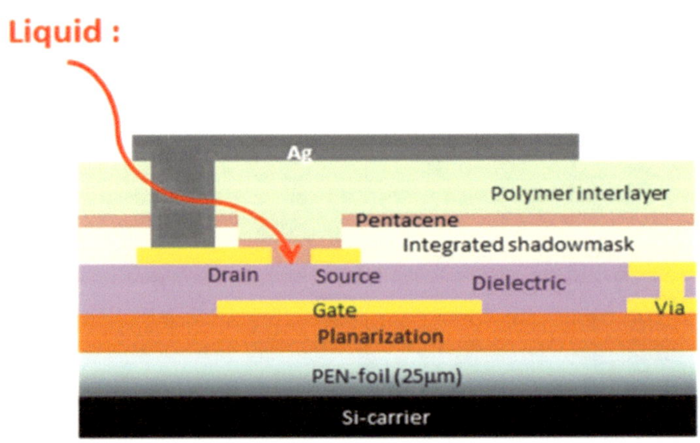

Fig. 12. Schematic presentation of the liquid impact on the OTFT during processing

However, for yield consideration, lift-off processes should be generally avoided and since the inherent limit to the metal thickness is approximately 50 nm, it would not be compatible with the non-transparent silver layer (200 nm) that is desired.

Another disadvantage of the lift off process is the long exposure of the wafer and OTFTs to the lift off solvent(s). It is proven by measuring mobilities of OTFTs that were encapsulated with different polymers and immersed for 1 minute in various solvents, that liquid chemicals used in general processing do have a negative impact on the mobility. Results of these tests are showed in Figure 13, using 2 types of polymer to protect the OTFTs from the chemicals.

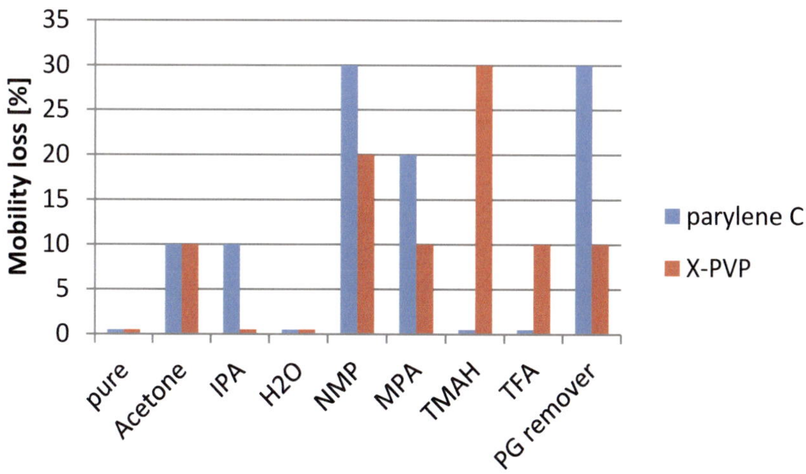

Fig. 13. Loss of mobility for OTFT after immersion in different process liquids for 1 min; IPA (isopropanol), NMP (n-methyl pyrolidone), MPA (methyl propyl acetate), TMAH (tetra methyl ammonium hydroxide), TFA (commercial Au etchant), PG remover (commercial photoresist stripper)

3.2 Wet etching

When looking at the option to pattern the anode layer with an etch process, the etching is performed after a silver layer has been deposited and patterned by photolithography using the inverted mask lay out of the lift off patterning. The etching itself can be done by wet etch or dry etching. If a wet etch process is used, a certain amount of reproducibility should be respected. To do this, the EDC system by Laurell Inc. was used, employing a commercial silver etchant. The system will spin the wafer while the etchant is sprayed over the wafer, causing the etching of the silver. The main issue with this system is to find a good rotation speed for the spinner. If the wafer rotates too slowly, the etchant arriving in the center will remain for too long and cause over etching here. If the resolution of the etched structures is monitored as shown in Figure 14, the ideal spin speed would be around 1500 rpm. At this equilibrium point, the overertch in the center and at the wafer edge was found to be the same. Adjusting the etch time leads to the desired amount of overetch, which is as close to zero as possible. Doing this however leaves a very small time window for the etching, resulting in a 7 second etch time for 200 nm of silver.

The main problem however of the wet etch technique is again the use of liquids. This can be seen by microscope inspection of the pixels after patterning. The swelling of the interlayer underneath will deform the pixel surface dramatically as can be seen in Figure 15, leading to pixel failures.

The liquid impact can also be observed in the performance of the transistors by comparing their mobilities before and after the silver has been patterned.

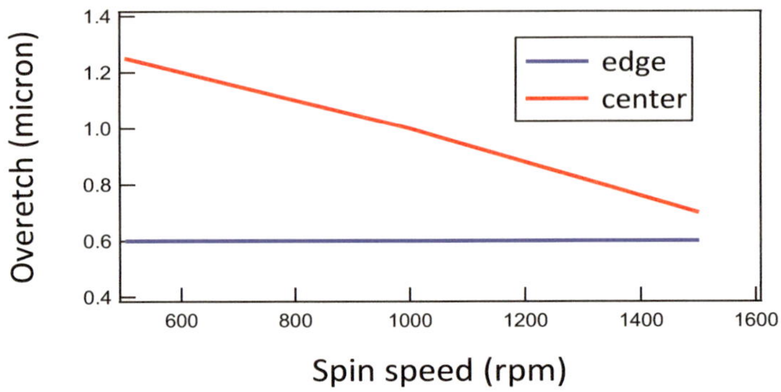

Fig. 14. Rotation speed versus overetch at edge and center of a 6 inch silver substrate

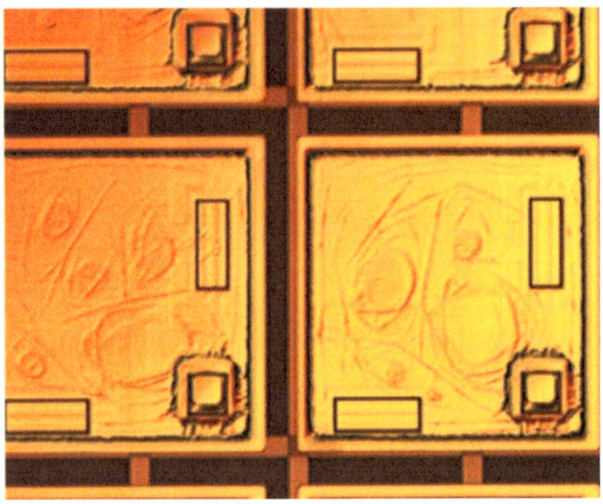

Fig. 15. Microscopy image of pixels deformed by swelling of the polymer interlayer

3.3 Dry etching

When looking to dry etching techniques, literature describes only few effective etching techniques for the removal of silver. These processes were typically used to pattern silver lines in circuit purposes and the use of an inductively coupled plasma (ICP) is the most common technique described by Lee *et al,* Jang *et al,* and Park *et al*. The disadvantage, however, of this technique is the likelihood of silver residues that will form a conductive film on the chamber wall, stopping the plasma. In literature, most of the etch recipes use a gas mixture in which argon and CF_4 are combined.

The purpose of this mixture is the fluorination of silver by CF_4, followed by argon sputtering in order to remove the silver salt.

$$Ag + F^- \rightarrow AgF \text{ and } AgF_2 \rightarrow \text{sputtered away by } Ar^+ \text{ ions}$$

We tried to produce a dry etch process running on a reactive ion etch (RIE) fab tool. At first, the effect of pure gasses present in the tool was tested to see their effect on a pristine silver film, deposited on a polymer coated wafer that could act as a dummy for the stack that will be used later. The gasses that can be used are argon, CF_4, oxygen and SF_6.

When using a pure argon gas plasma, the ionized argon atoms (Ar^+) will sputter on the silver surface, causing a rough surface, but without removal of silver. As a side effect of this sputtering, the temperature of the substrate increases too much to allow the use of argon when a foil substrate is used.

Fig. 16. Photo of the impact of pure argon etch on silver

Using an oxygen plasma, the reaction with silver is obvious to be observed and results in a highly affected surface. This etching will eventually remove all silver without generating too much heat, but the removal of silver is caused by a stress induced exfoliation. The oxygen will react with the silver and create silver oxides which have a certain amount of stress in them.

$$Ag + O^{2-} \rightarrow AgO \text{ and } Ag_2O$$

This stress will lead to the creation of flakes of silver oxide that will exfoliate from the surface because of their different expansion coefficient. The size of these flakes cannot be controlled and will result in low resolution if this gas only is used for the removal of silver. This observation has also been described in literature by Nguyen *et al.*

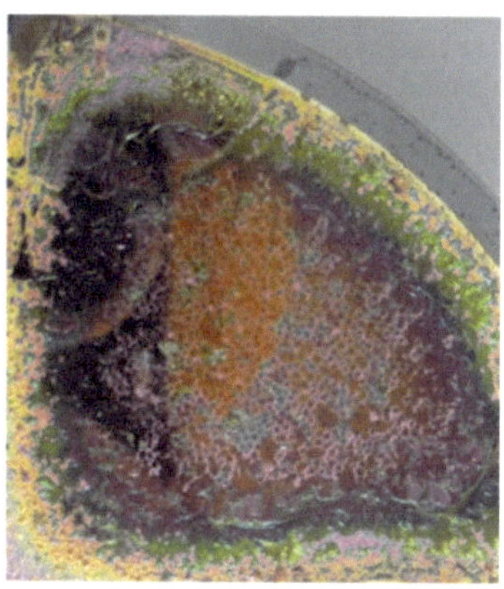

Fig. 17. Photo of the impact of pure oxygen etch showing the high affinity of oxygen to react with silver

Etching silver with a CF_4 plasma will result in a thin layer of silver fluoride which will be present on the top layer of the surface, but will act more similar to a passivation layer that prevents further fluorination of the metal. This is also why a mixture of CF_4 and argon is needed to remove this top layer such that a fresh silver layer is exposed to the CF_4 gas.

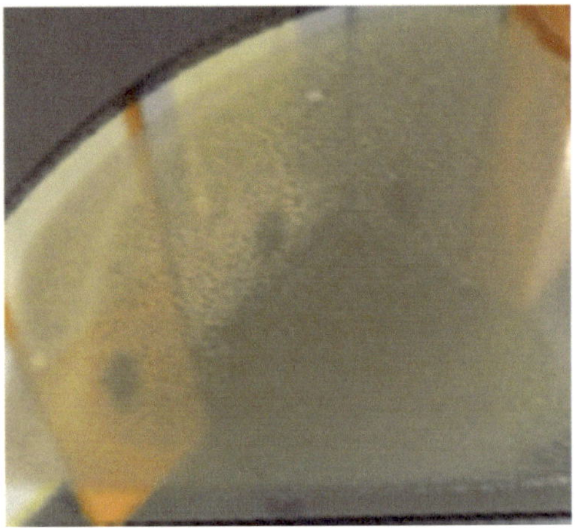

Fig. 18. Photo of the silver layer after plasma etch using pure CF_4 revealing a poor impact

The use of an SF_6 plasma will result in a gently affected silver layer in which probably some fluorination has occurred, but in which also sulfur depositions can be observed as yellow dots.

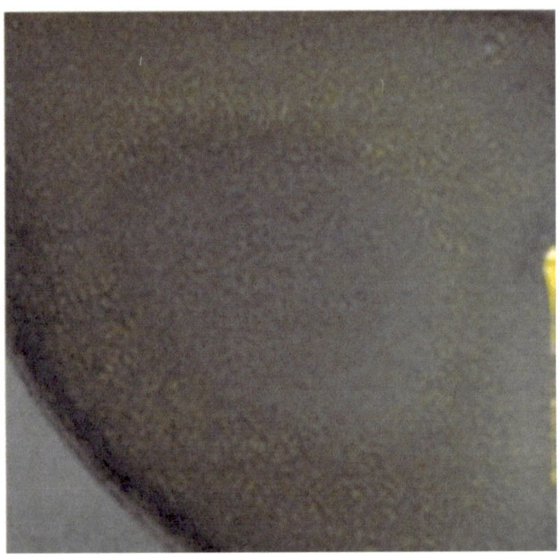

Fig. 19. The impact of pure SF_6 etch on a silver layer showing yellow (sulphur) dots in the surface

Since the use of a single gas appears not to be possible, a similar gas mixture as used in ICP appeared more useful. As a starting point, the same Ar/CF_4 mixture as described in literature can be tested, but will result in a non effective etch. Probably due to a low etch power (500 W) compared with ICP etch power (1500 W), this will not be suitable for RIE

tools. To get the mixture working, the energy needed to start the fluorination needs to be lowered. To do this, a certain amount of oxygen can be added to the Ar/CF_4 mixture, which will react first with the silver layer due to its favored reactivity. This will result in a silver oxide that will be easily accessible for the CF_4 gas to fluorinate it further into silver fluoride which can be sputtered away by the argon ions.
To understand what the right amount of oxygen in the Ar/CF_4 mixture is, samples with silver will be used again to verify the impact of different gas flows. In principle, the Ar/CF_4 mixture will be present in excessive amounts that will fluorinate immediately all silver oxide formed by the oxygen. Afterwards, the end product will be removed by the Ar^+ ions sputtering.

$$Ag + O^{2-} \rightarrow AgO \text{ and } Ag_2O$$

$$2\,AgO + CF_4 \rightarrow 2\,AgF_2 + CO_2 \rightarrow \text{removed by } Ar^+ \text{ ions}$$

When the amount of oxygen is too low, the silver will not etch completely and remains on the substrate.

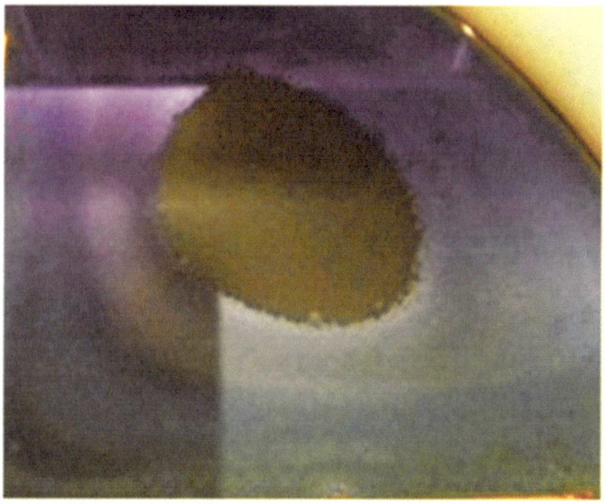

Fig. 20. Too low amount of oxygen, resulting in a not completed silver etch

Using too much oxygen will cause etching of the underlying polymer interlayer once the silver film has been removed.
If the right amount of oxygen is used in the mixture, the polymer interlayer will appear shiny as it was deposited before the silver deposition and all silver will be removed from the sample.
The compatibility of a dry etch recipe with that of the resist also requires optimization. The effect of a too high etch power will result in a carbonization of the resist and will not allow subsequent stripping of the resist with solvents. Also, the heating of the substrate will be reduced as the etch power is reduced.
If the etching would take too long, this will also affect the resist adversely, leading to the removal of the resist. This is mainly caused by the presence of oxygen in the gas mixture.
All the parameters were optimized for our tool and will of course depend from etch chamber to etch chamber. In our etch chamber, etch rates of ±13Å/s were obtained without affecting or removing a commercial photoresist used for the patterning of pixels.

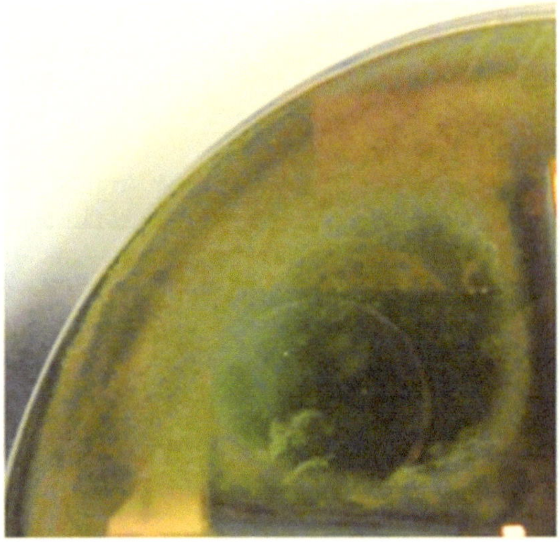

Fig. 21. Excess of oxygen in the gas mixture etching the polymer interlayer

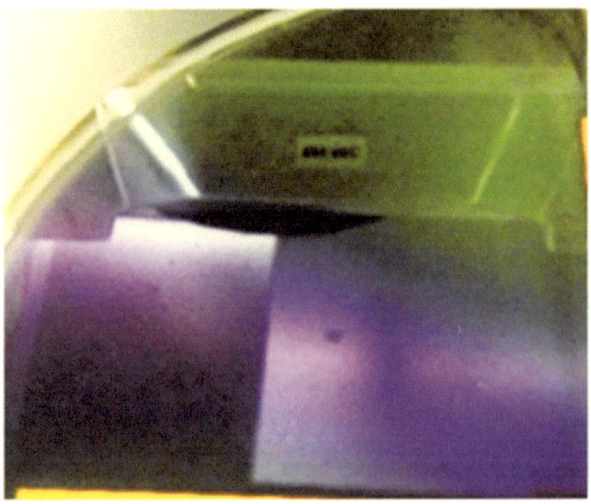

Fig. 22. Exposed and not affected polymer interlayer after complete silver etch

Fig. 23. Carbonized resist left on the silver structure after solvent treatment

Fig. 24. Removed photoresist caused by too long etch time

Fig. 25. Patterned pixels after stripping photoresist. Left: fully processed 6 inch wafer, right: detail of pixels in the QQVGA display

To confirm that the dry etch technique has not affected the performance of the OTFTs, electrical measurements were executed on several transistors on different wafers and mobilities were calculated. This resulted in an average mobility at the end of the process of 0.18 $cm^2/(Vs)$. When the final mobility is calculated for the OTFTs before and after the dry etch process step, 61% of the mobility has been preserved. This was also calculated for the wet etch technique, resulting in a final mobility 28% of the original mobility, showing the advantage of dry etch in terms of both transistor stability and resolution.

Furthermore, when comparing the transfer curves of the OTFTs depicted in Figure 26, it can be seen that there is no loss of uniformity over the processing steps, which is also one of the beneficial effects of the dry etch recipe.

After OLED depositions were finished, 1 cm^2 test pixels as depicted in Figure 27 show a high degree of uniformity of color indicating a smooth surface and giving an efficient light output of 83 cd/A for a green OLED and 37 cd/A for a red OLED.

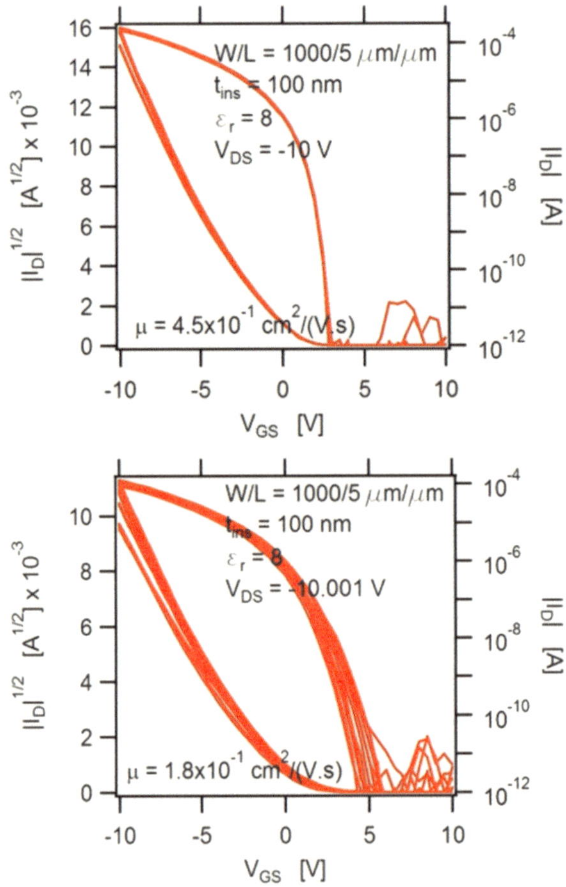

Fig. 26. Transfer curves of different OTFTs measured on a fully processed wafer starting after pentacene deposition (left) and anode etch (right)

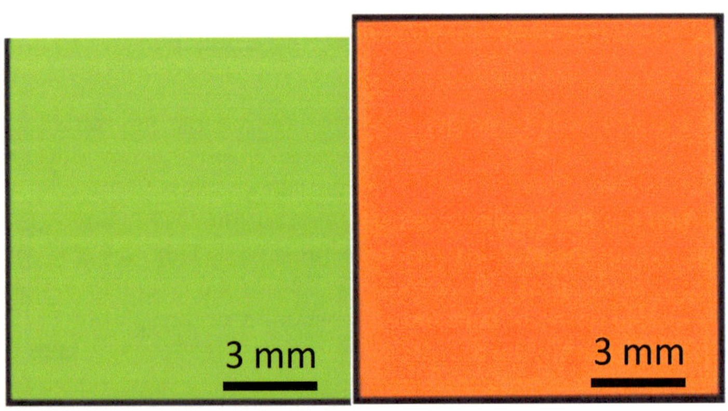

Fig. 27. Pictures of OLED test structures on a flexible substrate

When the active pixels are connected to a control box, the pixels in the display will switch on and off according to the commands of the controller. This allows us to activate or deactivate all pixels or switch on even or odd rows and/or columns or to make checker board patterns as is shown in Figure 28.

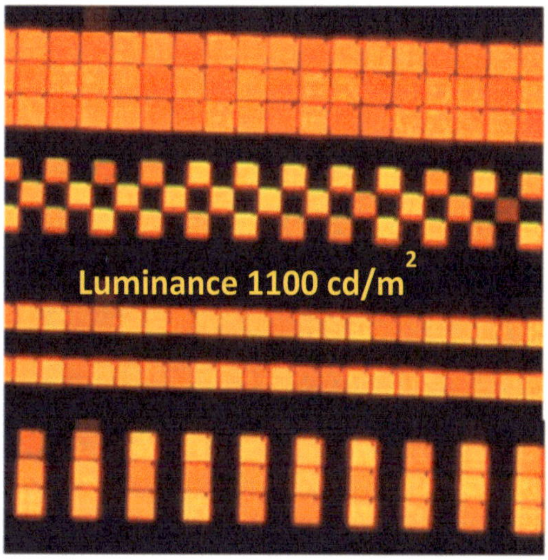

Fig. 29. Picture of different patterns created on the display by the active pixels

4. Conclusions

To tackle the challenges of a low temperature process for making organic backplanes for OLED displays, the use of different polymers in the interlayer, parylene as well as PVP, has been optimized in terms of their deposition, patterning and subsequent materials that will be deposited on them. Working transistors prove the quality of the materials and the efficiency of the process.

For the metal anode, a choice for silver was made mainly based on its excellent reflection and conductivity properties. The patterning of the silver anode is performed by dry etching using an optimized RIE process. This was developed due to the fact that other etch methods seemed to affect the whole stack adversely and the output of OTFT at the end.

Combining the experience and technology that has been gained over the research on the different interlayers and their appropriate processing techniques has led to the realization of high efficiency OLED pixels.

5. Acknowledgements

This work was realized in the framework of the EC project FLAME FP7 ICT-216546 in collaboration between imec, TNO/HOLST centre, Polymer Vision and Fraunhofer IPMS. I would like to thank K. Fehse for the simulated light output data for different anode materials and my colleagues at imec and in the HOLST centre for their support in this work.

6. References

Fukuda, K.; Yokota, T.; Kuribara, K.; Sekitani, T.; Zschieschang, U.; Klauk, H. & Someya, T. Thermal stability of organic thin-film transistors with self-assembled monolayers dielectrics, *Appl. Phys. Lett.*, Vol. 96, pp. 053302-1 - 053302-3 (2010)

Gelinck, G.H.; Huitema, H.; Edzer, A.; van Veenendaal, E.; Cantatore, E.; Schrijnemakers, L.; van der Putten, J. B. P. H. ; Geuns, T. C. T.; Beenhakkers, M.; Giesbers, J. B.; Huisman, B.-H.; Meijer, E. J.; Mena, B. E.; Touwslager, F. J.; Marsman, A. W.; van Rens, B. J. E. & de Leeuw, D. M. Flexible active-matrix displays and shift registers based on solution-processed organic transistors, *Nat. Mater.*, Vol. 3,pp. 106–110 (2004)

Huitema, E.; van Veenendaal, E.; van Aerle, N.; Touwslager F.J.; Hamers, J. & van Lieshout, P. Rollable displays – A technology development enabling breakthrough mobile devices, *SID 08 Digest*, pp. 927-930 (2008)

Hwang, M.; Lee, H.S.; Jang, Y.; Cho, J.H.; Lee, S.; Kim, D.H. & Cho, K. Effect of curing conditions of a poly(4-vinylphenol) gate dielectric on the performance of a pentacene-based thin film transistor, *Macromol. Res.*, Vol. 17, No. 6, pp. 436-440 (2009)

Jang, K. H.; Lee, W. J.; Kim, H. R. & Yeom, G. Y. Etching of copper films for thin film transistor liquid crystal display using inductively coupled chlorine-based plasmas, *Japn. J. Appl. Phys.*, Vol. 43, (No. 12), pp. 8300-8303 (2004)

Katsuhara, M.; Yagi, I.; Noda, M.; Hirai, N.; Yasuda, R.; Moriwaki, T.; Ushikura, S.; Imaoka, A. & Nomoto, K. A reliable flexible OLED display with an OTFT backplane manufactured using a scalable process, *SID Digest 09*, pp. 656–659 (2009)

Kawashima, N.; Kobayashi, N.; Yoneya, N.; Ono, H.; Fukuda, T.; Ohe, T.; Ishii, Y.; Nomoto, A.; Sasaki, M. & Nomoto, K. A high resolution flexible electrophoretic display driven by OTFTs with inkjet-printed organic semiconductor, *SID Digest 09*, pp. 25–27 (2009)

Lee, Y. J.; Park, S. D.; Song, B. K.; Kim, S. G.; Choe, H. H.; Hong, M. P. & Yeom, G. Y. Characteristics of Ag etching using inductively coupled Cl_2-based plasmas, *Jpn. J. Appl. Phys.*, Vol. 42, pp. 286-290 (2003)

Nguyen, P.; Zeng, Y. & Alford, T. L. *J.* Reactive ion etch of patterned and blanket silver thin films in Cl_2/O_2 and O_2 glow discharges, *Vac. Sci. Technol.*, Vol. 17, pp. 2204-2209 (1999)

Park, S. D.; Lee, Y. L.; Kim, S. G.; Choe, H. H.; Hong, M. P. & Yeom, G. Y. Etch characteristics of silver by inductively coupled fluorine-based plasmas, *Thin Solid Films*, Vol. 445, pp. 138-143 (2003)

Vicca, P.; Steudel, S.; Genoe, J. & Heremans, P. Adhesion promoting polymer interlayers for Ag layers
deposited in OLED processing *J. Adh. Sci. Technol.*, Vol. 24, pp. 1145-1151 (2010)

Vicca, P.; Steudel, S.; Smout, S.; Raats, A.; Genoe, J. & Heremans, P. A low-temperature-cross-linked poly(4-vinylphenol) gate-dielectric for organic thin film transistors, *Thin Solid Films*, Vol. 519, No. 1, pp. 391-393 (2010)

Yagi, I.; Hirai, N.; Noda, M.; Imaoka, A.; Miyamoto, Y.; Yoneya, N.; Nomoto, K.; Kasahara, J.; Yumoto, A. & Urabe, T. A full-color, top-emission AM-OLED display driven by OTFTs, *SID Digest 07*, pp. 1753–1756 (2007)

9

Transparent Conductive Oxide (TCO) Films for Organic Light Emissive Devices (OLEDs)

Sunyoung Sohn and Hwa-Min Kim
Catholic University of Daegu
Republic of Korea

1. Introduction

Transparent conducting oxide (TCO) thin films of In_2O_3, SnO_2, ZnO, and their mixtures have been extensively used in optoelectronic applications such as transparent electrodes in touch panels, flat panel displays (FPDs), and other future devices. The first chapter provides an introduction to the basic physics of TCO films and surveys the various topics and challenges in this field. It includes a description of the TCO materials used in some of the organic light emissive devices (OLEDs) that have been studied extensively to date, the performance of various OLEDs, and a brief outlook.

Chapter 2 focuses on TCO material development of p-type and n-type. Typical oxide kind of TCO materials consist of In_2O_3, SnO_2, and ZnO. These are applied as a TCO films with n-type semiconducting property according to highly doped dopants which acting as a carrier. Until today, in a n-type TCO materials, indium tin oxide (ITO) doped with SnO_2 of 10 wt.% in In_2O_3 has been widely commercialized. This is because the ITO film has high performance of both good electrical conductivity of ~10^{-4} Ω·cm and high transmittance of ~90% when the ITO film is coated on glass substrate. At present, In_2O_3-SnO_2 (ITO) films are most commonly used as TCO films, but they have ome disadvantages, such as high cost, instability, poor surface roughness, and toxicity in their further applications. And amorphous ITO film deposited at low temperature has low resistance under moist heat, which leads to a decrease in its conductivity and light transmittance. In addition, unfortunately, the price of Indium is dramatically increasing every day due to a mix-up between the supply and demand of raw materials by the exhaustion of Indium source. On the other hand, some zinc-based TCO materials have good optical and electrical properties comparable to the ITO films, as well as low cost, high stability, excellent surface uniformity, and good etching selectivity. The zinc-based TCO films are, therefore, regarded as promising substitutes for ITO film. In Chapter 3, the Indium-based and Zinc-based TCO Materials, and their electrical, optical, and structural properties will be discussed. Particularly, since more stringent specifications for TCO films have been required for realization of both higher resolution and larger screen size of FPDs, and preparation of high-quality TCO films at low temperature is very important to realize advanced optoelectronic devices.

Chapter 4 will introduce the new TCO materials, such as: organic conductors like poly(3,4-ethylenedioxy thiophene):poly(styrenesulphonicacid) (PEDOT:PSS), and the expanding field of nanomaterials including carbon nanotubes, nanoparticles, and composite materials combining one or more of these materials. For example, long metallic nanotubes have been found to have volume conductivities of ca. 700,000 S/cm, which is almost as conductive as pure copper. Moreover, carbon nanotubes (CNTs) because of their covalent bonding do not suffer from electromigration, which is a common problem that leads to failure in thin metal wires and films. It is the covalent bonds of CNTs that make them thermally stable and highly resistant to chemical damage. Therefore, these cheap and flexible transparent organic and nano conductors can be an appropriate substitution for conventional ITO in the next generation optoelectronic devices.

ITO films are very brittle and easily broken down by externally applied bending forces, while the ITO films are widely used as the transparent electrode for display device. Finally, Chapter 5 mainly examines how external deformation influences the mechanical stability of ITO thin films on flexible polymer substrates for flexible OLEDs (FOLEDs), typically in a bent state.

2. Classification of TCO materials

TCOs are very useful materials to transparent optoelectronics because they have unique features of optical properties in the visible light region such as the transparency over ~85% and optical band gap greater than 3 eV and controllable electrical conductivity such as carrier concentrations of at least 10^{20} cm^{-3} and resistivity of about 10^{-4} ohm·cm.(Kim et al., 2011) Nowithstanding their extraordinarily wide controllable conductivity range including that of semiconductor behavior, their applications are limited to transparent electrodes. It seems to us that the origin of this limited application is due to a lack of p-type conducting transparent oxide materials. TCO materials are naturally n-type degenerate semiconductors and the lack of a high quality p-type TCO always has been the main obstacle in front of the fabrication of a fully transparent complementary metal-oxide semiconductor(CMOS)-like devices. Although n-type TCO such as ZnO, SnO_2 and ITO are key components in a variety of technologies, p-type TCO are an emerging area with little work previous to four years ago. However, realization of good TCO could significantly impact a new generation of transparent electrical contacts for p-type semiconductors and organic optoelectronic materials and in conjunction with n-type TCOs could lead to a next generation of transparent electronics.

2.1 p-type TCO

Since the first report of a p-type TCO was NiO, In 1997, there was a report of transparent p-type conducting films of $CuAlO_2$ showing considerable improvement over NiO.(Sato et al., 1993; Kawazoe et al., 1997) Although the conductivity of 1 S·cm^{-1} was about three orders of magnitude smaller than that of n-type materials, the result was promising. Since the discovery of p-type conductivity in $CuAlO_2$, many Cu(I) based delafossites having transparency and p-type conductivity have been synthesized, such as $CuScO_2$, $CuYO_2$, $CuInO_2$, $CuGaO_2$, and $CuCrO_2$. Conductivity of the $CuInO_2$ film deposited under working oxygen pressure of 7.5 mTorr and 450 °C was reported as 2.8×10^{-3} S/cm. (Roy et al., 2003) And also, the dependence of the electrical conductivity of $CuInO_2$ films upon the deposition temperature was investigated. With increasing deposition temperature from room temperature to 600 °C, the conductivity increases and reaches a value of 5.8×10^{-2} S/cm for the film deposited at 400 °C temperature. Indeed, other structures have been identified that combine p-type conductivity and optical transparency in Cu(I) based materials, including $SrCu_2O_2$ and layered oxychalogenides (LaCuOS), although to date the p-type TCO with the highest conductivity is a delafossite (Mg doped $CuCrO_2$). Among various candidate materials, ZnO is one of the most important members of TCOs. Like the other members (e.g.: SnO_2, In_2O_3, IZO, and ITO), ZnO have been applied. Among the candidates of shallow acceptors, nitrogen is the most tried one due to its nearest-neighbor bond length of 1.88 Å that is similar to the Zn–O bond length of 1.93 Å. The p-type ZnO have been made by nitrogen using various deposition techniques like sputtering, chemical vapor deposition (CVD), metalorganic CVD (MOCVD), pulsed laser deposition (PLD) and spray pyrolysis (SP). (Huang et al., 2010) Growing p-ZnO was an important milestone in "Transparent Electronics", allowing fabrication of wide band gap p-n homo-junctions, which is a key structure in this field. It was anticipated that higher conductivity and optical transmission could be obtained by ZnO doped with N, F, P, Sb, and As. In this section, we discussed the necessary requirements in the electronic energy band structure and crystal structure with Cu based p-type TCO. The chemical formula of delafossites is AMO_2 in which A is the monovalent cation and M is a trivalent cation. Delafossites have a hexagonal, layered crystal structure: the layers of a cations and MO_2 are stacked alternately, perpendicular to the c-axis. As a class, p-type materials now include the copper-based delafossites $CuMO_2$.

Fig. 1 shows the schematic representation of the necessary electronic configuration of the cationic species. In this combination, considerable covalency can be expected for both the bonding and anti-bonding levels. The valence band edge shifts from the *2p* levels of oxygen ions to the anti-bonding levels because both the cation and anion have a closed shell electronic configuration. It should be noted that the localization nature of the valence band edge is greatly reduced by the modification. Cu and Ag have the appropriate d^{10} states for this purpose.(Sheng et al., 2006)

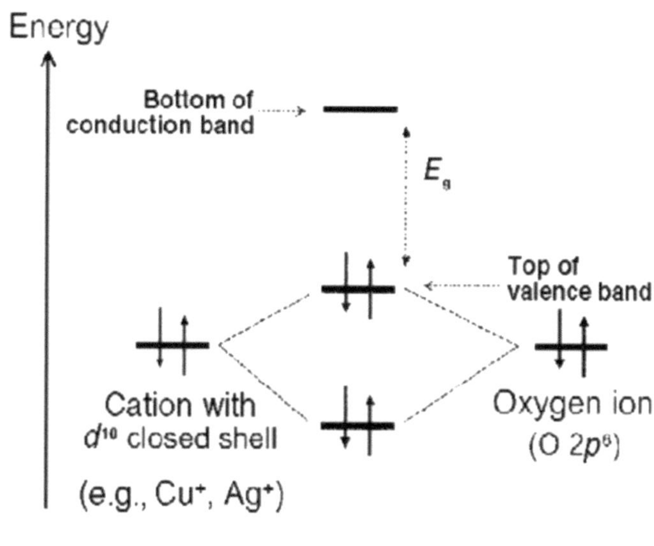

Fig. 1. Schematic of the chemical bond between an oxide ion and a cation that has a closed shell electronic configuration.

The metal states interact with some of the O *2p* states, which push up a more dispersive band above the non-bonding O *2p* or Cu *3d* states. Tetrahedral coordination of oxide ions is advantageous for p-type conductivity, as it acts to reduce the localization behavior of *2p* electrons on oxide ions. The valence state of the oxide ions can be expressed as sp^3 in this conformation. Since the Cu_2O has a rather small band gap(Eg) of 2.17 eV, we found that the

Eg of p-type TCO should be greater than 3.1 eV. Hence enlargement of the band gap would be another structural requirement for designing p-type TCOs, so that there is no absorption of visible photons. Two families of Cu based TCOs have been developed from this idea, $CuMO_2$(M = Al, Ga, In, Sc, Cr, Y, B, etc.) with the delafossite structure and the non-delafossite structure $SrCu_2O_2$, LaCuOCh (Ch = chalcogen). The band structural properties of these materials were calculated in detail by Nie et al., Robertson et al., and Ueda et al.(Sheng et al., 2006)

2.1.1 $CuBO_2$

A study by Snure and Tiwari has identified a new group 13 delafossite, $CuBO_2$, as a new p-type TCO.(Snure & Tiwari, 2007) In this study, a density functional theory (DFT) study are measured, and examining the detailed electronic structure of $CuBO_2$.(David et al., 2009) This group show conclusively that (i) the lattice parameters reported by Snure and Tiwari are not consistent with previous experimental trends and need to be reinvestigated, (ii) the valence band features of $CuBO_2$ are consistent with other delafossite p-type TCOs, (iii) the effective hole masses of the valence band maximum(VBM) are consistent with the reported good conductivity, and (iv) the predicted indirect band gap and optical band gap of $CuBO_2$ are 3.21 eV and ~5.1 eV, respectively. The GGA + U calculated bandstructure of $CuBO_2$ along the high symmetry lines taken from Bradley and Cracknell is shown in Fig. 2. The VBM is situated at the F point, while the conduction band minimum(CBM) lies at Γ, giving an indirect band gap of 1.94 eV, with the smallest direct band gap situated at Γ and measuring 3.21 eV.

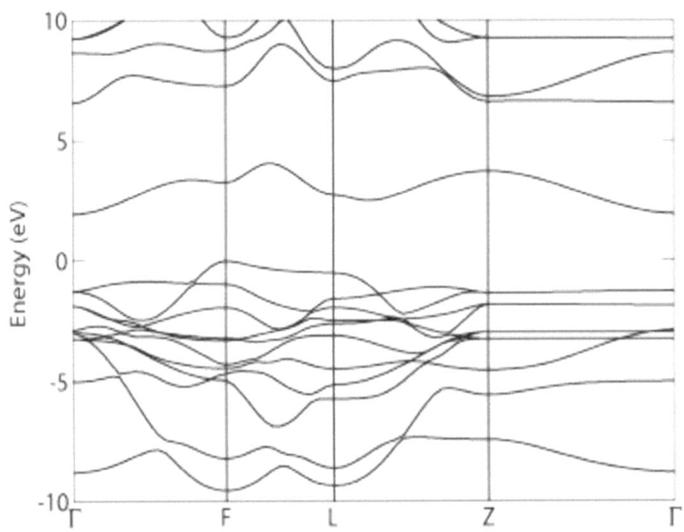

Fig. 2. GGA+U calculated bandstructure of $CuBO_2$. The top of the valence band is set to 0 eV.

To gain a deeper understanding of the band structure features, we have plotted projections of the wave function for the VBM at F and the CBM at the Γ point through a (001) plane containing Cu, B, and O atoms, labeled (b) and (c) in Fig. 3. (David et al., 2009) Anumerical breakdown of the states at the VBM shows that it contains ~67% Cu d character and ~31% oxygen p character, with B states effectively playing no role in the VBM makeup at F. This is further evidenced by the charge density plot of the VBM (Fig. 3b) which clearly shows d-like orbitals on the Cu ions and p-like orbitals on the O ions, with the absence of any density on the B states.

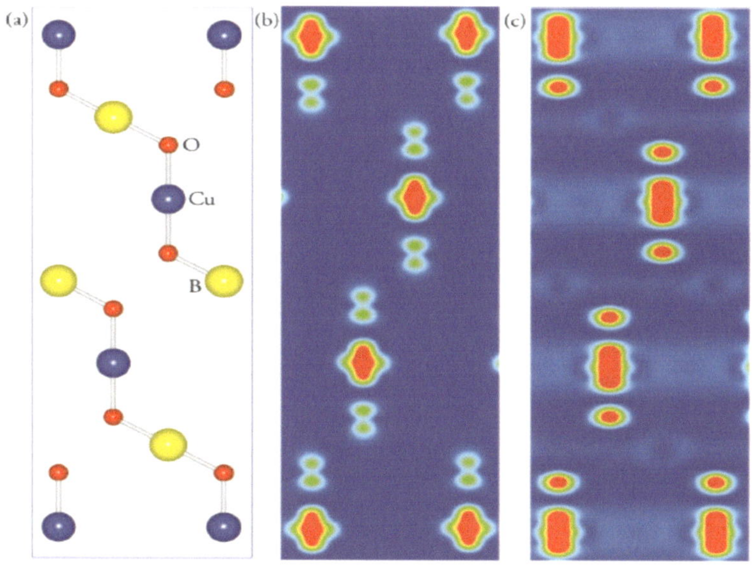

Fig. 3. Charge density contour plots showing the band edges of $CuBO_2$ through a (001) plane. (a) Structure of the cell in the (001) plane, (b) charge density of the VBM at F, and (c) charge density of the CBM at Γ, plotted from 0 eV(blue) to 0.003 eV(red) e·A^{-3}.

2.1.2 $CuAlO_2$

Kawazoe et al. first reported synthesis of $CuAlO_2$ das a p-type TCO material based on the copper I oxides such as $CuAlO_2$, $CuGaO_2$, and $SrCu_2O_2$ with chemical modulation of valence band.(Kwwazoe et al., 1997) In spite of several merits of $CuAlO_2$ as a p-type TCO, the main hurdle is its low electrical conductivity compared to the n-type TCO. Therefore, different methods such as high temperature solid-state reaction, hydrothermal method, ion exchanges, and sol-gel method etc. have been proposed to prepare $CuAlO_2$.(Ghosh et al., 2009) Each copper atom of $CuAlO_2$ is linearly coordinated with two oxygen atoms to form an O–Cu–O dumbbell unit placed parallel to the c-axis. Oxygen atoms of the O–Cu–O dumbbell link all Cu layers with the AlO_2 layers. After the report of p-type semiconducting, transparent $CuAlO_2$ thin film, a research field in device technology has emerged, called 'transparent electronics'. For the synthesis of $CuAlO_2$ thin films, the groups of Hosono, Gong, and Chattopadhyay used pulsed laser deposition, plasma-enhanced metalorganic chemical vapor deposition(PE-MOCVD), and dc sputtering, respectively.(Sheng et al., 2006) The electronic structures of $CuAlO_2$ were experimentally probed by normal/inverse photoemission spectroscopy(PES/IPES). The Fermi energy determined experimentally was set to zero in the energy scale in the three spectra, as shown in Fig. 4.

A band gap was observed between the valence band edge in the PES spectrum and the conduction band edge in the IPES spectrum. The band gap estimated was about 3.5 eV. The Fermi energy lies around the top of the valence band. The origin of the energy axis is the Fermi level which was determined using Au deposited on sample. These results means that

$CuAlO_2$ is a transparent p-type semiconducting material, which has excellent potential for use in optoelectronics device technology. Fig. 5 shows the emission current (I) versus macroscopic field(E) curve of $CuAlO_2$ thin films deposited on glass substrate as a function of the anode-sample separations(d).(Banerjee & Chattopadhyay, 2004)

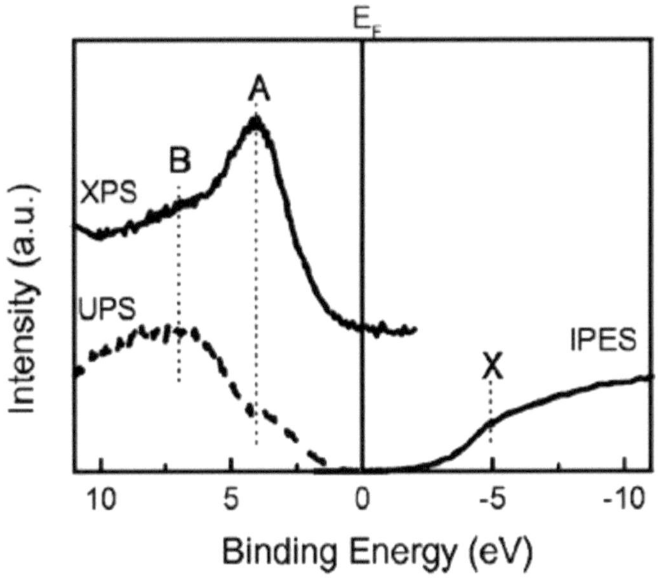

Fig. 4. Photoemission and inverse photoemission spectra of the valence and conduction region of $CuAlO_2$.

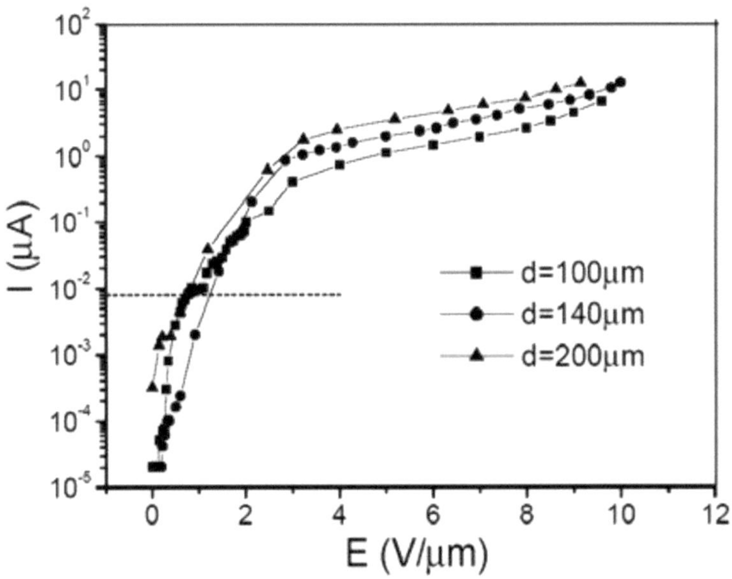

Fig. 5. Plot of emission current versus macroscopic field of $CuAlO_2$ thin film on glass substrate.

2.1.3 $CuGaO_2$

$CuGaO_2$ is a p-type TCO with E_g of ~3.6 eV. $CuGaO_2$ has a larger lattice constant of the a-axis, a = 2.98 A, than $CuAlO_2$ (a = 2.86 A). A polycrystalline thin film of $CuGaO_2$ prepared by rf sputtering method was obtained in an amorphous state by post-annealing for crystallization at 850 °C for 12 h under nitrogen atmosphere. And its activation energy was roughly estimated to be about 0.22 eV. The conductivity at room temperature was about 5.6×10^{-3} $S\cdot cm^{-1}$.(Sheng et al., 2006) The electrical conductivity, carrier (positive hole) density, and Hall mobility of epitaxial $CuGaO_2$ films at room temperature were respectively 6.3×10^{-2} $S\cdot cm^{-1}$, 1.7×10^{18} cm^3, and 0.23 $cm^2\cdot V^{-1}\cdot s^{-1}$, which were prepared on a-Al_2O_3(001) single-crystal substrates by PLD without post-annealing treatment and it were superior to those of the polycrystalline thin films of $CuGaO_2$. The films were grown epitaxially on the substrates in an as-deposited state. The films have high optical transparency (~80%) in the visible region, and the energy gap of $CuGaO_2$ for direct allowed transition was estimated to be 3.6 eV.(Ueda et al., 2001)

2.1.4 $CuInO_2$

The $CuInO_2$ system, which is copper-based delafossites, has been particularly interested because it can be doped both p-type (with Ca) and n-type (with Sn), allowing p–n homojunctions to be produced by applying doping of an appropriate impurity and tuning the deposition conditions.(Yanagi et al., 2001a, 2002b) On the other hands, no similar trend like $CuInO_2$ has ever been observed in any other semiconductors because it has the largest reported band gap of 3.9 eV. The conductivity of $CuInO_2$ films reported by Yangagi et al. shows about ~10^{-3} S cm^{-1} when it was grown by PLD from phase-pure $CuInO_2$ targets.(Yanagi et al., 2001) It has been smaller than that of the other p-type TCOs. However, Teplin et al. used $Cu_2In_2O_5$ as a target to deposit single phase undoped and Ca-doped $CuInO_2$ thin films because the oxygen-rich $Cu_2In_2O_5$ phase of Cu–In–O is easily prepared by solid-state synthesis in air.(Teplin et al., 2004)

2.1.5 $SrCu_2O_2$

$SrCu_2O_2$ (SCO) is one of the few non-delafossite p-type oxides. In some groups, the properties of SCO deposited under working oxygen pressure of 5.25×10^{-3} mTorr and 300 °C are measured. Roy et al. reported the performance of SCO films with copper formate ($Cu(CH_3COO)_2$) and strontium acetate ($Sr(CH_3COO)_2$), which were chosen as starting precursors.(Roy et al., 2003) Stoichiometric amounts of Copper formate and Strontium acetate were prepared as separate aqueous solutions and mixed just before application (otherwise copper acetate precipitated) and were sprayed on quartz substrates at ~250 °C. Non-doped and K-doped p-type SCO thin films were first deposited by Kudo et al.(Kudo et al., 1998) And the PLD technique has been used to fabricate Ca-doped SCO thin films on quartz glass substrates. In case of the synthesis of SCO, a p-type transparent conducting oxide by a chemical solution route as well as the conventional PLD method, for SCO by the chemical solution route, samples were made by spraying deposition on quartz substrates using an aqueous solution of copper formate and strontium acetate. The X-ray diffraction(XRD) spectra of as-deposited thin-film samples for different substrate temperatures had only single (202) peak of SCO films.(Sheng et al., 2006) In this case, as increase the deposition temperature, the intensity of the (202) peak was increased. And also, the transmittance of SCO films is observed to depend on deposition temperature. On increasing the deposition temperature, the average transmission was decreased. The $(ah\nu)^2$

versus *hv* plot is shown, from which the direct allowed band gap can be estimated; it was found to be about 3.2 eV for films deposited at 350 °C. The conductivity of the film deposited in room temperature was obtained as 8.2×10^{-2} S cm^{-1}, which is decreased as 5.4×10^{-2} S cm^{-1} deposited at 400 °C.(Roy et al., 2003)

2.2 n-type TCO

While the development of new TCO materials is mostly dictated by the requirements of specific applications, low resistivity and low optical absorption are always significant pre-requisites. There are basically two strategies in managing the task of developing advanced TCOs that could satisfy the requirements. The main strategy dopes known binary TCOs with other elements, which can increase the density of conducting electrons. More than 20 different doped binary TCOs were produced and characterized by n-type TCOs , of which ITO was preferred, while Al-doped ZnO (AZO) and Ga or Ga_2O_3-doped ZnO (GZO) come close to it in their electrical and optical performance. The TCOs have been intensively studied for their potential in optoelectronic applications, including for the manufacture of OLEDs. It has been well known that ITO is the most popular TCO, because of its high conductivity and transparency.(Kim et al., 1996) However, its chemical instability, toxic nature and high cost, combined with the diffusion of indium into surrounding organic materials, have stimulated efforts to find an alternative.(Kim et al., 2000)

Among these materials, one of most promising candidates is AZO which has sufficiently high conductivity and a transmittance of over 90% in the visible range, even in samples grown at room temperature.(Chem et al., 2000) Recently, it has been reported that optimized AZO films could replace ITO as an anode material in OLED applications. In comparison with ITO, AZO films are more stable in reducing ambient circumstance, more readily available, and less expensive. Because of these characteristics, AZO is often used as an anode material in photoelectronic devices such as solar cells, flat panel displays and OLEDs. (Yang et al., 1998; Ott & Chang, 1999; Wang et al., 2006; Deng et al., 1999) Previously, we reported that the AZO can be successfully adopted as a TCO on a flexible polymer polyethersulphone (PES) substrate for flexible OLED application.(Park et al., 2007) Because the anode materials in an OLED are in contact with organic molecules, both their surface chemical properties and their morphology affect the adhesion and alignment of molecules on the surface.(Guo et al., 2005)

Therefore, a microscopic understanding of wettability in solid surfaces is fundamentally interesting and practically valuable. Furthermore, the absorption of water on metal-oxide surfaces is an important subject in its own right, due to its crucial role in gas sensors, catalysis, photochemistry and electrochemistry. It is well known that the measurement of water contact angle (WCA) could reveal much useful information about characteristics of surface nanostructure and morphology.(Soeno et al., 2004) However, there have been few studies of the surface wettability of the transparent conducting oxide AZO thin film. AZO films with resistivity of ~$8.5.10^{-5}$ W·cm was reported by Agura et al.(Agura et al., 2003) An even lower resistivity was reported for GZO, ~8.1×10^{-5} W·cm.(Park et al. 2006) This *r* is very close to the lowest resistivity of ITO of 7.7×10^{-5} W·cm, with a free carrier density of 2.5×10^{21} cm^{-3}.(Ohta et al., 2000) The phase-segregated two-binary systems include ZnO-SnO_2, CdO-SnO_2, and ZnO-In_2O_3. In spite of the expectations, the electrical and optical properties of the two-binary TCOs were much inferior to those of ITO. Accordingly, the ternary TCO compounds could be formed by combining ZnO, CdO, SnO_2, $InO_{1.5}$ and $GaO_{1.5}$ to obtain Zn_2SnO_4, $ZnSnO_3$, $CdSnO_4$, $ZnGa_2O_4$, $GaInO_3$, $Zn_2In_2O_5$, $Zn_3In_2O_6$, and $Zn_4In_2O_7$.

However, since Cd and its compounds are highly toxic, the utilization of these TCOs is limited, though they have adequate electrical and optical properties. Other binary TCOs were synthesized from known binary TCOs and also from non-TCO compounds, such as In_6WO_{12} and the p-type $CuAlO_2$. All of the TCOs discussed above are n-type TCOs. In addition, p-type doped TCOs were also developed and could find interesting future applications, in particular as a new optoelectronic field like "transparent electronics". (Banerjee & Chattopadhyay, 2005)
The need to produce n-type TCOs with higher conductivity and better transmission, without relying on In, gave rise to research and development effort for new TCOs. Recently, mobility with more than twice that of commercial ITO was achieved in Mo-doped In_2O_3(IMO), and this material showed that the conductivity can be significantly increased with no changes in the optical transmittance upon doping of Mo.(Meng et al., 2001; Yoshida et al., 2004) Electronic band structure of IMO was investigated by Medvedeva, it was revealed that the magnetic interactions which had never been considered to play a role in combining optical transparency with electrical conductivity ensure both high carrier mobility and high optical transmittance in the visible range.(Medvedeva, 2006)
Recently, new thin film geometries were also explored by Dingle *et al.* in search of TCO films with higher conductivity.(Dingle et al., 1978) They showed that higher conductivity could be obtained by doping modulation, which spatially separates the conduction electrons and their parent impurity atoms (ions) and thereby reduced the effect of ionized and impurity scattering on the electron motion. Rauf used a zone confining process to deposit ITO with r = $4.4x10^{-5}$ W·cm and m = 10^3 cm^2/Vs.(Rauf, 1993) The highly and lowly doped regions were laterally arranged in the films, rather than vertically as in superlattice structures. A theoretical outline of a method to engineer high mobility TCOs was presented by Robbins and Wolden, based on the high mobility transistor structure discovered accidentally by Tuttle *et al.*(Robin & Wolden, 2003) The film should consist of alternating thin layers of two semiconductors. One layer provides a high density of carriers, while the second is a high mobility material. Electrons are supplied by the former and transported in the latter, mitigating the limitations of ionized impurity scattering.(Tuttle et al., 1989) The model of Robbins and Wolden assumes that the electrons move into the high mobility material in response to differences in electron affinity. However, the success of the proposed TCO design depends upon controlling the layer thickness at nano dimensions, (e.g. ~5 nm). In addition, this approach depends on having materials of excellent quality and compatible crystal structure in order to avoid problems related to interface defects. TCO materials with magnetic properties, which are ferromagnetic semiconductors with a Curie temperature well above room temperature, have also been explored recently, as they could be used for second generation spin electronics and as transparent ferromagnets. Ueda *et al.* reported that Co doped ZnO thin film ($Zn_{1-x}Co_xO$) with x = 0.05 - 0.25, had a large magnetic moment of 1.8mB per Co ion for x = 0.05. High-temperature ferromagnetism was subsequently found by other groups, with varying magnetic moments.(Ueda et al., 2001)

2.3 Indium-based TCOs

In fabricating OLED devices, ITO film among the transparent conductive oxide (TCO) films is widely used as an anode layer, because of its high transparency in the visible light range, low conductivity, and high work function (~ 4.8 eV). In majority of cases, a thin layer of a mixed ITO made of 9~10 mol % of tin oxide in indium oxide on a transparent substrate is used. However, conducting oxides such as pure tin oxide, Ga-In-Sn-O (GITO, 5.4 eV), Zn-In-Sn-O

(ZITO, 6.1 eV), Ga-In-O (GIO, 5.2 eV), and Zn-In-O (ZIO, 5.2 eV) films composed with In, Sn, Ga, Zn, and O components have particularly interesting transparency and conducting properties. They possess better characteristics than ITO such as a lager work function. Conducting polymer, TiN, and semitransparent at thicknesses that are suitable due to high conductivity as an electrodes. Besides, the FOLEDs has led to the utilization of the ITO or the organic conductors, such as; polyaniline(PANI) deposited on various plastic substrates of polyethylene terephthalate(PET), polyethylene naphthalate(PEN), polyimide(PI), polycarbonate(PC), polypropylene adipate, and acrylic polymer. The following paragraphs explain in detail the solutions found in the literature for realizing the anode. Among the many factors determining the performance of OLED devices, the interface between the organic hole transport layer (HTL) and the anode layer plays an important role in controlling the efficiency of the charge carrier injection into the emitting layer.(Li et al., 2005; Chan & Hong, 2004) The insertion of various thin insulating films, such as WO_3, NiO, SiO_2, ZrO_2, Ta_2O_5, and TiO_2, between the ITO anode and the HTL layer, was found to improve the performance of the OLEDs, which was explained by the energy level alignment or tunneling effect.(Qiu et al., 2003; Huang, 2003; Mitsui & Masumo, 2003; Lu & Yokoyama, 2003; Ishii et al., 1999; VanSlyke et al., 1996) In addition, the modification of the work function of the ITO surface was reported by doping it with Hf atoms using a co-sputtering technique or inserting a conducting oxide layer, i.e., IrO_x, which increases the work function of the ITO surface through dipole formation.(Chen et al., 2004; Kim & Lee, 2005)

Although many thin-film deposition techniques, such as sputtering or chemical vapor deposition, have been used to obtain an ultra-thin interfacial layer between the HTL and anode, these methods are not suitable for obtaining a high quality ultra-thin interfacial layer with a sub-nm range thickness. Recently, atomic layer chemical vapor deposition (ALCVD) has been widely used in many application areas which require precise thickness controllability and low structural defects, because the ALCVD process is based on surface adsorption- and saturation-controlled deposition kinetics. This results shows the effects of ALCVD treatment performed both at room temperature (RT) and at various temperatures up to the typical HfO_2 deposition temperature (300 °C) using tetrakis(ethylmethylamino) hafnium (TEMAH; $Hf[N(CH_3)C_2H_5]_4$) as the precursor on OLEDs. The binding and molecular structures of the HfO_x layer formed on the ITO surface were analyzed by X-ray photoelectron spectroscopy (XPS) and near-edge X-ray absorption fine structure (NEXAFS) spectroscopy.

Compared to the control sample without any treatment, the sample treated for 5 cycles at RT, which was referred to as RT-5C, exhibited significantly improved OLED performance, i.e., a decrease in turn-on voltage as depicted in the inset of Fig. 6(a) and an increase in brightness.

Because of the high current flow and subsequent increase in brightness mainly originating from the increased hole injection efficiency from the ITO anode into the organic layer, we found that the 5 cycle treatment at RT by the ALCVD process is an effective method of improving hole injection efficiency.(Shrotriya & Yang, 2005) However, when the number of ALCVD cycles was increased at RT, the turn-on voltage increased and the brightness decreased as compared to the control sample as shown in Fig. 6. This is believed to be caused by the formation of an insulating layer and the consequent retardation of the hole injection. With the increase of the deposition temperature, the electrical and optical characteristics of the OLEDs showed similar or worse performance when compared to the control sample as a function of deposition cycles as shown in Figs. 7 and 8.

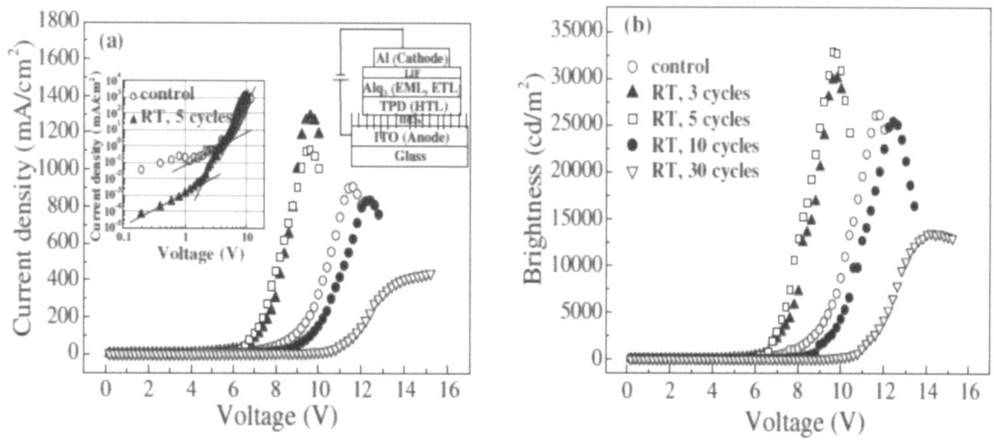

Fig. 6. (a) J–V, (b) B–V characteristics of OLEDs treated without and with ALCVD-HfO_x at room temperature as a function of deposition cycles.

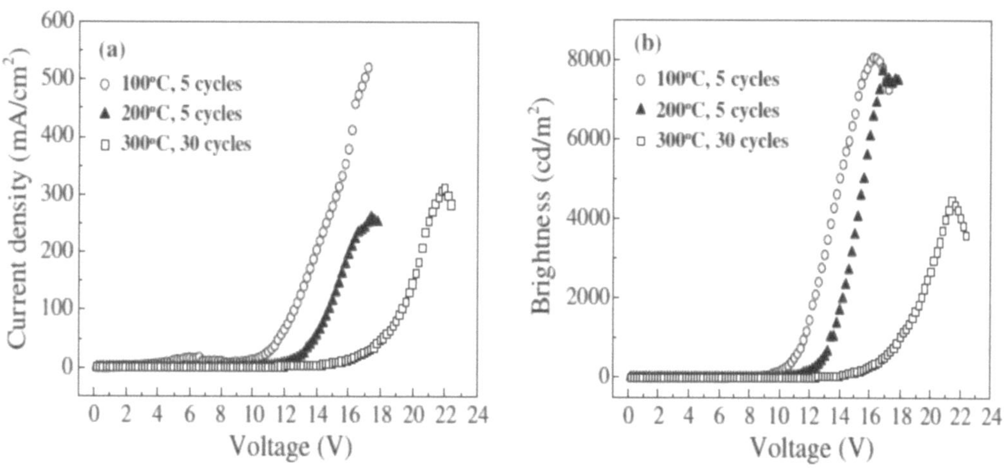

Fig. 7. (a) J–V, (b) B–V, and (c) power efficiency characteristics of OLEDs treated with ALCVD-HfOx at 100, 200, and 300 °C as a function of deposition cycles.

The Hf precursor formed Hf–O bonding having a high dipole moment with the underlying ITO network, and the subsequent bias-induced realignment of the anode Fermi level and highest occupied molecular orbital (HOMO) of the HTL lowered the band offset with the top HTL by modifying the work function of the ITO surface, as shown in Fig. 9.
However, the existence of the ultra-thin HfO_x layer did not retard the hole injection from the anode due to the tunneling effect. When the number of cycles or deposition temperature was increased, the peaks caused by the unoccupied hybridized orbitals of Hf and O appeared in the lower photon energy range, which confirmed the formation of an electrically insulating HfO_x layer. The O *K* edge NEXAFS spectra of the 300 °C-30 cycles sample were directly related to the oxygen p-projected density of states of ITO overlapped with that of HfO_2, which consists of the four unoccupied hybridized orbitals, Hf 5*d*+O 2*pπ*, Hf 5*d*+O 2*pσ*, Hf 6*s*+O 2*p*, and Hf 6*p*+O 2*p* of the HfO_2 film.(Cho et al., 2004) The formation

of a physically thick insulating HfO_x layer between the anode and HTL, as confirmed by the NEXAFS measurement, significantly deteriorated the OLED performance, as previously shown in Figs. 6 and 7.

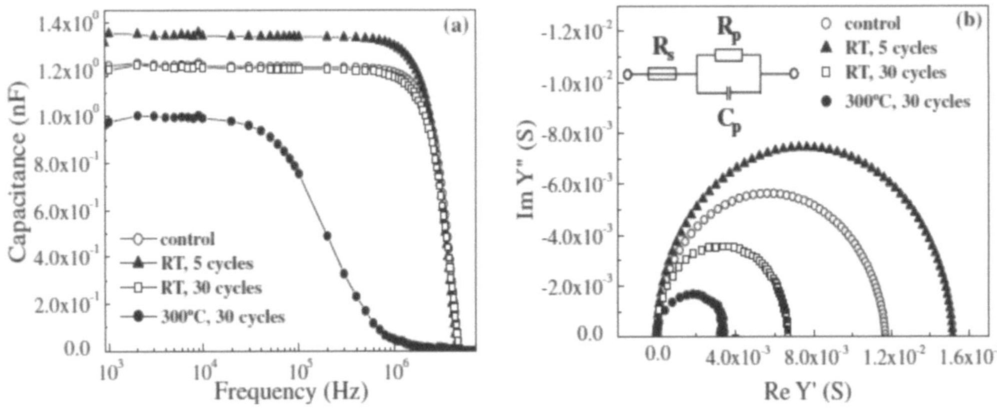

Fig. 8. (a) C–f characteristics and (b) Cole–Cole plots of OLEDs without and with surface treatment.

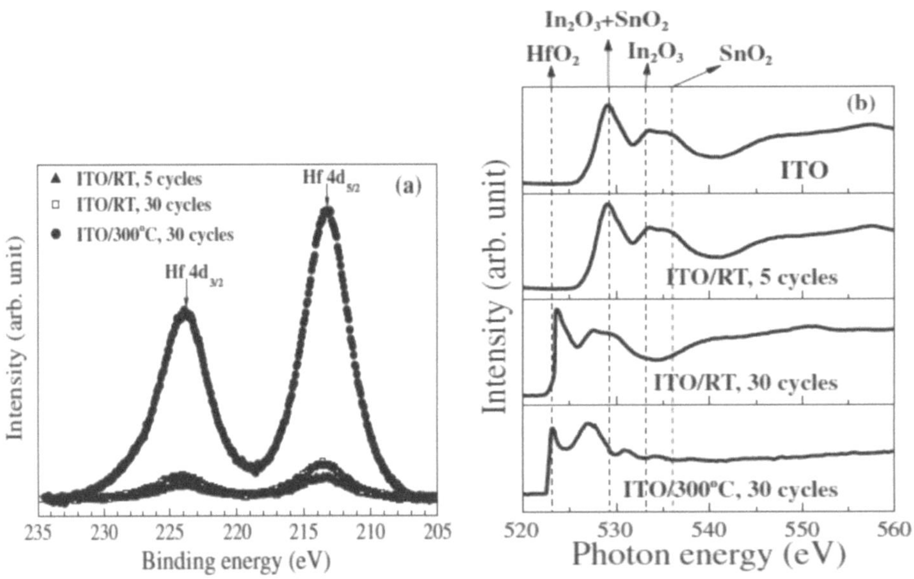

Fig. 9. (a) Hf 4d peak in the XPS peak spectra and (b) NEXAFS spectra of the O K edge features of the ITO surface as a function of TEMAH deposition conditions.

In order to measure the relative work function for a pristine ITO anode and ITO anodes modified by the different surface treatments of HfC_x, HfO_x and HfO_2 for 5 cycles, we carried out Kelvin probe current measurements as a function of the substrate bias by using the Kelvin probe microscopy(KPM) system in UHV conditions, as shown in Fig. 10.(Sohn, 2008) The applied bias(V_{App}) of the surface potential di_erence between the Kelvin probe tip and the pristine ITO substrate (V_{ITO}) under different HfC_x, HfO_x and HfO_2 treatment conditions were, respectively, shifted to -0.2, 0.4, -0.9 V, which resulted in improved or deteriorated device properties. The increased work function of the HfO_x-treated ITO anode reduced the barrier height for hole carrier injection in OLEDs compared to that of the HfC_x treatment without an oxidant or the HfO_2 treatment with a high deposition temperature.

And also, Sugiyama et al. suggested the three factors such as C-containing contaminants, the O/In ratio, and the In/Sn ratio for the increase of the ITO work function. In order to be utilized as excellent anode in OLEDs, however, the ITO film has to solve some problems such as formation of a defect region by diffusion of oxygen or In metal into the organic material layer, low transparency in the blue light range, and discord of the energy level alignment by difference between the ITO work function and HOMO level of a typical HTL. In order to increase the work function of ITO, a number of investigations were reported, such as the surface treatments under O_2, N_2, H_2, and N_2-H_2 condition, or the insertion of an anode interfacial layer with insulating wide band gap between the HTL and the anode.

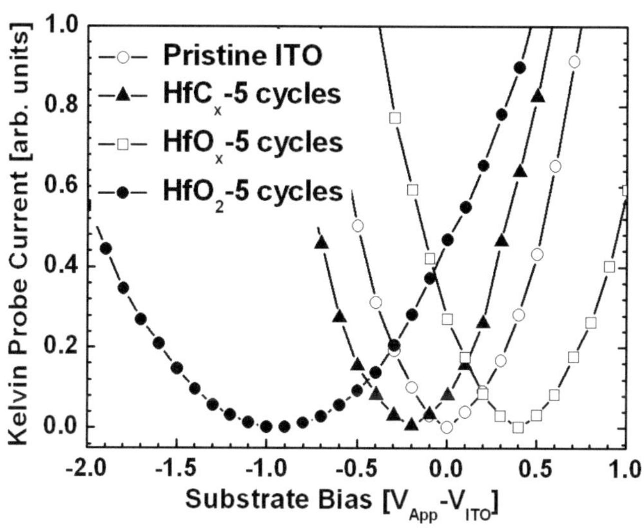

Fig. 10. Kelvin probe current for a pristine ITO anode and an ITO anode modi_ed by HfC_x, HfO_x and HfO_2 surface treatment for five cycles as a function of substrate bias. The applied bias (V_{App}) has been shifted by the surface potential difference between the Kelvin probes tip and the pristine ITO substrate (V_{ITO}).

The development of the TCO films such as Ga-In-Sn-O (GITO, 5.4 eV), Zn-In-Sn-O (ZITO, 6.1 eV), Ga-In-O (GIO, 5.2 eV), and Zn-In-O (ZIO, 5.2 eV) TCO films composed with In, Sn, Ga, Zn, and O materials. Especially, the oxygen plasma or UV ozone treatments on ITO surface can increase the work function of ITO and remove the carbon contamination of ITO surface. However, the improvement by oxygen plasma, widely used in OLEDs, is strongly dependent on processing conditions. Recently, Hung *et al.* reported that the polymerized fluorocarbon film formed on ITO surface can improve the charge carrier injection because it has a high ionization potential and relatively low resistance. The OLEDs with fluorocarbon/oxygen mixture showed the improved device performance with enhancing the holes injection by remove the carbon contamination on ITO surface and also accelerate the fluorine bonding directly to indium or tin on the ITO surface.

In order to determine the effect of the CF_x treatment, the conductance, capacitance, and impedance were respectively measured for the devices with and without the CF_x treatment in the frequency range of 10 Hz to 10 MHz for a zero bias voltage. In the low frequency region, the CF_x treated ITO anode had a higher capacitance than the device with the untreated ITO anode, which is related to the enhancement of carrier injection and space charge formed by the injected carriers.(Kim et al., 2008)

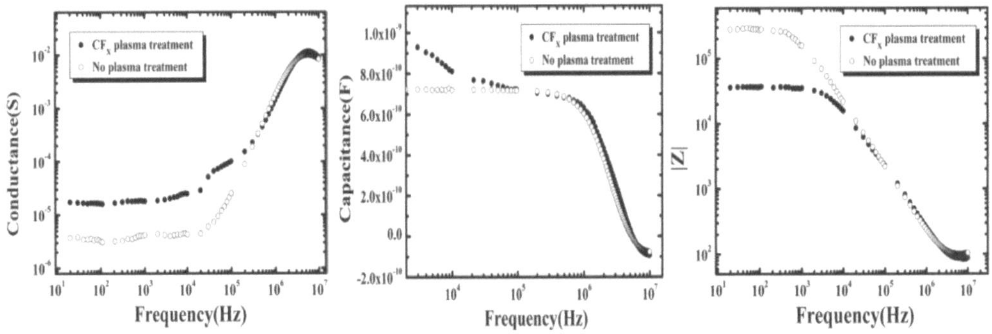

Fig. 11. Variation in conductance, capacitance, and impedance as a function of frequency in the device with and without the CFx plasma treatment.

2.4 Zinc-based TCOs without indium

The ITO is mostly used as a promising candidate material for TCO films due to many advantages, such as high conductivity (~10^{-4} Ω·cm), high transmittance (~85%) in visible light range, high uniformity, and high work-function (~4.8 eV).(Minami, 1999; Miyata, 1997; Shan, 2003; Yan, 1998) However, it was found that they have often been limited in their application because of the frequent necessity to optimize electrical, optical, and chemical properties for specialized applications. For example, the conventional ITO films have some substantial problem such relatively high deposition process (>300 °C) to get a low resistivity, drop of optical transmittance under H_2 plasma condition, and rising price due to indium exhaustion within a few years.(Han, 2001; Hirata, 1996; Honda, 1995; Minami, 1984; Park, 2006a, 2006b) The amorphous ITO film deposited at a low temperature has a low resistance to moist heat, which leads to a degradation in its conductivity and the light transmittance with time. Moreover, the chemical and electronic properties of ITO are far from optimum for current and future generation OLEDs. Drawbacks include deleterious diffusion of oxygen and In into proximate organic charge transporting/emissive layer, imperfect work function alignment with respect to typical HTL, HOMO level, and poor transparency in the blue region. For the purpose of improving TCO film properties, new materials consisting of ternary compound oxides based on ZnO were investigated. For example, In-doped ZnO (IZO), Al-doped ZnO (AZO), Ti-doped ZnO (TZO), and Si-doped ZnO (SZO) have been attracted, which are considerable attention as an alternative materials for ITO. Recently, zinc oxide or impurity (B, Al, Ga, In, and Zr) doped zinc oxide films have been investigated as alternate materials to ITO for OLEDs because zinc oxide is nontoxic, inexpensive and abundant. It is also chemically stable under exposure to hydrogen plasma that is commonly used for the fabrication of thin film transistor-liquid crystal display (TFT-LCD). Kim et al., investigated the Zr-doped ZnO (ZZO) thin film grown by PLD on glass substrates as a function of oxygen deposition pressure and film growth temperature for OLEDs.(Kim et al., 2003) For a 200-nm-thick ZZO film grown at 250°C in 1 mTorr, a resistivity of $5.6X10^{-4}$ Ω·cm and optical transmittance of 84% were measured. These results demonstrate that ZZO is a good anode material because the OLEDs fabricated on ZZO anodes exhibit external EL quantum efficiency comparable to a control device fabricated on commercial ITO.

Further increase of the good performance at ZnO based TCO films can be achieved through improving crystallinity by preparing single crystal or hetero-epitaxial ZnO films and/or increasing grain size in film by the post-annealing method. And the improvement of the electron mobility can be obtained by new composition materials by the addition of impurity dopants, such as Al, Ga, In, Ti and so on. Recently, TiO_2 has become the subject of many investigations for applications in optical coatings because of their good properties such as a high refractive index, high transparency, excellent water resistance, and thermal stability. However, since the conventional RF-magnetron sputter (RFS) system for TCO film deposition has consist with a system of the target and the substrate facing with each other, the particles with high energy such as γ-electrons, neutral Ar particles, and negative oxygen ions collide with the substrate. In this study, facing target sputtering (FTS) apparatus was designed to enhance the preciseness of manufactured thin film and the sputter yield rate with depositing film by forming higher density plasma in the electrical discharge space.(Kim, 2001; Noda, 1999, Nose, 1999) TiO_2-doped ZnO films, in comparison with the ZnO films doped with Group III elements, have more than one charge valence state. In this study, the electrical and optical properties of TiO_2-doped zinc oxide (TZO) films with various deposition thicknesses by FTS system were compared to those of the films made by conventional RFS method. For more details, the relations in the resistivity, carrier concentration and mobility, film density, and intrinsic stress in the films as a function of the deposition method with the FTS and conventional RFS system were analyzed.

The TZO films were deposited on slide glass substrates at RT by FTS and RFS methods, respectively. Target materials were made up of TiO_2 and ZnO powders with purity of 99.999 % that were calcined at 1000 °C in Ar atmosphere for 2 hours. The mixture with composition ratios was prepared for the target the composition ratios were selected as TiO_2 : ZnO = 2 : 98 weight percent (wt.%), and we will refer to the films deposited with the target as the TZO film. For FTS system, two circular targets with a size of 3 inch are located horizontally facing with each other, and more detail FTS structure was explained at previous report.(Kim, 2009) The applied RF-powers were respectively 120 W and 80 W for the film deposition using FTS and RFS system, the working pressure was set at 2×10^{-3} torr, and a pure Ar gas was used as discharge gas. Two circular targets with a size of 3 inch are located horizontally facing with each other, and Nd alloy permanent magnets of 4700 Gauss for plasma confining magnetic field was mounted to the back of the target, which was adjusted by variation of the distance between both two targets. In order to control the heat of the system caused by the ion bombardment of the cathode, cooling water was supplied. We investigated the process characteristics of the FTS apparatus under various deposition thicknesses compared with the film by RFS system. FTS system is a high-speed and low deposition temperature method, which arrays two sheets of targets facing each other. Inserts plasma is arresting magnetic field to the parallel direction of the center axis of both targets, discharged from targets and accelerated at the cathode falling area. Thus, this system is a plasma-free sputter method in which substrate is located at far from plasma. And also, the temperature on substrate during film deposition was much lower than that of the conventional sputtering method. And also,

the prepared films using FTS system as a function of the distance from center to edge has the uniform thickness. An ultra violet visible spectrophotometer (UV-VIS, Shimadzu Co.) was used to analyze the optical properties of the film such as transmittance and optical energy bandgap (E_{opt}). Crystallographic properties of the TZO films were analyzed by X-ray diffraction (XRD, Rikagu Co.) patterns by using the *Cu-Ka* (λ= 1:54Å) line. Surface morphology of the film was observed by a scanning electron microscopy (SEM, Hitachi Ltd.) and an atomic force microscope (AFM, Veeco Instruments Inc.), and the film thickness was measured by *a*-step. Electrical resistances and hall mobility of the films were measured by the Hall effect measurement system (HEM-2000, EGK Co.) using Van der paw method.
Fig. 12 shows the optical transmittance spectra of various TZO films prepared by respectively FTS and RFS system as a function of the film thickness. Under the same film thickness, the oscillation peaks by maximum and minimum points using a distributed Bragg reflector show a similar tendency. TZO thin films prepared by conventional sputtering and FTS method showed similar optical transmittance over 80 % in visible light range with baseline of glass substrate, which can applied in various optoelectronics like next generation FPDs, touch panel, and so on. The absorption edges of TZO films deposited by FTS method have been blue shifted compared with the film prepared by RFS method at same film thickness. It means that the optical band gaps were increased as shown in the inset of Fig. 12, which is attributed to Burstein-Mott effect due to the increase of carrier concentration by film density. In insertion of Fig. 12, the optical band gap E_{opt} of the TCO films were calculated by the Tauc's relation(Chowdhury, 2000; Tauc, 1974)

$$(ah\nu) = B(h\nu - E_{opt})^n \quad (2)$$

, where *a* is the absorption coefficient, is the energy of absorbed light, is the parameter connected with distribution of the density of states and *B* is the proportionality factor.
The TZO films by FTS system with various deposition thicknesses show the higher E_{opt} values than that of the films by RFS system, which is well correspond to the improvement of resistivity due to increase of carrier concentration of the films in Fig. 13. The ΔE_{opt} as the increase of optical bandgap by Burnstine Moss effect was as below:

$$\Delta E_{opt} = (\hbar^2/2m^*) \cdot (3\pi^2)^{2/3} \cdot N^{2/3} \quad (3)$$

, where $\hbar$ is Planck constant and m^* is effective mass.
Thus, the carrier concentration(N) is also increased when the optical bandgap is increased. The E_{opt} was increased from 3.4 to 3.5 eV at 100 nm film thickness when the TZO film was deposited by FTS system compared to those of the film prepared by the RFS system. The widening of the energy band gap with the TZO film could be due to the increase in the carrier concentration.
Fig. 13 shows the resistivity (left) and the carrier mobility (right) of the TZO films deposited by FTS and RFS system using Hall effect measurement. Usually, the resistivity (ρ) of film is in inverse proportion to film thickness, and relational expression is given by the equation:

$$\rho = R_s / t \quad (4)$$

, where R_s is the sheet resistance and t is the film thickness.
As shown in Fig. 13, the resistivity of TZO films is related to the carrier concentration and the Hall mobility. This indicates that the electrical conductivity of TZO films is due to the contribution from Ti^{4+} ions in the substitution sites of Zn^{2+} ions, interstitial atoms, and

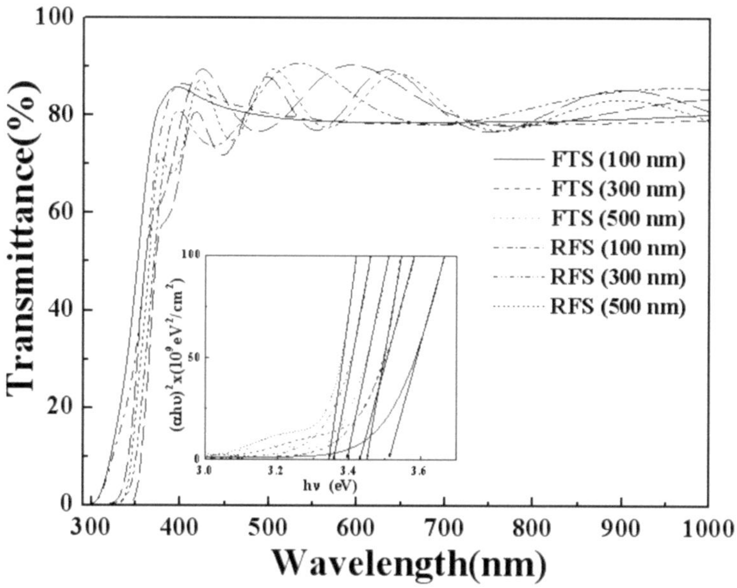

Fig. 12. Optical transmittance of TZO thin films deposited with deposition thickness of 500 nm on PEN substrate under various rf-power. The inset shows a plot of *ahν* vs. *hν* calculated from the optical transmittance spectra.

oxygen vacancies.(Chung, 2008) Finally, because the TCO film having both maximum conductivity and carrier mobility has a film density close to its theoretical density, it means that the electrical characteristics of the TCO films discussed here depend strongly on the grain size. More detail explanations will be discussed in Fig. 14 and 15.

However, the mean free paths were smaller than the grain size when the TCO films with the density showed lower than theoretical (not shown here). It is known that electron scattering at pores and voids within the grain is the major obstacle for electron conduction in the TZO films having a lower density. We thought that the scattering of the conduction electrons at the grain boundary during film deposition may be the major factor in determining the carrier mobility in TZO films. The electrical resistivity of TZO film deposited by FTS system showed about 5.0×10^{-4} Ω·cm, which was lower than that of the film made by conventional sputtering method with about 7.5×10^{-4} Ω·cm at 500 nm film thickness. We thought that the enhanced property of the film by FTS method was caused by the influence of the film density and/or the mean free path because the FTS used in this study is a high speed and low temperature sputter method that promotes ionization of sputter gas by screw-moving high-speed γ-electrons which array two sheets of targets facing each other. The generated plasma was arrested magnetic field to the parallel direction of the center axis of both targets, discharged from targets and accelerated at the cathode falling area. Therefore, the application parts of the FTS system will be extend because the FTS is a plasma-free sputter method in which the substrate is located apart from plasma.

Fig. 14 shows the XRD spectra of TZO films deposited by the FTS and the RFS with various film thicknesses. As increasing deposition thickness, the RF-sputtered films show the hexagonal wurtzite structure and has strong ZnO(002) peak of preferred orientation, together with relatively weak ZnO(103) peak.(Choi, 2005) It is notable that the intensity of ZnO(002) peak for the TZO films by the RFS system slightly increased with increasing deposition thickness. On the other hand, the intensity of the TZO films by the FTS system shows the weak (103) peak. It could be attributed to Ti atoms in TZO films. In Fig. 13, the resistivity of TZO films by both FTS and RFS system are significantly decreased while the deposition thicknesses are increase from 100 nm to 300 nm, however, and the value was almost saturated at 500 nm thickness. For the TZO film with optimum properties, we suggest that the crystallinity between (002) and (103) peaks was almost same in case of the film with 300 nm thickness as shown in Fig. 14(a). For more detail, the grain sizes of the films are calculated using Scherrer formula from XRD spectra as below:(Mardare, 2000)

$$D=(0.9\lambda)/(B\cos\theta) \quad (5)$$

where D is grain size, X-ray wavelength(λ) is using the Cu-Kα line(1.5405 Å), B is the full with at half maximum (FWHM) of (002) and (103) peaks, and θ is diffraction angle.

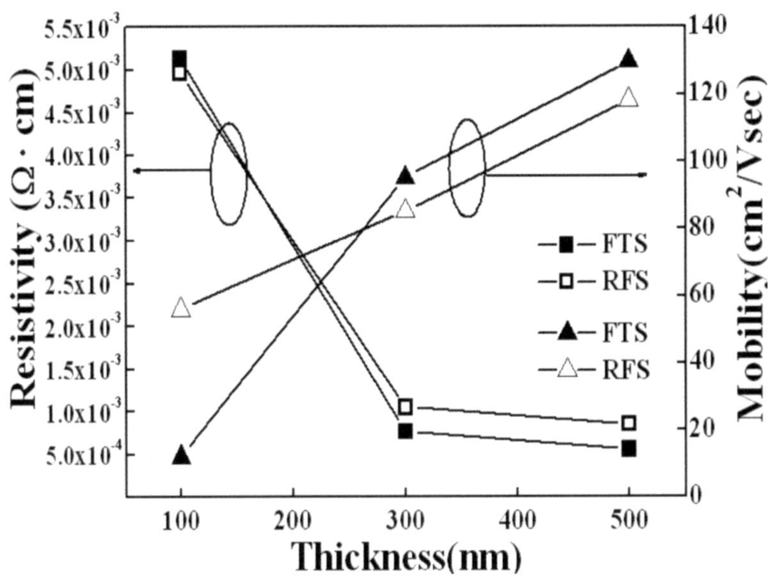

Fig. 13. Resistivity (left) and carrier mobility (right) of the TZO thin films as a function of the film thickness on glass substrate.

From the formula (4), the measured grain sized was varied from 105 nm to 155 nm. Thus, we thought that the film density was also improved as increase of the crystallinity as a function of the film thickness. The enhancement of the crystallinity and density in the TZO film can influenced on the conductivity of the film. The mean free path of the carrier as increase of the film density was also increased, which was resulted in increase hall mobility, as shown in Fig. 13.(Li et al., 2009)

Fig. 15 shows the SEM images of the TZO films with various deposition thickness using the FTS (a-c) and RFS (d-f) system. As increase film thickness, the grain sizes at both systems were proportionally increased. The film density was significantly improved while the deposition thicknesses are increase from 100 nm to 300 nm. However it was deteriorated at 500 nm thickness at both systems. The results are well agreed with the electrical properties in Fig. 13. The grain shapes of the TZO films by FTS system looks like the horizontal growth of (103) plane in Fig. 15(a)-(c). And also, the TZO films by RFS system in Fig. 15(d)-(f) looks

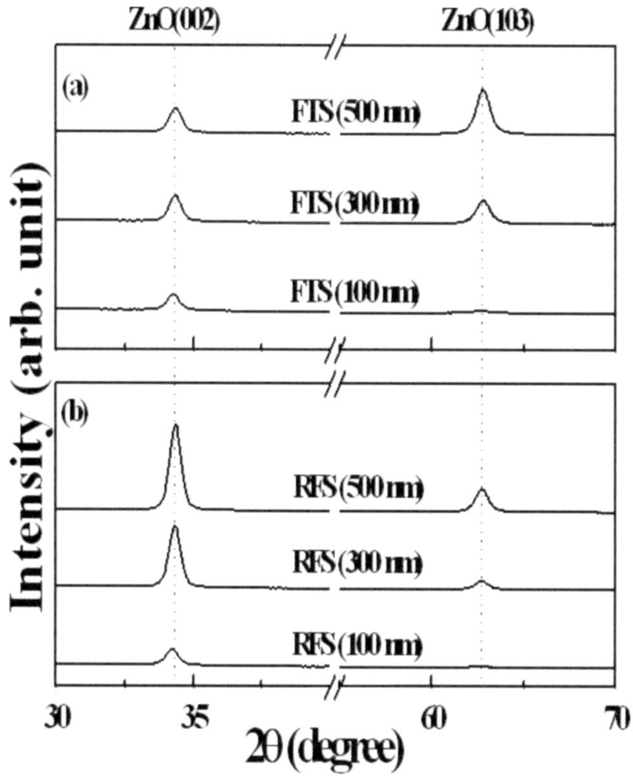

Fig. 14. X-ray diffraction patterns of the TZO thin films with thickness of 200 nm on PEN substrate under various rf-power.

like the shapes grown in vertical direction like (002) plane, as explained in Fig. 14. Generally, the thin films were condensed by vapor and atoms or molecular when the materials were deposited. At that time, some grains by the mobility decided as a function of the deposition methods and conditions were diffused, and then were moved in horizontal direction on the film surface.

However, if the applied energy for film deposition was low, the atoms with low mobility can't easily move in horizontal directions because it was frozen when the atoms were reached on the substrate. Therefore, the TZO films by FTS system shows relatively uniform surface morphology and the dense structures as shown in Fig. 16(a)-(c) while the films by the RFS system have open structures and rough surfaces contain with some pores between the grains in Fig. 16(d)-(f). The TZO film prepared by FTS system can reduce the damage on the films due to decrease the bombardment of high-energy particles such as gamma-electron. Thus, the surface roughness of the films by FTS system shows lower than that of the film by RFS system.

Among the films by FTS and RFS system, we suggest that the TZO film with 300 nm thickness prepared by FTS method can possible applied as promising substitutes for the conventional ITO film. Because the TZO film by FTS system was deposited by plasma-free sputter method with low temperature process and also has many advantages such as low resistance, high transmittance, uniform surface, cost effective production without indium component, and so on.

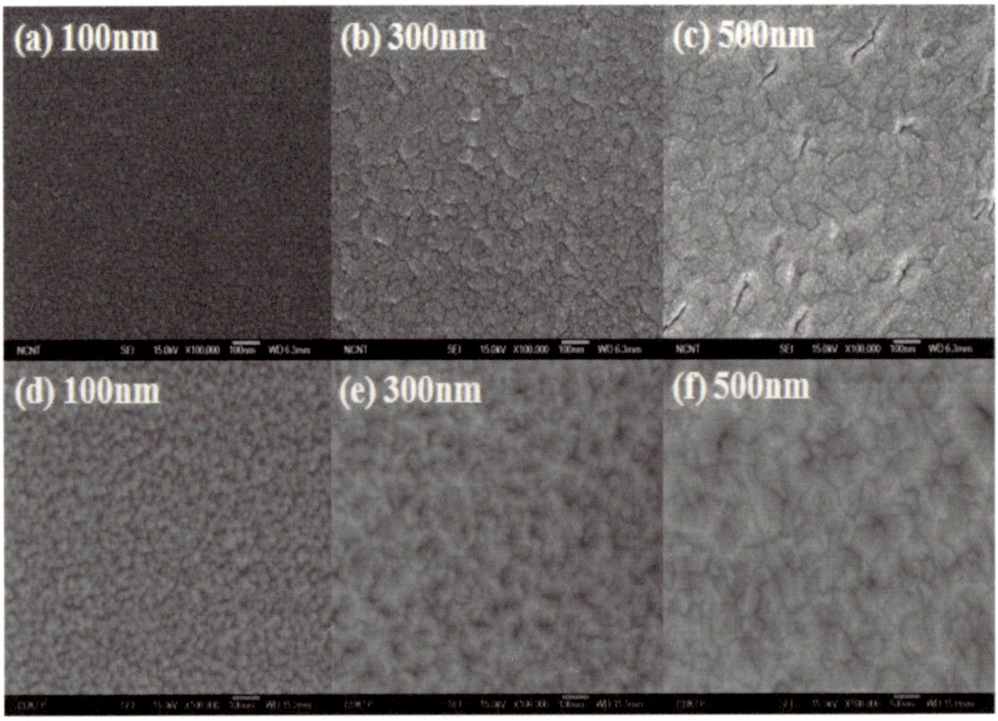

Fig. 15. SEM images of TZO films prepared by (a)-(c) FTS system and (d)-(f) RFS system with various deposition thickness.

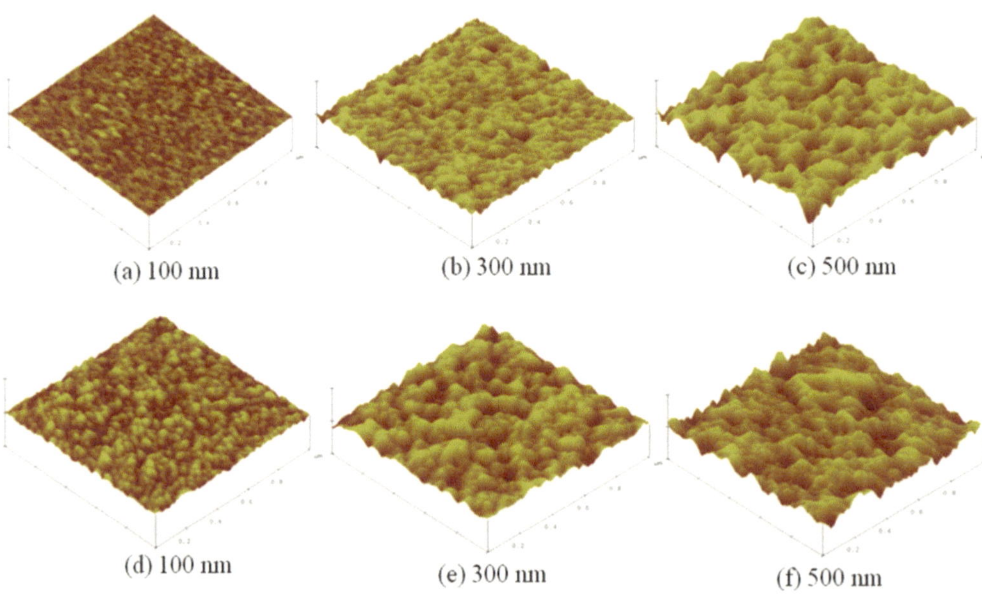

Fig. 16. AFM images of TZO films prepared by (a)-(c) FTS system and (d)-(f) RFS system with various deposition thickness.

2.5 Characterization of TCO films
2.5.1 Electrical properties

The purpose of the 4-point probe is to measure the resistivity of any semiconductor material. It can measure either bulk or thin film specimen, each of which consists of a different expression. The derivation will be shown in this tutorial. In a sheet resistance measurement, several resistances need to be considered, as shown in Fig. 17 (a). The probe has a probe resistance R_p. It can be determined by shorting two probes and measuring their resistances. At the interface between the probe tip and the semiconductor, there is a probe contact resistance, R_{cp}. When the current flows from the small tip into the semiconductor and spreads out in the semiconductor, there will be a spreading resistance, R_{sp}. Finally the semiconductor itself has a sheet resistance R_s. The equivalent circuit for the measurement of semiconductor sheet resistance by using the four-point probe is shown in Fig. 17. Two probes carry the current and the other two probes sense the voltage. Each probe has a probe resistance R_p, a probe contact resistance R_{cp} and a spreading resistance R_{sp} associated with it. However, these parasitic resistances can be neglected for the two voltage probes because the voltage is measured with a high impedance voltmeter, which draws very little current. Thus the voltage drops across these parasitic resistances are insignificantly small. The voltage reading from the voltmeter is approximately equal to the voltage drop across the semiconductor sheet resistance.

By using the four-point probe method, the semiconductor sheet resistance can be calculated:

$$R_s = F\,(V/I) \tag{6}$$

, where V is the voltage reading from the voltmeter, I is the current carried by the two current carrying probes, and F is a correction factor.

For collinear or in-line probes with equal probe spacing, the correction factor F can be written as a product of three separate correction factors:

$$F = F1 \cdot F2 \cdot F3 \tag{7}$$

$F1$ corrects for finite sample thickness, $F2$ corrects for finite lateral sample dimensions, and $F3$ corrects for placement of the probes with finite distances from the sample edges. For very thin samples with the probes being far from the sample edge, $F2$ and $F3$ are approximately equal to one (1.0), and the expression of the semiconductor sheet resistance becomes:

$$R_s = (\pi / \log 2)(V/I) \tag{8}$$

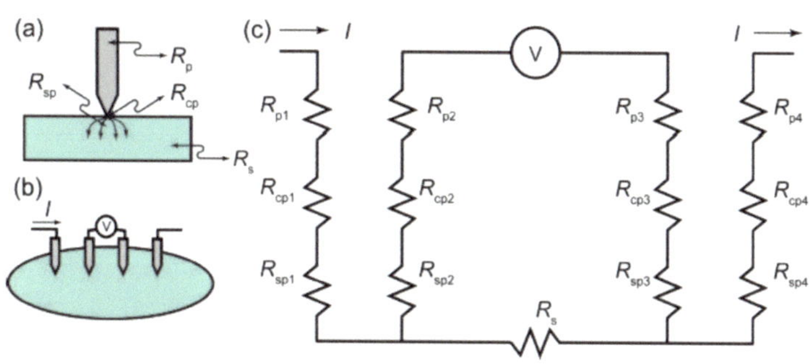

Fig. 17. 4-pin probe measurement of semiconductor sheet resistance.

The four-point probe method can eliminate the effect introduced by the probe resistance, probe contact resistance and spreading resistance. Therefore it has more accuracy than the two point probe method. For more detail the electrical property, the Hall measurement system is a complete system for measuring the resistivity, carrier concentration, and mobility of semiconductors. The system includes software with I-V curve capability for checking the ohmic integrity of the user made sample contacts. The systems can be used to characterize various materials including semiconductors and compound semiconductors (N Type & P Type) such as Si, Ge, SiGe, SiC, GaAs, InGaAs, InP, GaN, ZnO, TCOs, metals, etc., at both 300K and 77K (room temperature and liquid nitrogen temperature). An electric field from Fig. 18 and 19 is applied along the x-axis and a magnetic field is applied along the z-axis.

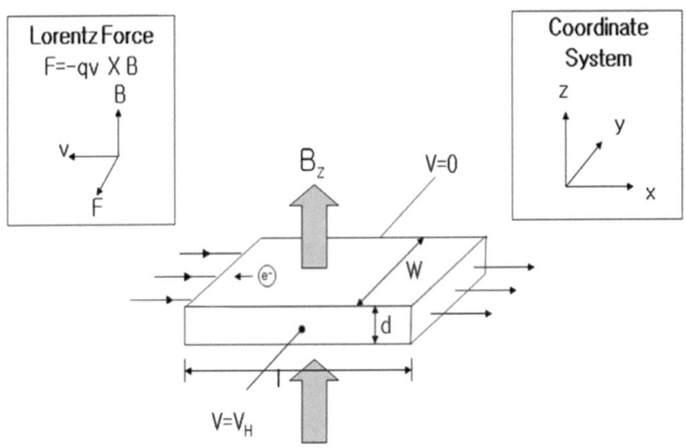

Fig. 18. Hall effects and Lorentz force.

For a p-type semiconductor sample, the Lorentz force due to the magnetic field exerts an average upward force on the holes flowing in the x-direction toward the positive y-axis which results in the accumulation of holes at the top of the sample that gives rise to a downward directed (y direction) electric field. The established electric field is called the Hall field and the voltage drop across the top and bottom of the sample is called the Hall voltage.

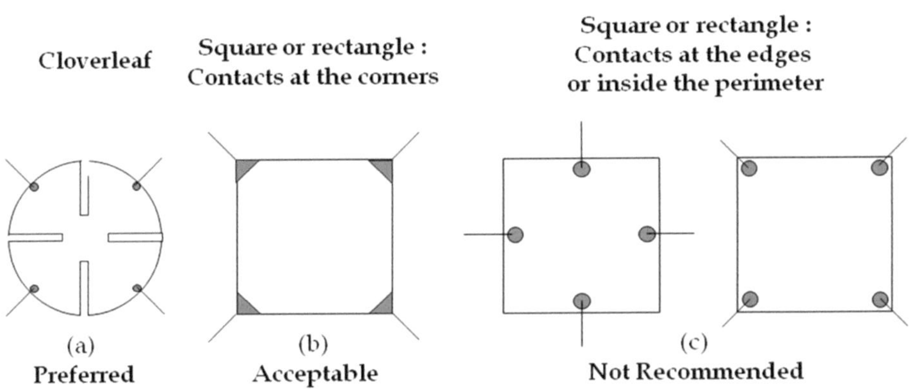

Fig. 19. Sample geometry using Van der Pauw method.

2.5.2 Optical properties

UV-VIS refers to absorption spectroscopy or reflectance spectroscopy in the ultraviolet-visible spectral region. This means it uses light in the visible and adjacent (near-UV and near-infrared (NIR)) ranges. A sample in a cuvette is exposed to light energy between 190 nm and 1000 nm. Spectrophotometry investigates the absorption of the different substances between the wavelength limits 190 nm and 780 nm (visible spectroscopy is restricted to the wavelength range of electromagnetic radiation detectable by the human eye, that is above ~360 nm; ultraviolet spectroscopy is used for shorter wavelengths). In this wavelength range the absorption of the electromagnetic radiation is caused by the excitation (i.e. transition to a higher energy level) of the bonding and non-bonding electrons of the ions or molecules. A graph of absorbance against wavelength gives the sample's absorption spectrum. Modern spectrophotometers draw this automatically. The measured spectrum is continuous, due to the fact that the different vibration and rotation states of the molecules make the absorption band wider. Certain parts of an organic molecule will absorbance some of this energy to create peaks on a spectrum for quantitative (primarily) and qualitative Analysis. The original UV-Vis specs were made as *DOUBLE-BEAM* units to correct for noise, drift and other Instabilities. Over 20 years ago, the well-known leaders in the analytical instrument markets; Beckman & Perkin-Elmer; began to focus on a line of "Stable-Beam" SINGLE-BEAM Instruments. Spectrophotometry is used for both qualitative and quantitative investigations of samples. The wavelength at the maximum of the absorption band will give information about the structure of the molecule or ion and the extent of the absorption is proportional with the amount of the species absorbing the light. Quantitative measurements are based on Beer's Law (also known as "Lambert-Beer Law" or even "Bouguer-Lambert-Beer Law") which is described as follows:

$$A = ec\,l \qquad (9)$$

, where A = absorbance [no units, because it is calculated as $A = \log_{10}(I_0/I)$, where I_0 is the incident light's intensity and I is the light intensity after it passes through the sample];
e = molar absorbance or absorption coefficient [in $dm^3\ mol^{-1}\ cm^{-1}$ units];
c = concentration (molarity) of the compound in the solution [in $mol\ dm^{-3}$ units];
l = path length of light in the sample [in cm units].

2.5.3 Structural properties

Atomic force microscope(AFM) provides a 3D profile of the surface on a nanoscale, by measuring forces between a sharp probe (<10 nm) and surface at very short distance (0.2-10 nm probe-sample separation). The probe is supported on a flexible cantilever. The AFM tip "gently" touches the surface and records the small force between the probe and the surface. The probe is placed on the end of a cantilever (which one can think of as a spring). The amount of force between the probe and sample is dependent on the spring constant (stiffness) of the cantilever and the distance between the probe and the sample surface. This force can be described using Hooke's Law:

$$F = -k\,x \qquad (10)$$

, where F is force, k is spring constant, and x is cantilever deflection.
If the spring constant of cantilever (typically ~0.1-1 N/m) is less than surface, the cantilever bents and the deflection is monitored. This typically results in forces ranging from nN(10^{-9}) to μN(10^{-6}) in the air. If the tip was scanned at a constant height, a risk

would exist that the tip collides with the surface, causing damage. Hence, in most cases a feedback mechanism is employed to adjust the tip-to-sample distance to maintain a constant force between the tip and the sample. Traditionally, the sample is mounted on a piezoelectric tube that can move the sample in the z direction for maintaining a constant force, and the x and y directions for scanning the sample. Alternatively a 'tripod' configuration of three piezo crystals may be employed, with each responsible for scanning in the x, y and z directions. This eliminates some of the distortion effects seen with a tube scanner. In newer designs, the tip is mounted on a vertical piezo scanner while the sample is being scanned in X and Y using another piezo block. The resulting map of the area z = f(x,y) represents the topography of the sample.
The AFM can be operated in a number of modes, depending on the application. In general, possible imaging modes are divided into static (also called contact) modes and a variety of dynamic (or non-contact) modes where the cantilever is vibrated.
Contact Mode AFM: (repulsive VdW) When the spring constant of cantilever is less than surface, the cantilever bends. The force on the tip is repulsive. By maintaining a constant cantilever deflection (using the feedback loops) the force between the probe and the sample remains constant and an image of the surface is obtained from Fig. 20.

Fig. 20. Schematic of contact mode AFM.

Intermittent Mode (Tapping): The imaging is similar to contact. However, in this mode the cantilever is oscillated at its resonant frequency. The probe lightly "taps" on the sample surface during scanning, contacting the surface at the bottom of its swing. By maintaining constant oscillation amplitude a constant tip-sample interaction is maintained and an image of the surface is obtained from Fig. 21.

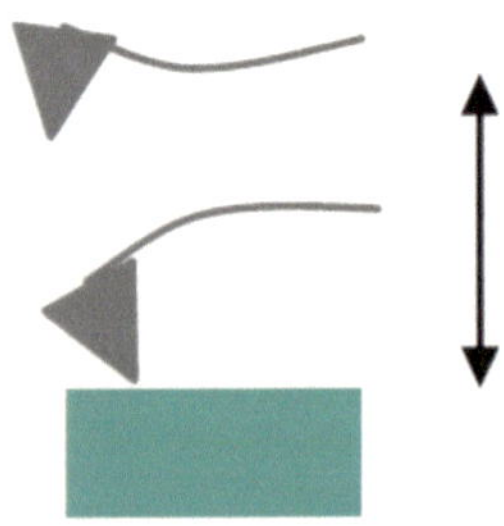

Fig. 21. Schematic of tapping mode AFM with oscillation amplitude of 20-100 nm.

Non-contact Mode: (attractive VdW) The probe does not contact the sample surface, but oscillates above the adsorbed fluid layer on the surface during scanning. (Note: all samples unless in a controlled UHV or environmental chamber have some liquid adsorbed on the

Film	Thickness/ nm	Sheet resistance/(Ω·□$^{-1}$)	Work function/eV	Transmittance(%) (at 1.54 μm)
ITO(12 Ω/□)	~180	~12	4.5±0.1	35
ITO(100 Ω/□)	~35	~100	4.4±0.1	85
ZnO	~400	~10000	—	88
AZO	~130	~400	5.1±0.1	30
PANI	~110	~800	4.8±0.1	39

Table 1. Relevant properties of transparent conducting films on glass substrates.

Dopant-polymer/ solvent[b]	$\sigma(\Omega cm)^{-1}$	%QE at 7 V	L (cd/m^2)	L (cd/m^2)
			at 7 V fresh	200 h at 7 V
PSS-PAni/H_2O	0.1	1.4	4000	30 w/ITO [c]
PSS-PAni/NMP	0.1	1.0	1500	
PAPS-PAni/H_2O	discontinuous	1.1	1200	
APS-PAni/FA	100	0.4	3000	
			1000 w/Au	50 w/Au
CSA-PAni/FA	200	0.3	1800	
PSS-PEDT/H_2O	2	1.3	4000 w/ITO	30 w/ITO
			300 w/Au	35 w/Au
			180 none	60 none

Table 2. Performance of PLEDs with various polymer anodes ~100 nm MEH-PPV.

The two types of conducting polymers studied were PAni and polyethylenedioxythiophene (PEDT). The PAni materials were doped with two different polymer dopants, polystyrenesulfonic acid and polyacrylamidopropanesulfonic acid, and two monomer dopants, amidopropanesulfonic acid and camphor sulfonic acid, for comparison. Average lifetime behaviors on samples with and without ITO are shown in Fig. 24.(Carter et al., 1997) In general, the brightest diodes with the higher current densities decayed the most rapidly; however, the diodes with the ITO in the device structure continued to decay more rapidly even when the current densities were below that of the devices without ITO in the structure. In the next 200 h, the light output in non-ITO devices decayed less than 20% while the ITO-based diodes lost nearly another order of magnitude. This effect is also observed if the diodes are aged at a much lower dc voltage ~4 V and current density. These results indicate that the long-term device failure is accelerated by the presence of ITO, caused by photo-oxidation of the light-emitting polymer via oxygen evolved from the ITO. The mechanism for the short-term aging is currently under further investigation.

2.6.2 Nanometals

Since Pt has a very high work function of about 5.6 eV, it could strongly enhance hole injection. However, in order to use for TCO films, Pt must be very thin to be transparent, and it would be deposited on, e.g., the conventional ITO. Malliaras et al. have shown that a

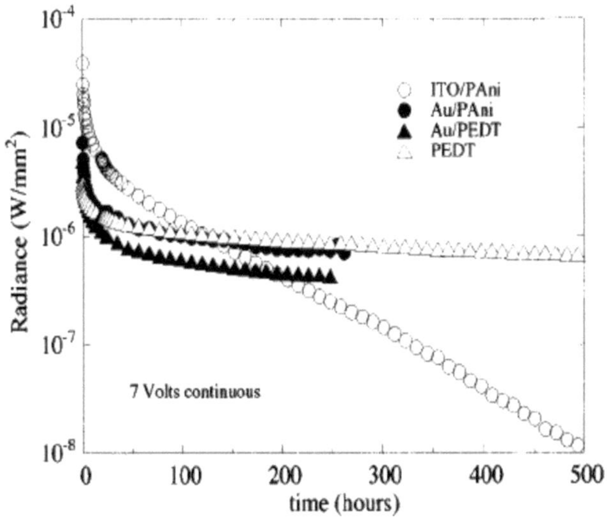

Fig. 24. Radiance lifetime studies for different PLED device structures. Devices without ITO in the device structure have improved long lifetime behavior. 1 W/mm^2=7.3X10^7 cd/m^2.

thin layer (≤10 Å) of Pt on ITO enhances hole injection by up to a factor of 100 relative to the uncoated ITO. In order to investigate the properties of the multilayer TCO, a sandwich structure of ITO(50 nm)/Au(5 nm)/ITO(45 nm) (IAI) multilayer films compared with conventional ITO film were analyzed on glass substrates without intentional substrate heating for application such as OLEDs, STN-LCDs, gas sensor, solar cells, and so on. The electrical properties of the IAI and ITO films evaluated with Hall Effect measurements were measured. Although ITO/Au/ITO films have a lower mobility than ITO single-layer films, they have a lower resistivity as shown in Table 3.(Kim et al., 2007) The decrease in the resistivity of the IAI films may be caused by the increase in the carrier density of the IAI films, which results from the presence of the Au interlayer. However, the ITO/Au/ITO films have a lower mobility than the ITO single-layer films, which suggesting that the two interfaces between the ITO and Au film may act as a barrier to carrier movement.

	ITO	ITO/Au/ITO
Carrier density	1.8	220
Mobility	140.8	47.4
resistivity	40.2	0.5

Table 3. The comparison of the carrier density (× $10^{19}/cm^3$), mobility (cm^2/Vs) and resistivity (× 10^{-4} Ω·cm) of the ITO and ITO/Au/ITO films.

Transmittance and sheet resistance values at ITO/Ag/ITO structure are found mainly dependent on the Ag film thickness; whereas the wavelength range at which the maximum transmittance is controlled by the ITO film thickness.(Guillen et al., 2009) Lowest sheet resistance at ITO/Ag/ITO structure have been obtained below 6 ohm/sq at Ag film thickness above 10 nm and ITO layers thickness in the 30-50 nm range. Choi *et. al* investigated the effects of the post-annealing temperature on the TCO film properties with ITO/Ag/ITO structure.(Choi et al., 1999) The properties of multilayer films, especially the optical and electrical properties, depended dominantly on the characteristics of Ag film. The morphology and structure of very thin Ag film were sensitive to the deposition temperature. The Fig. 25 shows the optical transmission spectra of Ag films deposited for 70 s at 25, 100 and 200°C, which shows various transmittances from 20% to 80% in visible light range. The spectrum of 200°C shows the lowest transmittance. The color of the film was dark violet, but the others were blue. Light scattering by the rough surface reduced the transmittance. Substrate heating during Ag film deposition led to the same effect on the multilayer film. When Ag film deposition temperature was 200°C in the preparing process of multilayer, both visible transmittance and conductivity decreased.

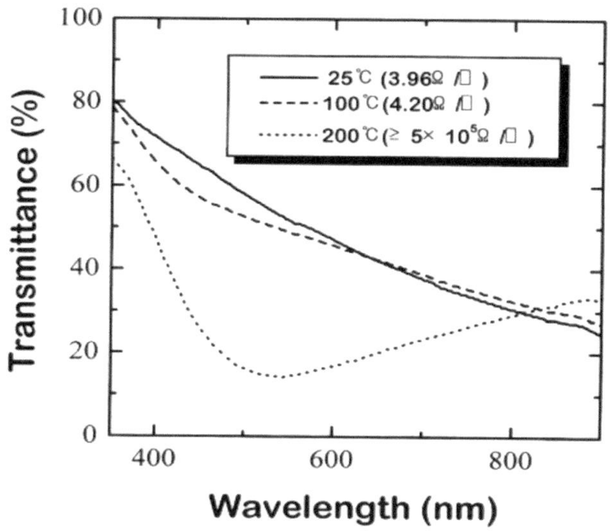

Fig. 25. Optical transmission spectra and sheet resistance of Ag thin films deposited on glass at 25, 100, and 200°C. Deposition time was 70 s.

Their micrographs of these films are shown in Fig. 26. Because substrate heating causes low nucleation density, we obtained island structure films at 200 °C, where a continuous film had not formed yet.

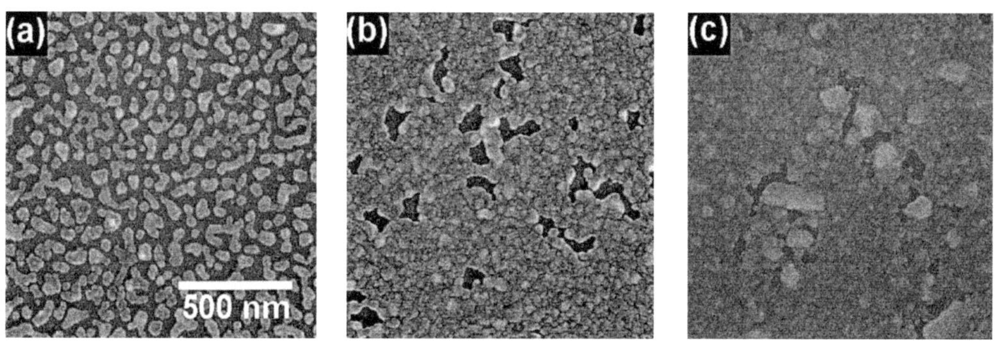

Fig. 26. SEM micrographs of Ag thin ®lms deposited on glass at (a) 25, (b) 100, and (c) 200 °C Deposition time was 70 s.

2.6.3 Carbon nanotubes and graphene

Electrode materials for the most important properties are very high conductivity. Use materials with high conductivity and relatively small amount when using the material to lower the price, and the concentration of material to help penetration of the transparent electrode is connected directly. The following requirements are important to low processing temperatures, low manufacturing cost and a uniform printing properties, adhesion with the substrate, such as external friction by abrasion, weathering and chemical resistance of various organic solvents, etc. are needed. The conductive ink materials which is currently used or has been studied, give a similar examples, such as; conductive polymer solution, liquid dispersed in metal nano-particles, carbon nanotubes(CNT) composite materials, as shown in Fig. 27.

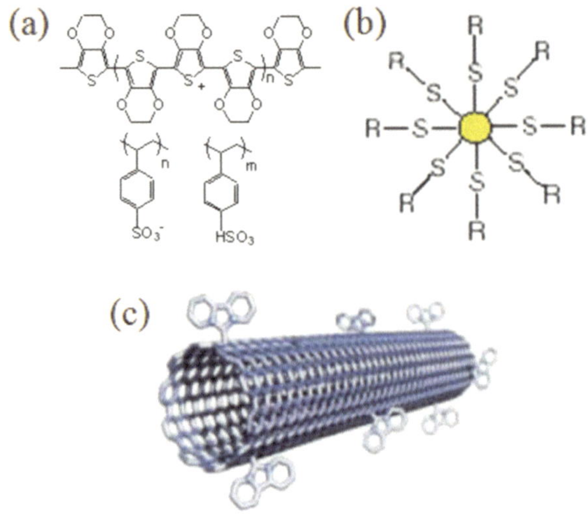

Fig. 27. Conductive ink materials. (a) PEDOT:PSS, (b) metal nano particle, and (c) surface modified carbon nanotube.

Metal nano-particles have enough stocks as high conductivity. But, to remove the dispersion agent that the wrists are used to distribute these materials, relatively high firing temperatures(>150 °C) and the expensive production costs are required. Therefore, there can be a little more low-cost manufacturing process for the development of distributed and can lower the sintering temperature is necessary to develop the dispersion agent. Conductive polymer dispersion in high and low process temperature, the easy one because the fairness, conductivity is very low(1~10 S/cm) compared to the metal nano-particles. Therefore, the research on polymer material with high conductivity is needed. Because the CNT dispersion solutions are difficult to disperse in common solvents, Modified CNT surface and dispersed in a solvent such as water and ink are used. CNT is lower than the relatively high conductivity of metal nanoparticles (100~1000 S/cm) and low processing temperature (100 °C), but because of being researched in recent long-term stable dispersion of CNT getting more difficult, and studies are needed.

Dai Nippon Printing (DNP) Co., Ltd. and Fujifilm Co. have developed a new method, which can directly coating the metal grid using fine pattern printing method by Ag nano-ink. Fig. 28(a) shows the TCO electrode of a nano-mesh type developed in DNP Co., Ltd. and

Fujifilm Co. Because the films of DNP Co., Ltd. can form the uniform pattern at only necessary parts, it can be reduced the unnecessary deposition and etching process during ITO fabrication. And also, Fujifilm Co. of Fig. 28(c) has developed a new type of electrode, which can control the resistivity from 0.2 Ω/cm^2 to thousands Ω/cm^2 using Ag pattern of mesh type on PET substrate and organic/inorganic conductive materials. U.S. venture cambrios Co. released the transparent conductive ink as a wet coating, which contains soluble metal nano-wire, as shown in Fig. 28(b). The image shows the dispersed shape on a substrate at Ag nanowires fabricated in Cambrios Co. uniformly distributed wire of the nano-scale film can form the film as a type of mesh, has excellent permeability and high resistivity. And also, it has advantage which can be directly coated at room temperature. And also, Toray and Cimma Co. in Fig. 28(d) developed the transparent electrode with transmittance over 80% using self-alighning mechanism of siver nano-particles.

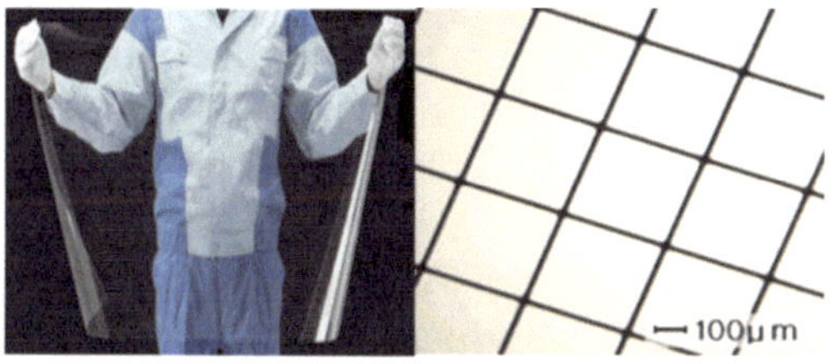

(a) Dai Nippon Printing Co., Ltd.

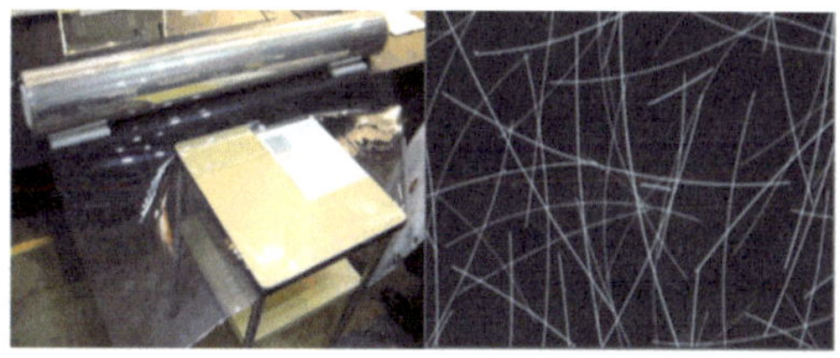

(b) Cambrios Co.

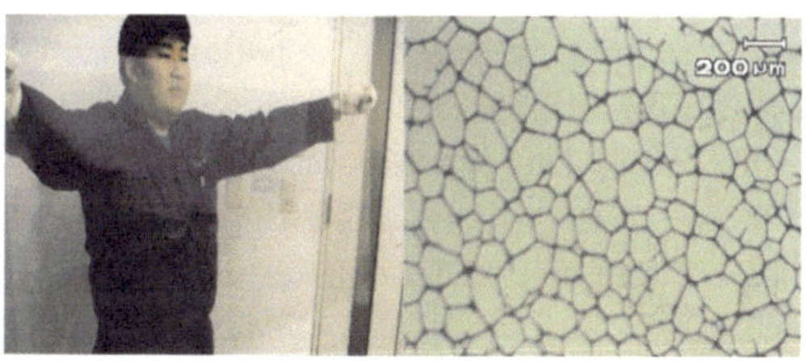

(c) Fujifilm Co.

(d) Toray-Cimma Co.

Fig. 28. Transparent electrodes of nano-mesh type.

2.7 TCOs for flexible OLEDs

In most case, OLEDs have been traditionally fabricated on glass substrate, however, there OLEDs have several disadvantage for certain applications such as portable communication display, because the glass substrate is very fragile, heavy and non-flexible. Flexible OLEDs (FOLEDs) using the plastic substrate are growing attention, because a plastic substrate can be overcome disadvantages of a glass substrate. These devices provide the ability to conform, bend or roll-up display into any shape.(Gu et al., 1997) This means that FOLEDs may be laminated onto an automotive windshield, or an aircraft cockpit et al. However, the low thermal stability of a plastic substrate (T_g=80 °C) is difficult the process for making transparent conducting electrode, indium tin oxide (ITO), and TFT. These materials are necessary for the high temperature ≥ 200 °C. Conventional FOLEDs are fabricated the polymer materials.(Gustaffson et al., 1993; He & Kanicki, 2000) Y. Zhang and S. R. Forrest were suggested that thermal deposited organic thin films have the general flexibility property by the van der waals force between aromatic atoms-to-atoms.(Zhang

& Forrest, 1993) Since there results were reported, G. Gu shows that a conventional small molecule organic materials can be successfully fabricated FOLEDs.(Gu et al., 1997) As FOLEDs attempted cyclic bending test, the electrode layer consist of an inorganic materials, such as ITO and metal cathode, had cracked surface because an inorganic layers are brittle materials. Under mechanical stresses, micro crack and propagation of existing pinholes will be significantly reductive such as the contact properties between an organic and electrode layer. In this work, we report that FOLEDs in the sequence of ITO, organic materials, and aluminum(Al) deposited on the PET substrate by using low temperature process. The current density and brightness property of FOLEDs were investigated as a function of the radius of variation bending test. To investigate change of the surface in an inorganic electrode after the mechanical flexibility test, scanning electron microscope (SEM) was used.

For fabricating the FOLEDs, an ITO, N,N'-diphenyl-N,N'-bis(3-methylphenyl)-1,1'-diphenyl-4,4'-diamin (TPD), tris-(8-hydroxyquinoline) aluminum (Alq_3), and Al were used as an anode, HTL, emitting material layer(EML), and cathode, respectively. The ITO electrode was deposited on top of the PET substrate by RF-magnetron sputtering at room temperature. The ITO target is an alloy of In_2O_3 (90 %) and SnO_2 (10 %) by weight, with 99.99 % purity. The base pressure of the sputtering chamber was 5×10^{-6} torr, and the sputter deposition pressure of the ITO film was 6×10^{-3} torr with Ar and O_2 flows regulated by mass flow controller at 25 sccm and 1 sccm, respectively. The r. f. sputtering power was 50 W, resulting in a deposition rate of 2.8 nm/s. The sheet resistance of 150 nm thick-deposited the ITO film is 50 Ω/sq, and its transmission was 90 % at the wavelength of 513 nm. To optimized properties of the sputter deposited electrode, ITO surface was treated by O_2 plasma at 2×10^{-1} torr for 60 sec.(Ishii et al., 2000) The organic materials TPD and Alq_3 were deposited as thickness of 60 nm and 40 nm, respectively, at the deposition rates of 1.0-1.5 Å/s on the PET/ITO substrate. After the deposition of organic layer, the Al cathode layer was deposited as the thickness of 100 nm at the deposition rates 10 Å/s by thermal evaporation. The performance of FOLEDs was measured after repeated bending each 100 times per samples at radius as a function of various bending test range. To analyze electrical property and brightness intensity of FOLEDs, we measured by Keithly 2400 electrometer, Si photodiode and Keithly 485 picoammeter. The sheet resistance of ITO after the bending test were measured by four-point probe, and surface morphology of ITO, organic films, and Al layer after bending test were observed by SEM(FEI company, XL30 ESEM-FEG).

Fig. 29(a) and 29(b) shows the change of current density-voltage (J-V) and brightness-voltage (B-V) characteristics for radius of curvature. The bent FOLEDs at radius of 19 mm, 16 mm, 13 mm, 10 mm, and non bending were referred to as FOLED(19), FOLED(16), FOLED(13), FOLED(10), and FOLED(0), respectively. In Fig. 29(a), the turn-on voltage of FOLED(0), FOLED(19), FOLED(16), FOLED(13), and FOLED(10) were observed at 9 V, 10 V, 10 V, 15 V, and 17 V, respectively. As the given voltage, the current density and the brightness of bent FOLEDs were substantially lower than those of the control device. The J-V and B-V properties of FOLED(19) were analogous with FOLED(0). FOLED(16) showed similar current density to FOLED(19), however, this device inferior maximum bright intensity (1440 cd/m^2) than FOLED(0) (1900 cd/m^2). Further decrease of the radius to 13 mm and 10 mm resulted in reduce J-V and B-V characteristics of FOLEDs.

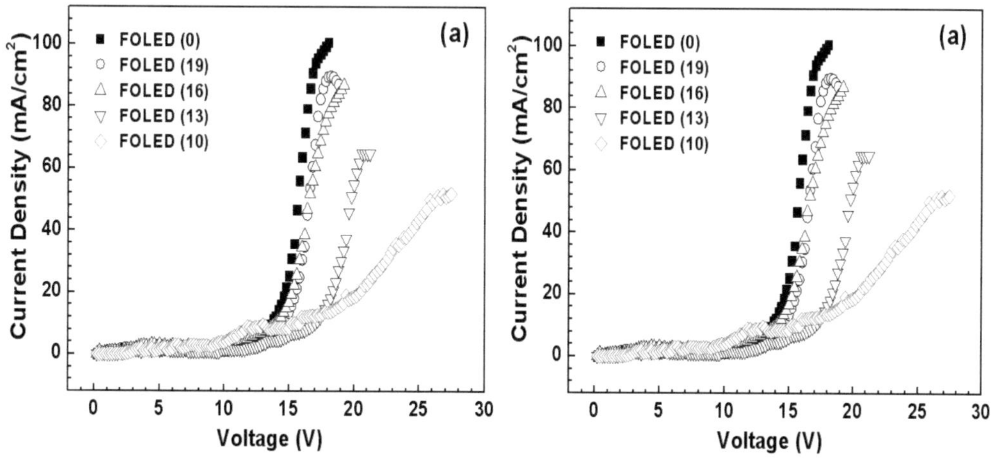

Fig. 29. (a) Current density versus voltage (J-V) and (b) brightness ver. voltage (B-V) curves of the bent FOLEDs at radius of 19 mm, 16 mm, 13 mm, 10 mm, and non bending.

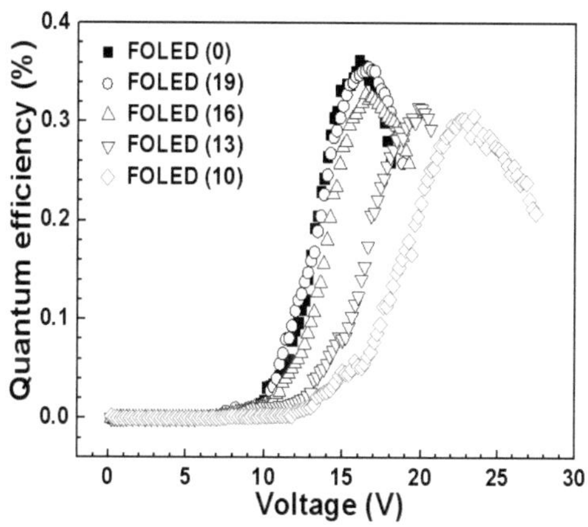

Fig. 30. Quantum efficiency ver. voltage (I-V) curves of the bent FOLEDs at radius of 19 mm, 16 mm, 13 mm, 10 mm, and non bending.

Fig. 30 shows external quantum efficiency of bent FOLEDs. FOLED(0), FOLED(19), and FOLED(16) showed maximum quantum efficiency of ~ 0.35 % at 16 V. FOLED(13) and FOLED(10) have lowered quantum efficiency of 0.31 % and 0.30 % at 20 V and 23 V, respectively, than the other FOLEDs. As the radius of bending test decreased, the reduction of device properties, such as current density, brightness, and quantum efficiency is thought that the contact property between the layers was decreased because of mechanical stress by the bending test of FOLEDs. To measure changes of the sheet resistance after the bending test of ITO films, four point probe was observed (Fig. 31). The ITO films were deposited on the PET substrate and then ITO coated substrates were operated bending test (100 times) with various range of 19 mm, 16 mm, 13 mm, and 10 mm. The bent ITO with 19 mm showed analogous sheet resistance (62 Ω/□), with the control ITO (50 Ω/□). However, the sheet resistance of bent ITO increased as the radius of bending test decreased when the radius of bending test ≤ 16 mm. Since increased the sheet resistance of bent ITO, FOLED(13) and

FOLED(10) were thought to have a lower electric and brightness properties than FOLED(0). Although, the sheet resistance of 16 mm bent ITO is 230 Ω/□, this value is sufficiently small for using electrode in OLEDs.(Gu et al., 1996)

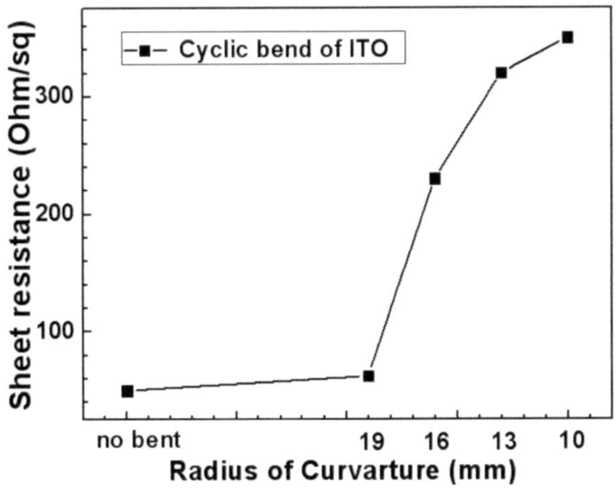

Fig. 31. Changes of the sheet resistance after the bending test of ITO films.

Fig. 32 showed the SEM images of the ITO surface as a function of the variation radius of curvature. The crack phenomenon of bent ITO was appeared from the lowered radius of bending test ≤ 16 mm. The ITO for the bending test at 13 mm and 10 mm radius showed the rough surface and a large amount of crack from the whole area. In our case, the bent ITO at 16 mm, 13 mm, and 10 mm have the hasty reductive properties, such as sheet resistance and surface morphology.

In this result, the reasons for reduced properties of bent FOLEDs are follows. Firstly, the bent FOLEDs have lowered device characteristics, such as increased driving voltage and decreased luminance property, because of the increase resistance of bent ITO by a bending test in radius of below 16 mm.(Chen et al., 2002) Secondly, many attempts have been focused on interface property between an organic and electrode layer. The OLEDs using the ITO anode with smooth surface have shown superior device properties, as lower turn-on voltage and higher luminescent efficiency, because these devices are improved contact property. In this study, the bent FOLEDs have shown inferior device performance because rough surface of the bent ITO was decreased contact property in the interface between the TPD organic layer and ITO electrode.(Kwon et al., 2002) In conclusions, we fabricated FOLEDs with an ITO anode, a TPD hole transport layer, an Alq_3 emitting layer, and an Al cathode deposited on the PET substrate and studied FOLEDs characteristics after bending test at various radiuses of 10 mm, 13 mm, 16 mm, and 19 mm. The performance of FOLEDs with lowered radius (≤ 16 mm) was decreased the device properties, and increased the sheet resistance of bent ITO. These devices showed the crack phenomena and rough surface in the ITO and Al inorganic layers. In our experiment, the optimum radius of bending test was 19 mm. When FOLEDs was bent at 19 mm radius, inorganic layer, ITO and Al, cannot show the crack phenomena. The electrical property and brightness efficiency of FOLED(19) were similar with the control device. In this result was suggested that the performance of the bent FOLEDs was affected significantly by the crack phenomenon of an inorganic layer and increased sheet resistance of bent ITO.

Fig. 32. SEM images of the ITO surfaces (a) without bending and with bending as a function of the variation radius of curvatures in (b) 16 mm, (c) 13 mm, and (c) 10 mm.

3. Conclusion

In fabricating OLED devices, ITO film among the TCO films is widely used as an anode layer, because of its high transparency in the visible light range, low conductivity, and high work function, etc. However, indium in ITO has a tendency to diffuse into the emissive polymer layer under device operation, which may in turn influence the quantum efficiency and lifetimes of OLEDs. In addition, it is known that the performance of ITO-based polymer LEDs is highly dependent on the chemical condition of the ITO electrode, which is affected, at least in part, by the particular method used to clean the ITO prior to device fabrication. Therefore, various TCOs, such as ZnO based TCO or conducting polymer, nanometal, and carbon nanotube, etc. have been investigated and can be applied in OLEDs and/or other optoelectronic devices. For example, ZnO has the advantages, such as the absence of toxicity, low cost, and good thermal stability. ZnO films with a hexagonal wurtzite structure have a wide optical energy band gap (around 3.3 eV). However, the electrical properties of undoped ZnO films are subjected to stoichiometric deviations resulting from oxygen vacancies and interstitial zinc atoms. In order to improve this deficiency, many workers have researched how the electrical properties of future TCO films are influenced by doping or new material development.

4. Acknowledgment

This research was financially supported by the Second Stage of Brain Korea 21 Project and the Ministry of Knowledge Economy(MKE), Korea Institute for Advancement of Technology(KIAT) and Dae-Gyeong Leading Industry Office through the Leading Industry Development for Economic Region.

5. Reference

Agura, H.; Suzuki, H.; Matsushita, T.; Aoki, T. & Okuda, M. (2003). Low Resistivity Transparent Conducting Al-doped ZnO Films Prepared by Pulsed Laser

Deposition. *Thin Solid Films*, Vol.445, No.2, (December 2003), pp. 263-267, ISSN 0040-6090

Banerjee, A.N. & Chattopadhyay, K.K. (2005). Recent Developments in The Emerging Field of Crystalline p-Type Transparent Conducting Oxide Thin Films. *Progress in Crystal Growth and Characterization of Materials*, Vol.50, No.3, (September 2005), pp. 55-105, ISSN 0960-8974

Carter, S. A.; Angelopoulos, M.; Karg, S.; Brock, P. J. & Scott, J. C. (1997). Polymeric Anodes for Improved Polymer Light-Emitting Diode Performance, *Applied Physics Letters*, Vol.70, No.1, (April 1997), pp. 2067-2069, ISSN 0003-6951

Chan, I. M. & Hong, F. C. (2004). Improved Performance of The Single-Layer and Double-Layer Organic Light Emitting Diodes by Nickel Oxide Coated Indium Tin Oxide Anode. *Thin Solid Films*, Vol.450, No.2, (October 17 2003), pp. 304-311, ISSN 0040-6090

Chen, M.; Pei, Z.; C, Sun.; L.S, Wen. & X, Wang. (2000). Surface Characterization of Transparent Conductive Oxide Al-Doped ZnO Flms. *Journal of Crystal Growth*, Vol.220, No.3, (February 2000), pp. 254-262, ISSN 0022-0248

Chen, T. H.; Liou, Y.; Wu, T. J. & Chen, J. Y. (2004). Enhancement of Organic Light-Emitting Device Performances with Hf-Doped Indium Tin Oxide Anodes. *Applied Physics Letters*, Vol.85, No.11, (April 2004), pp. 2092-2094, ISSN 1077-3118

Chen, Z.; Cotterell, Brian. & Wang, Wei. (2002). The Fracture of Brittle Thin Films on Compliant Substrate in Flexible Displays. *Engineering Fracture Mechanics*, Vol.69, No.5, (February 2002), pp.597-603, ISSN 0013-7944

Cho, M.-H.; Chang, H. S.; Cho, Y. J.; Moon, D. W.; Min, K.-H.; Sinclair, R.; Kang, S. K.; Ko, D.-H.; Lee, J. H.; Gu, J. H. & Lee, N. I. (2004). Change in Chemical State and Thermal Stability of HfO_2 by The Incorporation of Al_2O_3. *Applied Physics Letters*, Vol.84, No.4, (May 2003), pp. 571-573, ISSN 1077-3118

Choi, B.G.; Kim, I. H.; Kim, D.H.; Lee, K.S.; Lee. T.S.; Cheong, B.; Baik, Y.-J. & Kim, W.M. (2005) Electrical, Optical and Structural Properties of Transparent and Conducting ZnO Thin Films Doped with Al and F by Rf Magnetron Sputter. *Journal of The European Ceramic Society*, Vol.25, No.12, (March 2005), pp. 2161-2165, ISSN 0955-2219

Choi, K.; Kim, J.; Lee, Y. & Kim, H. (1999). ITO/Ag/ITO Multilayer Films for The Application of A Very Low Resistance Transparent Electrode. *Thin Solid Films*, Vol.341, No.1-2, (March 1999), pp. 152-155, ISSN 0040-6090

Chowdhury, F.-U.-Z. & Bhuiyan, A.H. (2000). An Investigation of The Optical Properties of Plasma-Polymerized Diphenyl Thin Films. *Thin Solid Films*, Vol.360, No.1-2, (February 2000), pp. 69-74, ISSN 0040-6090

Chung, J.-L.; Chen, J.-C. & Tseng, C.-J. (2008). Preparation of TiO_2-Doped ZnO Films by Radio Frequency Magnetron Sputtering in Ambient Hydrogen-Argon Gas. *Applied Surface Science*, Vol.255, No.5, (July 2008), pp. 2494-2499, ISSN 0169-4332

Custaffson, G.; Treacy, G. M.; Klavertter, Y. C.; Colaneri, N. & Heeger, A. J. (2003). The "Plastic" Led: A Flexible Light-Emitting Device Using A Polyaniline Transparent Electrode. *Synthetic Metals*, Vol.57, No.1, (February 2000), pp. 4123-4127, ISSN 0379-6779

David, O.; Scanlon, Aron, W. & Graeme, W. W. (2009). Understanding the p-Type Conduction Properties of the Transparent Conducting Oxide $CuBO_2$: A Density

Functional Theory Analysis. *Chemistry of Materials*, Vol.21, No.19, (July 2009), pp. 4568–4576, ISSN 0897-4756

Deng, Z.B.; Ding, X.M. & Lee, S.T. (1999). Enhanced Brightness and Efficiency in Organic Electroluminescent Devices using SiO_2 Buffer Layers. *Applied Physics Letters*, Vol.74, No.15, (December 1998), pp. 2227-2229, ISSN 0003-6951

Dingle, R.; Störmer, H.-L.; Gossard, A.C. & Wiegmann, W. (1978). Electron Mobilities in Modulation-Doped Semiconductor Heterojunction Superlattices. *Applied Physics Letters*, Vol.33, No.7, (July 1978), pp.665-667, ISSN 0003-6951

Ghosh, C. K.; Popuri, S. R.; Mahesh, T. U. & Chattopadhyay, K. K. (2006). Preparation of Nanocrystalline $CuAlO_2$ through Sol-Gel Route. *Journal of Sol-Gel Science and Technology*, Vol.52, No.1 (March 2009), pp. 75-81, ISBN 0928-0707

Gu, G.; Bulović, V.; Burrows, P. E. & Forrest, S. R. (1996). Transparent Organic Light Emitting Devices. *Applied Physics Letters*, Vol.68, (March 1996), pp. 2606-2608, ISSN 0003-6951

Guan, H.-S.; Gheng, C.-H.; Li, W.-C.; Geng, D.-F.; Fan, Z.-Q.; Chang, Y. C.; Zhao, W.; Guo, Z.-Q. & Du. G.-T. (2009). Influence of Transparent Anode on Luminescent Performance of Near-Infrared Organic Light-Emitting Diodes. *Chemical Research in Chinese Universities*, Vol.25, No.6, (November 2009), pp. 786-789, ISSN 1005-9040

Guillen, G. & Herrero, J. (2009). Transparent Conductive ITO/Ag/ITO Multilayer Electrodes Deposited by Sputtering at Room Temperature, *Optics Communications*, Vol.282, No.4, (November 2008), pp. 574-578, ISSN 0030-4018

Gu, G.; Burrows, P. E.; Forrest, S. R. & Thompson, M. E. (1997). Vacuum-Deposited, Nonpolymeric Flexible Organic Light-Emitting Devices. *Optics Letters*, Vol.22, No.3, (1997), pp. 172-174, ISSN 0146-9592

Guo, Z.G.; Zhou, F.; Hao, J.C. & Liu, W.M. (2005). Stable Biomimetic Super-Hydrophobic Engineering Materials. *Journal of the American Chemical Society*, Vol.127, No.45, (July 2005), pp. 15670-15671, ISSN 0002-7863

Han, J.; Mantas, P.Q. & Senos, A.M.R. (2001). Effect of Al and Mn Doping on the Electrical Conductivity of ZnO. *Journal of European Ceramic Society*, Vol.21, No.10-11, (2001), pp. 1883~1886, ISSN 0955-2219

He, T. & Kanicki, J. (2000). High-Efficiency Organic Polymer Light-Emitting Heterostructure Devices on Flexible Plastic Substrates. *Applied Physics Letters*, Vol.76, No.6, pp. 661-663, ISSN 0003-6951

Hirata, G.A.; McKittrick, J.; Cheeks, T.; Siqueiros, J.M.; Diaz, J.A.; Contreras, O. & Lpez, O.A. (1996). Synthesis and Optoelectronic Characterization of Gallium Doped Zinc Oxide Transparent Electrodes. *Thin Solid Films*, Vol.288, No.1-2, (November 1996), pp. 29-31, ISSN 0040-6090

Honda, S.; Tsujimoto, A.; Watamori, M. & Oura, K. (1995). Oxygen Content of Indium Tin Oxide Films Fabricated by Reactive Sputtering. *Journal of Vacuum Science & Technology A: Vacuum, Surfaces, and Films*, Vol.13, No.3, (May 1995), pp. 1100-1103, ISSN 0734-2101

Huang, C. J. (2003). Silicon Dioxide Buffer at Anode/Polymer Interface for Enhanced Brightness and Efficiency of Polymer Light-Emitting Diode. *Journal of Materials Science Letters*, Vol.22, No.20, (May 2003) pp. 1423-1425 ISSN 1573-4811

Huang, Y.-J.; Liu, C.-W.; Chu, S.-Y. & Lo, K.-Y. (2010) The Formation of p-Type ZnO Films by Thermal Diffusion from the Low Energy, High Dose Phosphorus-Implanted Si

Substrate. *Journal of The Electrochemical Society*, Vol.157, No.4, (December 2009), pp. H435-H437, ISBN 0013-4651

Ishii, H.; Sugiyama, K.; Ito, E. & Seki, K. (1999). Energy Level Alignment and interfacial Electronic Structure at Organic/Metal and Organic/Organic Interfaces. *Advanced Materials*, Vol.11, No.11, (June 1999), pp. 605-625, ISSN 1521-4095

Kawazoe, H.; Yasukawa, M.; Hyodo, H.; Kurita, M.; Yanagi, H. & Hosono, H. (1997). p-Type Electrical Conduction in Transparent Thin Films of $CuAlO_2$. *Nature*, Vol.389, No.6654, (June 1997), pp. 939-942, ISBN 0028-0836

Kim, H.-M.; Bae, K. & Sohn S. (2011). Electronic and Optical Properties of Indium Zinc Oxide Thin Films Prepared by Using Nanopowder Target. *Japanese Journal of Applied Physics*, Vol.50, No.4, (June 2010), pp. 045801-045805, ISBN 0021-4922

Kim, H. & Sohn, S. (2009). Preparation of SiO_x and SiO_xN_y Films on PEN using Facing Target Sputtering System. *ECS Transactions*, Vol.19, No.9, (May 2009), pp. 9-16, ISSN 1938-6737

Kim, H.; Gilmore C.; Piqu'e, A.; Horwitz, J.; Mattoussi, H.; Murata, H.; Kafari, Z. & Chrisey, D. (1999). Electrical, Optical, and Structural Properties of Indium-Tin–Oxide Thin Films for Organic Light-Emitting Devices. *Journal of Applied Physics*, Vol.86, No.11, (June 1999), pp. 6451-6461, ISSN 0021-8979

Kim, H.; Gilmore, C.; Horwitz, J.; Piqu'e, A.; Murata, H.; Kushto, G.; Schlaf, R.; Kafafi, Z. & Chrisey, D. (2000). Transparent Conducting Aluminum-Doped Zinc Oxide Thin Films for Organic Light-Emitting Devices. *Applied Physics Letters*, Vol.76, No.3, (January 2000), pp. 259-261, ISSN 0003-6951

Kim, H.; Horwitz, J.S.; Kim, W.H.; Qadri, S.B. & Kafafi, Z.H. (2003). Anode Material Based on Zr-Doped ZnO Thin Films for Organic Light-Emitting Diodes. *Applied Physics Letters*, Vol.83, No.18, (September 2003), pp. 3809-3811, ISSN 0003-6951

Kim, H.; Kim, H.; Lee, J.; Lee, K.; Yi, J.; Oh, S.; Sohn, S.,; Jung, D.; Jang, S. & Chae, H. (2008). Admittance spectroscopic analysis of organic light emitting diodes with the CF_x plasma treatment on the surface of indium tin oxide anodes. *Thin Solid Films*, Vol.516, No.7, (August 2008) pp.1370, ISSN 0040-6090

Kim, K. H.; Son, I. H.; Song, K. B.; Kong, S. H.; Keum, M. J.; Nakagawa, S. & Naoe, M. (2001). Thin Film Properties by Facing Targets Sputtering System. *Applied surface Science*, Vol.169-170, No.1, (January 2001), pp. 410-414, ISSN 0169-4332

Kim, S. Y. & Lee, J.-L. (2005). Effect of Thin Iridium Oxide on The Formation of Interface Dipole in Organic Light-Emitting Diodes. *Applied Physics Letters*, Vol.87, No.23, (September 2005), pp. 232105-232107, ISSN 1077-3118

Kim, Y.; Park, J.; Choi, D.; Jang, H.; Lee, J.; Park, H.; Choi, J.; Ju, D.; Lee, J. & Kim, D. (2007). ITO/Au/ITO Multilayer Thin Films for Transparent Conducting Electrode Applications. *Applied Surface Science*, Vol.254, No.5, (June 2007), pp. 1524-1527, ISSN 0169-4332

Kudo, A.; Yanagi, H.; Hosono, H. & Kawazo, H. (1998). $SrCu_2O_2$: A p-Type Conductive Oxide with Wide Band Gap. *Applied Physics Letters*, Vol. 73, (May 1998), pp. 220-222, ISSN 0003-6951

Kwon, S. H.; Paik, S. Y. & Yoo, J. S. (2002). Electroluminescent Properties of MEH-PPV Light-Emitting Diodes Fabricated on The Flexible Substrate. *Synthetic Metals*, Vol.130, No.1, (August 2002), pp. 55-60, ISSN 0379-6779

Li, C.; Furuta, M.; Hiramatsu, T.; Furuta, H. & Hirao, T.(2009). Effects of Substrate on The Structural, Electrical and Optical Properties of Al-Doped ZnO Films Prepared by Radio Frequency Magnetron Sputtering. *Thin Solid Films*, Vol.517, No.11, (November 2008), pp. 3265-3268, ISSN 0400-6090

Li, J.; Yahiro, M.; Ishida, K.; Yamada, H. & Matsushige K. (2005). Enhanced Performance of Organic Light Emitting Device by Insertion of Conducting/Insulating WO_3 Anodic Buffer Layer. *Synthetic Metals*, Vol.151, No.2, (May 2005), pp. 141-146, ISSN 0379-6779

Lu, H. T. & Yokoyama, M. (2003). Enhanced Emission in Organic Light-Emitting Diodes using Ta_2O_5 Buffer layers. *Solid-State Electronics*, Vol.47, No.8, (December 2002), pp. 1409-1412, ISSN 0038-1101

Mardare, D.; Tasca, M.; Delibas, M. & Rusu, G.I. (2000). On The Structural Properties and Optical Transmittance of TiO_2 R.F. Sputtered Thin Films. *Applied surface Science*, Vol.156, No.1-4, (2000), pp. 200-206, ISSN 0169-4332

Medvedeva, J. E. (2006). Magnetically Mediated Transparent Conductors: In_2O_3 Doped with Mo. *Physical Review Letters*, Vol.97, No.8, (August 2006), pp. 086401-086404, ISSN 0031-9007

Minami T.; Nanto, H. & Takata, S. (1985). Highly Conductive and Transparent Aluminum Doped Zinc Oxide Thin Films Prepared by RF Magnetron Sputtering. *Thin Solid Films*, Vol.124, No.1, (February 1985), pp. 43-47, ISSN 0040-6090

Minami, T. (1999) Transparent and conductive multicomponent oxide films prepared by magnetron sputtering. *Journal of Vacuum Science & Technology A: Vacuum, Surfaces, and Films*, Vol.17, No.4, (July 1999), pp. 1765-1772, ISSN 0734-2101

Mitsui, A. & Masumo , K. (2003). Effect of Zirconium Oxide Undercoat on Microstructure and Properties of Tin-Doped Indium Oxide Film for Organic Light Emitting Devices. *Thin Solid Films*, Vol. 442, No.1-2, (September 2003), pp. 140-144, ISSN 0040-6090

Miyata, T.; Minami, T.; Shimokawa, K.; Kakumu, T. & Ishii, M. (1997). New Materials Consisting of Multicomponent Oxides for Thin-Film Gas Sensors, *Journal of the Electrochemical Society*, Vol.144, No.7, (February 1997), pp. 2432-2436, ISSN 0013-4651

Nose, M.; Nagae, T.; Yokota, M.; Saji, S.; Zhou, M. & Nakada, M. (1999). Electrical and Colorimetric Properties of TiN Thin Films Prepared by DC Reactive Sputtering in a Facing Targets Sputtering (FTS) System. *Surface and Coatings Technology*, Vol.116-119, No.1, (September 1999), pp. 296-301, ISSN 0257-8972

Ohta, H.; Orita, M.; Hirano, M.; Tanji, H.; Kawazoe, H. & Hosono, H. (2000). Highly Electrically Conductive Indium-Tin-Oxide Thin Films Epitaxially Grwn on Yttria-Stabilized Zirconia(100) by Plused-Laser Deposition. *Applied Physics Letters*, Vol.76, No.19, (March 2000), pp. 2740-2742, ISBN 0003-6951

Ott, A.W. & Chang, R.P.H. (1999). Atomic Layer-Controlled Growth of Transparent Conducting ZnO on Plastic Substrates. *Materials Chemistry and Physics*, Vol.58, No.2, (November 1998), pp. 132-138, ISSN 0254-0584

Park, J.-M.; Hong, J.-S.; Kim, J.-J.; Park, S.-H.; Kim, H.-M. & Ahn, J.-S. (2006). Bending Effects of Indium-Zinc Oxide Thin Films Deposited on Polyethylene Terephthalate Substrate by Radio Frequency Magnetron Sputtering. *Journal of the Korean Physical Society*, Vol.48, No.6, (June 2006), pp.1530-1533, ISSN 0374-4884

Park, J. H.; Cho, Y. C.; Shin, J. M.; Cha, S.-Y.; Cho, C. R.; Kim, H. S.; Yoon, S. J.; Jeong, S.-Y.; Park, S. E.; & Lim, A.-R. (2007). A Study of Transparent Conductive Aluminum-Doped Zinc Oxide Fabricated on a Flexible Polyethersulphone (PES) Substrate. *Journal of the Korean Physical Society*, Vol.51, No.6 (December 2007), pp. 1968-1972, ISSN 0374-4884

Park, J. M.; Kim, J. J.; Kim, H. M.; Kim, J. H.; Ryu, S. W.; Park, S. H. & Ahn, J. S. (2006). Substrate Effects on The Characteristices of $(In_2O_3)_{1-x}(ZnO)_x$ Films. *Journal of the Korean Physical Society*, Vol.48, No.6, (June 2006), pp. 1624-1627, ISSN 0374-4884

Park, S.-M.; Ikegami, T. & Ebihara, K. (2006). Effects of Substrate Temperature on the Properties of Ga-doped ZnO by Pulsed Laser Deposition. *Thin Solid Films*, Vol.513, No.1-2, (December 2005), pp. 90-94, ISSN 0040-6090

Qiu, C.; Xie, Z.; Chen, H; Wong, M. & Kwok, H. S. (2003). Comparative Study of Metal or Oxide Capped Indium Tin Oxide Anodes for Organic Light-Emitting Diodes. *Journal of Applied Physics*, Vol.93, No.6, (January 6 2003), pp. 3253-3258, ISSN 1089-7550

Rauf, I.A. (1993). Low Resistivity and High Mobility Tin-Doped Indium Oxide Films. *Materials Letters*, Vol.18, No.3, (September 1993), pp. 123-127, ISSN 0167-577X

Robins, J. J. & Wolden, C.A. (2003). High Mobility Oxides: Engineered Structures to Overcome Intrinsic Performance Limitations of Transparent Conducting Oxides. *Applied Physics Letters*, Vol.83, No.19, (September 2003), pp. 3933-3935, ISSN 0003-6951

Roy, B.; Ode, A.; Readey, D.; Perkins, J.; Parilla, P. Tepliin, C.; Kaydanova, T.; Miedaner, A.; Curtis, C.; Martinson, A.; Coutts, T.; Ginley, D. & Hosono, H. (2003). Towards High Performance p-Type Transparent Conducting Oxides. *National Center for Photovoltaics and Solar Program Review Meeting*, Denver, Colorado, March 24-26, 2003, NREL reported NREL/CP-520-33595

Sato, H.; Minami, T.; Takata, S. & Yamada, T. (2003). Transparent Conducting p-Type NiO Thin Films Prepared by Magnetron Sputtering. *Thin Solid Films,* Vol. 236, No. 1-2, (December 1993), pp. 27-31, ISSN 0040-6090

Shan, F.K.; Shin, B.C.; Kim, S.C. & Yu, Y.S. (2003). Optical Property and Aging Effect of ZnO Thin Films. *Journal of the Korean Physical Society*, Vol.42, No.94, (April 2003), pp. S1174-1177, ISSN 0374-4884

Sheng, S.; Fang, G.; Li, C.; Xu, S. & Zhao, X. (2006). p-Type Transparent Conducting Oxides. *Physica Status Solidi (a)*, Vol.203, No.8, (April 2006), pp. 1891-1900, ISSN1862-6300

Shrotriya, V. & Yang, Y. (2005). Capacitance-Voltage Characterization of Polymer Light-Emitting Diodes. *Journal of Applied Physics*, Vol.97, No.5, (December 2004), pp. 54504-54509, ISSN 1089-7550

Snure, M. & Tiwari, A. (2007). $CuBO_2$: A p-Type Transparent Oxide. *Applied Physics Letter,* Vol.91, No.9 , (July 2007), pp. 092123-092125, ISBN 0003-6951

Sohn, S.; Lee, J.; Park, K.; Jung, D.; Kim, H. & Yi, J. (2008). Enhanced Performance of an Organic Light-Emitting Device by Using an Oxidant Component during a Surface Reaction Process by Using Atomic Layer Chemical Vapor Deposition. *Journal of the Korean Physical Society*, Vol. 53, No.6, (June 2008), pp. 3416-3421, ISSN 0374-4884

Takiji, S.; Kohei, I. & Seimei, S. (2004). Ultra-water-repellent surface: fabrication of complicated structure of SiO_2 nanoparticles by electrostatic self-assembled films.

Applied Surface Science, Vol.237, No.3, (September 2004), pp. 539-543, ISSN 0169-4332

Teplin, C.; Kaydanova, T.; Young, D.; Perkins, J. & Ginley, D. (2004), A Simple Method for the Preparation of Transparent p-Type Ca-doped $CuInO_2$ Films: Pulsed-Laser Deposition from Air-Sintered Ca-Doped $Cu_2In_2O_5$ Targets. *Applied Physics Letters*. Vol.85, No.17, (October 2004), pp. 3789-3791, ISSN 0003-6951

Tuttle, G.; Kroemer, H. & English, J. H. (1989). Electron Concentrations and Mobilities in AlSb/InAs/AlSb Quantum Wells. *Journal of Applied Physics*, Vol.65, No.12, (February 1989), pp. 5239-5242, ISSN 0021-8979

Ueda, K.; Tabata, H. & Kawai, T. (2001). Magnetic and Electric Properties of Transition-Metal-Doped ZnO Films. *Journal of Applied Physics*, Vol.79, No.7, (August 2001), pp. 988-990, ISSN 0021-8979

Ueda, K.; Hase, T.; Yanagi, H.; Kawazoe, H.; Hosono, H.; Ohta, H.; Orita, M. & Hirano, M. (2001). Epitaxial Growth of Transparent p-Type Conducting $CuGaO_2$ Thin Films on Sapphire (001) Substrates by Pulsed Laser Deposition. *Journal of Applied Physics*, Vol.89, No.3, pp. (July 2000), pp. 1790-1793, ISSN 1089-7550

VanSlyke, S. A.; Chen, C. H. & Tang, C. W. (1996). Organic Electroluminescent Devices with Improved Stability. *Applied Physics Letters*, Vol.69, No.15, (August 1996), pp. 2160-2162, ISSN 1077-3118

Wang, R.-C.; Liu, C.-P.; Huang, J.-L. & Chen, S.-J. (2006). Single-Crystalline AlZnO Nanowires/Nanotubes Synthesized at Low Temperature. *Applied Physics Letters*. Vol.88, No.2, (August 2005), pp. 023111-023113, ISSN 0003-6951

Yan, Y.; S.J, P.; Dai, J.; Chang, R.P.H.; Wang, A. & Marks, T.J,. (1998). Polytypoid Structures in Annealed In_2O_3–ZnO Films, *Applied Physics Letters*, Vol.73, No.18, (August 1998), pp. 2585-2587, ISSN 0003-6951

Yanagi, H.; Hase, T.; Ibuki, S.; Ueda, K. & Hosono, H. (2000). Bipolarity in Electrical Conduction of Transparent Oxide Semiconductor $CuInO_2$ with Delafossite Structure. *Applied Physics Letters*, Vol.78, No.11, (January 2001), pp. 1583-1585, ISSN 0003-6951

Yanagi, H.; Kawazoe, H.; Kudo, A.; Yasukawa, M. & Hosono, H. (2000). Chemical Design and Thin Film Preparation of p-Type Conductive Transparent Oxides. *Journal of Electroceramics*, Vol. 4, (June 2000), pp. 407-414, ISSN 1385-3449

Yanagi, H.; Ueda, K.; Ohta, H.; Orita, M, Hirano, M. & Hosono, H. (2002). Fabrication of All Oxide Transparent p-n Homojunction using Bipolar $CuInO_2$ Semiconducting Oxide with Delafossite Structure. *Solid State Communication*, Vol.121, No.1, (December 2001), pp. 15-17, ISSN 0038-1098

Yang, M.; Xiliang, Y.; Huaxian, C.; Jie, S.; Yiming, J.; Zhuangjian, Z. & Zhongyi, H. (2001). A New Transparent Conductive Thin Film In_2O_3:Mo. *Thin Solid Films*, Vol. 394, No.2, (August 2001), pp. 219-223, ISSN 0040-6090

Yang, T.L.; Zhang, D.H.; Ma, J.; Ma, H.L. & Chen, Y. (1998). Transparent Conducting ZnO:Al Films Deposited on Organic Substrates Deposited by R.F. Magnetron-Sputtering. *Thin Solid Films*, Vol.326, No.1, (August 1998), pp. 60-62, ISSN 0040-6090

Yoshida, Y.; Wood, D. M.; Gessert, T. A. & Coutts, T. J. (2004). High-Mobility Sputtered Films of Indium Oxide Doped with Molybdenum. *Applied Physics Letters*. Vol.84, No.12, (January 2004), pp. 2097-2099, ISSN 0003-6951

Zhang, Y. & Forrest, S. R. (1993). Mechanisms of Quasiepitaxial Ordering at Molecular Thin Film Interfaces. *Physical Review Letters*, Vol.71, No.17, (October 1993), pp. 2765-2768, ISSN 0031-9007

10

Micro-Cavity in Organic Light-Emitting Diode

Young-Gu Ju
Department of Physics Education, Kyungpook National University
Korea

1. Introduction

The study on micro-cavity in organic light-emitting diode(OLED) demands understanding the theory of multi-layer films. It is because OLED is basically an optical device and its structure consists of organic or inorganic layers of sub-wavelength thickness with different refractive indices. When the electron and holes are injected through the electrodes, they combine in the emission layers emitting the photons. These photons will experience the reflection and transmission at each interface and the interference will determine the intensity profile. The light reflected at the interfaces or the metallic electrode returns to the emission layer and affects the radiation efficiency. In optical terminology, OLED belongs to a micro-cavity being comprised of multi-layers. Therefore, before studying the cavity effect of OLED, we better begin with the theory of multi-layer film theory in optics. This theory is well explained in most of textbook dealing with optics since it relates to optical coatings and lasers(Fowles, 1975; Born & Wolf, 1989). In this section, a brief review will be given for the purpose of self-containment, which will be especially helpful for the beginners.

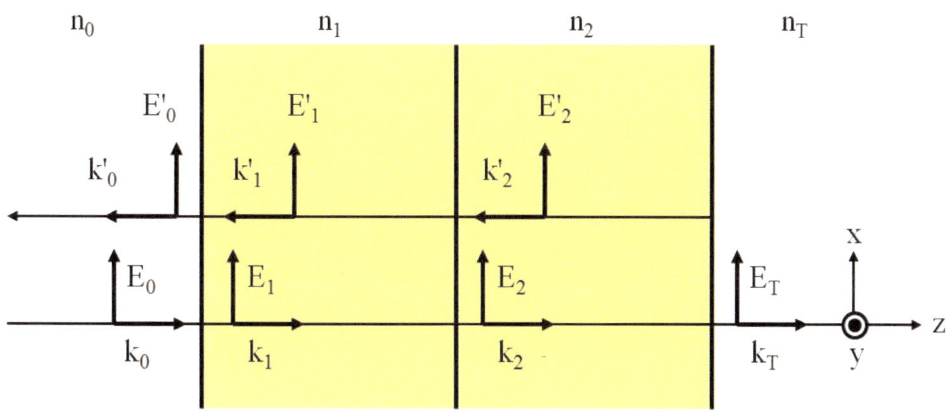

Fig. 1. The schematic diagram of multi-layer and the electric fields inside the film

The four layers and the electric fields are displayed in Fig. 1. For the time being, the four layers are called as the zero-th, the first, the second and the final layer, respectively. A plane wave designated E_0 with propagation vector of k_0 is normally incident on the first layer from the zero-th layer and generates the reflected electric field designated by E_0' with k_0'.

The first medium also contains two electric fields E_1 and E_1'. E_1 and E_1' represents the electric fields measured at the first interface between the zero-th medium and the first medium. E_2 and E_2' also represents the electric fields travelling in positive and negative z-direction, which is also measured at the second interface. The final layer has only one electric field E_t since it doesn't have a reflection.
In terms of geometrical optics, the reflection occurs infinite times between the interfaces. However, in this type of analysis, all those rays are summed up into two electric fields in each medium. In other words, the electric fields propagating in positive and negative z-direction already take into account of multiple reflections. In this configuration, the final amplitude of the electric field is obtained by applying the boundary condition at each interface. The boundary conditions used in the calculation requires that the E field component and H-field component parallel to the interface should be continuous at the interfaces. If these boundary conditions apply to the interface between the first medium and the second medium, the following equations hold.

$$E_1 e^{ikl} + E_1' e^{-ikl} = E_2 + E_2' \tag{1}$$

$$H_1 e^{ikl} + H_1' e^{-ikl} = H_2 + H_2' \tag{2}$$

The electric E_1 has the extra phase factor of exp(ikl) since it propagate further in z direction by the layer thickness l. In addition, H field satisfies the following relation with E field.

$$\vec{H} = \frac{1}{\mu\omega}\vec{k} \times \vec{E} \tag{3}$$

The propagation vector k is equal to the product of refractive index n and the propagation vector k_0 in vacuum. The amplitude of H field is proportional to E field and refractive index n since the k_0, permeability μ and angular frequency ω do not depend on the medium when the medium is non-magnetic. In this manner, Eq. 2 becomes Eq. 4.

$$n_1 E_1 e^{ikl} - n_1 E_1' e^{-ikl} = n_2 E_2 - n_2 E_2' \tag{4}$$

In general, the subscript denoting the layer number can be replaced by i. For instance, the number 1 can be changed into i and the number 2 can be i+1, representing i-th and (i+1) th layer, respectively. The two equations with the new indices are expressed in Eq. 5, and Eq. 6.

$$E_i e^{ik_i l_i} + E_i' e^{-ik_i l_i} = E_{i+1} + E_{i+1}' \tag{5}$$

$$n_i E_i e^{ik_i l_i} - n_i E_i' e^{-ik_i l_i} = n_{i+1} E_{i+1} - n_{i+1} E_{i+1}' \tag{6}$$

The two special cases for this general iteration formula occur at the first interface and the last one. The first layer and the final layer have zero thickness so that exponential factors become 1. Moreover, the final layer doesn't have the reflected wave so that the second term on the right hand side of equation should disappear. The conversion of the equations into the neater forms can be done using a matrix.

$$\begin{pmatrix} e^{ik_i l_i} & e^{-ik_i l_i} \\ n_i e^{ik_i l_i} & -n_i e^{-ik_i l_i} \end{pmatrix} \begin{pmatrix} E_i \\ E_i' \end{pmatrix} = \begin{pmatrix} 1 & 1 \\ n_{i+1} & -n_{i+1} \end{pmatrix} \begin{pmatrix} E_{i+1} \\ E_{i+1}' \end{pmatrix} \tag{7}$$

$$\begin{pmatrix} E_i \\ E_i' \end{pmatrix} = \begin{pmatrix} e^{ik_i l_i} & e^{-ik_i l_i} \\ n_i e^{ik_i l_i} & -n_i e^{-ik_i l_i} \end{pmatrix}^{-1} \begin{pmatrix} 1 & 1 \\ n_{i+1} & -n_{i+1} \end{pmatrix} \begin{pmatrix} E_{i+1} \\ E_{i+1}' \end{pmatrix} = A_{i+1} \begin{pmatrix} E_{i+1} \\ E_{i+1}' \end{pmatrix} \tag{8}$$

$$\begin{pmatrix} E_0 \\ E_0' \end{pmatrix} = A_1 A_2 \cdots A_N \begin{pmatrix} E_T \\ 0 \end{pmatrix} \tag{9}$$

From the matrix formula, we can see more clearly the evaluation process of the electric field in the multi-layer films. If we set the final goal to finding the reflectivity of the films when looking on the first layer, the problem is equivalent to evaluating the ratio of E_0' to E_0. According to Eq. 8, E_0 fires the next sequence by multiplying A_1 and (E_1, E_1'). Since A_1 can be obtained using the index and the thickness of the layer, the problem is finding out E_1 and E_1'. The E_1 and E_1' can be calculated by finding out E_2 and E_2'. In this way, the iteration continues until it reaches the final layer. At the final layer, E_t and E_t' should be given somehow. The solution is rather simple. E_t' is 0 and E_t can be any value, for example, 1. The matrix equations express E_0 and E_0' as a ratio to E_t anyway. However, the reflectivity is again the ratio of E_0' to E_0 so that the absolute value doesn't affect the answer no matter what value we choose for E_t. If someone insists on the absolute value of E_0 and E_0', then the E_0 and E_0' are evaluated in the unit of E_t. Since E_0 is given as an input value, the E_t is adjusted to make E_0 equal to the given input value. This E_t also can be used to get the absolute value for E_0'.

2. Programming the multi-layer equations in MATLAB

In this section, we present a MATLAB code to calculate the electric field and the reflectivity inside the multi-layers. This approach will provides a realistic view on the field profile inside the cavity and how it affects the optical property by giving a numerical value in a concrete example. Among many programming languages, MATLAB is chosen since it is a high level language. Programmers usually don't have to worry about the details of matrix manipulation, complex variable, graphics and so on. From an educational viewpoint, the code in MATLAB is easy to explain.

```
( FILE: MLay.m )
clear all;
global Nm;
global rfr;
global MA;
global Efl;
wvl = 0.5;  % in um
rfr = [1.0 2.5 1.5 2.5 1.5 2.5 1.0];
thick = wvl./(4*rfr);
Nm = length(rfr);
thick(1) = 0;
thick(Nm) = 0;
kv = 2*pi*rfr/wvl;
MA = @(k) [exp(i*kv(k)*thick(k)), exp(-i*kv(k)*thick(k));
rfr(k)*exp(i*kv(k)*thick(k)), -rfr(k)*exp(-i*kv(k)*thick(k))];
Efl=[];
```

```
vEf = Ef(1);
[thl2, Efl2, intfl2] = EProf(Efl,kv,thick);
Instl2=abs(Efl2).^2;
plot(thl2, Instl2, 'r-', thl2(intfl2), Instl2(intfl2), 'bO');
R = abs(vEf(2)/vEf(1))^2
```

The program consists of three script files. The codes of "MLay.m" are presented above. The lines of "Ef.m" and "EProf.m" are also presented in the following. The input parameters such as wavelength, refractive index and thickness of the layers are set in the first program. The variables corresponding to these input parameters are "wvl", "rfr" and "thick". It is worthy to note that the first layer and the final layer are the zero-th layer(incident medium) and the transmission medium as described in Fig. 1. The thicknesses of those two layers are set to 0. The thicknesses of the layers are set to be a quarter wavelength as a starting example. These values can be changed complying with the user's need. The other variation is found in the indexing of the array variable. The initial index of the array in MATLAB is 1 instead of 0, which is different from Eq. 8. The main calculation is performed in the definition of "MA" and the "Ef(1)". Ef(1) calls the function routine defined in the separate script file "Ef.m". It is an execution of iteration formula Eq. 8. When it reaches the final layer, it is given the final numerical value 1.0 for the electric field at the final medium and returns to the previous function calls, consecutively. All the readers have to do is to put these three files in a MATLAB current folder and run "MLay.m".

```
( FILE: Ef.m )
function [z]=Ef(k)
global Nm;
global rfr;
global MA;
global Efl;
if(k == Nm)
z=[1.0; 0.0];
else
z=inv(MA(k))*[1 1;rfr(k+1) -rfr(k+1)]*Ef(k+1);
end
Efl = [transpose(z); Efl];
end
( FILE: EProf.m )
function [thl2, Efl2, intfl2] = EProf(Efl, kv, thick)
Ndiv = 10;
Nm = length(thick);
cth0 = cumsum(thick);
Efl2 = [Efl(1,1)+Efl(1,2)];
thl2 = [cth0(1)];
intfl2 = [1];
for n=2:(Nm-1)
cth = [0:thick(n)/Ndiv:thick(n)];
Efl_tmp = Efl(n, 1)*exp(i*kv(n)*cth)+Efl(n, 2)*exp(-i*kv(n)*cth);
Efl2 = [Efl2, Efl_tmp];
thl2 = [thl2, cth0(n-1)+cth];
intfl2 = [intfl2, length(thl2)];
```

```
end
Efl2 = [Efl2, [Efl(Nm,1)+Efl(Nm,2)]];
thl2 = [thl2, cth0(Nm-1)+thick(Nm)];
end
```

Finally, the "EProf.m" is in charge of calculating the field profile at the interfaces and the between them. The E field is assumed to vary in accordance with the plane wave solution. In other words, the electric field amplitudes evaluated at the interfaces are used to give the value between them. The results are plotted using "plot" function and the reflectivity of the multi-layers is calculated from the E_0 and E_0'.

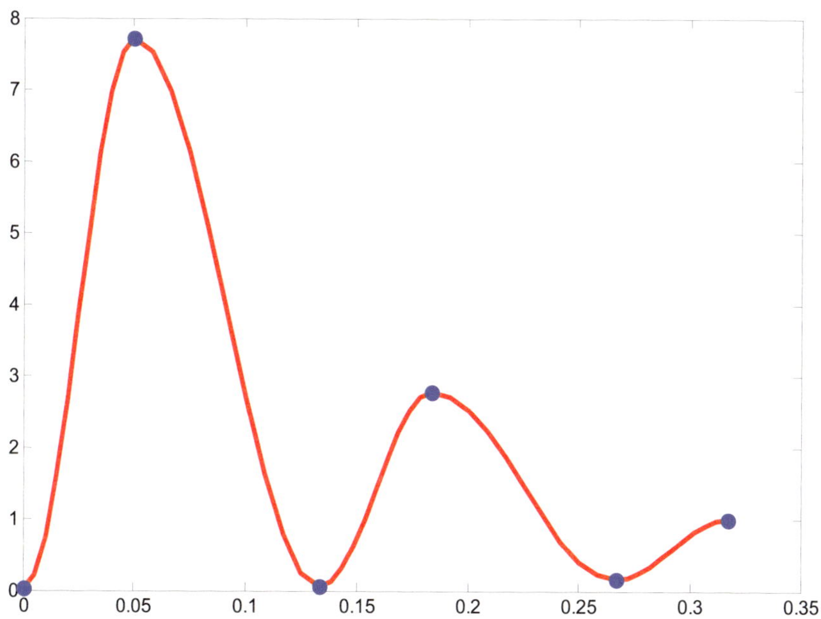

Fig. 2. The output from the execution of the program "MLay.m" is displayed. R = 0.9204. The refractive index and the thickness for input are

"rfr = [1.0 2.5 1.5 2.5 1.5 2.5 1.0];"
"thick = wvl./(4*rfr);"
"thick(1) = 0;"
"thick(Nm) = 0;".

The result of execution is shown as in Fig. 2. In this simulation, the layer structure is comprised of air, 2.5 pairs of TiO2/SiO2, and air from the incident medium. The final reflectivity of the layers is about 92 %. The thickness of each layer is a quarter wavelength to obtain the high reflection. The circular marks represent the position of the interfaces so that the reigon between the marks corresponds to a single layer. Since the incident medium and the final medium have zero thickness, the number of layers seen in Fig. 2 is only 4 instead of 6. The peak intensity decreases as the light penetrates the layers.

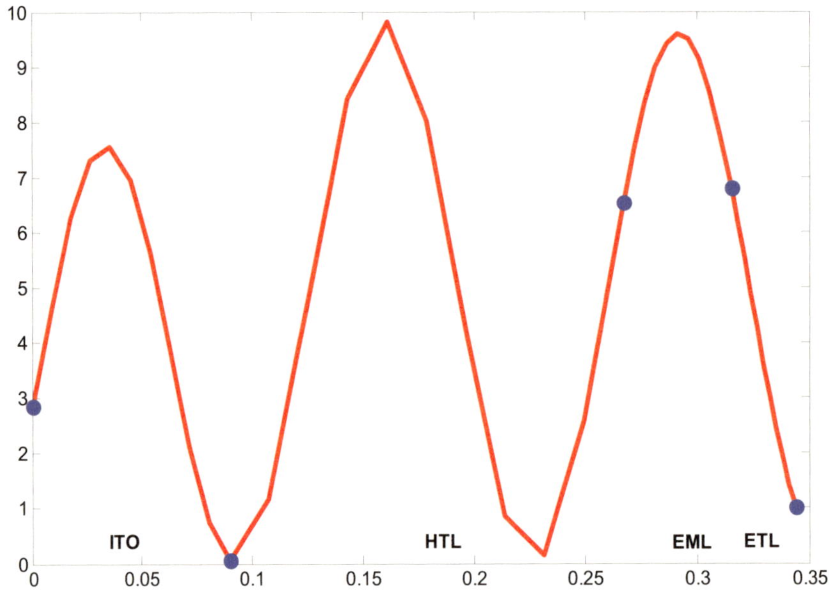

Fig. 3. The output from the execution of the program "MLay.m" is displayed. R = 0.8706. The refractive index and the thickness for input are

"rfr = [1.5 2.13 1.87 1.94 1.75 0.644+5.28i];"

"thick = [0 0.09 0.177 0.048 0.029 0];"

In order to analyze of the real OLED device, the refractive index and the thickness are replaced by those of the OLED. The layers used in the simulation are overcoat, indium tin oxide(ITO), hole transport layer(HTL), emission layer(EML) and aluminum cathode, respectively. The title of each layer is typed in as an annotation at the bottom of the plot between the circular marks, which represent the boundaries of each layer. The light is supposed to be incident on the overcoat layer first.

In the design of the OLED cavity, the position of the emission layer is important since it affects the emission efficiency of the device. As in Fig. 3, the emission layer occupies the peak position of the intensity profile. It means that the optical density of state is high over the emission layer and the electron transition rate is also high. If the emission layer is placed around the node of the intensity profile, the spontaneous emission is suppressed rather than enhanced. Quantum mechanics states that the radiation probability is proportional to the electron transition probability and the optical density of states(Gasiorowicz, 1984). If the two atoms have the same transition probability, but have different optical density of states, which generally depends on the position of the atom inside the cavity, the resultant transition rate comes to be determined by the optical density around the atom. This kind of enhancement is often observed in the design of a micro-cavity laser such as vertical-cavity surface-emitting laser(VCSEL). Therefore, the OLED engineer should pay attention to controlling the thickness of the layers so that the emission layer is positioned at the peak of the intensity, otherwise, he will lose optical efficiency.

Eq. 8 can be modified into a bit more complicated form in order to handle oblique incidence angle, polarization, dispersion, user-friendly interface, and so on. But, all these variations stick around the same boundary conditions under different circumstances. The generalization of the program is left for the fun of the reader.

3. Tunable micro-cavity in OLED

In the former section, the MATLAB program calculates the electric field profile inside the conventional OLED device. The multi-layer structure naturally forms the cavity through the reflections at the interfaces of the cathode, organic layers, anode, and so on. The bandgap difference between the materials causes the index differences and this index mismatch accompanies reflections. Therefore, it is very hard to get rid of the reflections at the interfaces and the cavity effect inside the OLED. In this section, we present the method of how to more actively use this interference in order to control the emission wavelength of the white OLED.

The design of a strong cavity can be realized by reinforcing the bottom mirror. In the conventional OLED, the anode usually consists of a transparent conductive oxide like indium zinc oxide(IZO) and it plays a role of bottom mirror. The higher index contrast between IZO and the neighboring layers increases the reflectivity at the interfaces. Since Al cathode already keeps high reflectance over the visible wavelength, increasing the reflectivity of the bottom mirror can strengthens the cavity effect. At first thought, a metallic layer at the bottom can be used to make a strong cavity. However, metals usually have a very high imaginary index, which means strong absorption. In this case, the light generated inside cannot escape the cavity unless the metallic layer is very thin. As matter of fact, very thin layer of silver can forms the strong cavity in OLED. However, the thin layer is not easy to deposit maintaining the uniformity of the device.

In this article, we present the results in which distributed Bragg reflector(DBR) comprises the bottom mirror of the cavity. The DBR is simply a periodic stack of dielectrics with different refractive index. The thickness is usually a quarter wavelength to increase the reflectivity. The schematic diagram of the micro-cavity consisting of DBR is displayed as in Fig. 4. Compared to the structure used in the simulation for Fig. 3, the bottom side has three additional layers of IZO/SiO2/IZO. Although these three layers seem to lack the number of layers comparing to the DBR in VCSELs, this number of layers are adequate for controlling the emission property of the OLED.

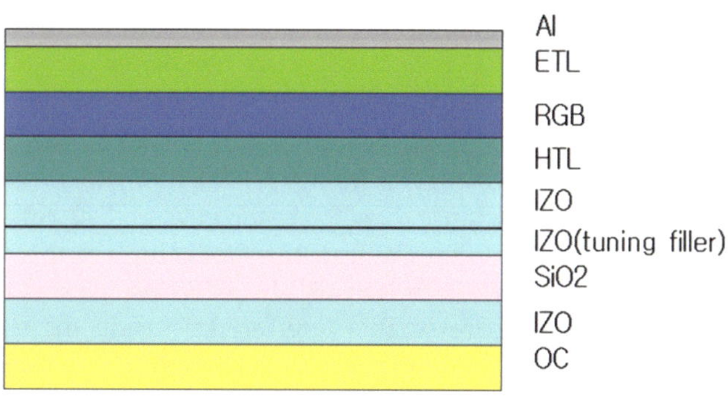

Fig. 4. The structure of strong cavity used for controlling the emission wavelength of white OLED.

It is worthy to note that the thickness of the second IZO layer is variable to tune the resonant wavelength. In general, the resonant peak moves depending on the thickness of the cavity. In a conventional cavity, the central layer such as the emission layer in this structure is adjusted to change the resonance. However, the emission layer in OLED is kind of sensitive layer which should be untouched due to the optimization of the emission characteristics. Instead, the extra tuning layer is added to the anode layer to affect the cavity length. This technique is often used to make a tunable VCSEL(Ju et al, 2000).
The analysis of the cavity effect needs the information on the spectral reflectance. The program used in the previous section should be modified to calculate the reflectivity over a range of wavelength instead of a single wavelength. The modified code is shown below.

```
( FILE: MLay_a2.m )
clear all;
global Nm;
global rfr;
global MA;
global Efl;
wvli=0.4;
wvlf=0.7;
wvlwd=0.002;
rfr = [1.5 2.1+0.016i 1.5 2.1+0.016i 2.13 1.87 1.94 1.75 0.644+5.28i];
thick = [0 0.073 0.089 0.040 0.09 0.117 0.048 0.029 0];
Nm = length(rfr);
thick(1) = 0;
thick(Nm) = 0;
wvll = [wvli:wvlwd:wvlf];
Rcurve=[];
for n=1:length(wvll)
wvl=wvll(n);
kv = 2*pi*rfr/wvl;
MA = @(k) [exp(i*kv(k)*thick(k)), exp(-i*kv(k)*thick(k));
rfr(k)*exp(i*kv(k)*thick(k)), -rfr(k)*exp(-i*kv(k)*thick(k))];
vEf = Ef(1);
Rcurve = [Rcurve; [wvl*1000, abs(vEf(2)/vEf(1))^2]];
end
plot(Rcurve(:, 1), Rcurve(:,2));
grid on;
xlabel('wavelength(nm)');
ylabel('reflectivity');
```

The outcome of the program is displayed in Fig. 5. This modelling assumes that the tuning layer is 40 nm IZO. The reflectance has its resonance at 540 nm. The rapid drop of reflectivity at the center indicates that the transmission is high at this wavelength. This is a typical behavior of a cavity made with two highly reflecting mirrors. Although the mirror reflects the incident light, the constructive interference inside the cavity under resonant condition makes the transmission very high. Although the simulation calculates the reflectivity curve, it also provides the information about the spectral emission. The transmittance or the spectral emission can be obtained by subtracting reflectance from 1. Therefore, the spontaneous emission other than the resonant wavelength is suppressed by the cavity. If the emission layer contains the RGB emission layer as in white OLED, the resonant wavelength determines the emission color of the pixel. As for 40 nm IZO tuning layer, the cavity allows the pixel to emit 540 nm with spectral width of 50 nm, which is narrower than the natural spectral width of the green emission in OLED. In other words, the use of cavity enhances the color purity of the emission.

For the most part, OLED includes color filter(CF) layer to reduce the spectral width and improve color gamut. However, the reduction of spectral width comes through the absorption of the light energy outside the spectral window of the pigment. It suffers energy loss for the sake of enhancement in color purity. On the contrary, the spectral narrowing induced by cavity effect doesn't suffer energy loss since it originates from the change of optical modes inside the cavity, not the absorption process. Therefore, the cavity in OLED increases the transmission efficiency through CF by narrowing the spectral width without energy loss.

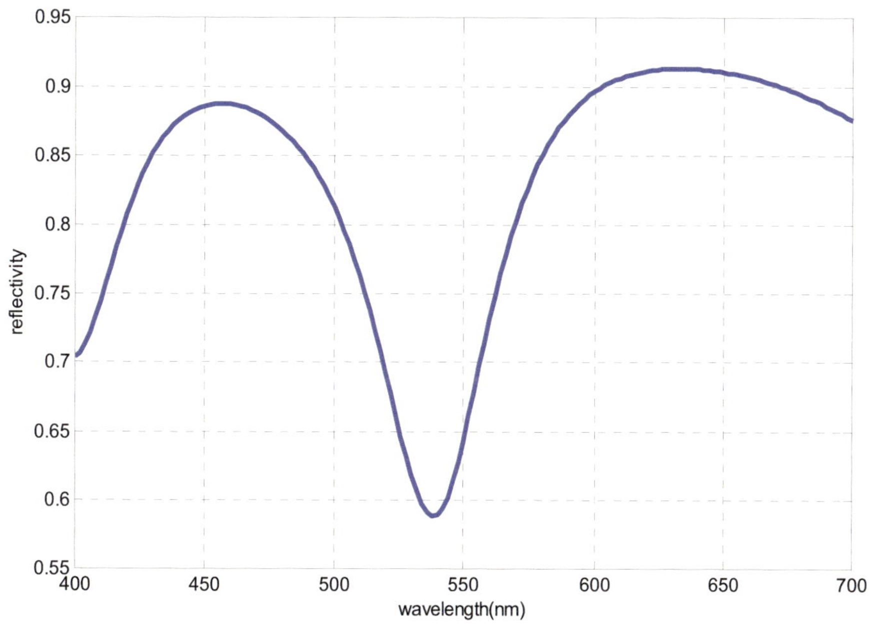

Fig. 5. The reflectivity curve is calculated when the tuning layer is IZO and 40 nm thick.

Furthermore, the resonance in this DBR cavity varies as a function of the thickness of the tuning layer. As the thickness of the tuning layer increases, the resonance shifts toward long wavelength. When the tuning IZO layer is 100 nm, the resonance goes from 540 nm to 620 nm and the secondary resonance appears at 460 nm as seen from Fig. 6. The thicker tuning layer provides the way to narrow the spectral width of blue and red emission. It means that one tuning layer can create two resonances at the same time. In this way, two types of tuning layers can make the interference filter for RGB and enhance the optical efficiency in OLED. The reflectivity curves with different tuning layer and thickness are shown in Fig. 7. It shows that SiNx also function as a tuning layer but with smaller spontaneous enhancement due to its smaller index.

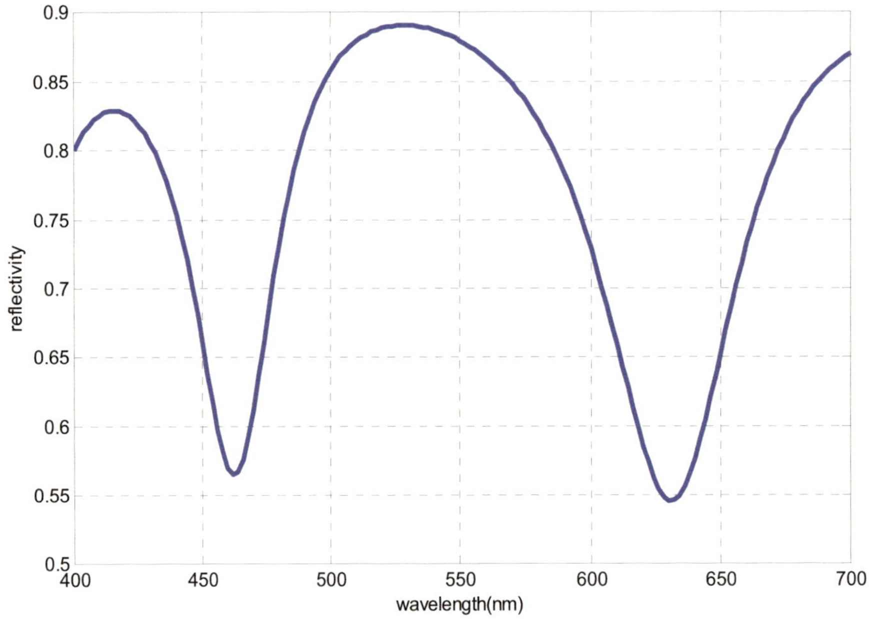

Fig. 6. The reflectivity curve is calculated when the tuning layer is IZO and 100 nm thick.

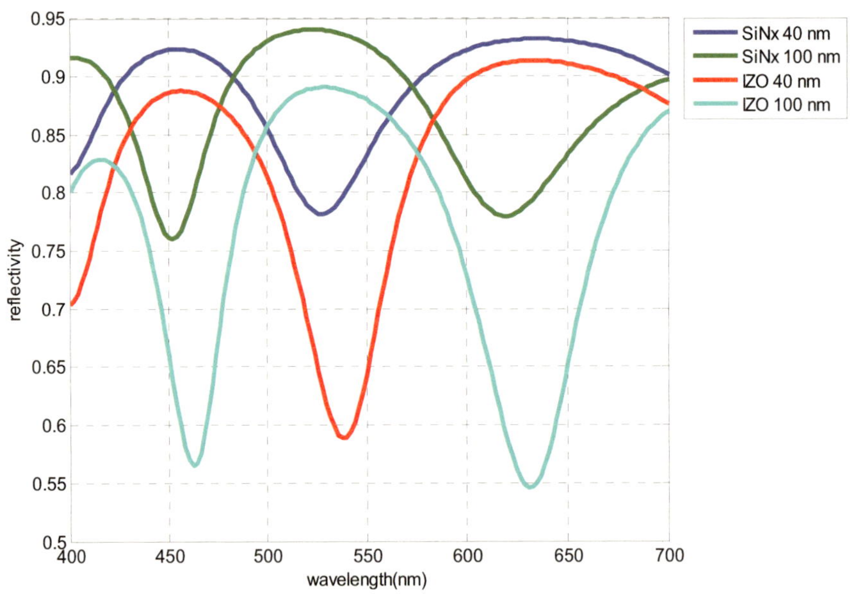

Fig. 7. The reflectivity curves are plotted. The tuning layers are (a) 40 nm SiNx, (b) 100 nm SiNx, (c) 40 nm IZO, and (d) 100 nm IZO, respectively.

Based on this idea, the practical OLED devices were fabricated and demonstrated(Lee et al, 2009a). The schematic diagram of OLED structure with DBR cavities and tuning layers are illustrated in Fig. 8. We fabricated RGBW AMOLED panels with the above optical designs and the conventional CF used in LCDs (color gamut = 72%). The panels showed a color gamut in the range of 100–110% NTSC. On the other hand, panels with no micro-cavity design had a color gamut of only 75%. Another benefit was increased light output through the CF. With micro-cavity designs, the CF transmission ratio increased to 40% from 27%. The light output from R, G, and B subpixels increased by about 50%.

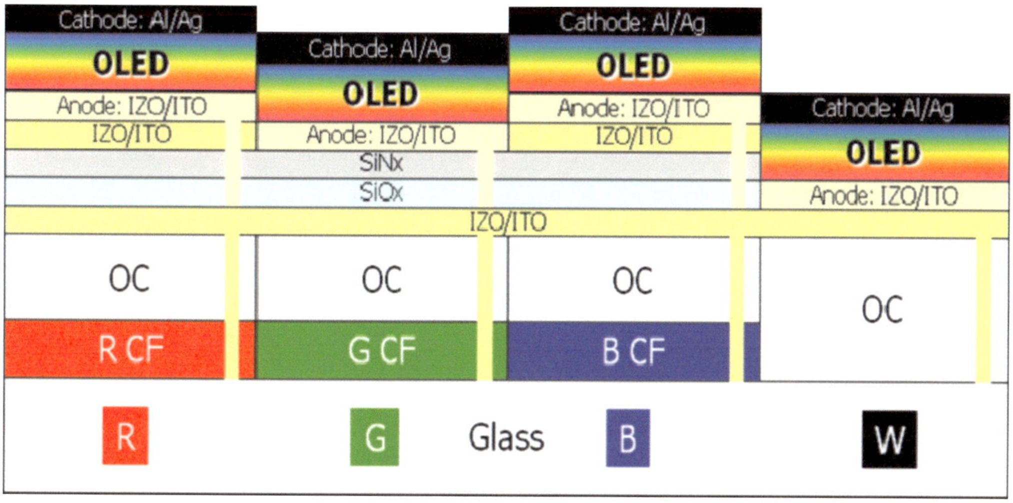

Fig. 8. A micro-cavity design of a RGBW bottom-emitting AMOLED. OC stands for overcoat. The RGB subpixels have DBR (IZO or ITO and SiOx), a filter common for RGB (SiNx), and another filter (IZO or ITO) for R and B subpixels. The W subpixels do not have DBR in order to avoid the spectral modification and dependence on the viewing angle.

4. FDTD analysis for OLED(Lee et al, 2009b)

As matter of fact, the multi-layer theory deals with one dimensional problem since it assume that the physical situation doesn't change for the translation movement in the plane of the layers. This assumption is quite good for the conventional OLED device. The functional layers of submicron thickness are deposited on large area glass, whose lateral dimension is virtually infinite compared to the thickness. However, suppose the OLED structure is made on the curved surface or the wavy substrate for some reason. From this point, the problem becomes two dimensional or three dimensional one, which is not within the scope of multi-layer theory. Then, what kinds of theoretical tools are available to the designer? Even though ray tracing is most frequently used algorithm to calculate the optical property of the system, it lacks capability of handling interference and sub-wavelength feature like thin film. FDTD can be a good option although it takes much time to compute the electric field inside the cavity. The computation time increases rapidly as the feature size and the dimension increases. Therefore, the analysis should be carefully planned to reduce the computation time and obtain the goal of calculation at the same time. In this section we present an example of analyzing the undulated cavity of the OLED structure using FDTD method.

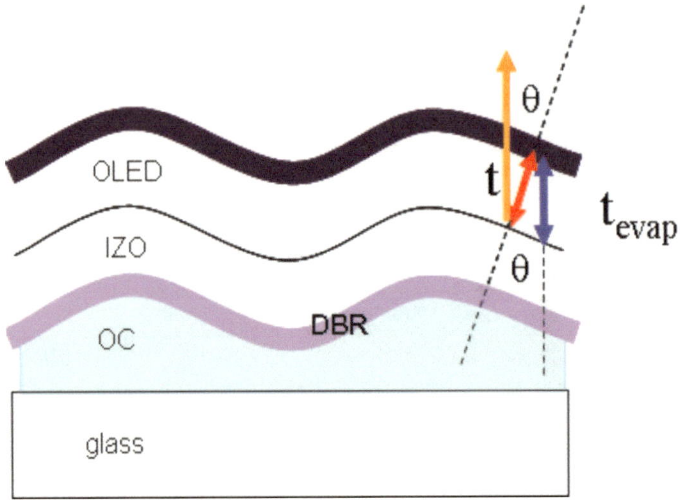

Fig. 9. Principles of tuning resonance in undulated micro-cavity

The main purpose of the undulated micro-cavity is to modify spectrum with the change of undulation profile. The physical mechanism of tuning resonance in undulated cavity is illustrated in Fig. 9. When the OLED layers are deposited over the undulated profile of the overcoat (OC) layer, the thickness of evaporation normal to substrate is uniform over the whole device. However, the thickness normal to the slope is proportional to cosθ if the angle between the slope and the substrate is θ. In addition, the emission is not normal to the slope, which leads to the blue shift of resonance peak proportional to the cosθ. As a result, the resonant wavelength of an undulated micro-cavity is proportional to $\cos^2\theta$.

The angle of slope is controlled by varying the amount of UV exposure in the lithography process. More UV exposure causes a deeper trench in OC layer after development process. The reflow process smoothen the rectangular profile into a curved surface. In general, the angle of the slope increases as the UV exposure increases. The experimental results are displayed in Fig. 10. The emission peak has changed with the increase of UV exposure time. The red curve represents the micro-cavity tuned to the R and B pixel showing the strong peaks around the red and blue wavelength. The increased UV exposure moves the peak to green wavelength which is optimized for G pixel in OLED display as seen in the brown curve.

In a real device, the UV exposure is controlled by the opening size of the mask pattern. Since the opening size of the mask pattern can be set to different values for RGB pixels, the resonance frequency of each pixel can be tuned simultaneously by one step photolithography. It greatly alleviates the processing burden which typically comes with the micro-cavity devices used for multi-colors.

In order to analyze the resonance shift and other optical properties of an undulated micro-cavity, the FDTD method and the permittivity profile as shown in Fig. 11 were used. In this structure, the micro-cavity largely consists of the organic layers sandwiched between DBR and a cathode. From the bottom-most layer, DBR layers are comprised of 70 nm SiNx, 90 nm SiO2, 35 nm SiNx, 90 nm indium zinc oxide (IZO). The organic layers are 120 nm hole transport layer, 50 nm emission layer, 30 nm electron transport layer. Lastly, the thickness of Aluminum cathode is 200 nm on the top. The region below distributed Bragg reflector (DBR) is assumed to be filled with OC layer. The permittivties used for SiNx, SiO2, IZO, hole transport layer, emission layer, electron transport layer and Alumium are 3.61, 2.25, 4.23, 3.74, 3.51, 2.94 and 0.06+I 3.50 respectively. The undulation profile is modeled using a sine wave form. The period and the height of the sine wave determine the angle of the slope in the undulated micro-cavity.

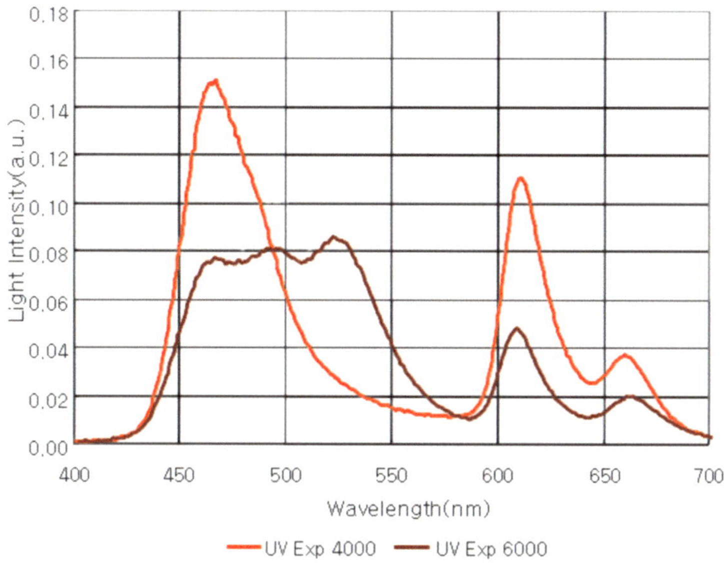

Fig. 10. Emission spectra of white OLED with variation of UV exposure time. The red line and brown line corresponds to 4 s and 6 s of exposure time, respectively.

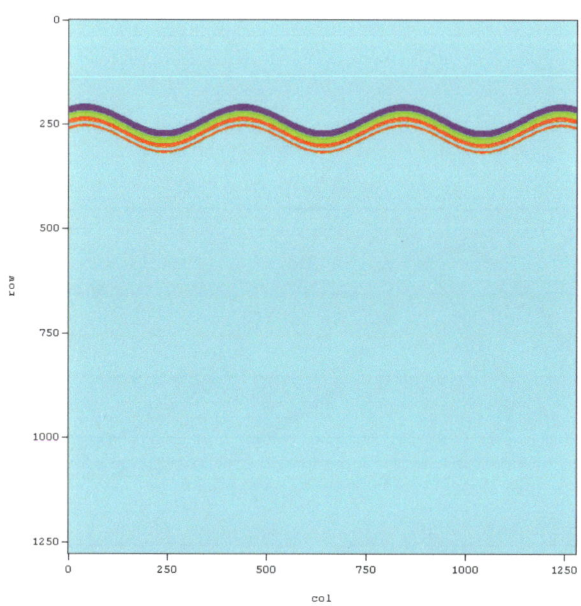

Fig. 11. Permittivity profile of undulated micro-cavity in OLEDs. Period = 5.0 μm, height = 0.6 μm.

For measuring the spectrum change, the dipole is allowed to oscillate for a short time. Since the short pulse acts as a broad band source, it can be a good light source to characterize the frequency response of the micro-cavity. The detector apart from the dipole source collects the wave emitted at a certain angle. The Fourier transformed output of the collected wave is divided by the Fourier transformed input to give frequency response at each wavelength. In this way, the spectrum change of the micro-cavity can be analyzed.

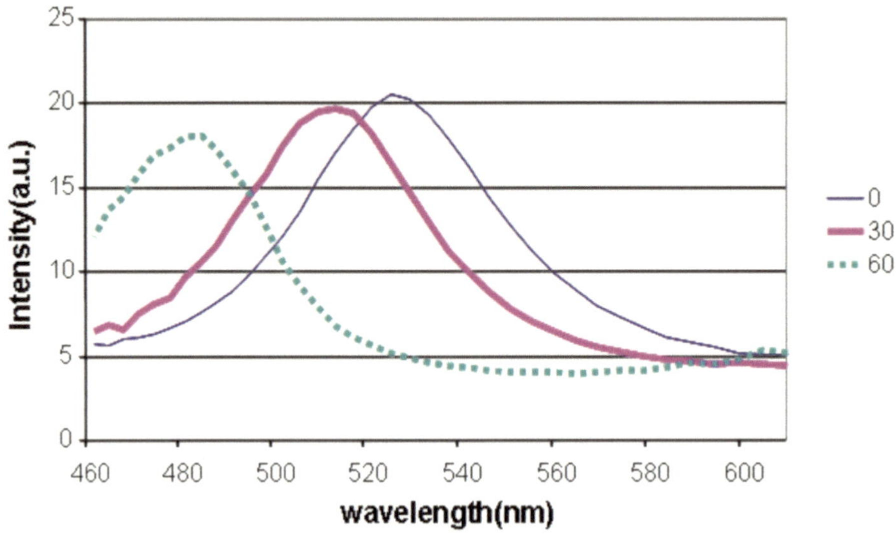

(a)

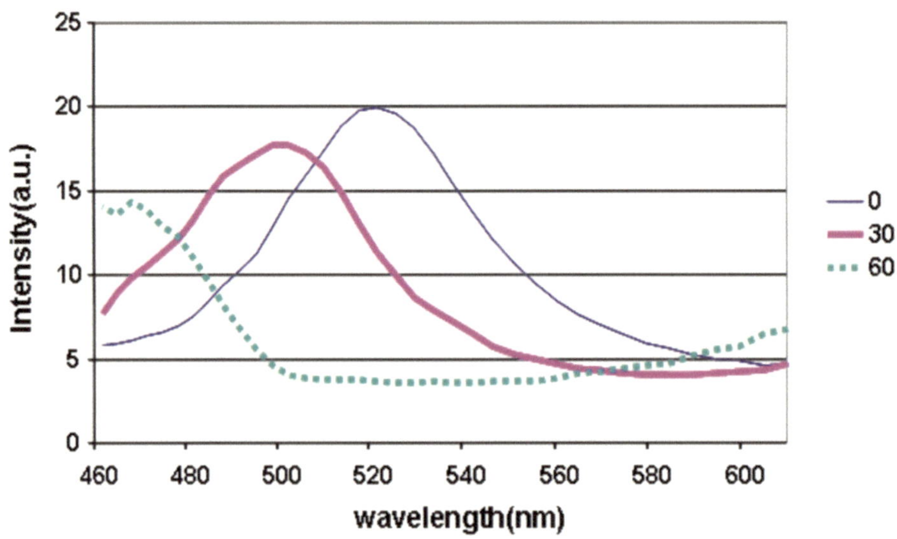

(b)

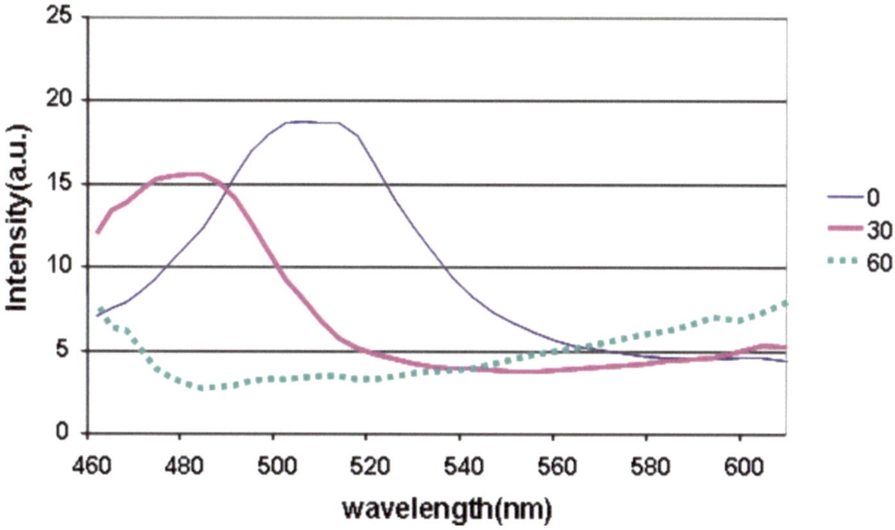

(c)

Fig. 12. Spectrum of dipole emission in undulated micro-cavity with various heights and emission angle; Period = 5.0 μm, height = (a) 0.0 μm, (b) 0.2 μm, and (c) 0.4 μm, emission angle = 0° (solid), 30° (thick solid), 60° (dotted).

When the dipole is placed in the middle of the slope of the undulation profile, the spectrum shows blue shifts as the height increases as seen in Fig. 12. It agrees with the experimental results observed in the real device(Lee et al, 2009a). The dipole is placed in the middle of the slope since the slope region is the brightest part of the device. Thermal evaporation of organic material results in uniform thickness in the vertical direction. It means that the thickness normal to the surface is the smallest near the slope. The reduced thickness of organic layers also reduced the electrical resistance in the part, which leads to the current crowding near the slope. Therefore, locating the dipole in the middle of slope is reasonable to explain the blue shift of the spectrum. Otherwise, the dipole oscillation near peak or valley gives less change to the resonance peak. The resonance peak at normal angle shifts from 525 nm to 510 nm when the height changes from 0.0 μm to 0.4 μm. Since the angle of the slope is roughly 9 degrees at the height of 0.4 μm, the amount of blue shift may be explained by $\cos^2\theta$ dependence. As explained before, this angle dependency can be ascribed to the layer thinning due to the slanted evaporation and oblique resonance condition, each of which has $\cos\theta$ dependency.
The internal physical mechanisms of undulated micro-cavity that controls the emission wavelength in OLEDs was investigated. The finite-difference-domain method is applied to a previously manufactured OLED design featuring optical structure on a wavy over-coat layer. The emission spectrum shows blue shift of 15 nm when the height of undulation changes from 0.0 μm to 0.4 μm with the period fixed at 5.0 μm. The blue shift is also observed in the experiment and the amount of shift in the simulation complies with $\cos^2\theta$ dependence.

5. Acknowledgment

This work was supported by Samsung Electronics Corporation. This work was supported by National Research Foundation of Korea Grant funded by the Korean Government(2009-0071253).

6. References

Born M and Wolf M, *Principles of Optics*, 6th Ed., ISBN 0-02-946146-4, p.55 (1989)

Fowles G R, *Introduction to Modern Optics*, 2nd Ed., ISBN 0-03-089404-2, Chapter 4 (1975).

Gasiorowicz S, Quantum physics, Chapter 22 (1984)

Ju Y G, Lofgreen D, Fiore A, Hu S Y, Hegblom E, Louderback D, Sjolund O, Huntington A, Coldren L A, "Densely packed pie shaped vertical-cavity surface-emitting laser array incorporating a tapered one-dimensional wet oxidation", *IEEE Photonics Technology Letters*, vol.12, no. 5, p.462-464, May 30 (2000)

Lee B W, Ju Y G, Hwang Y I, Lee H Y, Kim C W, Lee J S, and Souk J H, "Micro-cavity design of bottom-emitting AMOLED with white OLED and RGBW color filters for 100% color gamut", *Journal of the Society for Information Display*, Vol. 17, Issue 2, pp. 151-157 (2009a) Feb.

Lee B W, Ju Y G, "Analysis of undulated micro-cavity in organic light-emitting diodes", *Optical and quantum electronics*, vol. 41, Issue 8, pp. 627-633 (2009b)

11

Fast-Response Organic Light-Emitting Diode for Interactive Optical Communication

Takeshi Fukuda[1] and Yoshio Taniguchi[2]
[1]*Department of Functional Materials Science, Saitama University*
[2]*Shinshu University*
Japan

1. Introduction

In recent years, many types of electronic equipment have come into wide use in our lives. Especially, mobile phones and personal computers have been widely used by many people, and this fact causes the drastically change of our lives. In addition, we can connect global networks using mobile phones and personal computers, and we can get much information in a short time without moving. Nowadays, several mobile networks are widely used, such as, Bluetooth, ultra wideband, ZigBEE, and so on. Furthermore, the global computer networks will be used unconsciously without thinking the connection in near future, and many researchers demonstrated unique concepts of intuitive interface modules. (Morrison et al., 2005; Wilson et al., 2007; Mignonneau et al., 2005) To realize an intuitive interface module between real the world and the global computer network, we proposed the free space visible optical communication system utilizing organic light-emitting diodes (OLEDs) as a transceiver module and organic photo-diodes (OPDs) as a receiver module, as shown in Fig. 1. In this system, we can get information from the OLED by touching the emitting area, and the emitting area of the OLED is large enough to connect without the precious alignment between the OLED and the OPD.

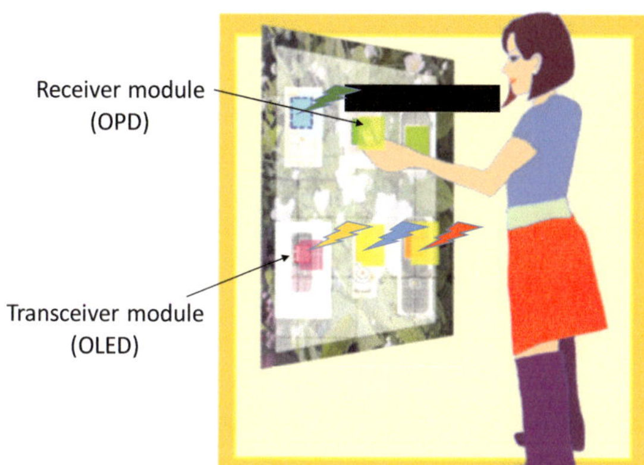

Fig. 1. Concept of the interactive visible optical communication system using the OLED and the OPD with the high response speed.

By now, several research groups demonstrated OLEDs and OPDs with the high response speed for the novel application of the optical communication, and the response speed of more than Mbps has been achieved by optimizing the device structure. (Shimada et al., 2006; Morimune et al., 2006) The reported optical communication system consists of an optical fiber to transmit optical signals generated from the OLED to the OPD. In generally, a core diameter of the multimode optical fiber is several 100 μm. (Koike, 2008) Even though the optical signal reaches far from the OLED, the high accuracy alignment between the OLED/OPD and the optical fiber is necessary to achieve the efficient optical communication. Furthermore, the emitting area of the OLED and the receiving area of the OPD can be controlled by changing the deposition areas of electrodes, which sandwiches organic layers. Therefore, we have proposed that the free space optical data transmission is suitable for the next generation visible optical communication system due to the alignment-less connection. The visible light of the OLED announces the connection point, and everyone can get optical information by touching the visible light using the receiver module (OPD). Moreover, OLEDs and OPDs can be fabricated by printing processes, resulting in the low-fabrication cost and the flexible devices. (Mori et al., 2003; Ooe et al., 2003)

OLEDs have attracted a great deal of public attention as visible light sources of flat panel displays and lightings. In recent years, several breakthroughs have led to significant enhancements of performances in OLEDs, such as the improvement in the charge-carrier balance, (Tsutsui, 1997) the low-work function electrode material, (Parker, 1994) the efficient injection of the electron from a metal cathode to an adjacent organic layer by inserting an electron injection layer (EIL), (Kido et al., 1998; Hung et al., 1997; Stößel et al., 2000; Kin et al., 2006) the high carrier mobility of electron/hole transport materials, (Ichikawa et al., 2006; Uchida et al, 2001) the high efficiency fluorescence and phosphorescence emitting materials. (Tang et al., 1987; Adachi et al., 2001; Cao et al., 1999; Xu et al., 2003) In the case of the visible optical communication system, the response speed is an important factor for the practical application. The reported cutoff frequency of the output power, which indicates the response speed, has been achieved up to 25 MHz for the OLED with a small area of 300 μm circle. (Kim et al., 2006) However, the large emitting area of the OLED is necessary for our proposed institutive visible optical communication system.

We investigated the response speed of the OLED by changing device parameters, such as the device area (capacitance of the organic layer), the fluorescence lifetime of the organic emitting material, (Fukuda et al., 2007) the thickness of hole/electron transport layers (HTL and ETL) corresponding to the carrier transport time from the electrode to the EML, the energy gap at a metal/organic interface (Fukuda et al., 2007), the combination of the host-guest materials used as the emitting layer (EML) (Fukuda et al., 2009), and the effect of the hole blocking layer (Fukuda et al., 2007). In this chapter, we show the experimental result of the fast response OLED. Then, we investigated the organic-inorganic hybrid device using ZnS as the ETL (Fukuda et al., 2008a). This is because that the response speed of the OLED is limited by the low electron mobility of the organic ETL material, and ZnS has higher electron mobility than organic materials. Finally, we demonstrated the intuitive optical communication system utilizing the OLED as a transceiver. In this system, we succeeded in the transmission of the pseudo-random signal with 1 Mbps and the movie files with 230 kbps, when the pen-type photo-diode is touched the emitting area of the OLED.

2. Limiting factor of the response speed of the OLED

The conventional OLED consists of a transparent anode, several organic layers, and a metal cathode, as shown in Fig. 2. The each organic layers are called as the hole injection layer (HIL), the HTL, the EML, the HBL, and the HTL. The names of these organic layers indicate their functions of the operation mechanism. When the voltage is applied between the transparent anode and the metal cathode, holes and electrons (carriers) are injected into the organic layers, respectively. Then, these injected holes and electrons transport into the HTL and the ETL, respectively. Finally, the carriers recombine into the EML, resulting in the generation of light. The generated light comes out from a transparent anode and a transparent substrate. That is to say the response speed of OLEDs is limited by the time from the applying voltage to the generation of light caused by the carrier recombination. We examined the details of these processes and the method to improve the response speed of the OLED.

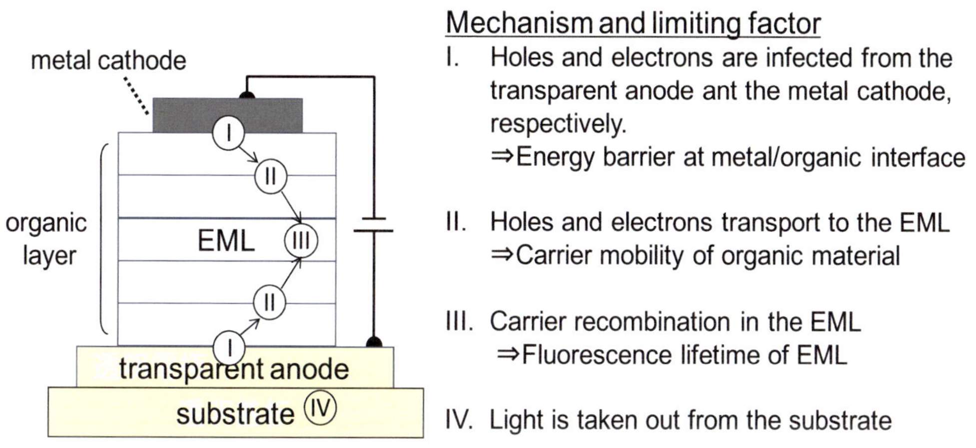

Fig. 2. Cross sectional view of the conventional OLED structure and limiting factors of the transmission speed of the OLED.

3. Fabrication process of the OLED and the experimental setup to estimate the response speed of the OLED

The fabrication process of the OLED is described in the following sentence. OLEDs were fabricated on glass substrates covered with a patterned indium tin oxide (ITO) anode. The thickness of the ITO layer was 150 nm. The prepared glass substrates were cleaned in deionized water, detergent, and isopropyl alcohol sequentially under ultrasonic waves, and then treated with oxygen plasma for 5 min. Next, several organic layers, an EIL and a metal cathode were thermally deposited successively using a conventional vacuum deposition system at a base pressure of below 5.0 x 10^{-6} Torr. Deposition rates were maintained at 0.8-1.0 Å/s for both the HTL and the ETL, 5.0 Å/s for both the EML and the metal cathode, and 0.1-0.2 Å/s for the EIL as determined using a quartz crystal monitor.

To evaluate the response speed of the OLED, we measured the relative EL intensity as a function of the frequency of an applied sine wave voltage. Figure 3 shows the schematic configuration of the experimental setup. The sine wave and bias voltages were applied to the OLED using a programmable FM/AM standard signal generator (KENWOOD, SG-7200)

and a DC power supply (ISO-TECH, IPS-3610D), respectively. The amplitude of the sine wave voltage was controlled using an attenuator (Furuno Electric, VHF-STEP) and a high speed amplifier (ARF Japan, ARF-15237-25). In addition, several resistances and capacitances were used to reduce the frequency dependence of the amplitude of the applied sine wave voltage, as shown in Fig. 3.
The generated light wad guided into a plastic optical fiber (Moritex, PJR-FB250) with the diameter of 250 μm. Then, the output EL intensity was observed using an avalanche photodiode (Hamamatsu Photonics, S5343) and an oscilloscope (Yokogawa Electronic, DL-1740). The frequency dependence of EL intensity was measured by changing the modulation frequency of the sine wave voltage from 100 kHz to 10 MHz. In addition, the rise and decay times of output EL intensity were also measured while applying a pulse voltage with a width of 1 μs to investigate the transient properties of the OLED. The rise and decay times were defined as the times required for the optical response to change from 10% to 90% and from 90% to 10% of the maximum EL intensity, respectively. We also measured the luminance-current density-voltage characteristics of the OLED using a source measure unit (Hewlett-Packard, HP4140B) and a luminance color meter (Topcon, BM-7).

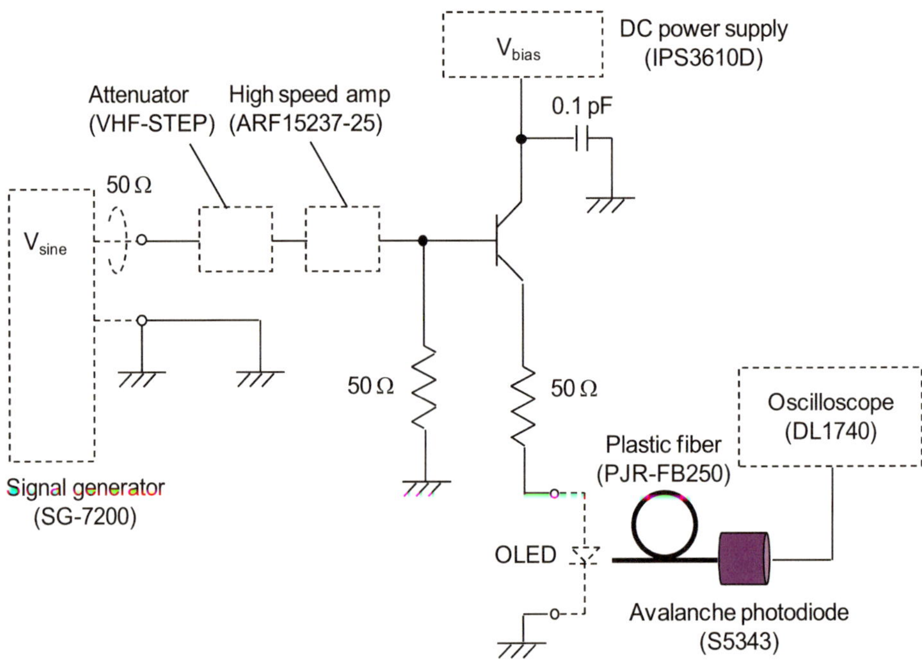

Fig. 3. Cross sectional view of the conventional OLED structure and limiting factors of the transmission speed of the OLED.

4. Response speed of the OLED

4.1 Device area (capacitance of the organic layer)

The conventional OLED consists of several organic layers with a total thickness of less than 200 nm due to the low carrier mobility of organic materials, resulting in the large capacitance of an emitting area. The capacitance of an emitting area is well known to affect pulse voltage-

transient current characteristics, and the large capacitance of the organic layer causes the long decay time of the transient current while applying a pulse voltage. (Wei et al., 2004) Therefore, the lower capacitance, corresponding to the smaller emitting area, is required for the high response speed of OLEDs. By now, previous papers demonstrated that the response speed of the OLED increases by reducing the capacitance of the emitting area. (Kajii et al., 2002a)
To investigate the influence of the emitting area on the response speed of the OLED, we used 4,4'-bis[*N*-(1-napthyl)-*N*-phenyl-amino]-biphenyl (α-NPD) as the HTL, 6,11,12-tetraphenyltetracene (rubrene) as the dopant in the EML, and tris(8-hydroxyquinoline) aluminium (Alq_3) as the EML and the ETL. Figure 4 shows molecular structures of used organic materials. The device structure is ITO 150 nm/α-NPD 40 nm/rubrene:Alq_3(0.5wt%) 20 nm/Alq_3 40 nm/LiF 0.4nm/MgAg (9:1) 150 nm/Ag 20nm for the device A and ITO 150 nm/α-NPD 40 nm/Alq_3 60 nm/LiF 0.4nm/MgAg (9:1) 150 nm/Ag 20nm for the device B. In addition, the emitting area was changed from 0.2 to 1.5 mm^2 to investigate the influence of the emitting area on the response speed of the OLED.

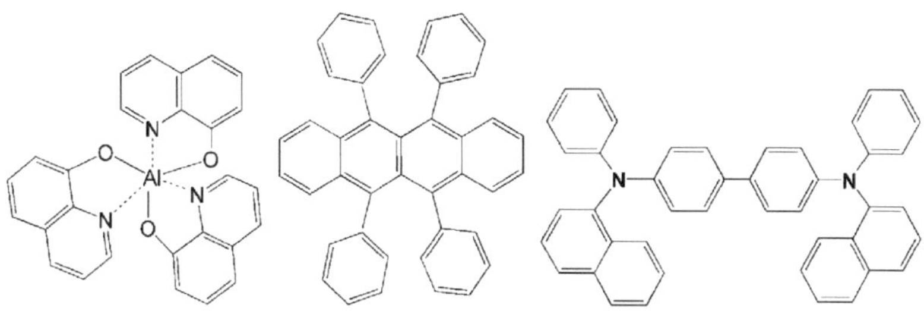

Fig. 4. Molecular structures of organic materials (α-NPD, rubrene and Alq_3)

Figure 5 shows the relative output EL intensity as a function of modulation frequency for the two OLEDs, that is, devices A and B with rubrene doped Alq_3 and Alq_3 as EMLs, respectively. The sine wave voltage was 7 V and the bias voltage was 5 V. Here, the EL intensities at various modulation frequencies are normalized with respect to the EL intensity at a frequency of 100 kHz. It was observed that the relative EL intensity of device A with the rubrene doped Alq_3 EML is higher than that of device B, which has the Alq_3 EML. This result indicates that the device A has a higher response speed than the device B. This result can be explained by the fluorescence lifetime of the EML. (Kajii et al., 2002b) The fluorescence lifetime of rubrene doped Alq_3 (0.5wt%) and non-dope Alq_3 were 10 ns and 16 ns, respectively. (Fukuda et al., 2007b) Therefore, the response speed of the OLED was improved by doping rubrene in the EML.
The cutoff frequency of the device A with the emitting area of 1.2 mm^2 was 4.0 MHz, and the 2-times faster cutoff frequency (8 MHz) was achieved when the emitting area was 0.2 mm. The cutoff frequency corresponds to the responses speed of the OLED; therefore, this result indicates that the response speed of the OLED was improved with decreasing capacitance of the emitting area. In the case of the institutive optical communication system, the large emitting area is important factor to connect between the OLED and the OPD. Therefore, the response speed of the OLED is necessary to improve by optimizing other device parameters.

4.2 Thickness of hole/electron transport layers (carrier injection time)

In generally, the carrier mobility of organic materials is much lower than that of inorganic materials. This fact causes the long decay time from the carrier injection to the generation of

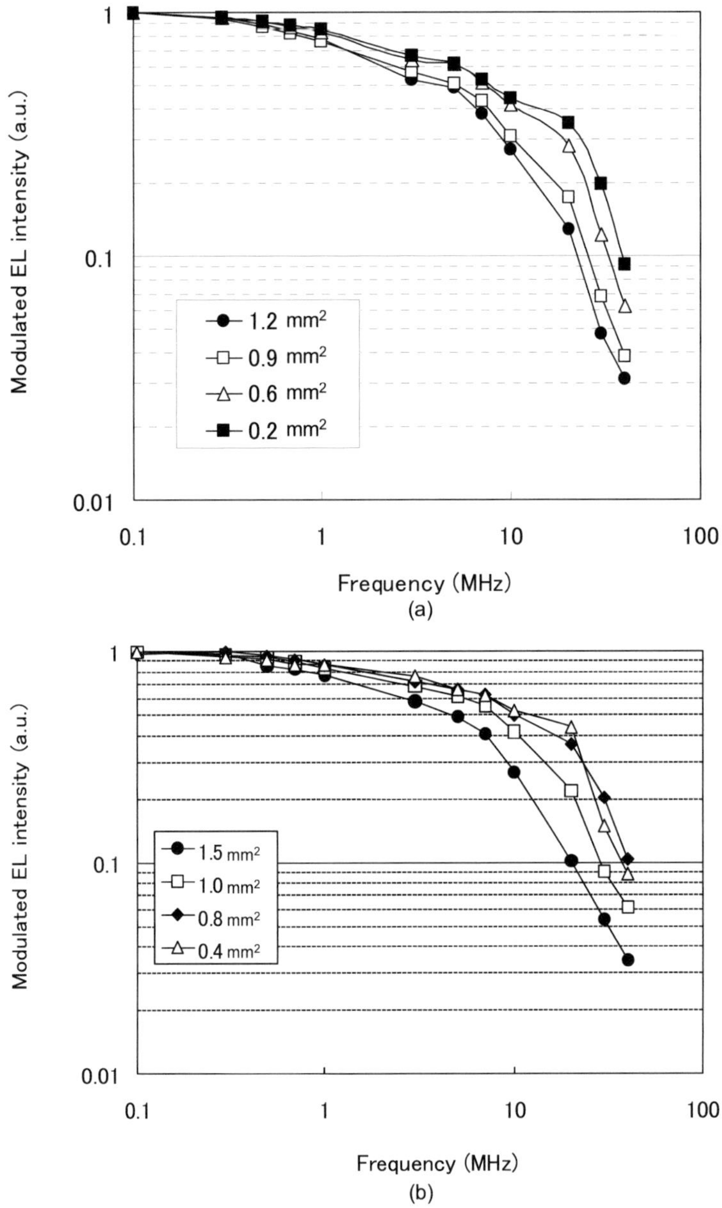

Fig. 5. Relative EL intensity while applying the sine wave voltage for (a) the device A with rubrene:Alq_3 and (b) the device B with Alq_3 as EMLs.

light, resulting in the slow response time of the OLED. In addition, the carrier transport time from the electrode to the EML is related with the thickness of HTL and the ETL. Here, we show the relationship between the thicknesses of the HTL/ETL and the response speed of the OLED. (Fukuda et al., 2007e)

The device structure is ITO 150 nm/α-NPD 40 nm/rubrene:Alq_3(0.5wt%) 20 nm/Alq_3 10-40nm/LiF 0.4nm/MgAg (9:1) 150 nm/Ag 20nm. Active areas were decided as the sandwiched region of ITO/MgAg, and those of all the devices were fixed at 1 mm^2. The detail of the measurement is described in the above-mentioned section.

Figure 6(a) shows the relationship between the applied pulse voltage and the rise time of output EL intensity of OLEDs with different thicknesses in the ETL. The thicknesses of the ETLs were 10 nm, 20 nm, 30 nm, and 40 nm. As clearly shown in Fig. 6(a), the rise time decreased with decreasing thickness of the ETL. The electron injection time is calculated from the electron mobility, the thickness, and the applied electric field. The electron mobility of Alq_3 is about 10^{-5} cm^2/Vs (Barth et al., 2001). Therefore, we can estimate the electron injection time of 450 ns, 250 ns, 150 ns, and 50 ns for OLEDs with thicknesses in 40 nm, 30 nm, 20 nm, and 10 nm, respectively. The measurement results of the rise times were longer than the estimated electron injection times. These differences are considered to be caused by the energy gap at metal/organic interface and the capacitance of the organic layer. In addition, the decay time was also reduced with decreasing thickness of the ETL. In addition, the decay time shown in Fig. 6(b) also decreased with decreasing thickness of the ETL. This result indicates that the carrier injection time mainly affect the decay time of the output EL intensity while applying the high speed pulse voltage.

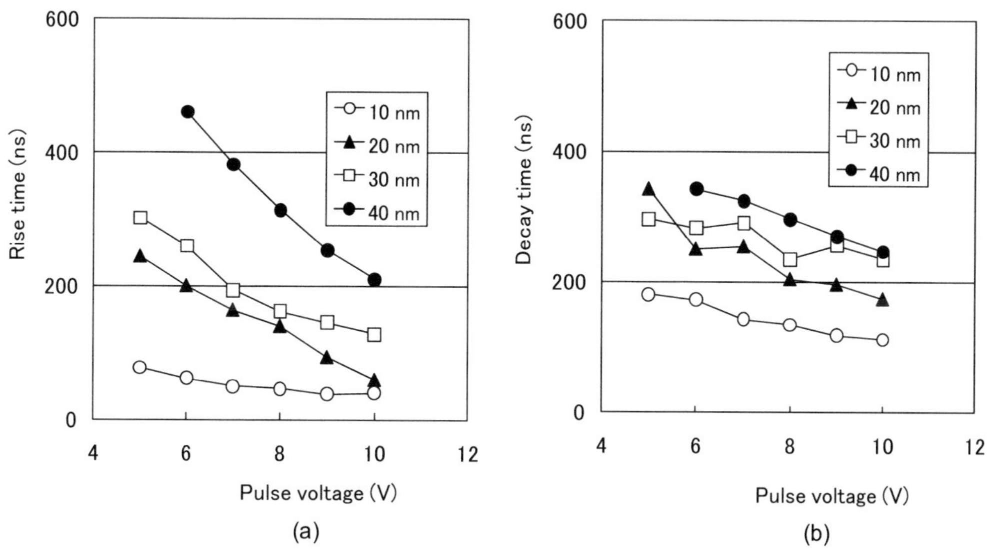

Fig. 6. Influence of the pulse voltage on (a) rise and (b) decay times of the OLEDs with different thickness of the ETL (Alq_3). (Fukuda et al., 2007e)

Figure 7 shows the relative EL intensity of OLEDs with different thicknesses of the ETLs when the sine wave voltage was applied to the device. The sine wave voltage was 8 V and the bias voltage was 5 V. The relative EL intensity at the high frequency region increased with decreasing the thickness of the ETL. This result indicates that the response speed increased with decreasing thickness of the ETL, which corresponds to the electron travelling length from the metal cathode to the EML.

On the other hand, the rise time was little influenced by the thickness of the HTL ranged from 20 nm to 40 nm, as shown in Figs. 8(a). The device structure was ITO 150 nm/α-NPD 20-40 nm/rubrene:Alq_3(0.5wt%) 20 nm/Alq_3 10nm/LiF 0.4nm/MgAg (9:1) 150 nm/Ag

20nm. Active areas were decided as the sandwiched region of ITO/MgAg, and those of all the devices were fixed at 1 mm2.

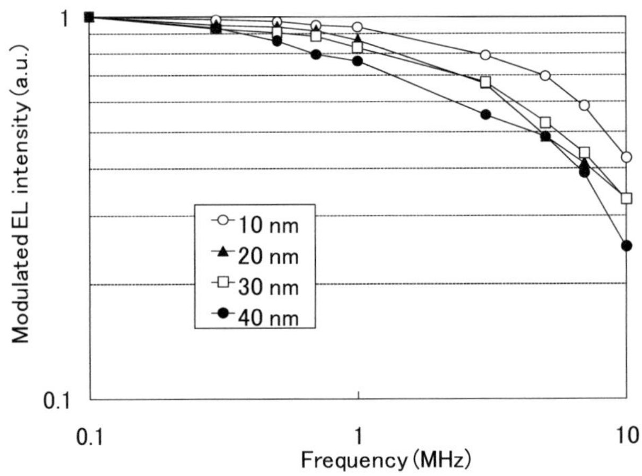

Fig. 7. Relative EL intensity while applying the sine wave voltage for OLEDs with different thicknesses of the ETLs. (Fukuda et al., 2007e)

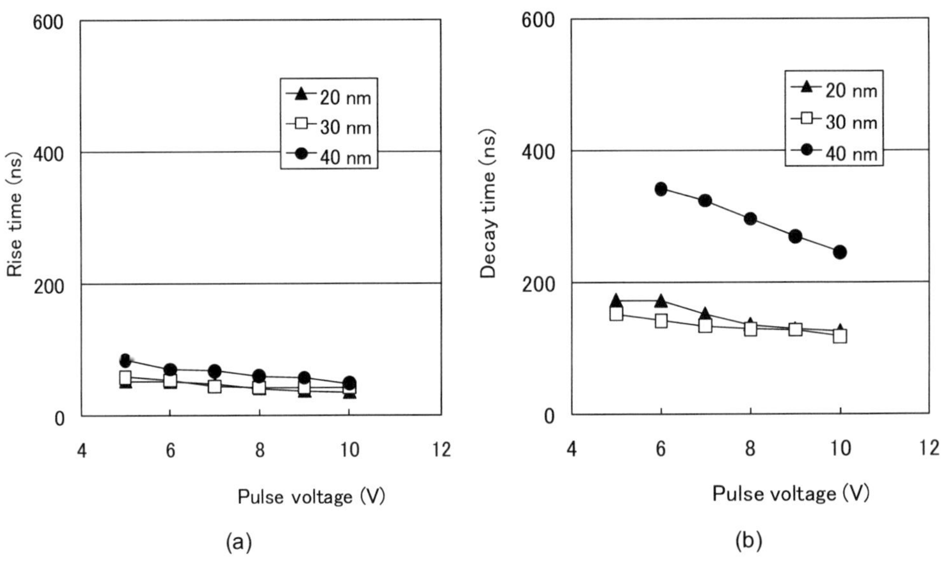

Fig. 8. Influence of the pulse voltage on (a) rise and (b) decay times of the OLEDs with different thickness of the HTL (α-NPD). (Fukuda et al., 2007e)

The thickness of the ETL (Alq_3) was 10 nm, and the response speed of the OLED was almost same for all the devices. This is because that the electron mobility of Alq_3 is much lower than the hole mobility of α-NPD. These experimental results indicate that the thickness of the ETL mainly limits the response speed of OLEDs owing to the low electron mobility of Alq_3 used as the ETL (Fukuda et al., 2007e).

4.3 Energy gap between metal the cathode and the adjacent organic layer

In generally, holes and electrons (carriers) are injected from an anode and a cathode, respectively. The injection efficiency of carriers is defined by the energy level difference between an electrode and an adjacent organic layer (Kampen et al., 2004). Therefore, the low energy gap at the electrode/organic interface is necessary to realize efficient carrier injection and to reduce the drive voltage of OLEDs. By now, many researchers have investigated, such as the surface treatment of the indium tin oxide (ITO) layer used as a transparent anode (Nüesch et al., 1998; Hatton et al., 2001), the low work function metal cathode, (Parker, 1994) and hole/electron injection layers at the electrode/organic interface. (Kido et al., 1998; Hung et al., 1997; Stößel et al., 2000; Kin et al., 2006) Especially, the metal/organic interface has a large energy gap, and Schottky barrier is formed at the metal/organic interface. As a result, the efficiency of injecting electrons into an organic layer form a metal cathode is low, and the high drive voltage is necessary. Furthermore, the large energy gap at metal/organic interface causes the decrease in the response speed of the OLED (Ichikawa et al., 2003; Fukuda et al., 2007d). In addition, the carrier injection efficiency at the organic/organic interface is also important factor for high speed OLEDs. (Fukuda et al., 2007c)

The thicknesses of the organic layers are 40 nm for α-NPD, 20 nm for rubrene-doped Alq_3, and 30 nm for Alq_3. In addition, we employed three species of metal cathodes of 100 nm thickness, namely, Ca/Al, Al, and MgAg (9:1 w/w)/Ag for devices C, D and E, respectively. To investigate the effects of an inserted EIL, we fabricated a similar set of OLEDs using a thin 8-hydroxyquinolinato lithium (Liq) layer with thickness of 0.4 nm as an EIL. We also used Ca/Al, Al and MgAg (9:1 mass ratio)/Ag as metal cathodes for devices F, G, and H, in which Liq was inserted between the metal cathode and the ETL. The current efficiency of the OLEDs with Liq is less sensitive to a change in EIL (Liq) thickness than that of OLEDs with the conventional EIL material of LiF, resulting in their suitability for mass production. (Zheng et al., 2005). The active areas of all the OLEDs were fixed at 1 mm^2.

Figure 9(a) shows the relationship between the relative EL intensity and the frequency of the applied sine wave voltage for the three OLEDs (devices C, D, and E). The sine wave voltage was 7 V and the bias voltage was 5 V. Here, the EL intensities at various frequencies are normalized with respect to the EL intensity at a frequency of 100 kHz. It was observed that the relative EL intensity of device C with the Ca/Al cathode was higher than those of devices D and E, which have Al and MgAg/Ag cathodes, respectively. The relative EL intensity at the high frequency region corresponds to the response speed of the OLED. Therefore, this result indicates that device C has a higher response speed than devices D and E. The cutoff frequency of device C was 8.5 MHz, while those of devices D and E were 1.3 and 4.2 MHz, respectively.

Figure 9(b) shows the influence of the barrier height at the metal/organic interface on the cutoff frequency. Here, the barrier height was calculated to be the difference between the work function of the metal cathode and the LUMO level of Alq_3 used as the ETL. The LUMO level of Alq_3 was 3.1 eV and work functions of metal cathodes were 3.0, 4.3, and 3.6 eV for Ca, Al, and MgAg, respectively. Therefore, the barrier heights were estimated to be 0.1, 1.2, and 0.5 eV for devices C, D, and E, respectively. The cutoff frequency increased with decreasing barrier height, which affects the efficiency of injecting electrons into the organic layer from the metal cathode. The cutoff frequency relates the response speed of the OLED; therefore, the response speed increases with decreasing barrier height at the metal/organic interface.

Figure 10 shows the relationship between the frequency of sine wave voltage and the relative EL intensity for the three EIL (Liq)-inserted OLEDs, that is, devices F, G, and H with

Ca/Al, Al, and MgAg/Ag as metal cathodes, respectively. The response speed of the OLED also increased when the low-work function metal electrode was used for the EIL-inserted OLED. The cutoff frequency of device F was observed to be about 11.2 MHz, while those of devices G and H were approximately 6.7 and 8.8 MHz, respectively. By comparing Fig. 9(a), we found that the cutoff frequency increased by inserting Liq layer for all the devices with the different cathode materials. Here, Li has low work function of 2.9 eV, and thus appears to be a good candidate for injecting electrons into the Alq_3 layer. It is known that diluted Li-metal alloys can act a cathode and exhibit much better transient characteristics than a pure metal cathode. (Zheng et al., 2005).

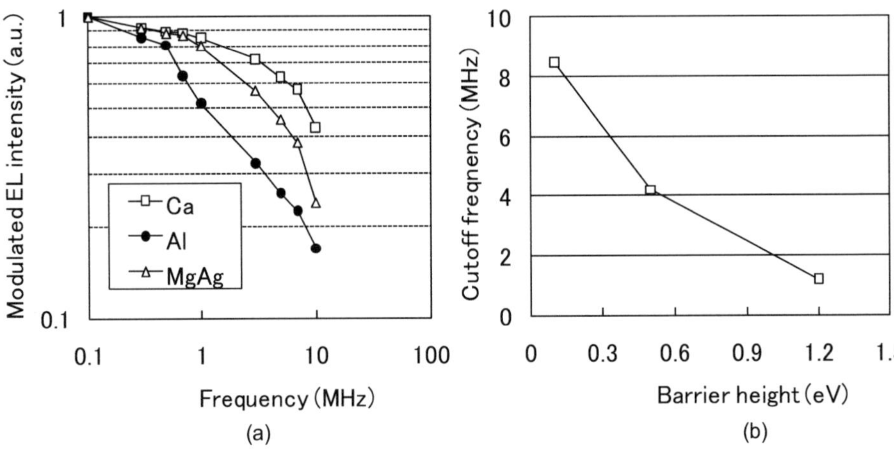

Fig. 9. (a) Frequency dependence of relative EL intensity for devices C, D, and E with Ca, Al, amd MgAg as metal cathodes, respectively. (b) Relationship between cutoff frequency and barrier height at metal cathode/Alq_3 interface. The cutoff frequency was calculated from the experimental result in Fig .9(a). (Fukuda et al., 2007c)

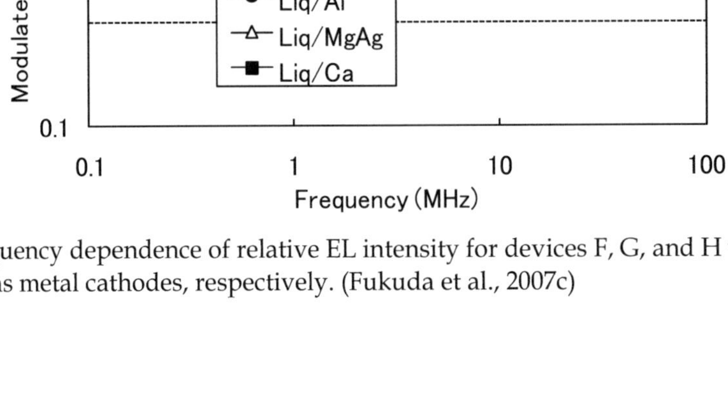

Fig. 10. Frequency dependence of relative EL intensity for devices F, G, and H with Ca, Al, and MgAg as metal cathodes, respectively. (Fukuda et al., 2007c)

4.4 Influence of fluorescence lifetime of EML and response speed of OLED

The fluorescence life time of organic emitting materials is important factor to determine the response speed of the OLED. (Kajii et al., 2002b) Therefore, we investigated the direct influence of the fluorescence lifetime on the response speed of the OLED. (Fukuda et al., 2007b)

We fabricated organic neat films on glass substrates by a conventional thermal evaporation system at a base pressure of below 5.0x10^{-6} Torr. The glass substrates were cleaned in deionized water, detergent, and isopropyl alcohol sequentially under ultrasonic waves, and then treated with 50 W oxygen plasma for 5 minutes just before use. Finally, the following 10 species of organic materials were deposited on glass substrates, and molecular structures of these organic materials are shown in Fig. 11. The used organic materials were 1,4-bis[2-[4-[*N,N*-di(*p*-tolyl)amino]phenyl]vinyl]benzene (DSB), 2-(4-tert-butylphenyl)-5-(4-biphenylyl)-1,3,4-oxadiazole (PBD), (3-(2-benzothiazolyl)-*N,N*-diethylumbelliferylamine (coumarin 6), 4,4'-(bis(9-ethyl-3-carbazovinylene)-1,1'-biphenyl (BCzVBi), 4,4-bis(2,2-ditolylvinyl)biphenyl (DPVBi), α-NPD, 4,4'-bis[9-dicarbazolyl]-2,2'-biphenyl (CBP), Alq$_3$, doped Alq$_3$, rubrene doped Alq$_3$, Pyrromethen 567 doped Alq$_3$, 4-(dicyanomethylene)2-methyl-6-(julolidin-4-yl-vinyl)-4H-pyran (DCM2), 4,7-diphenyl-1,10-phenanthroline (BPhen), Bis-(2-methyl-8-quinolinolate)-4-(phenylphenolato)aluminum (BAlq), Perylene, and rubrene.

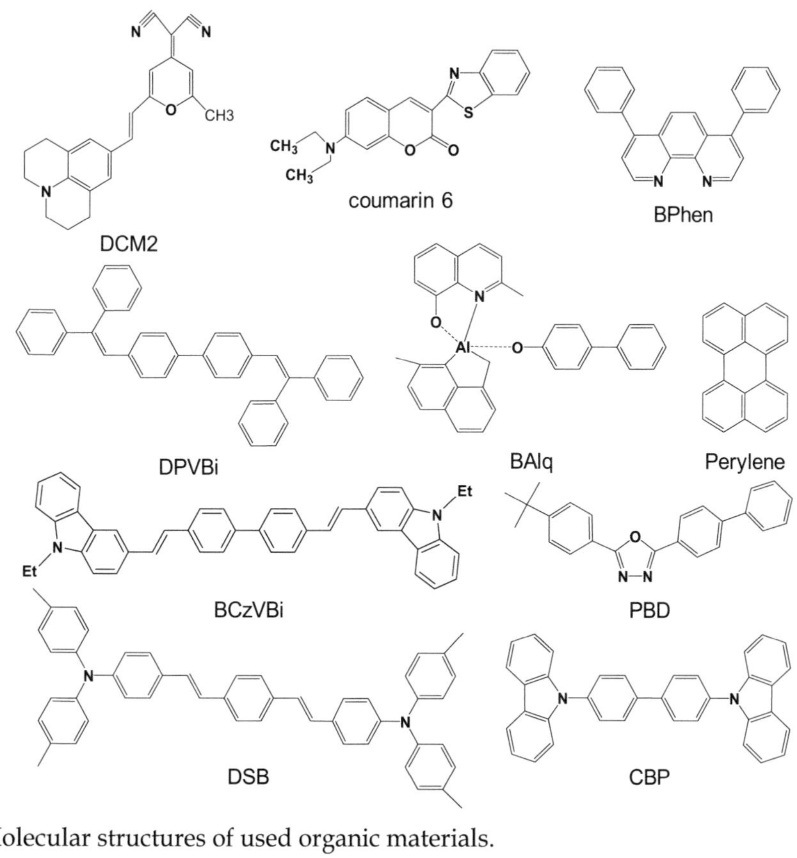

Fig. 11. Molecular structures of used organic materials.

After deposition of organic neat films, we measured fluorescence lifetimes of all the organic films by a femtosecond pulse laser (THALES Laser, Bright). After passing through the second harmonic generator, the center wavelength and the pulse width of the femtosecond pulse laser were 390 nm and 112 femtosecond, respectively. All the organic films radiated photoluminescences (PLs), when the femtosecond pulse laser was irradiated. The radiated PL was captured with a spectrometer and a streak camera (Hamamatsu Photonics, A5760), then time-resolved PL spectra were measured. Finally, Mono-exponential fitting was employed to derive the FL from the measured time-resolved PL intensity.

The frequency dependence of PL intensity was measured to investigate the direct relationship between the cutoff frequency of PL intensity and the fluorescence lifetime of the organic neat film. A schematic configuration of an experimental setup is shown in Fig. 12. The organic neat film was excited by the violet laser diode (NDHV220APAE1-E, Nichia corp.). The center wavelength of the excited violet laser diode was 405 nm, and the all the organic neat film absorbs the excited light. In addition, the violet laser diode was operated by a high-frequency sine wave voltage utilizing the programmable FM/AM standard signal generator (SG-7200, KENWOOD). And then, PL intensity was observed using the avalanche photo diode (S5343, Hamamatsu Photonics), which was located perpendicular to the optical axis of the laser diode, as shown in Fig. 12.

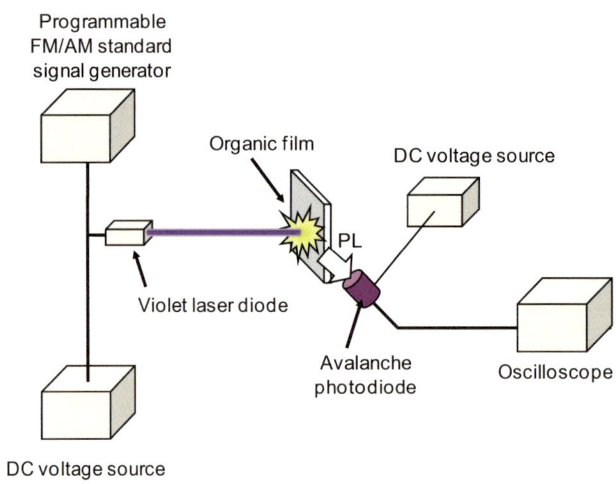

Fig. 12. Schematic configuration of the experimental setup to estimate the influence of the relative PL intensity while irradiating the violet laser diode.

As a result, the frequency dependence of PL intensity was estimated by changing the modulation frequency of the violet laser diode. Moreover, PL spectra were measured by the spectrophotometer (USB 2000, OceanOptics Company) also located perpendicular to the optical axis of the laser diode.

Figure 13(a) shows the influence of PL intensity on the frequency of the violet laser diode for two organic materials, DSB and Alq_3. For both organic films, PL intensity decreases with increasing frequency of the violet laser diode due to the decay time of the PL. This experimental result showed that cutoff frequencies were 160 MHz and 20 MHz for DSB and Alq_3, respectively. The difference of the cutoff frequency can be explained by the fluorescence lifetime of the organic material. The fluorescence lifetimes of DSB and Alq_3 were 0.2 ns and 16.0 ns, respectively. Therefore, the long fluorescence lifetime Alq_3 of causes the decreased cutoff frequency.

Figure 13(b) shows the relationship between the cutoff frequency of PL intensity and the fluorescence lifetime of the organic emitting material. This result is a consequence of fluorescence lifetimes without the influences of the capacitance and the carrier mobility, which are known to affect the response speed of the OLED. Therefore, we can estimate the direct influence of the fluorescence lifetime on the response time of the OLED. The transient characteristic of PL is strongly dependent on the fluorescence lifetime, and the response is considered to increase utilizing the short fluorescence lifetime of the organic material as a light-emitting layer of OLEDs. The highest cutoff frequency of PL intensity can reach about 160 MHz using one substituted phenyl/vinyl compound, DSB, of which the fluorescence lifetime was 0.2 ns.

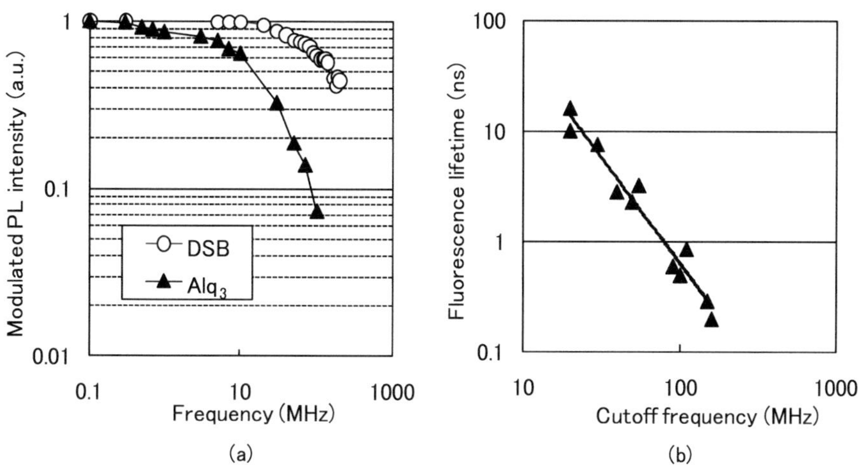

Fig. 13. (a) Relationship between the frequency of the irradiated violet laser diode and the relative PL intensity for DSB and Alq_3 neat films. (b) Influence of the cutoff frequency of the experimental result in Fig. 13(a) on the fluorescence lifetimes of organic neat films. (Fukuda et al., 2007b)

4.5 Combination of host-guest materials in EML

In the previous chapter, the fluorescence lifetime of the EML is important factor to realize the fast response speed of the OLED. In addition, the efficient energy transfer from the host material to the guest material is also key parameter for the increased response speed. (Fukuda et al., 2009)

To investigate the response speed of OLEDs with different combinations of host-guest materials, we fabricated three devices, referred as devices I, J, and K. The guest materials of devices I, J, and K were DSB, DPVBi, and BCzVBi, respectively. The device structure was α-NPD 40 nm/EML 20 nm/bathocuproine (BCP) 10 nm/Alq_3 20 nm/LiF 0.4 nm/MgAg (9:1 w/w) 100 nm/Ag 50 nm. Three organic emitting materials were choosed as DSB (device I), DPVBi (device J), and BCzVBi (device K) doped with CBP at 0.5 wt.%, respectively Emitting areas of all the devices were fixed at 1 mm^2.

Figures 14(a) and (b) show rise and decay times of the output EL intensity while applying the pulse voltage with the duration of 1 μs. The bias voltage was fixed at 6 V and the pulse voltage was ranged from 5 to 10 V. Both rise and decay times decreased with increasing pulse voltage due to the high carrier mobility at the high electric field. In addition, the rise times of devices I (DSB), J (DPVBi), and K (BCzVBi) were 58, 345, and 257 ns at the pulse voltage of 5 V, respectively. The measured rise times were larger than the decay time of all the devices. In generally, the large capacitance of organic layers limits the response speed of OLEDs owing to the large emitting area and the thinness of organic layers compared to semiconductor emitting devices. (Kajii et al., 2002a) However, the rise time is same as the decay time when only the capacitance affects the response speed of the OLED. Therefore, we can conclude that the rise time of optical response is primarily associated with the carrier dynamics between the applying voltage and the generation of light. Furthermore, the fluorescence lifetime of DSB:CBP, DPVBi:CBP, and BCzVBi:CBP neat films at the concentration of 0.5 wt.% were 1.4, 1.6, and 0.8 ns, respectively. These values were estimated time-resolved PL intensity using a femtosecond pulse laser with the center wavelength of 390 nm and a streak camera. Such short fluorescence lifetimes were assumed to give little effect on response speed of device. (Fukuda et al., 2007b)

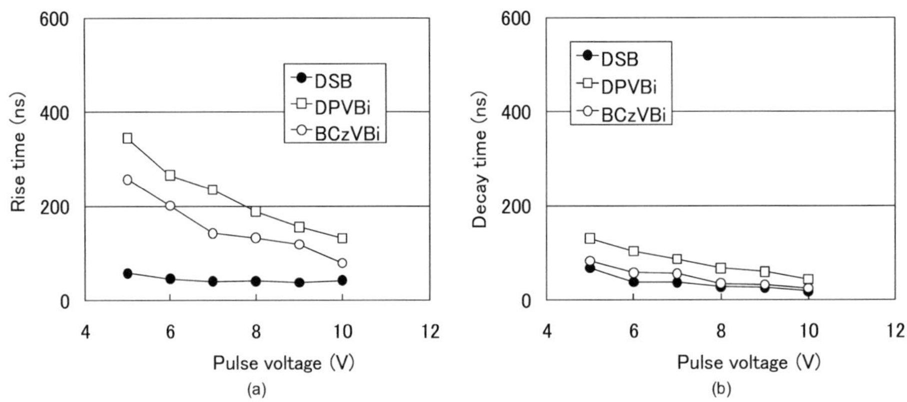

Fig. 14. (a) Rise and (b) decay times of the output EL intensity while applying the pulse voltage for three kinds of devices with different guest materials. (Fukuda et al., 2009)

Figure 15 shows absorption spectra of organic neat thin films of guest materials (DSB, DPVBi, and BCzVBi) and the PL spectrum of the CBP neat film. The PL spectrum of CBP showed the peak wavelength at 411 nm, and the PL intensity rapidly decreased in both shorter and longer wavelengths. The guest materials had peculiar absorption bands at the violet wavelength region. The peak wavelengths of absorption spectra of DSB, DPVBi, and BCzVBi were 418, 354, and 372 nm, respectively. Based on previous researches, the energy transfer efficiency of dye-doped OLEDs depends on the overlap integral of the emission spectrum of the host material and the absorption spectrum of the guest material. (Dexter, 1953; Eisenthal et al., 1953). The measured spectral overlap was different from each combination of the host-guest system, and the largest spectral overlap was achieved using DSB as a guest material. Therefore, efficient Förster energy transfer from the host material to the guest material is considered to be realized in the case of DSB doped CBP. As a result, the response speed of the OLED was also improved using DSB.

4.6 Organic-inorganic hybrid device

In the previous section, we showed that the low electron mobility of the ETL prevents the improved response speed of the OLED. (Fukuda et al., 2007e) In this section, we

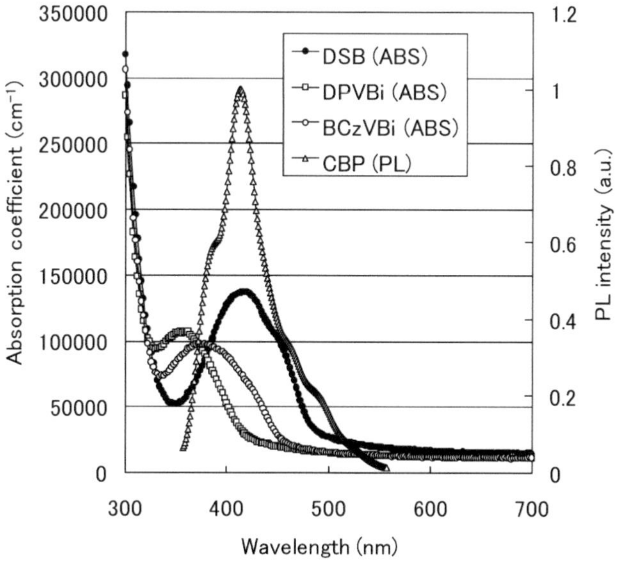

Fig. 15. Absorption spectra of DSB, DPVBi, and BCzVBi neat films and the PL spectrum of the CBP neat film. (Fukuda et al., 2009)

demonstrated organic-inorganic hybrid light-emitting diode, of which ZnS was used as the ETL. The ZnS layer has higher electron mobility compared to the organic electron transport material. Therefore, higher response speed can be realized compared to the OLED even though the emitting area is large. (Fukuda et al., 2008a) This fact indicates that such a device can be applicable for the institutive visible optical communication system.

Each layer consisted of α-NPD as an HTL, CBP doped with 0.5 wt% BCzVBi as an EML, and ZnS (device L) and Alq_3 (device M) as ETLs. Here, the fluorescence lifetime of BCzVBi was 0.6 ns, and it was short enough to realize the fast response speed. The thicknesses were 40 nm for α-NPD, 20 nm for BCzVBi doped CBP, ZnS, and Alq_3. Finally, LiF (0.4 nm) and MgAg (9:1 w/w) were evaporated on the top of the ETL layer. The active areas of all the OLEDs were fixed at 1 mm^2.

Figure 16(a) shows the relationship between the frequency of an applied sine wave voltage and the output relative EL intensity of the devices L and M, which consisted of ZnS and Alq_3 as ETLs, respectively. The relative EL intensity of the organic-inorganic hybrid device (device L) showed higher response speed compared to the OLED (device M). This result indicates that the low electron mobility of Alq_3 causes the low response speed. On the other word, we can realize the increased response speed utilizing the ZnS layer with high electron mobility as the ETL. This is because the response speed of the OLED is limited by the low electron mobility of organic electron transport materials, and the electron mobility of ZnS is higher than that of Alq_3

Figure 16(b) shows the influence of the sine wave voltage on the cutoff frequency for devices L and M. By comparing the devices L and M, we found that the cutoff frequency was influenced by the applied voltage for only the device L with ZnS. The low drive voltage of the OLED has been required for all the applications, such as mobile phones, flat panel displays, general lightings, and visible optical communications. This result indicates that the ZnS-ETL is important technique to improve both the response speed and the drive voltage.

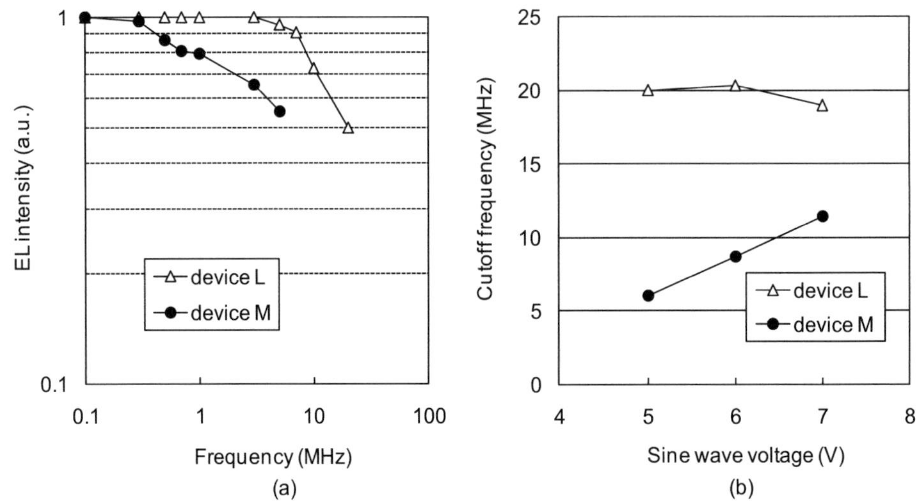

Fig. 16. (a) Relative EL intensity as a function of the applied sine wave voltage for two device with ZnS (device L) and Alq3 (device M) used as ETLs. (b) Relationship between the cutoff frequency and the sine wave voltage. (Fukuda et al., 2008a)

5. Intuitive visible data communication system with OLED as transceiver

5.1 Experimental

In this section, we show a demonstrator of the institutive visible optical communication system utilizing the OLED as an electro/optic converter. This system consisted of the transceiver module with the OLED and the pen-type receiver module with the semiconductor photo diode at a point, as shown in Fig 16. When the point of the pen-type receiver module approaches the emitting area of the OLED, you can get information from the OLED. Furthermore, the emitting area was 2 mm x 2mm, and the many people can touch without thinking the precious alignment between the pen-type receiver module and the OLED.

The fabrication process and the experimental results are discussed as bellows. We deposited cupper-phthalocyanine (CuPc) as a hole injection leayr, α-NPD as a HTL, rubrene in Alq_3 as an EML, Alq_3 as an ETL, and LiF as an EIL subsequently, upon the ITO-coated glass substrate. The device structure is glass substrate/ITO 150 nm/CuPc 10 nm/α-NPD 40 nm/1.0wt% rubrene:Alq_3 20 nm/Alq_3 40 nm/LiF 0.5 nm/Al 150 nm. Since the degradation of organic layers is caused by humidity and oxygen, the device was deposited by employing the conventional thermal evaporation at 6.0×10^{-6} Torr without breaking the vacuum. Then, the fabricated device was encapsulated under nitrogen atmosphere using UV-curable adhesives and cavity glass lids.

The inset of Fig. 16 shows the inside of the pen-type receiver module. We used driver IC (MAXIM, MAX749CSA) to apply the modulated pulse voltage. By applying the pulse voltage, the modulated optical signal generated from the emitting area of the OLED. In addition, the pen-type receiver module consisted of the photo-diode at a point, the comparator (NEC, μPC271G2), the operational amplifier (Linear Technology, LT1192 S8), and many electric parts (resistors, capacitor, and mechanical switch).

Pseudo electric signals were applied to the OLED to demonstrate the data transmission between the OLED and the photo diode. The amplitude and the clock frequency of pseudo signals were 4 V and 1 Mbps, respectively. In addition, the bias voltage of 4 V was also applied to the OLED by a DC voltage source. Then, the pen-type receiver was approached the emitting area of the OLED, as shown in Fig. 17. Finally, the output optical signal was changed the electric signal by a photo-diode, and the time-resolved output power was measured by an oscilloscope.

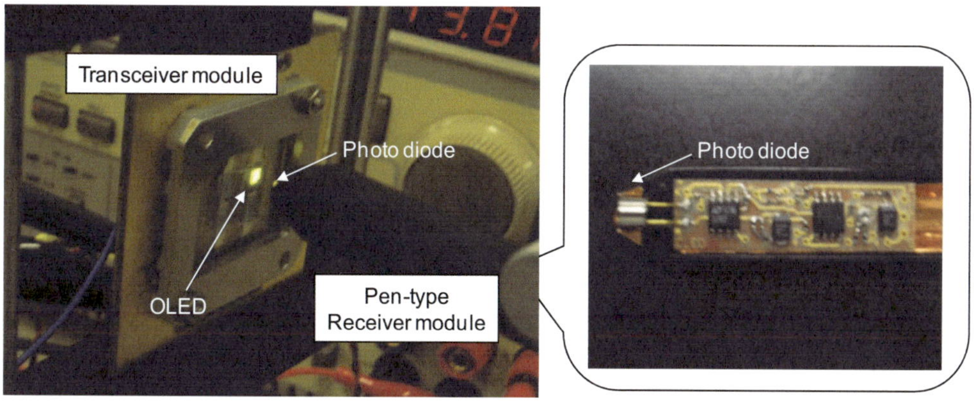

Fig. 17. (a) Demonstrator of the intuitive visible optical communication system, which consists of the OLED as a transceiver module and the pen-type receiver module. The inset shows the inside of the pen-type receiver module. (Fukuda et al., 2008b).

5.2 Result and discussions

Figure 18 shows the input electrical signal and the output optical signal as a function of the time. The output optical signal was received using the pen-type receiver module when the

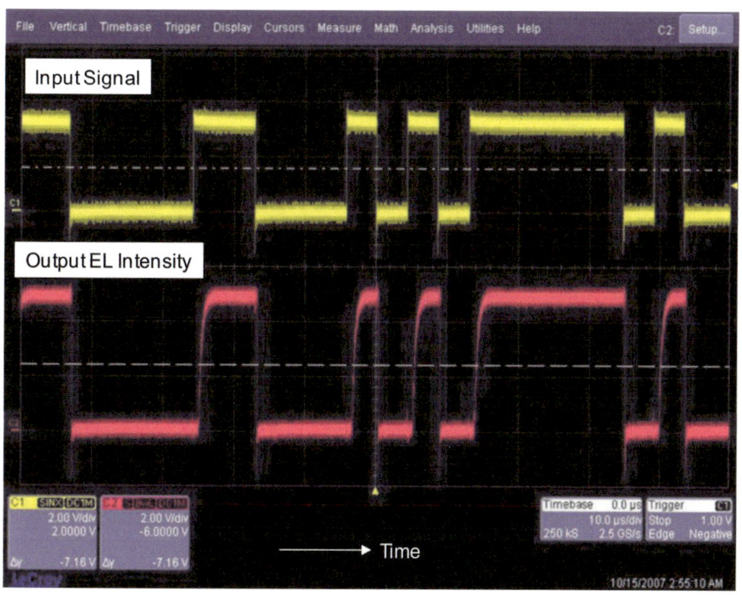

Fig. 18. Input electrical signal (yellow) and the output optical signal (pink) as a function of the time. The frequency of the input electrical signal was 1 Mbps.

pseudo-random signals were applied to the OLED. Transmission speed of pseudo-random signals was 1 Mbps. As clearly shown in Fig. 18, the rise time is larger than the decay time. This is because that the injection time of carriers from electrodes to the EML is long. However, we can realize the error-free data transmission at a speed of 1 Mbps using the transceiver module with the OLED.

6. Conclusion

In this chapter, we demonstrated fast-response OLEDs for the intuitive visible optical communications. We successfully achieved the more than 20 MHz by optimizing device parameters, such as the emitting area, the thickness of the carrier transport layer, the metal cathode, the fluorescence lifetime of the emitting material, the combination of the host-guest material, and the semiconductor ETL. Finally, we also demonstrated the demonstrator of the institutive optical data transmission system using the OLED as the transceiver. If the pen-type receiver module is touched the emitting area of the OLED, we can get the pseudo-random signals without thinking the precious alignment between the OLED and the pen-type receiver module.

7. Acknowledgment

The authors would like to thank Ms. K. Tanaka, Mr. K. Hashizume, Mr. H. Ohashi, and Mr. T. Hanawa of Nokia Research Center, Nokia Japan Co. Ltd. for their advices to practical applications of institutive visible optical communications. The authors also would like to thank Mr. H. Hosoya, Mr. Fujimaki, Mr. M. Ohashi, Mr. K. Asano, Mr. K. Azegami, Mr. Y. Terada, Mr. K. Ichii, and Mr. H. Kannno of Fujikura Ltd. for their help. The authors also would like to thank Prof. M. Ichikawa, Prof. B. Wei, Miss. T. Okada, Mr. E. Suto to do experiment and discussion. The part of this work was supported by Fujikura Ltd. And CLUSTER (the second stage) of Ministry of Ministry of Education, Culture, Sports, Science and Technology, Japan.

8. References

Adachi, C., Baldo, M.A., Thompson, M.E. & Forrest, S.R. (2001). Nearly 100% internal phosphorescence efficiency in an organic light emitting device, *J. Appl. Phys.*, Vol.90: 5048-5051.

Barth, S., Müller, P., Riel, H., Seidler, P.F. & Rieb, W. (2001). Electron mobility in tris(8-hydroxyquinoline)aluminum thin films determined via transient electroluminescence from single- and multilayer organic light-emitting diodes, *J. Appl. Phys.*, Vol.89: 3711-3719..

Cao. Y., Parker, I. Yu, G., Gang, Z. & Heeger, A. (1999). Improved quantum efficiency for electroluminescence in semiconducting polymers, *Nature*, Vol.397: 414-416.

Dexter, D. (1953). A theory of sensitized luminescence in solids, *J. Chem. Phys.*, Vol.21: 836-850.

Eisenthal, K.B., Siegel, S. (1964). Influence of Resonance Transfer on Luminescence Decay, *J. Chem. Phys.*, Vol.41: 652-655.

Fukuda, T., Ohashi M., Wei, B., Okada, T., Ichikawa, M. & Taniguchi, Y. (2007). Transient response of blue organic electroluminescence devices with short fluorescence lifetime of substituted phenyl/vinyl compound as an emissive layer, *Opt. Lett.*, Vol.32: 1150-1152.

Fukuda, T., Okada, T., Wei, B., Ichikawa, M. & Taniguchi, Y. (2007). Transient property of optically pumped organic film of different fluorescence lifetimes, *Appl. Phys. Lett.*, Vol.90: 231105.

Fukuda, T., Okada, T., Wei, B., Ichikawa, M. & Taniguchi, Y. (2007). Influence of carrier-injection efficiency on modulation rate of organic light source, *Opt. Lett.*, Vol.32: 1905-1907.

Fukuda, T., Wei, B., Ichikawa, M. & Taniguchi, Y. (2007). Enhanced Modulation Speed of Tris(8-hydroxyquinoline)aluminium-Based Organic Light Source with Low-Work-Function Electrode, *Jpn. J. Appl. Phys.*, Vol.46: 7880-7884.

Fukuda, T., Wei, B., Ichikawa, M. & Taniguchi, Y. (2007). Effect of Hole and Electron Injection Time on Modulation Speed of Organic Light-Emitting Diode, *Abstract of the 13th microoptics conference 2007*, pp.154-155, Kagawa, Japan, Oct. 2007.

Fukuda, T., Okada, T., Wei, B., Ichikawa, M. & Taniguchi, Y. (2008). Fast-response hybrid organic-inorganic light-emitting diode, *Phys. Status Sol.: Rap. Res. Lett.*, Vol.2: 290-292.

Fukuda, T. & Taniguchi, Y. (2008). Fast response organic light-emitting diode for visible optical communication, *Proceedings of SPIE*, Vol.6899: 68990K-1-83990K-13.

Fukuda, T., Wei, B., Ichikawa, M. & Taniguchi, Y. (2009). Transient characteristics of organic light-emitting diode with efficient energy transfer in emitting material, *Thin Solid Films*, Vol.518: 567-570.

Hatton, R.A., Day, S.R., Chesters, M.A. & Willis, M.R. (2001). Organic electroluminescent devices: enhanced carrier injection using an organosilane self assembled monolayer (SAM) derivatized ITO electrode, *Thin Solid Films*, Vol.394: 292-297.

Hung, L.S., Tang, C.W. & Mason, M.G. (1997). Enhanced electron injection in organic electroluminescence devices using an Al/LiF electrode, *Appl. Phys. Lett.*, Vol70: 152-154.

Ichikawa, M., Amagai, J., Horiba, Y., Koyama, T. & Taniguchi, Y. (2003). Dynamic turn-on behavior of organic light-emitting devices with different work function cathode metals under fast pulse excitation, *J. Appl. Phys.*, Vol.94: 7796-7800.

Ichikawa, M., Kawaguchi, T., Kobayashi, K., Miki, T., Furukawa, K., Koyama, T. & Taniguchi, Y. (2006). Bipyridyl oxadiazoles as efficient and durable electron-transporting and hole-blocking molecular materials, *J. Mater. Chem.*, Vol.16: 221-225.

Kajii, H., Tsukagawa, T., Taneda, T. & Ohmori, Y. (2002). Application of Organic Light Emitting Diode Based on the Alq_3 Emissive layer to the Electro-Optical Conversion Device, *IEICE Trans. Electron.*, vol.E85-C: 1245-1246.

Kajii, H., Tsukagawa, T., Taneda, T., Yoshino, K., Ozaki, M., Fujii, A., Hikita. M., Tomaru, S., Imamura, S., Takenaka, H., Kobayashi, J., Yamamoto, F. & Ohmori, Y. (2002). Transient Properties of Organic Electroluminescent Diode Using 8-Hydroxyquinoline Aluminum Doped with Rubrene as an Electro-Optical Conversion device for Polymeric Integrated devices, *Jpn. J. Appl. Phys.*, Vol.41: 2746-2747.

Kampen, T., Bekkali, A., Thurzo, I., Zahn, D.R.T., Bolognesi, A., Ziller, T., Carlo, A.D. & Lugli, P., (2004). Barrier height of organic modified Schottky contacts:theory and experiment, *Appl. Surf. Sci.*, Vol.234: 313-320.

Kido, J. & Matsumoto, T. (1998). Bright organic electroluminescent devices having a metal-doped electron-injecting layer, *Appl. Phys. Lett.*, Vol.73: 2866-2868.

Kim, J.-S. Kajii, H. & Ohmori, Y. (2006). Characteristics of optical response in red organic light-emitting diodes using two dopant systems for application to the optical link devices, *Thin Solid Films*, Vol.499: 343-348.

Kin, Z., Yoshihara, K., Kajii, H., Hayashi, K. & Ohmori, Y. (2006). Effects of CsF/Metal Interface on Electron Injection in Polymer Light-Emitting Diodes, *Jpn. J. Appl. Phys.*, Vol.45: 3737-3741.

Koike, Y. (2008). Microoptics and Photonics Polymer, Jpn. J. Appl. Phys., Vol.47: 6629-6634.

Mignonneau, L. & Sommerer, C. (2005). Designing emotional, metaphoric, natural and intuitive interfaces for interactive art, edutainment and mobile communications, *Computers & Graphics*, Vol.29: 837-851.

Mori, K., Ning, T., Ichikawa, M., Koyama, T. & Taniguchi, Y. (2003). Organic Light-Emitting Devices Patterned by Screen Printing, *Jpn. J. Appl. Phys.*, Vol.39: L942-L944.

Morimune, T., Kajii, H. & Ohmori, Y. (2006). High-Speed Organic Photodetectors Using Heterostructure with Phthalocyanine and Perylene Derivative, *Jpn. J. Appl. Phys.*, Vol.45: 546-549.

Morrison, G.D. (2005). A Camera-Based Input Device for Large Interactive Displays, *IEEE Comput. Grap. Appl.* Vol.23: 52-57.

Nüesch, F., Kamaraś, K. & Zuppiroli, L. (1998). Protoned metal-oxide electrode for organic light emitting diodes, *Chem. Phys. Lett.*, Vol.283: 194-200.

Ooe, M., Satoh, R., Naka, S., Okada, H. & Onnagawa, H. (2003). Painting Method for Organic Electroluminescent Devices, *Jpn. J. Appl. Phys.*, Vol.42: 4529-4534.

Panchaphongsaphak, B., Burgkart, R. & Riener, R. (2007). Three-Dimensional Touch Interface for Medical Education, *IEEE Trans. Info. Tech. BioMed.* Vol.11: 251-263.

Parker, I.D. (1994). Carrier tunnelling and device characteristics in polymer light-emitting diodes, *J. Appl. Phys.*, Vol.75: 1656-1666.

Shimada, H., Yanagi, J., Matsushita, Y., Naka, S., Okada, H. & Onnagawa, H. (2006). Organic Multifunction Diodes Operable for Emission and Photodetection Modes, *Jpn. J. Appl. Phys.*, Vol.45: 3750-3753.

Stößel, M., Staudigel, J., Steuber, F., Blässing, J. Simmerer, J. Winnacker, A., Neuner, H., Metzdorf, D., Johannes, H.-J. & Kowalsky, W. (2000). Electron injection and transport in 8-hydroxyquinoline aluminium, *Synth. Met.*, Vol.111-112: 19-24.

Tang, C.W. & VanSlyke, S.A. (1987). Organic electroluminescent diodes, *Appl. Phys. Lett.*, Vol.51: 913-915.

Tsutsui, T. (1997). Recent progress of molecular organic electroluminescent materials and devices, *MRS Bulletin*, Vol.22: 39-45.

Uchida, M., Izumisawa, T., Nakano, T., Yamaguchi, S., Tamao, K. & Furukawa, K. (2001). Structural Optimization of 2,5-Diarylsiloles as Excellent Electron-Transporting Materials for Organic Electroluminescent Devices, *Chem. Mater.*, Vol.13: 2680-2683.

Wei, B., Furukawa, K., Amagai, J., Ichikawa, M., Koyama, T. & Taniguchi, Y. (2004). A dynamic model for injection and transport of charge carriers in pulsed organic light-emitting diodes, *Semicond. Sci. Technol.*, Vol.19: L56-L59.

Xu, Q., Ouyang, J., Yang, Y. Ito, T. & Kido, J. (2003). Ultrahigh efficiency green polymer light-emitting diodes by nanoscale interface modification, *Appl. Phys. Lett.*, Vol.83: 4695-4697.

Zheng, X., Wu, Y., Sun, R., Zhu, W., Jiang, X., Zhang, Z. & Xu, S. (2005). Efficiency improvement of organic light-emitting diodes using 8-hydroxy-quinolinato lithium as an electron injection layer, *Thin Solid Films*, Vol.478: 252-255.

12

Effect of High Magnetic Field on Organic Light Emitting Diodes

Toshihiro Shimada
Hokkaido University
Japan

1. Introduction

This chapter aims at reviewing magnetic field effects (MFE) in organic light emitting diodes (OLED) with an emphasis on our study under high magnetic field up to 9 T. This subject includes organic spintronics in general, which is a hot subject attracting many researchers recently. Since singlet-triplet conversion in excitons is critically important in the current efficiency of OLEDs, spintronics aspects of OLEDs should be studied in detail. However, due to the difficulty in the fabrication of stable devices, the number of the researches has been limited.

We have found two things up to now, by making very stable OLEDs and measuring them under high magnetic fields. (1) Efficiency of OLEDs decreases quadrically with the magnetic field up to 6 T and the rate of decrease becomes smaller between 6T and 9T. (2) Minority carrier conductivity decreases linearly with the magnetic field, whereas that of majority carrier is almost constant. (3) Anomalous behaviors (large magnetoresistance etc.) are only seen in bipolar injection, which agrees with previous reports.

Although the mechanisms behind these findings have not been clarified yet, some hypotheses have been made with the analogy with MFE on chemical reactions. In principle, MFE on charge transport and recombination has similarity with chemical reaction under magnetic field, and the terminology and concept should be parallel between these subjects. We will try to combine the current knowledge of organic charge transport, OLEDs, spintronics and MFEs of chemical reaction to make a unified picture of these issues.

2. Spins in organic devices

2.1 Spins in OLEDs

The roles of spins in OLEDs and other organic devices are discussed well but have not been clarified quantitatively in experiments. We will review it in terms of OLED efficiency first. Figure 1 shows the schematic mechanism of OLED with the emphasis on the spin states. Since most of the organic semiconductors are used as intrinsic, the charges are transported via HOMO (highest occupied molecular orbital, in the case of holes) or LUMO (lowest unoccupied molecular orbital, in the case of electrons) of organic semiconductor molecules, usually by hopping in amorphous devices. Electrons and holes finally meet in one luminescent molecule and make excited states or excitons. The important thing is that there are two kinds of excitons, namely singlet excitons and triplet excitons. Although singlet

excitons can be relaxed radiatively, triplet excitons cannot emit light in ordinary materials due to the spin selection rule. Since the charges injected from electrodes are not spin polarized unless spintronics techniques were used, the spin polarizaion statistics is singlet : triplet = 1 : 3. The emission efficiency of OLEDs are governed by this factor and it is well known that incorporation of heavy atoms (Pt, Ir etc.) in the luminescent dye molecule greatly alleviate this burden via intersystem crossing (Baldo et al. 1999). Since the chemical synthesis of the luminescent molecules with heavy atoms is not fully developed and the heavy atoms are costly, other methods such as applying magnetic field to OLED (Kalinowski 1997) or mixing magnetic nanoparticles in the device (Hu et al. 2006, Sun et al. 2007) have been attempted. These approaches uses MFEs on carrier injection, transport and recombination, which are related with spintronics of organic semiconductors.
Pure MFE without using magnetic electrodes has been studied. Experimentally, the reports on MFE of OLEDs without ferromagnetic component qualitatively agree with each other, *i.e.*, steep increase in efficiency (2~10%) in the low magnetic field (< 100mT) and gradual decrease in the higher magnetic field. Although the behavior in the low magnetic field region is intensively studied and complicated phenomena including magnetoresistance are being elucidated (Bobbert et al., 2007; Davis & Bussmann, 2004; Desai et al., 2007a, 2007b; Hu & Wu, 2007; Kalinowski, 1997; Kalinowski et al., 2003, 2004; Lei et al., 2009; Liu et al., 2009; Odaka et al., 2006; Sakaguchi et al., 2006; Shakya et al., 2008; Shemg et al. 2007), very few experiments in relation to the MFE on organic semiconductors have been performed under high magnetic field larger than 2T (Reufer et al. 2005).

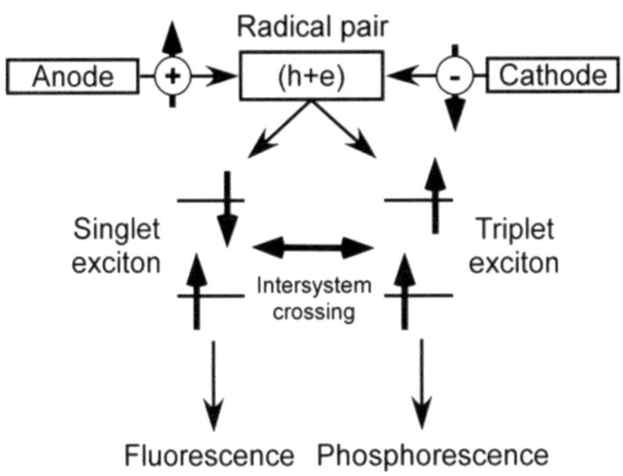

Fig. 1. Role of spins in organic light emitting diodes (OLEDs).

2.2 Organic spintronics

Spintronics study is now extended to all kinds of semiconductor materials. Organic semiconductors are not the exception. Since organic semiconductors consist of light elements such as carbon, hydrogen, oxygen and nitrogen, lifetime of spin polarized carriers might be long in organic semiconductors. After the proposal of this concept (Dediu et al. 2002), many papers have been published on the spin injection and transport in organic semiconductors. Most of the researches have been focused on performance of spin valves and magnetoresistance of organic semiconductors. A spin valve is a two terminal device consisting of a non-magnetic layer sandwiched by two different magnetic electrodes. The coersive forces of two electrodes are different and spin-polarized carrier injection and scattering makes characteristic magnetic field dependence of the device characters (I-V curve). Various organic semiconductors have been attempted in the device structures, and the large difference in the device resistance (magnetoresistance; MR) are achieved depending upon the spin orientation of the magnetic electrodes. At first MR was only substantially observed at low temperatures, but recently great MR at room temperatures are frequently reported. A variation of this research is MR measurement of mixture of magnetic nanoparticles and organic semiconductors. Some samples were prepared by codeposition of magnetic metals (cobalt etc.) and organic semiconductors. MR corresponding to the magnetization of magnetic nanoparticles (Sakai et al. 2006, Miwa et al. 2007) can be observed and its origin has been elucidated by x-ray magnetic circular dichromism (Matsumoto et al. 2009, Zhang et al. 2010).

An important topic related to the subject in the following is MR of devices without magnetic (or spin polarized) materials. Strong increase in conductance is observed in organic semiconductors when weak magnetic field (~ 100 mT) is applied. It is becoming a consensus that this high MR is only observed with bipolar injection, i.e., both of electrons and holes are injected to one layer, as shown in the following experiment (section 4.5).

3. Magnetic field effect in chemical reactions

First, we will follow up the current understanding of MFE in chemical reactions. Some of the chemical reaction change their reaction rate under magnetic field. In a simplified picture, those reactions proceed via intermediate state whose energy can be altered by the magnetic field. The energy difference can be due to the Zeeman effect on spin triplet state, which does not work on the spin singlet state. Therefore the reaction path between the singlet to or from the triplet can change the reaction rate. It must be noted that the Zeeman energy is too small to alter the reaction path in a single molecule. This is because the energy difference between spin singlet and spin triplet is very large (0.1~1eV) compared to the Zeeman energy (< 10 meV) under easily achievable magnetic field. Therefore it is considered that the intermediate state to which magnetic field can affect is a "radical pair", in which an anion radical and a cation radical are placed closely and about to transfer charges. Those radicals have unpaired spins and thus spin triplet and spin singlet states exist. The energy difference between the singlet and the trpilet is very small because the spin-spin interaction is small due to the large distance belonging to different (but adjacent) atoms (or molecules or ions) and can be comparable with the Zeeman energy. MFE on chemical reaction rate comes from the radical pairs.

The MFE on chemical reaction rates are complicated and are classified as follows. The dependence of these effects on the magnetic field is schematically shown in Fig. 2.

(a) Hyper fine coupling (hfc) mechanism: This effect is caused by the interaction between nuclear spin and external magnetic field. hfc mechanism causes the increase of reaction rate in low magnetic field.

(b) Δg mechansim: The difference of g-factor between anion and cation makes the Larmor frequency of the spins of the radicals different under magnetic field. Difference of frequency changes the relative orientation of radical spins, leading to change of the ratio of interconversion of singlet / triplet radical pairs. In the study of the radical pair in solution, it is known that the concentration of singlet excitons decreases (ΔY) in proportion to $B^{1/2}$ by Δg mechanism.

$$\Delta Y = -\frac{m}{p}\left(\frac{\pi \Delta g \mu_B B}{2}\right)^{1/2}, \quad (1)$$

where m, p, μ_B and B are, time constant of dissipation of radical pairs, collision probability between radical pairs, Bohr magneton and magnetic field, respectively.
(c) triplet-triplet anhilation (TTA): TTA is the collision reaction between triplet excitons to make singlet. It occurs when the density of the excited states created from radical pairs is high and they are migrating. This effect is rarely importent in MFE of solution chemistry but becomes important in solid state devices.

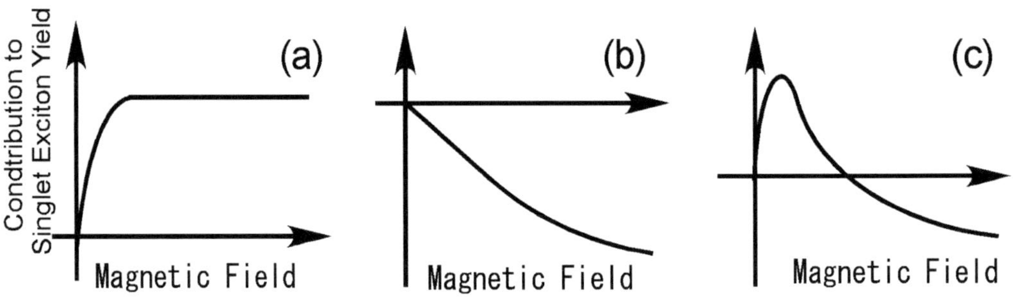

Fig. 2. Magnetic field effect of (a) hfc (b) Δg and (b) TTA.

The contribution of these mechanisms in the actually observed MFE is still under active argument. The characteristic magnetic field strength which gives the inflection points in (a)-(c) greatly differs from each other as discussed in the following. It is expected that the contributions can be elucidated by measuring the device properties in the wide range of the magnetic field. We therefore started experimental study described in the following section.
It should be noted here that all of the above mentioned mechanisms exhibit *concave* curves of the emission efficiency as a function of external magnetic field (B).

4. Effect of high magnetic field on organic light emitting diodes

In the previous sections, we have reviewed the MFEs on charge transport in organic semiconductor devices including OLEDs and chemical reaction kinetics. Since some of the exciton-related effects saturate at relatively low magnetic field (~ 1T), it is expected that the contribution of the above mechanisms in MFE will be separated if high magnetic field is applied. In this section, we present our experimental study of MFE on OLEDs under high magnetic field (Goto *et al.*, 2010). It seems that the results cannot be explained by the known exciton-related mechanisms (Δg, hfc, TTA etc.), and the origin is discussed based on the transport characteristics.

4.1 Preparation of compact and stable OLEDs for the measurement in high magnetic field

Since the sample space of the high field magnet is small, the sample must be as compact as possible, while maintaining the stability to warrant the reliable measurement. We made fluorescent devices and phosphorescent devices. The structure of fluorescent device is shown in Fig. 3(a). An indium tin oxide (ITO) coated glass substrate (Aldrich) with a sheet resistivity of 8-12 Ω/square was used as the substrate. 100nm N',N'-Di(naphthalene-1-yl)-

N,N′ dipheyl-benzidine (α-NPD) and 100nm Tris-(8-hydroxyquinolino) aluminum (Alq) were deposited successively as the hole transporting layer and emitting & electron transporting layer, respectively. Then a cathode was deposited, which consisted of a 2 nm Cs layer followed by 150 nm of Al. The ITO substrate was cleaned by ultrasonicating in ethanol and acetone. Following this, the ITO was treated in ozone for 20 min.. The deposition of the organic layers (α-NPD and Alq, Luminescence Technology Corporation) was performed using Knudsen-cells in a vacuum chamber with a base pressure during evaporation of ~10^{-7} Torr. Cs was deposited with alkali metal dispenser (SAES Getters). The deposition rate of organic materials was about 0.1 nm / s, which was measured by calibrated quartz crystal microbalances. The structure of phosphorescence device is shown in Fig. 3(b). Doping of 5% $Btp_2Ir(acac)$ in CBP was performed by controlling the evaporation rate by monitoring the quartz crystal microbalances.

The sample OLEDs and unipolar devices were transferred from the deposition chamber to glove box filled with dry N_2 without exposing them to air. The electrical connection to the OLED was made using thin Cu wire with In contact. Then the OLED sample was sealed in a glass box (made of O.D. 20 mm x t 3 mm pyrex tube and two t 0.1mm glass plates) using photo-hardening epoxy (Threebond 3124) together with a zeolite desiccant (Shinagawa Kasei Co. LTD). These sealing process was essential to obtain stable devices.

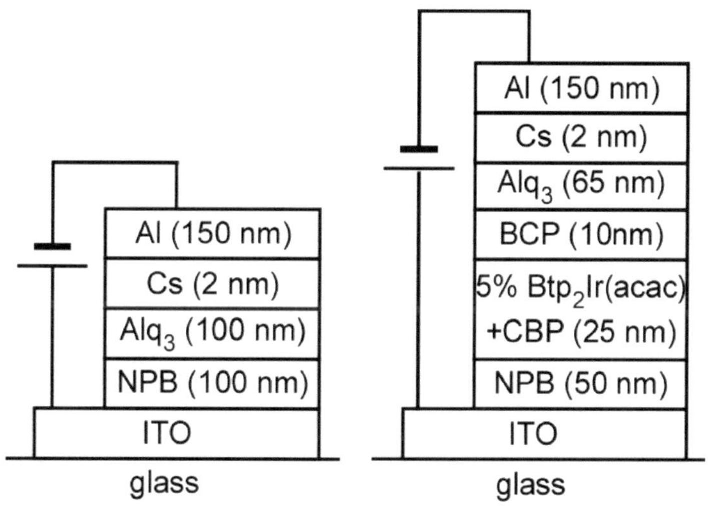

Fig. 3. Structure of (a) fluorescence and (b) phosphorescence devices

4.2 Measurement under magnetic field

MFE was measured at 300K in superconducting magnet using Physical Property Measurement System (PPMS; Quantum Design). The magnetic field was perpendicular to the device plane. The magnetic field was increased from 0 T to 9 T and then was decreased from 9 T to 0 T in order to check the temporal changes. The results are shown after confirming pure MFE is observed, unless stated otherwise. The emission intensity was measured with photon counter H7155-21 (Hamamatsu) in magnetic shielding made of thick iron plates and cylinders. The shielding of photon counter was tested and it was confirmed that there was no magnetic field dependence on its output. The bias was applied by the Keithley 6487 picoammeter / voltage source in constant voltage mode.

4.3 Results of fluorescent OLED

First we show the characteristics of the fluorescent device without applying the magnetic field. Figure 4(a) shows the emission intensity and current of the fluorescent OLED as a function of voltage, together with those of a phosphorescent device (Fig.4(b)). It is reported that TTA in Alq-based fluorescent OLEDs occurs when the current density is larger than 100 mA/cm^2 (Kondakov 2007) and some of our measurement exceeds this limit. However, since the magnetic field dependence of TTA appears only at low temperatures (Lei et al. 2009, Liu et al. 2009), we consider we can neglect contribution from TTA in the present measurement at 300 K.

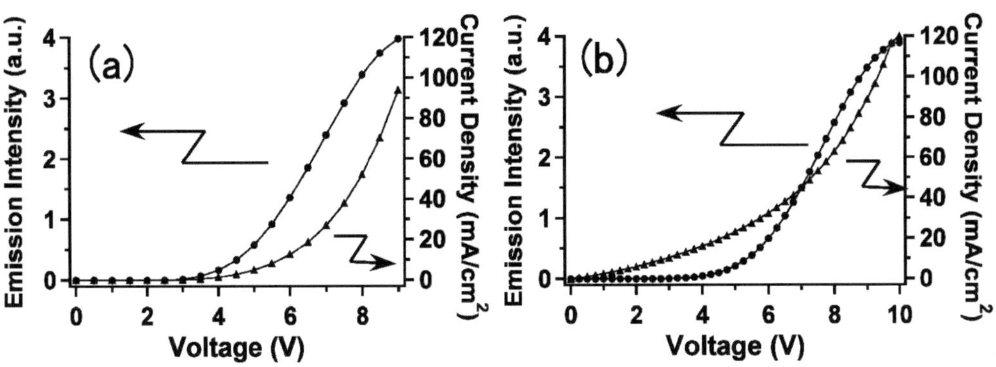

Fig. 4. I-V and light emitting properties of (a) fluorescent and (b) phosphorescent devices without magnetic field.

The emission intensity, current and emission efficiency of the same device under the magnetic field are shown in Fig. 5. The emission intensity and current show different magnetic field dependence on the sweep direction (0 -> 9T / 9 -> 0T) (Fig 5 (a)(b)). However, their ratio, i.e. emission efficiency, does not show the hysteresis as shown in Fig. 5(c). It means that the hysteresis comes from the charge injection process from the electrodes to the emission layer. We found that the "hysteresis" is dependent on both of the magnetic field and the time from the start of the current flow. The time dependence is probably due to the bias stress on the device, but the magnetic field dependence might be related with MFE of the trap / detrap processes.

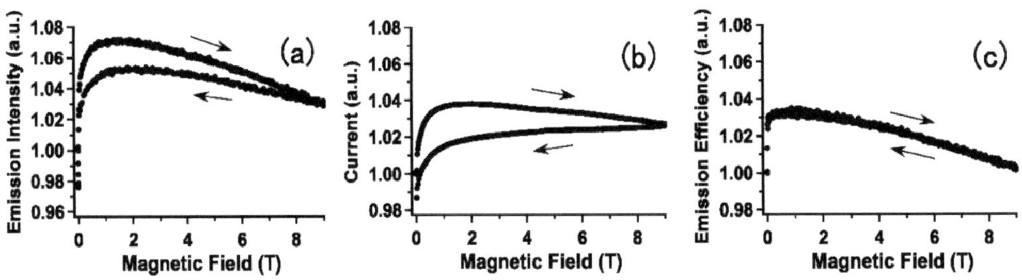

Fig. 5. Magnetic field effect of the fluorescent device (normalized at zero field). ((a) Emission Intensity, (b) Current, (c) Emission Efficiency at 4V). The arrows show the sweep direction.

The emission efficiency (and also the emission intensity and the current) increase steeply as a function of B when it was less than 0.02 T as reported in the literature, and gradually decreases as B was further increased. It should be noted that the decrease in the mid~ high B region is *convex* function, which cannot be explained by widely accepted behavior of hfc, Δg and TTA mechanisms which exhibit *concave* behavior against B. We will discuss this point later.

In order to see dependence on B more clearly, we re-plotted Fig. 5(c) as a function of B^2. Figure 6(a) clearly shows the linear decrease of the fluorescent efficiency against B^2 in the range of 0.1 T~ 6.5 T. On the other hand, by re-plotting Fig 5(c) as a function of $B^{1/2}$ (Fig. 6(b)), it is noticed that the decrease of the emission efficiency shows $B^{1/2}$ dependence in the range of 6.5 T ~ 9 T.

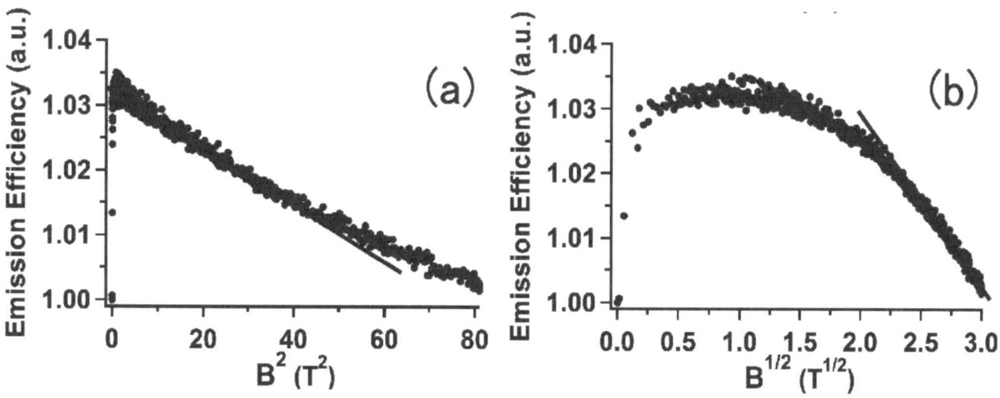

Fig. 6. Normalized emission efficiency of the fluorescent device plotted as a function of (a) B^2 (b) $B^{1/2}$.

4.4 Results on phosphorescent OLED

Figure 7 shows the MFE on the emission efficiency of the phosphorescent OLED. In contrast to the results of the fluorescent device, it did not show the magnetic field dependence. Although we changed the driving voltage (4V, 6V, 8V, 10V), the magnetic field dependence did not appear.

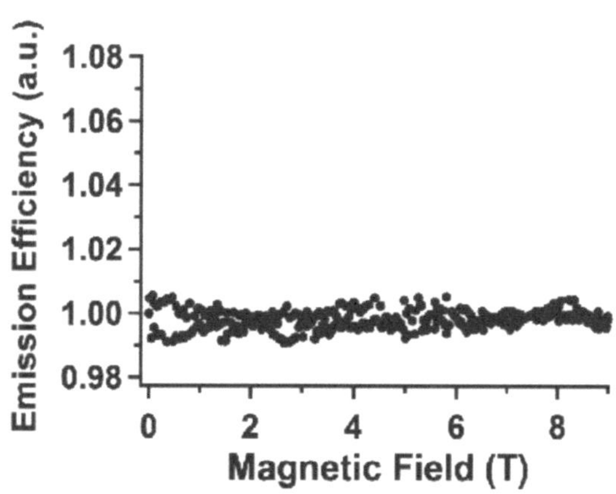

Fig. 7. Magnetic field effect of the phosphorescent device (normalized at zero field). The result at 10V is shown.

4.5 Magnetoconductance measurement of unipolar devices

In order to investigate the charge balance factor which might influence the EL efficiency, we measured the magnetoresistance of the majority and minority carriers in α-NPD and Alq by making the unijunction devices with different work function electrodes (Au and Cs). All of the devices showed Ohmic *I -V* characteristics in the measured range (-10~10V). The results of MFE on the current at constant voltage are shown in Figs. 8(a)-(d). The voltages were chosen to give current in the range of 10~100 μA and the results are shown after normalization at zero field. It is easily noticed that the MFEs on the conductivity of the majority carriers (holes in α-NPD and electrons in Alq) are negligible, whereas linear decreases in the conductivity was observed for the minority carriers (electrons in α-NPD and holes in Alq). Since the *I-V* characteristics are Ohmic, it shows the carrier mobility values of the minority carriers decrease linearly as a function of the magnetic field.

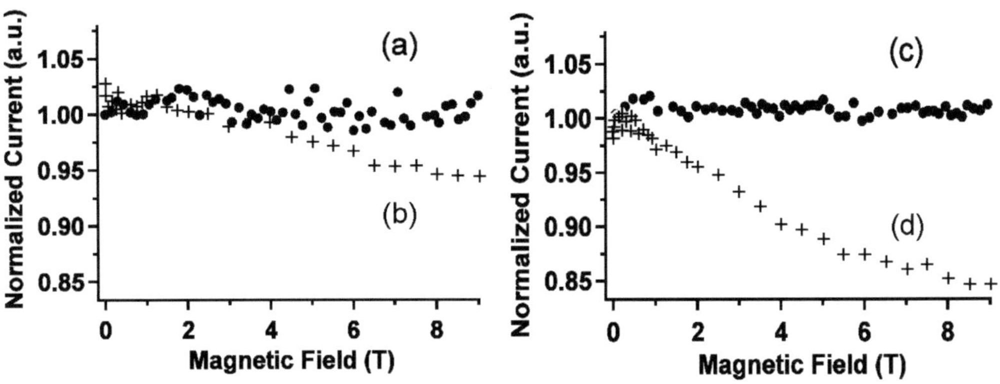

Fig. 8. Magnetic field effect of conductance of unipolar devices (normalized at zero field). (a)Au/α-NPD/Au (holes) (b)Au/Cs/α-NPD/Cs/Au (electrons) (c)Au/Cs/Alq/Cs/Au (electrons) (d)Au/Alq/Au (holes)

Also it should be noted that steep increase around zero field was not observed in unipolar devices, which agrees with previous reports (Yusoff, 2009).

4.6 Discussions – the origin of field dependence

We found the decrease in the fluorescent efficiency in organic EL devices proportional to B^2 in the range of 0.1~6.5T. Such dependence has not been reported to the authors' knowledge. The EL efficiency (η_{ext}) is given by the following:

$$\eta_{ext} = \alpha \times \Phi_{PL} \times \Phi_{exciton} \times \gamma \tag{2}$$

Here, α, Φ_{PL}, $\Phi_{exciton}$, γ are the light extraction efficiency, quantum efficiency of organic material, exciton formation efficiency, and carrier balance factor, respectively.

α is related with the magnetic field via Faraday /Kerr effects with interference. The Faraday rotation of the non-magnetic and thin organic layers is not significant even at 9T and interference would not change as a function of magnetic field. Thus we can neglect α. The fluorescent in optically excited organic dyes in the magnetic field has been studied in detail (Katoh & Kotani 1992), but B^2 dependence in high magnetic field was not reported in the literature. Therefore, our result cannot be explained by Φ_{PL}. Since all known mechanism of

MFE on $\Phi_{exciton}$ gives concave dependence on B as mentioned earlier, we here tentatively rule out the contribution from $\Phi_{exciton}$ as the main mechanism of the present B^2 dependence. The remaining factor in eq. (2) is the charge balance factor. We examined various models to relate the MFE on the charge balance factor, and the following is the only model that can barely explain the B^2 dependence. We observed that the conductance of the minority carrier changes linearly as a function of magnetic field as shown in Fig. 8. We considered various models based on our observation that the mobility of the minority carrier (μ) depends upon the magnetic field (B) as

$$\mu = \mu_0 (1 - aB) \tag{3}$$

where μ_0 and a are materials dependent constants. Since the EL intensity is determined only by the charge balance if the carrier recombination rate is proportional to the radiation (Scott et al. 1997), eq. (3) gives the fluorescent efficiency linearly related with B^1. However, our experiments showed that the current remains almost constant as a function of the magnetic field in the mid ~ high B range. We have found that we can deduce the B^2 dependence from eq. (3) with additional two assumptions. The assumptions are as follows. (i) The emission region is very narrow and only the recombination at this region contributes to the emission. This is reasonable because Alq layer (200nm) is much thicker than the thickness of emission region of ordinary devices. The interfacial mixing and the damage caused by the electrode formation might also justify this assumption. (ii) The hole current (J_h) and electron current (J_e) are balanced in the emission region when $B = 0$. Since the mobility of the majority carrier is smaller in Alq than α-NPD (Kepler et al. 1995, Nguyen et al. 2007), it is believed that the recombination and the emission occurs in Alq. This assumption is also reasonable because the device characteristics (Fig. 4), in which turn-on voltage of current is almost the same as that of emission intensity, show that this device has good carrier balance.

Since the current is constant as a function of B, $J_h + J_e$ is constant. From the assumption (ii), $J_h = J_e$ at the emission region when B=0. The emission region is in Alq and Eq.(3) becomes

$$J_h = J_0(1-aB), \tag{4}$$

where J_0 is a constant.

Because J_h+J_e is constant, J_e at the emission region can be written as

$$J_e=J_0(1+aB). \tag{5}$$

The recombination rate is proportional to J_hJ_e

$$J_hJ_e = J_0^2 (1-aB)(1+aB) = J_0^2 (1-a^2B^2) \tag{6}$$

and the decrease of the emission proportional to B^2 is explained.

We understand that the carrier transport of unipolar devices are not the same as the bipolar devices, for example, charge injection at the electrodes might be strongly involved in the minority carriers. However, our finding of linear MFE on minority carriers has not been reported and no theoretical prediction has been made to the author's knowledge. There might exist other mechanisms which also explain these results, but we hope the present result and discussions may stimulate the study of MFE on organic semiconductors and devices. Although TTA is not likely to work at room temperature and hfc will saturate at

relatively low magnetic field, ΔG works at high magnetic field region and might be cooperative with other factors.
There is another mechanism which might explain the B^2 dependence, although it is not consistent with the results on the unipolar devices (eq.(3)). It is theoretically predicted that the decrease of the conductivity proportional to B^2 is characteristics of magnetoresistance of hopping transport and that it levels off to the B^1 dependence when the magnetic field is high (Kepler et al. 1995). If it is applicable to the carriers in the emission layer of an OLED, the charge balance of the device will change in the same manner as the conductance and the emission decrease proportional to B^2 will be observed. The difficulty of this model is that we did not see such magnetoconductance in unipolar devices with organic single layers.
In the range beyond 6.5T, we observed $B^{1/2}$ dependence (Fig. 6(b)). In the study of the radical pair in solution, it is known that the density of singlet excitons decreases in proportion to $B^{1/2}$ by Δg mechanism (Sakaguchi & Hayashi 1995). The magnetoconductance of the minority carriers shown in Fig. 8 start to saturate in the range beyond 6.5 T. It might be the reason why $B^{1/2}$ dependence due to Δg mechanism start to appear in this region.
We did not find observable MFE in phosphorescent OLEDs (Fig.7). This result indicates that intersystem crossing occurs so fast in RPs and excited state molecules that Larmor precession in Δg mechanism does not affect the recombination kinetics. It also suggest that the charge balance effect discussed above does not come into play. It will be interesting to measure the magnetoconductance of organic semiconductors doped with phosphorescent dyes.

5. Conclusion

We reviewed recent studies on organic spintronics and MFE on chemical reactions in relation to the MFE on OLEDs. We measured EL efficiency of fluorescent and phosphorescent OLEDs in the magnetic field up to 9T and in the fluorescent device we found quadratic decrease as a function of the magnetic field between 0.1 ~ 6.5T. We also measured magnetoconductance of unipolar devices and observed that only minority carriers show significant magnetoconductance decreasing linearly with the magnetic field (15% at 9T in Alq). $B^{1/2}$ dependence in the range beyond 6.5T can be explained by MFE on the density of singlet exciton caused by Δg mechanism. In contrast, we did not find any MFE in the phosphorescent devices.

6. Acknowledgment

The author is grateful to the collaboration and discussions with Mr. Yuichiro Goto, Mr. Takuya Noguchi, Mr. Utahito Takeuchi, Dr. Kunitada Hatabayashi, Dr. Yasushi Hirose, Prof. Takehiko Sasaki, Prof. Tetsuya Hasegawa (all at the University of Tokyo). He would like to thank Prof. Takayuki Uchida (Tokyo Polytechnic University) for valuable information on the fabrication of OLED.

7. References

Baldo, MA, Lamansky, S., Burrows, PE, Thompson, ME & Forrest, SR (1999) Very high-efficiency green organic light-emitting devices based on electrophosphorescence, *Appl Phys Lett* 1999, 75, 4-6.

Bobbert, P.A.; Nguyen, T.D., van Oost, F.W.A., Koopmans, B. & Wohlgenannt, M. (2007) Bipolaron Mechanism for Organic Magnetoresistance, *Phys. Rev. Lett.* 99 (2007) 216801/1-216801/4.

Davis, A.H. & Bussmann, K. (2004) Large magnetic field effects in organic light emitting diodes based on tris(8-hydroxyquinoline aluminum) (Alq3)/N,N'-Di (naphthalen-1-yl)-N,N'diphenyl-benzidine (NPB) bilayers , *J. Vac. Sci. Technol. A* 22 (2004) 1885-1891.

Dediu, V., Murgia, M., Matacotta, F.C., Taliani, C., Barbanera, S. (2002) Room temperature spin polarized injection in organic semiconductor, *Solid State Commun.*, 123, 181-184(2002)

Desai, P.; Shakya, P., Kreouzi, T. & Gillin, W.P. (2007) , The role of magnetic fields on the transport and efficiency of aluminum tris(8-hydroxyquinoline) based organic light emitting diodes, *J. Appl. Phys.* 102 (2007) 073710/1-073710//5.

Desai, P.; Shakya, P., Kreouzis, T. & Gillin, W.P. (2007) Magnetoresistance and efficiency measurements of Alq3-based OLEDs, *Phys. Rev. B* 75 (2007) 094423/1-094423/5.

Goto, Y., Noguchi, T., Takeuchi, U., Hatabayashi, K., Hirose, Y., Uchida, T., Sasaki, T., Hasegawa, T., Shimada, T., (2010), High magnetic field effect in organic light emitting diodes, *Org. Elec.* 11 (2010) 1212-1216.

Hayashi, H.& Sakaguchi, Y. (2005) Magnetic field effects and CIDEP due to the d-type triplet mechanism in intra-molecular reactions , *J. Photochem. Photobiol. C : Photochem. Reviews*, 6 (2005) 25-36.

Hu, B. & Wu, Y. (2007) Tuning magnetoresistance between positive and negative values in organic semiconductors, *Nature Mat.* 6 (2007) 985-991.

Kalinouski, J; (1997). in Organic Electroluminescent Materials and Devices, Gordon Publishers (Eds. Miyata, S. & Nalwa, H.S.), New York, 1997. ISBN 2919875108

Kalinouski,J., Cocchi, J., Virgili, D., Marco, P.D. & Fattori, V. (2003). Magnetic field effects on emission and current in Alq3-based electroluminescent diodes, *Chem. Phys. Lett.* 380 (2003) 710-715.

Kalinowski, J., Cocchi, M., Virgili, D., Fattori ,V.& Marco, P.D. (2004). Magnetic field effects on organic electrophosphorescence, *Phys. Rev. B* 70 (2004) 205303.

Katoh, R. & Kotani, M. (1992) Fission of a higher excited state generated by singlet exciton fusion in an anthracene crystal, *Chem. Phys. Lett.* 196 (1992) 108-112.

Kepler, R. G.; Beeson, P. M., Jacobs, S. J., Anderson, R. A., Sinclair, M. B., Valencia, V. S. & Cahill, P. A. (1995), Electron and hole mobility in tris (8 - hydroxyquinolinolato - N1,O8) aluminum, *Appl. Phys. Lett*, 66 (1995) 3618-3620.

Kondakov, D.Y. (2007) Characterization of triplet-triplet annihilation in organic light-emitting diodes based on anthracene derivatives, *J. Appl. Phys.* 102 (2007) 114504/1-114504/5.

Lei, Y.L., Zhang, Y., Lui, R., Chen, P., Song, Q.L. & Xiong, Z.H. (2009), Driving current and temperature dependent magnetic-field modulated electroluminescence in Alq3-based organic light emitting diode, *Organic Electronics* 10 (2009) 889-894.

Liu, R., Zhang, Y., Lei, Y.L., Chen, P. & Xiong, Z.H. (2009) Magnetic field dependent triplet-triplet annihilation in Alq3-based organic light emitting diodes at different temperatures, *J. Appl. Phys.* 105 (2009) 093719/1-093719/5.

Matsumoto, Y., Sakai, S. Takagi, Y., Nakagawa , T., Yokoyama, T. , Shimada, T., Mitani, S., Naramoto, H., Maeda, Y. , X-ray absorption spectroscopy and magnetic circular dichroism in codeposited C60–Co films with giant tunnel magnetoresistance, *Chem. Phys. Lett.* 470 (2009) 244–248.

Miwa, S., Shiraishi, M., Tanabe, S., Mizuguchi, M., Shinjo, T., Suzuki, Y. (2007) Tunnel magnetoresistance of C60-Co nanocomposites and spin-dependent transport in organic semiconductors, *Phys. Rev. B* 76 (2007) 214414/1-214414/7.

Nguyen, N. D.; Schmeits, M. & Loebl, H. P. (2007) Determination of charge-carrier transport in organic devices by admittance spectroscopy: Application to hole mobility in α-NPD, *Phys. Rev. B* 75 (2007) 075307/1-075307/13.

Odaka, H.; Okimoto Y., Yamada, T., Okamoto, H., Kawasaki, M. & Tokura, Y. (2006). Control of magnetic-field effect on electroluminescence in Alq3-based organic light emitting diodes, *Appl. Phys. Lett*, 88 (2006) 123501/1-123501/3.

Reufer, M.; Walter, M.J., Lagoudakis, P.G., Hummel, A.B., Kolb, H.S., Roskos, H.G., Scherf, U. & Lupton, J.M. (2005) Spin-conserving carrier recombination in conjugated polymers, *Nature Mat.* 4 (2005) 340-346.

Sakaguchi, Y.; Iwasaki, Y., Osasa T., Asahi, M. & Matsumura M. (2006). Fractions of singlet and triplet excitons generated in organic light-emitting devices based on a polyphenylenevinylene derivative, *Phys. Rev. B* 74 (2006) 195209/1-195209/8.

Sakaguchi, Y. & Hayashi, H. (1995), Magnetic field effects on the photodissociation reaction of triphenylphosphine in non-viscous homogeneous solutions , *Chem. Phys. Lett.* 245 (1995) 591-597.

Sakai,S., Yakushiji, K., Mitani, S., Takanashi, K. Naramoto, H. Avramov, P.V., Narumi K., Lavrentiev, V., Maeda, Y. (2006)., Tunnel magnetoresistance in Co nanoparticle/Co–C60 compound hybrid system *Appl. Phys. Lett.* 89 (2006) 113118/1-113118/3.

Scott, J.C.; Karg, S. & Carter, S.A. (1997) Bipolar charge and current distributions in organic light-emitting diodes, *J. Appl. Phys.* 82 (1997) 1454-1460.

Shakya, P.; Desai, P., Somerton, M., Gannaway, G., Kreouzis, T. & Gillin, W.P. (2008) , The magnetic field effect on the transport and efficiency of group III tris(8-hydroxyquinoline) organic light emitting diodes, *J. Appl. Phys.* 103 (2008) 103175/1-103175/5.

Shemg, Y.; Nguyen, T.D., Veeraraghavan, G.& Mermer, O. (2007) Effect of spin-orbit coupling on magnetoresistance in organic semiconductors, *Phys. Rev. B* 75 (2007) 035202/1-035202/7.

Sun, C.J., Wu, Y., Xu, Z., Hu, B., Bai, J. Wang, J. P. & Shen, J. (2007) Enhancement of quantum efficiency of organic light emitting devices by doping magnetic nanoparticles , *Appl. Phys. Lett.* 90 (2007) 232110/1-232110/3.

Yusoff, A.R.B.M., da Silva, W.J., Serbena, J P. M. , Meruvia, M. S. & Hümmelgen, I.A. (2009) Very High Magnetocurrent in Tris-(8-hydroxyquinoline) Aluminum-Based Bipolar Charge Injection Devices - *Appl. Phys. Lett.* 94 (2009) 253305/1-253305/3.

Zhan, Y., Holmström, E., Lizárraga, R., Eriksson, O., Liu, X., Li, F., Carlegrim, E., Stafström, S. & Fahlman, M., (2010),Efficient Spin Injection Through Exchange Coupling at Organic Semiconductor/Ferromagnet Heterojunctions, *Adv. Mater.*, 22 (2010) 1626.